AF686236

Renewable and Sustainable Energy: Current State and Prospects

Renewable and Sustainable Energy: Current State and Prospects

Editors

Bartlomiej Iglinski
Michał Bernard Pietrzak

MDPI • Basel • Beijing • Wuhan • Barcelona • Belgrade • Manchester • Tokyo • Cluj • Tianjin

Editors
Bartlomiej Iglinski
Faculty of Chemistry,
Nicolaus Copernicus
University in Toruń,
Toruń, Poland

Michał Bernard Pietrzak
Department of Statistics and
Econometrics, Faculty of
Management And Economics,
Gdansk University of Technology,
Gdansk, Poland

Editorial Office
MDPI
St. Alban-Anlage 66
4052 Basel, Switzerland

This is a reprint of articles from the Special Issue published online in the open access journal *Energies* (ISSN 1996-1073) (available at: https://www.mdpi.com/journal/energies/special_issues/renewable_and_sustainable_energy).

For citation purposes, cite each article independently as indicated on the article page online and as indicated below:

LastName, A.A.; LastName, B.B.; LastName, C.C. Article Title. *Journal Name* **Year**, *Volume Number*, Page Range.

ISBN 978-3-0365-5231-6 (Hbk)
ISBN 978-3-0365-5232-3 (PDF)

© 2022 by the authors. Articles in this book are Open Access and distributed under the Creative Commons Attribution (CC BY) license, which allows users to download, copy and build upon published articles, as long as the author and publisher are properly credited, which ensures maximum dissemination and a wider impact of our publications.
The book as a whole is distributed by MDPI under the terms and conditions of the Creative Commons license CC BY-NC-ND.

Contents

energies

MDPI

Editorial

Renewable and Sustainable Energy: Current State and Prospects

Bartłomiej Igliński [1] and Michał Bernard Pietrzak [2,*]

[1] Faculty of Chemistry, Nicolaus Copernicus University in Toruń, Gagarina 7, 87-100 Toruń, Poland; iglinski@chem.umk.pl
[2] Faculty of Management and Economics, Gdańsk University of Technology, Narutowicza 11/12, 80-233 Gdańsk, Poland
* Correspondence: michal.pietrzak@zie.pg.gda.pl

Citation: Igliński, B.; Pietrzak, M.B. Renewable and Sustainable Energy: Current State and Prospects. *Energies* **2022**, *15*, 4735. https://doi.org/10.3390/en15134735

Received: 9 June 2022
Accepted: 27 June 2022
Published: 28 June 2022

Publisher's Note: MDPI stays neutral with regard to jurisdictional claims in published maps and institutional affiliations.

Copyright: © 2022 by the authors. Licensee MDPI, Basel, Switzerland. This article is an open access article distributed under the terms and conditions of the Creative Commons Attribution (CC BY) license (https://creativecommons.org/licenses/by/4.0/).

1. Introduction

The last two decades of the twentieth century represented a period of above-average, systematic growth of formal and informal interdependencies between economies of different countries and between world markets. The intensity, magnitude, and diversity of these interdependencies have never been recorded before in economic history, and the market transformations taking place have been referred to in the literature as the process of world globalization. Over the next twenty years of the 21st century, the dynamic and systematic development of globalization processes has progressed to such an extent that, in fact, a very high level of interdependence has been achieved in every sector of the economy [1]. This means that national economies for the last forty years have faced functioning in new economic conditions, forced mainly by the developing globalization processes. This has mainly contributed to a significant increase in the socio-economic development [2,3] and the associated enrichment of the populations of most economies and worldwide changes in the labor market [4]. In addition, there have been new trends in consumer attitudes [5,6], where environmental issues have begun to play a dominant role in consumer decision-making.

It should be emphasized that the progressing globalization processes have contributed to a significant increase in the level of foreign direct investment, especially in less developed countries, and thus to an increase in investment expenditures in enterprises. In addition, over the last 20 years, there has been a hitherto unseen increase in the level of innovation and the number of innovations implemented [7,8]. All this, combined with the high competitiveness of economies, has created the possibility of introducing modern technologies, including the use of energy from renewable energy sources (RES) [9]. The appearance of further innovations in the production of renewable energy, a significant reduction in the cost of production facilities and a change in consumer attitudes combined with a significant increase in the level of household wealth have resulted in the fact that every consumer in the world has the opportunity to produce renewable energy. It can be concluded that a new branch of the world economy, the RES has emerged in the 21st century and is one of the fastest-growing sectors in world economies. On the other hand, the global change of consumer attitudes gave the opportunity to implement the idea of sustainable development, both in its economic, environmental, social, and institutional aspects [10,11]. It should be emphasized that in the European countries that are members of the European Union, the process of implementation of sustainable development goals is fully institutionalized, and the member countries are obliged to adapt their institutions and legislation to the adopted Community-wide sustainable development strategies [12].

However, within the dynamically developing processes of globalization and civilization changes caused by them, including above-average economic growth, a key problem turned out to be the systematic increase of energy demand. The developed global supply chains proved to be insufficient, as they led to overexploitation of non-renewable energy sources and to the destruction of the natural environment. This means that there was a need for an intensive energy transition at the global level [13–15], which would be based on the development of new, alternative ways of energy production and consumption and on the gradual replacement of non-renewable energy sources with renewable or low-carbon sources [16,17]. It should be emphasized that the transformation, by taking into account the idea of sustainable development, leads to further fundamental changes in almost all sectors of the economy, as well as the way of life of the societies of most countries.

2. A Short Review of the Contributions in This Issue

There is a need to discuss the issue of energy transformation and the related problem of Renewable and Sustainable Energy Production. The answer is to prepare a special issue entitled "Renewable and Sustainable Energy: Current State and Prospects", which contains 14 published articles. The articles successively deal with the content related to the issue of sustainable development and the problem of achieving the goals of sustainable development. Subsequent articles then address issues related to RES mix production, different approaches to renewable energy production (from national-level production approaches to local level production, to a city level, and even to individual household level), and issues related to energy investment and renewable energy consumption, where attention has been focused on both businesses and households.

Cheba and Bąk in the article [18] undertook to present the relationship between the ecological efficiency of production and the 7th Agenda for Sustainable Development 2030. Both these areas indicate the relationship between the economy and the natural environment, emphasizing changes in the field of energy use. The multi-criteria taxonomy and selected taxonomic methods were used to investigate the relationship between these two areas. Despite the ongoing attempts to equalize the levels of development between individual EU countries in many strategic areas, they remain very diverse. Can it be assumed that the next steps taken by the European Commission will eliminate these differences? According to the authors, it is not entirely possible. In subsequent studies, the authors also plan to expand the scope of research, the methods used to study the relationship between these areas, and qualitative and quantitative techniques using, for example, cognitive mapping.

Kludacz-Alessandri and Cygańska in the work [19] pointed out that nowadays the company's reputation depends on corporate social responsibility (CSR). CSR has a positive effect on the company's financial performance (FP). The authors examined whether the financial results influence the adoption of corporate social responsibility in companies in the energy sector. It turned out that the measure of profit before interest, the measure of return on assets (ROA), and taxation (EBIT) were much higher among companies implementing a CSR strategy. Enterprise Value to Earnings Before Interest, Depreciation, EBITDA, and taxes was lower among companies that adopted CSR.

In developing countries, decentralized renewable energy systems such as mini-solar grids (MG) play and will play an important role. Stritzke and Jain [20] believe that RES in developing countries is facing major technical, financial, and social challenges in terms of sustainable development. The research conducted by the authors aimed to understand the sustainable development of RE MG in a developing context based on an integrated assessment of the technical, financial, and social dimensions of the exploitation of the ME through empirical data from community surveys on energy consumption from Uganda and Zambia and two other in-depth case studies MG from Zambia. The authors concluded that the complex ecosystem of the rural community is the most important determinant of the sustainable development of the ME. There should be an appropriate match between the tariffs of the ME and the affordable price for consumers.

On the other hand, Piekut [21] presented an analysis of sustainable development and fuel and energy transformation in the household sector in 2004–2019 in the EU. The subject of the research was various sources of RES used by households, i.e., primary solid biofuels, charcoal, solar systems, geothermal technologies, biogas, biodiesel, bioethanol, and heat pumps. In the analyzed period of 2004–2019, there was an absolute and relative increase in the use of RES in the household sector.

It should be emphasized that for most countries the problem of the production of the RES becomes a priority in the policy concerning the development of the economy. RE plays a key role, especially in EU policy, which assumes that in 2050 the share of the RE is to increase to 50%, and 80% of electricity is to be generated from low-emission sources. The authors of the article [22] analyzed the changes in the use of RES for electricity production from 2005 to 2019 in the EU countries. The k-mean and the Gini coefficient were used in the study. The research confirmed that the EU countries, in line with the assumptions of the energy policy, increased both the share of RES in energy production, especially electricity, and increased the use of RES. Individual EU countries differed in terms of the use of the RES for the production of electricity. This means that the energy transformation in each of the EU countries proceeds in a different way. EU countries with similar problems should undertake joint actions with regard to the Community's internal policy, technological development, and energy production. Programs promoting the purchase and use of RES installations should be launched/continued. Moreover, cooperation between individual countries in the field of RES should be increased, such as joint research, joint projects, or joint support systems.

As already mentioned, the EU aims to create sustainable, low-carbon economies based on the RES. This also applies to the new EU member states. In order to be successful, the new EU member states must carry out quick and effective changes in the energy sector. Wałachowska and Ignasiak-Szulc in the work [23] presented in the article evaluation of new EU member states in terms of diversification of renewable energy countries. Ward's method was used for the analysis. The obtained results can be used in countries of comparable specificity to undertake activities of a similar nature with regard to internal energy production, technological development, or a common energy policy.

Poland as an EU member state should decarbonize the economy and become "climate neutral" by 2050. In the case of Poland, it is very difficult, as currently, about 80% of energy comes from hard coal and lignite. The country's energy transformation is openly opposed by miners or some energy engineers. Several programs supporting the development of the RES have been introduced in Poland. One of the most important ones is "My Electricity", which, depending on the edition and investment, is subsidized from 3000 to 20,500 PLN. In the article [24] authors believe that the development of prosumer photovoltaics in Poland is important, but it will not replace coal-fired power plants. More research is needed on the ecological energy mix of Poland. The most important goal of the research was to make a proper review of the energy sector, with particular emphasis on technologies that can be used as ecological systems of distributed energy production, and to outline scenarios for the development of the sector The authors used the Delphi method supported by the Computer Assisted-Web Interview (CAWI) technique, Desk research, and the Weighted SWOT analysis. The obtained results showed that despite some disadvantages, it is the photovoltaic systems that will be the fastest-growing energy sector in Poland. Additionally, technologies will be developed on the basis of dispersed systems of biomass and biogas use.

In recent years, a transformation of energy towards RES has been observed in many countries. Huterski et al. [25] undertook to assess the level of development of electricity production from RES using one of the methods of multivariate comparative analysis (WAP)—a taxonomic measure of Hellwig's development. A total of 28 countries were surveyed, including Great Britain and 27 countries of the present EU. Panel models were used to describe the relationship between the share of RES electricity production in total electricity production and GDP per capita, electricity production from water, wind, solar, and biogas per person, and countries' public energy expenditure as a percentage of GDP. It was found that rich countries are much easier to invest in the RES than in countries that have recently joined the EU.

Rapid economic development implies increased production and consumption of energy. As a result, conventional sources are no longer sufficient, and their extraction and combustion cause a large burden on the environment and climate. In connection with the above, many countries have decided to transform their energy towards a low- and zero-emission economy. In the EU, the development of a "green economy" has become a strategic goal in the fight against climate change. The systematic development of the RES leads to the improvement of the energy security of a given country and the entire EU.

Another very important problem is the issue of socio-economic development, innovativeness and RES at the regional level [26–29]. In the paper [30] authors took up the problem of carrying out energy transformation in the example of Pomerania Voivodeship (Poland). In this regard, the current status, potential, and development prospects of the RES in the Pomerania region are presented. Additionally, a PEST analysis was performed for the renewable energy sector. The calculated RES potential indicates that Pomerania could become energy self-sufficient on the basis of RES. It was concluded that not only this Voivodeship but also the whole of Poland is characterized by a high potential of the RES [14].

In the next article [31] authors also developed the subject of RES in Poland for the Greater Poland Voivodeship. It is important because the EU Member States are obliged to implement the adopted Community Energy Strategy, which was defined under the European Green Deal. Energy transformation is to be based largely on the diversification of energy sources used, with a predominance of the RES. The authors asked themselves whether, based on the available technologies of the RES mix, it is possible to decarbonize Greater Poland Voivodeship. The research consists in determining the energy potential of RES in Greater Poland Voivodeship based on the methods of the geographic information system (GIS). The GIS methods were chosen because they allow for spatial positioning of surface, linear, and RES potential structures, thus ensuring high accuracy of the obtained estimates. The authors concluded that the technical potential of the RES in the Greater Poland Voivodeship is higher than the current consumption of electricity and heat. It should be added that the Greater Poland Voivodeship is one of the regions dependent on coal in Poland, which has already prepared a structured plan for a just transformation towards clean energy technologies.

The paper [32] deals with the problem of using renewable energy at the city level. Sidełko dealt with an innovative approach to municipal waste management in the example of the commune of Koszalin (Poland). The author proposed the Waste Processing Energy Recovery model which is a universal solution for provinces and cities. The waste balance includes waste from the selective collection, mixed municipal and commercial waste as well as sewage sludge from the municipal sewage treatment plant. The developed model is based on the functioning of four facilities. Every day, this system produces 5519 m^3 of gas and high-energy waste fuel for combustion in the amount of 82.2 tons. The proposed energy recovery from waste is 754 kWh/inhabitant/year.

The overall energy mix is made up of many sources of renewable energy. Therefore, it is also important to consider the individual selected energy sources in detail. Every year the share of wind energy (EC) in the energy mix of many countries increases. It is clean energy and more and more competitive in terms of prices for energy from burning fossil fuels. Matching the appropriate statistical distribution to the wind speed (WS) data is crucial in analyzing and estimating the EC potential. In the paper [33] the efficient global optimization (EGO) technique to fit the statistical distribution to WS data were proposed, and the technique performance was compared to the genetic algorithm (GA), simulated annealing (SA), and differential evolution (DE). On the basis of Weibull parameters, the authors obtained the potential of the EC and the potential annual revenues for Gdańsk (Pomerania Voivodeship, Poland). The conducted research has shown that urban wind turbines can be installed in the city with virtually no restrictions. Installed on the roofs of shopping malls, office buildings, or houses, they would partially cover the electricity needs of these buildings.

In Poland, a new form of settlement of investments in improving energy efficiency is the formula of involving an energy service company (ESCO). Kurowska-Pysz and Kunikowski in the paper [34] showed that many entities in Poland still lack sufficient knowledge on this subject. The problem discussed in the article concerned the conditions for applying the ESCO formula (investment financing model with the participation of a specialized company) to support enterprises and local government units in the development of energy and energy projects. Research questions were asked to analyze the following issues: sources of knowledge and reasons for interest in the ESCO formula, activities and other factors that may increase or decrease the interest in the ESCO formula, and attractiveness of alternative instruments for financing energy projects. The research problem was solved by means of the triangulation of research methods: empirical qualitative research (desk research analysis and one of the foresight methods (plate expert), individual in-depth interviews, CAWI questionnaire, and focus questionnaire). It was noticed that there is a lack of knowledge among enterprises and local government units about the ESCO formula. Hence the most important conclusion: education of enterprises and local government units in the field of energy efficiency. The authors recommend strengthening the energy market and supporting ESCO companies.

In the last article [35] authors considered the issue of management of electricity consumption in manufacturing companies. This is to allow enterprises to optimally control the costs of electricity in the current times of pandemic, political crises, and energy transformation toward the RES. A method of analysis and management of electricity consumption in enterprises based on simulation modeling was proposed. The model takes into account energy consumption, production order execution time, machine load, and employee overtime. The obtained results show that it is possible to determine the level of power available for the process execution and its impact on the production volume and execution time. In the event that the available capacity was reduced by half, the order fulfillment time increased by nearly 25% and an increase in energy consumption by nearly 15%.

3. Conclusions

Summarizing the content of the articles in the special edition "Renewable and Sustainable Energy: Current State and Prospects", it should be stated that the energy transformation processes will systematically develop and will result in further, dynamic development of the RES sector. The process of producing renewable energy is particularly important, because the RES mix allows for the satisfaction of energy needs in the region, and even nationwide. Further development of RES means an increase in the energy security of a given country, which will increasingly use its own RES. This is of great importance especially at the present time, where, in the face of the war in Ukraine, many countries have given up importing fossil fuels from Russia and are looking for energy solutions in integrated RES. Undoubtedly, the RES sector needs additional determinants of development. One such factor is initiatives in the form of the creation of startups focused on the production, storage and distribution of renewable energy [36–38]. Another issue is to subsidize already existing successful companies in the RES sector by going public and raising new funds from the capital market. Here, the most important thing is the right moment for a company to go public (IPO) [39–42], since on the right moment depends upon the success of the debut in the form of sale of the majority of shares and their good price.

Funding: This research received no external funding.

Conflicts of Interest: The authors declare no conflict of interest.

References

1. Rees, W.E. Globalization, trade and migration: Undermining sustainability. *Ecol. Econ.* **2006**, *59*, 220–225. [CrossRef]
2. Dreher, A. Does globalization affect growth? Evidence from a new index of globalization. *Appl. Econ.* **2006**, *38*, 1091–1110. [CrossRef]
3. Skare, M.; Porada-Rochoń, M. Financial and economic development link in transitional economies: A spectral Granger causality analysis 1991–2017. *Oecon. Copernic.* **2019**, *10*, 7–35. [CrossRef]
4. Gajdos, A.; Arendt, L.; Balcerzak, A.P.; Pietrzak, M.B. Future trends of labour market polarisation in Poland. *Perspect. Trans. Bus. Econ.* **2020**, *19*, 114–135.

5. Jankiewicz, M.; Pietrzak, M.B. Assessment of trends in the share of expenditure on services and food in the Visegrad Group member states. *Int. J. Bus. Soc.* **2020**, *21*, 977–996. [CrossRef]
6. Piekut, M. Patterns of energy consumption in Polish one-person households. *Energies* **2020**, *13*, 5699. [CrossRef]
7. Szopik-Depczyńska, K.; Kędzierska-Szczepaniak, A.; Szczepaniak, K.; Cheba, K.; Gajda, W.; Ioppolo, G. Innovation in sustainable development: An investigation of the EU context using 2030 agenda indicators. *Land Use Policy* **2018**, *79*, 251–262. [CrossRef]
8. Nowak, P. Cooperation of enterprises in innovative activities on the example of Polish regions. *Equilib. Quart. J. Econ. Econ. Policy* **2021**, *16*, 839–857. [CrossRef]
9. Roszko-Wójtowicz, E.; Grzelak, M.M. Macroeconomic stability and the level of competitiveness in EU member states: A comparative dynamic approach. *Oeconomia Copernic.* **2022**, *11*, 657–688. [CrossRef]
10. Szopik-Depczynska, K.; Cheba, K.; Bąk, I.; Stajniak, M.; Simboli, A.; Ioppolo, G. The study of relationship in a hierarchical structure of EU sustainable development indicators. *Ecol. Indic.* **2018**, *90*, 120–131. [CrossRef]
11. Balcerzak, A.P. Quality of institutions in the European Union countries. Application of TOPSIS based on entropy measure for objective weighting. *Acta Polytech. Hung.* **2020**, *17*, 101–122. [CrossRef]
12. European Union. Directive 2009/28/EC of the European Parliament and of the Council of 23 April 2009 on the promotion of the use of energy from renewable sources and amending and subsequently repealing Directives 2001/77/EC and 2003/30/EC. *Off. J. Eur. Union* **2009**, *140*, 16–62.
13. Lin, M.-X.; Liou, H.M.; Chou, K.T. National energy transition framework toward SDG7 with legal reforms and policy bundles: The case of Taiwan and its comparison with Japan. *Energies* **2020**, *13*, 1387. [CrossRef]
14. Pietrzak, M.B.; Igliński, B.; Kujawski, W.; Iwański, P. Energy transition in Poland—assessment of the renewable energy sector. *Energies* **2021**, *14*, 2046. [CrossRef]
15. Gielen, D.; Boshell, F.; Saygin, D.; Bazilian, M.D.; Wagner, N.; Gorini, R. The role of renewable energy in the global energy transformation. *Energy Strat. Rev.* **2019**, *24*, 38–50. [CrossRef]
16. Chovancová, J.; Tej, J. Decoupling economic growth from greenhouse gas emissions: The case of the energy sector in V4 countries. *Equilib. Q. J. Econ. Econ. Policy* **2020**, *15*, 235–251. [CrossRef]
17. Grosse, T.G. Low carbon economy policy in Poland: An example of the impact of europeanisation. *Equilib. Quart. J. Econ. Econ. Policy* **2011**, *6*, 9. [CrossRef]
18. Cheba, K.; Bąk, I. Environmental production efficiency in the European Union countries as a tool for the implementation of goal 7 of the 2030 agenda. *Energies* **2021**, *14*, 4593. [CrossRef]
19. Kludacz-Alessandri, M.; Cygańska, M. Corporate social responsibility and financial performance among energy sector companies. *Energies* **2021**, *14*, 6068. [CrossRef]
20. Stritzke, S.; Jain, P. The sustainability of decentralised renewable energy projects in developing countries: Learning lessons from Zambia. *Energies* **2021**, *14*, 3757. [CrossRef]
21. Piekut, M. The consumption of renewable energy sources (RES) by the European Union households between 2004 and 2019. *Energies* **2021**, *14*, 5560. [CrossRef]
22. Matuszewska-Janica, A.; Żebrowska-Suchodolska, D.; Ala-Karvia, U.; Hozer-Koćmiel, M. Changes in electricity production from renewable energy sources in the European Union countries in 2005–2019. *Energies* **2021**, *14*, 6276. [CrossRef]
23. Wałachowska, A.; Ignasiak-Szulc, A. Comparison of renewable energy sources in 'New' EU Member States in the context of national energy transformations. *Energies* **2021**, *14*, 7963. [CrossRef]
24. Senkus, P.; Glabiszewski, W.; Wysokińska-Senkus, A.; Cyfert, S.; Batko, R. The potential of ecological distributed energy generation systems, situation, and perspective for Poland. *Energies* **2021**, *14*, 7966. [CrossRef]
25. Huterski, R.; Huterska, A.; Zdunek-Rosa, E.; Voss, G. Evaluation of the level of electricity generation from renewable energy sources in European Union countries. *Energies* **2021**, *14*, 8150. [CrossRef]
26. Ključnikov, A.; Civelek, M.; Fialova, V.; Folvarčná, A. Organizational, local, and global innovativeness of family-owned SMEs depending on firm-individual level characteristics: Evidence from the Czech Republic. *Equilib. Quart. J. Econ. Econ. Policy* **2021**, *16*, 169–184. [CrossRef]
27. Prokop, V.; Kotkova Striteska, M.; Stejskal, J. Fostering Czech firms? innovation performance through efficient cooperation. *Oecon. Copernic.* **2021**, *12*, 671–700. [CrossRef]
28. Rogers, J.; Simmons, E.; Convery, I.; Weatherall, A. Public perceptions of community-based renewable energy projects. *Energy Policy* **2008**, *36*, 4217–4226. [CrossRef]
29. Zoellner, J.; Schweizer-Ries, P.; Wemheuer, C. Public acceptance of renewable energies: Results from case studies in Germany. *Energy Policy* **2008**, *36*, 4136–4141. [CrossRef]
30. Igliński, B.; Flisikowski, K.; Pietrzak, M.B.; Kiełkowska, U.; Skrzatek, M.; Zyadin, A.; Natarajan, K. Renewable energy in the Pomerania Voivodeship—institutional, economic, environmental and physical aspects in light of EU energy transformation. *Energies* **2021**, *14*, 8221. [CrossRef]
31. Igliński, B.; Pietrzak, M.B.; Kiełkowska, U.; Skrzatek, M.; Gajdos, A.; Zyadin, A.; Natarajan, K. How to meet the Green Deal objectives—is it possible to obtain 100% RES at the regional level in the EU? *Energies* **2022**, *15*, 2296. [CrossRef]
32. Sidełko, R. Application of technological processes to create a unitary model for energy recovery from municipal waste. *Energies* **2021**, *14*, 3118. [CrossRef]

33. Aydin, O.; Igliński, B.; Krukowski, K.; Siemiński, M. Analyzing wind energy potential using efficient global optimization: A case study for the city Gdańsk in Poland. *Energies* **2022**, *15*, 3159. [CrossRef]
34. Kurowska-Pysz, J.; Kunikowski, G. The ESCO formula as support for public and commercial energy projects in Poland. *Energies* **2021**, *14*, 8098. [CrossRef]
35. Smagowicz, J.; Szwed, C.; Dąbal, D.; Scholz, P. A simulation model of power demand management by manufacturing enterprises under the conditions of energy sector transformation. *Energies* **2022**, *15*, 3013. [CrossRef]
36. Gorączkowska, J. Enterprise innovation in technology incubators and university business incubators in the context of Polish industry. *Oecon. Copernic.* **2020**, *11*, 799–817. [CrossRef]
37. Zinecker, M.; Skalická, M.; Balcerzak, A.P.; Pietrzak, M.B. Business angels in the Czech Republic: Characteristics and a classification with policy implications. *Econ. Res.-Ekonomska Istraživanja* **2021**, *16*, 273–298. [CrossRef]
38. Zinecker, M.; Skalická, M.; Balcerzak, A.P.; Pietrzak, M.B. Identifying the impact of external environment on business angel activity. *Econ. Res.-Ekonomska Istraživanja* **2021**. [CrossRef]
39. Meluzín, T.; Balcerzak, A.P.; Pietrzak, M.B.; Zinecker, M.; Doubravský, K. The impact of rumours related to political and macroeconomic uncertainty on IPO success: Evidence from a qualitative model. *Transform. Bus. Econ.* **2018**, *2017*, 148–169.
40. Meluzín, T.; Zinecker, M.; Balcerzak, A.P.; Doubravský, K.; Pietrzak, M.B.; Dohnal, M. The timing of initial public offerings: Non-numerical model based on qualitative trends. *J. Bus. Econ. Manag.* **2018**, *19*, 63–79. [CrossRef]
41. Meluzín, T.; Zinecker, M.; Balcerzak, A.P.; Pietrzak, M.B. Why do companies stay private? Determinants for IPO candidates to consider in Poland and the Czech Republic. *East. Eur. Econ.* **2018**, *56*, 471–503. [CrossRef]
42. Meluzín, T.; Zinecker, M.; Balcerzak, A.P.; Pietrzak, M.B.; Doubravský, K. Institutional Settings and their Impact on the IPO Activity: An Exploratory Study Based on Qualitative Modelling. *Acta Polytech. Hung.* **2021**, *18*, 215–235. [CrossRef]

energies

MDPI

Article

Application of Technological Processes to Create a Unitary Model for Energy Recovery from Municipal Waste

Robert Sidełko

Faculty of Civil Engineering, Environmental and Geodetic Sciences, Koszalin University of Technology Poland, Śniadeckich St. 2, 75-453 Koszalin, Poland; robert.sidelko@tu.koszalin.pl; Tel.: +48-94-347-85-00

Abstract: The subject matter of this paper is the functioning of a highly effective waste management system. Assumptions of the Energy Recovery Waste Processing (ERWP) model, being a universal solution for towns and regions irrespective of their population, are presented here. The result of simulations illustrating the energetic potential of municipal waste stored and processed in biological and physicochemical processes are also presented. Calculations were performed for the municipality of Koszalin (Poland), with a population of 106,000. Mixed household and commercial waste, organic waste, waste from selective collection and sewage sludge from a municipal wastewater treatment plant were considered in the waste mass balance. Empirical equations and unit coefficients describing the energetic efficiency of particular processes originating from the author's own research work as well as from the results available from the scientific literature were used in the calculations. The developed ERWP model is based on the functioning of four objects constituting a comprehensive technical infrastructure, i.e., biological stabilisation in air condition (BSAC), mechanical treatment plant (MTP), cogeneration system plant (CSP) and gas production plant (GPP) where two independent modules operate, namely, dry/wet methane fermentation (DMF and WMF). Each day, this system generates highly energetic refuse-derived fuel (RDF) for combustion in amounts of 82.2 t for CSP and 127.3 t for GPP, generating 5519 m^3 of gas/d. The value of the energy contained in such generated gas and in waste making up an alternative fuel is 1027.4 GJ, which is equivalent to 285.4 MWh. It should be noted that the creation of a waste management system based on the ERWP model assumptions fulfills the criteria of energetic recycling and allows for recovery of energy in the form of gas and heat equivalent to 79,917.6 MWh/a, i.e., 754 kWh/inhabitant/a.

Keywords: waste management; energy recovery; model of energy recovery; biogas; fermentation; combustion

Citation: Sidełko, R. Application of Technological Processes to Create a Unitary Model for Energy Recovery from Municipal Waste. *Energies* **2021**, *14*, 3118. https://doi.org/10.3390/en14113118

Academic Editor: Paolo Defilippis

Received: 26 April 2021
Accepted: 25 May 2021
Published: 26 May 2021

Publisher's Note: MDPI stays neutral with regard to jurisdictional claims in published maps and institutional affiliations.

Copyright: © 2021 by the author. Licensee MDPI, Basel, Switzerland. This article is an open access article distributed under the terms and conditions of the Creative Commons Attribution (CC BY) license (https://creativecommons.org/licenses/by/4.0/).

1. Introduction

This paper pertains to municipal waste that may have a negative impact on the natural environment and human health [1–3]. Waste management is a global problem originating from economic development [4–7]. Globalisation processes have contributed, through the systematic increase in the interlinkage between various markets and in numerous aspects of economic and social life, to the establishment of a new institutional order, new institutions and legal and economic solutions [8,9]. Such processes also promote increases in outlays on investment projects [10,11], boosting the level of innovation and competitiveness of economies [12–16]. The fact that in the last thirty years an economic convergence has occurred for the majority of countries is indicated in the literature [17,18]. The economic convergence processes contributed to permanent changes in consumers' attitudes and awareness, particularly in terms of sustainable consumption when considering concerns about energy consumption and the natural environment [19–23]. Energetic transformation embraces the majority of economies, including agriculture, in which use of biocomponents produced from organic waste plays a significant role [24,25]. This means that waste management is equally important for big metropolitan areas and small rural communities [26,27]. An observed trend of the use of waste as a source of valuable raw materials requires development of a functioning waste management model, taking into consideration the possibilities and needs of a given region [28].

In the article, practical and theoretical aspects of the municipal waste elimination for the city of Koszalin based on an effective functioning waste management system will be considered. The main objective of this paper is the presentation of the ERWP model to be used as a complementary waste management system. The effectiveness of a system based on the ERWP model has been discussed in relation to Koszalin city (Poland).

It should be underlined that the volume of waste generated worldwide increases systematically, yet a significant part of it is not covered by the system, thus causing quantifiable losses in the economy and natural environment [29,30]. In 2018 in the European Union, 2538 million t of municipal waste was generated, and 55% was recycled [31]. The remainder was put in landfills, causing their degradation to a variable degree [32]. The absence of proper industrial infrastructure for waste processing, mainly in the eastern EU countries, is still the main reason why 40 ÷ 45% of waste does not go to local waste treatment systems, resulting in wasting of valuable sources of recyclable and energetic materials [33,34]. This results in an increase in costs of economic system functioning that is different depending on the solutions adopted [35]. In Poland, almost 42% of generated municipal waste is directly put into landfills [36].

The ERWP model takes into account a number of conditions, among which the most important ones are: waste type, physicochemical properties, morphological composition and volume. Irrespective of the values of the above parameters, such systems are always based on the application of well-known unit processes, such as: screening, separation, biological treatment, dehydration, thermal transformation and storage. The proper compilation of processes that make up a complementary system depends on the adopted objective that is contingent upon financial and technical possibilities. Such an objective can be, for example, the recovery of valuable waste components within the material or organic or energetic recycling framework [37]. Ultimately, the objective can also be rendering such waste harmless through its storage in controlled conditions. However, taking into account waste composition and its physicochemical properties, this method is economically ineffective and, importantly, has a negative impact on the natural environment [38,39].

2. Potential of Municipal Waste to Energy Production—Review

The introduction indicates the need for the construction of a waste management system for any local government unit. The main task of such a system is to use waste as a source of raw materials including raw materials for energy production. The volume of energy generated from waste depends, first and foremost, on the methods of processing, allowing for the recovery of energy in variable forms, generally in the form of heat generated in the combustion process [40,41]. Waste can be, as a source of energy, a significant element of the local energetic balance [42]. A good example here is the municipality of Copenhagen, which adopted in 2011 a strategy for development until 2025, which will eliminate the use of coal as the energetic raw material [43]. The share of energy generated from municipal waste in Copenhagen's energetic mix will finally be approximately 40%.

In 2018, 12.5 million t of mixed municipal waste was stored in Poland. Most of the waste, i.e., 9971.2 thousand t, was generated in households, which made up 83% of total generated waste. The remaining part of the waste, i.e., 1997.5 thousand t of household and commercial types, collected from the servicing of municipal infrastructure and entrepreneurs, amounted to 17%. Analysis of the morphological composition of waste delivered to plants using methane-biological processing technology (MBT), based on results of research work performed in 20 plants located in Poland, shows a great potential for broadly understood recovery [44]. Classification of particular waste components in terms of their use leads to distinguishing the following groups: Group No I—recyclable materials (glass, metals, synthetic materials qualified for recycling), Group No II—waste having high energetic value (paper, cardboard, textiles, composites and synthetic materials not qualified for recycling, as well as wood), Group No III—biodegradable waste (BIO1, BIO2, BIO3) and

Group No IV—waste classified as useless or dangerous. Percentage shares of particular waste components in mixed waste as well as in fractions separated with an 80–90 mm mesh screen, as average values obtained from research work performed in various plants in Poland, are shown in Table 1.

Table 1. Percentage shares of waste components and classification into four different groups.

No.	Components	Mixed [(1)]	>80–90 mm [(1)]		<80–90 mm [(1)]	
		$\%_{(w/w)}$	$\%_{(w/w)}$	Group, No	$\%_{(w/w)}$	Group, No
1	Kitchen (BIO1)	25.1	0.9	III	12.7	III
2	Park, garden (BIO2)	0.3	0.6	III	0.3	III
3	Organic (BIO3)	0.6	5.9	III	34.7	III
4	Wood	0.6	1.3	II	0.6	II
5	Paper	14.6	22.4	I	9.8	II
6	Plastic	14.1	32.6	I	5.7	II
7	Glass	8.6	1.4	I	11.1	I
8	Textiles	39	12.7	II	0.6	II
9	Metals	2	1.4	I	1.3	I
10	Multicomponent	3.6	6.5	II	1.5	II
11	Other	9.5	14.2	IV	21.7	IV
	<20 mm	17.7	-	-	-	-

[(1)] Own study.

Waste possessing values of recyclable materials (Group No I) recovered for mixed waste during the manual separation process can be obtained, as a rule, exclusively from the oversize fraction separated in, e.g., rotary drum sifters featuring mesh not less than 8 cm; the total content is 37.2%. A method of using Group No II, defined as RDF, considering its high calorific value of 18–24 MJ/kg, is combustion [44]. The application of screening and mechanical separation of mixed waste resulting in energetic raw material in RDF form allows, in the extreme case, for an increase in the calorific value from 8.4 to 25.0 MJ/kg [45]. A calorific value exceeding 11 MJ/kg guarantees, in principle, energetic efficiency of combustion or gasification exceeding 65%, which allows for classification of the process as energy recovery [46]. According to the International Energy Agency, the calorific value of waste to be used in combustion processes should not be, for process profitability, below 7942 kJ/kg [47]. In Germany, one of the biggest EU economies, the volume of RDF separated in MTP installations increased from 31% in 2006 [48] to 34.2% in 2017 [49]. A significant criterion for the application of the available techniques of thermal transformation of mixed municipal waste is relatively high humidity, which reduces their calorific value. The use of more advanced techniques such as, for example, pyrolysis or gasification, requires higher calorific value of waste; this is associated with a necessity to apply proper methods of batch preparation [50,51]. Unfortunately, pyrolytic installations of an industrial scale used for waste processing are unreliable, which has been proved by plants closing shortly after being put into operation. For example, THERMOSEL 2002 (opened)/2006 (closed), DBA 2001/2010, EDDITH 2002/2009 and Schwel-Brenn 1997/2000 [52,53].

Group No III comprises biodegradable waste, i.e., waste that can be subjected to biological gasification in the methane fermentation process. During decomposition of organic matter under controlled anaerobic conditions, biogas is generated, which contains flammable components, including methane. The share of biogas production in European Union countries makes up 136.6 million tons of oil equivalent [54]. The content of methane, depending on the raw material, is 50–75% and the calorific value is, on average, 22 MJ/m^3 [55,56]. In the case of the organic fraction separated from municipal waste, the yield of biogas volume in the plants that use the methane fermentation process in low hydration conditions (DFI) reaches 339 m^3 CH_4/Mg organic matter [57]. A possibility to generate biogas, resulting in a reduction in waste processing costs will be, in future, the primary factor deciding the selection of this biological waste processing method. Waste that is also suitable for the methane transformation process is sewage sludge (SS), originating from municipal wastewater treatment plants. In 2018 in Poland, 640,000 t of dry matter from sewage sludge was generated. This is equivalent to approx. 25 million tons of mechanically dehydrated sludge to 80% humidity on average [36]. The volume of methane generated during the fermentation process carried out in separate chambers in high-hydration conditions is 0.19–0.24 m^3/kg organic matter [58]. The application of the thermal method for the disposal of sewage sludge requires a reduction in its humidity to at least 10%. This means that the consumption of energy for mechanical dehydration, then water evaporation from sludge with an initial humidity of 98%, is very high. Other sorts of waste, which due to their properties may constitute a raw material for energy generating processes in various forms, are generated in populated areas; they comprise, among other things, waste from selective collection, including biowaste, organic waste from green area cultivation and biodegradable waste from production and foodstuff processing as well as flammable packaging waste with low value as recyclable material.

The diversity of municipal waste means that optimisation of the system of energy generation from waste should take into account not just waste fuel properties but also its morphological features. This also means there is a necessity to apply various methods for the preparation of the raw material earmarked for the generation of energy in the form of heat, electricity and gas. Using (i) available techniques and (ii) applying the principle of cooperation between the waste generating and processing entities [59], an ERWP model was developed; it allows for assessment of the volume of energy generated from the processing of amassed waste independently of its volume, type and specific features. Empirical equations describing the impact of selected variables on the volume of generated energy and values of empirical indicators describing energetic efficiency of particular processes, achieved both in industrial plants as well as used in scientific research, were used for setting the energy balance. ERWP reflects the circular economy idea, which promotes the maximum usage of available raw materials in line with a rule that waste becomes a raw material for the next production cycle [60,61].

Waste-to-energy (WtE) plants are an integral part of the circular economy strategy in the treatment of non-recyclable waste. Waste with a high potential for thermal gasification or biogas production is converted into heat and electric energy. The ERWP model takes into account two of the six defined trends of the WtE strategy, i.e., more gasification plants offering commercial-scale operations and a push to use organic waste to replace natural gas [62].

3. Materials and Methods

Organisational and urban structures that are suitable for the creation of comprehensive energy generation systems using various sources, including municipal waste, are big cities and communal special purpose associations. The benefits originating from implementation of waste processing procedures contained in the ERWP model were analysed based on municipal waste produced in the municipality of Koszalin located in the northern part of Poland in the Pomerania region. The simulation was performed for mixed waste, biodegradable waste, waste from selective collection and mechanically dehydrated sewage sludge from the municipal wastewater treatment plant. Based on the structural research results, using the ERWP model assumptions, volumes of waste flowing between particular plants were determined; on these grounds, the volumes of waste for making energetic raw material were fixed. This model allows for comprehensive assessment of the waste economy system's efficiency, including definition of the waste volume being directed to raw materials recycling or materials recycling. The aspect associated only with the energetic potential of waste processed in the plants that use the combustion and methane fermentation techniques is presented herein.

3.1. Composition of Municipal Waste

Structural research work on mixed waste was performed at the Regional Waste Management Plant in Sianów, which has a mechano-biological waste processing installation. The research work comprised the performance of a morphological analysis of two fractions that were mechanically separated in a rotary drum sifter with 90 mm screen mesh. Samples of waste from both fractions amounting to approximately 100 kg were taken for morphological examination. Tests were performed three times. From fractions defined as oversized (fr. > 90 mm) and subscreen (fr. < 90 mm), nine waste components were separated, i.e., biodegradable waste, plastics, paper and cardboard, glass, metals, textiles, wood, composite materials and other. Percentage shares of particular components were determined by their weight and comparison of the partial results to the entire mass of the sample. The shares, which were calculated in that way, indicated percentage by weight ($\%_{(w/w)}$). The percentage shares of the oversized and subscreen fractions in the mixed waste stream separated in the rotary drum sifter were determined from the mass balance of MBTP's mechanical part [44]. The components separated in both fractions were combined, based on the classification criterion for a given group, thus creating Group No II and Group No III. From both groups, samples were taken for physicochemical tests.

3.2. Physicochemical Analyses

Apart from the samples taken from Groups No II and No III, which were composed of the components separated at the structural examination stage, samples of mechanically dehydrated sewage sludge taken from the municipal wastewater treatment plant in Koszalin were also subjected to physicochemical tests. The research included recording dry mass (*dm*) according to PN-EN 14346 and organic matter (*om*) according to PN-EN 15169: 2011. The analysis of C and N content was accomplished by Elementar, VarioMax CN. About 10 mg of dried powdered homogenous sample was used to determine the percentage of carbon and nitrogen. The measurement uncertainty of both analytes was the same, i.e., ±0.5%.

3.3. Model for Energy Waste Management

Taking into account the method of waste storage, four primary sorts of municipal waste generated in populated areas can be distinguished, i.e., mixed waste, waste from selective collection, biowaste and sewage sludge originating from a municipal wastewater treatment plant. The ERWP model is based on the cooperation of four industrial objects responsible for performing various tasks connected with processing of the above defined sorts of waste (Figure 1). The abbreviations used are presented in Table 2.

Table 2. List of abbreviations.

BOF	Biodegradable Organic Fraction	BHDS	Biological Half-Digested Sludge
BSAC	Biological Stabilisation in Air Condition	CSP	Cogeneration System Plant
DMF/WMF	Dry/Wet Methane Fermentation	GPP	Gas Production Plant
GPI	Gas Pretreatment Installation	HHV/LHV	Higher/Lower Heating Value
HM	Harmful Materials	MBTP	Mechanical Biological Treatment Plant
MTP	Mechanical Treatment Plant	RDF	Refuse-Derived Fuel
RM	Recyclable Materials	SRM	Separated Row Materials
SS	Sewage Sludge		

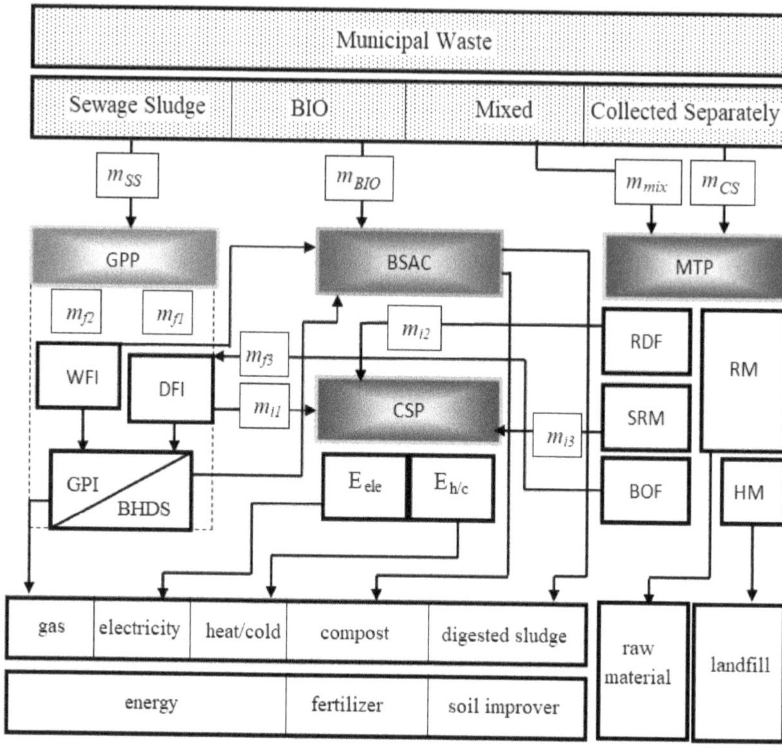

Figure 1. Block diagram of the ERWP model.

The first object is the mechanical treatment plant (MTP), to which mixed waste is directed. The MTP technique is based mainly on the screening as well as manual/mechanical separation processes. The second plant is the cogeneration system plant (CSP). In this object, thermal transformation of waste using a highly efficient combustion process will be carried out. The third plant is the gas production plant (GPP) processing biodegradable waste using two types of methane fermentation performed in high (WMF module) and in low water content conditions in the so-called dry fermentation process (DMF module). The fourth plant is the biological stabilisation in air condition (BSAC), i.e., the use of an intensive biological material aeration process under controlled conditions (composting). The key issue qualifying determination of the entire system's energetic potential is the mass balance of waste flowing between particular plants and the balance of energy generated in unit processes provided in the ERWP model.

3.4. Mass Balance and Energy Balance

In setting the mass balance for raw materials used in energy generation, the volume of waste streams resulting from the splitting of waste mass originating from MTP and GPP technology—m_{f1}, m_{f2}, m_{f3}, m_{i1}, m_{i2}, m_{i3}—must be calculated (Figure 1). In order to do this, it is necessary to determine volumes of waste supplied to the system—m_{SS}, m_{BIO}, m_{mix}, m_{CS}—and know the results of the examination of waste morphological composition. The method of assessment of particular waste stream masses is shown in Table 3.

Table 3. Mass balance equations.

Mass	Equation [(1)]	Eq. No.	Assumption [(1)]
m_{f1}	$m_{f3}(W_{BOF} - W_{DF})/(W_{DF} - W_{SS})$	(1)	$m_{DF} \cdot W_{DF} = \sum(m_i \cdot \cdot \cdot W_i)$
m_{f2}	$m_{SS} - m_{f1}$	(2)	part of *SS* to *WMF* process
m_{f3}	$m_{mix}(\%_{fr>90\ mm} \cdot \%_{No\ III} + \%_{fr<90\ mm} \cdot \%_{No\ III})0.01$	(3)	Group No III of both fractions
m_{i1}	$\Delta m\ (m_{f3} + m_{f1})$	(4)	Δm, as a result of *om* reduction
m_{i2}	$m_{mix}(\%_{fr>90\ mm} \cdot \%_{No\ II} + \%_{fr<90\ mm} \cdot \%_{No\ II})0.01$	(5)	Group No II of both fractions
m_{i3}	$m_{CS} \cdot \cdot \cdot 0.4 + m_{BIO} \cdot \cdot \cdot 0$	(6)	worn out furniture, doors, ... : *Cs*, wood and branches: *BIO*

(1) Own study.

Two energy-generating processes have been provided in the system based on the ERWP model. These are methane fermentation in high- and low-hydration conditions and thermal waste transformation through combustion. The result of the oxidation of organic compounds in the municipal waste combustion process in the presence of oxygen is the liberation of thermal energy amounting to 10 MJ/kg (LHV) [63]. To calculate the parameters allowing for assessment of the energy potential, equations describing the relationship between the higher heating value (HHV) and lower heating value (LHV) as variables dependent on the contents of selected elements: HHV, LHV = $f(C, H, N, O, S)$, can be used [64]. The elements are determined by application of elemental analysis and their contents are indicated as percentages of dry mass. Mutual relationships described in the literature are expressed by multiple regression equations having the general form of $Y = a + b_1 \cdot X_1 + \ldots\ldots + b_n \cdot X_n$. The equations obtained are then approximated by application of the least squares method and in the majority of cases they have a high coefficient of determination (R^2), which shows a good match of the estimated function (Table 4).

Table 4. Equations used to determine HHV and LHV in MJ/kg depending on the chemical composition.

Materials	Equation	R^2	References
Over sieve fraction of MSW	HHV = 0.3491C + 1.1783H + 0.1005S − (0.1034O + 0.0151N + 0.0211A)	0.85	[65]
Sewage Sludge	HHV = 0.2322C + 0.7655H − 0.072O + 0.0419N+ 0.0698S + 0.0262Cl + 0.18814P	-	[66]
Biomass materials	HHV = 0.328C + 1.4306H + 0.0929S − 0.0237N − (1 − A/100)· (40.11H/C) + 0.3466	-	[67]
CR- sifted ballast after composting of MSW and MTR-rejected after manual separation	$LHV_d = HHV - (212.2H + 0.8(O + N))$ $LHV_w = HHV_d - ((1 - 0.01M) - 24.43M)$	-	[48]

Usage of the equations contained in Table 3 is associated with a necessity to analyse contents of the elements making up the independent variables. Arriving at a reliable result requires (i) application of the same methodology of preparation of the analytical sample and (ii) use of raw material procured in the same way as that used for the development of the given equation. Condition (ii) originates from the limitation of the impact of the so-called discreet variables that are not taken into account in the equations in Table 3. Finally, to estimate HHV, an equation was used which takes into account, as independent variables, contents of various materials in the waste mix making up the energetic material [68], in the following form:

$$HHV = 0.0535(F + 32.6 \cdot \cdot \cdot CP) + 0.3722 \cdot PLR,\ MJ/kg \tag{7}$$

Particular values in Formula (7) indicate contents of: *F*—bio fraction, *CP*—cardboard and paper, *PLR*—plastic, leather and rubber in the dry waste mixture expressed in $\%_{(w/w)}$.

Approximation of the results based on determination of the *HHV* value, being a dependent variable for sewage sludge samples after the methane fermentation process with variable organic substance contents, allowed for the development of a model for which coefficient R^2 was 0.87 [69]:

$$HHV = 0.2132 \cdots Z_{om}, \text{MJ/kg} \quad (8)$$

Z_{om} in Formula (8) indicates contents of organic matter in % *dm*.

The higher heat value of wood waste was assessed from the following equation [70]:

$$HHV = 0.4373C - 1.6701, \text{MJ//kg} \quad (9)$$

Gas yield from methane fermentation is described, in practice, by two indicators. The first one is the rate of biogas production (GPR), counted as a quotient of the daily volume of generated gas and reactor volumetric capacity expressed in $m^3_{gas}/m^3_{reactor}/d$. This indicator is used directly to fix the calculated flow rate in the installation and to select proper biogas-processing devices. The second indicator is the unit gas production (GP), calculated by division of the daily generated gas volume by the daily load of organic matter, which shows the volume of biogas generated from raw material mass, i.e., $m^3_{gas}/Mg\ om$. GP is used, first and foremost, for assessment of energetic potential associated with economic analysis. Fermentation of the organic fraction, depending on the participation of other co-materials, gives the value of GP = 222 ÷ 350 $dm^3\ CH_4/kg\ om$. [71]. Table 5 shows GP values obtained from examination of the fermentation process using various raw materials both in low- and high-hydration conditions.

Table 5. The indicators used to determine WFI and DRI yield of gas production.

Raw Materials	T °C	M %	GP $dm^3/kg\ dm$	$CH_4\ \%_{(vol)}$	HRT d	References
SS	36	95.5	56.5	-	28	[72]
SS	35	98.4	64.7	72.3	27	[73]
SS	35	90	171	-	4 ÷ 59	[74]
SS	55	82.5	164 ÷ 233	-	30	[75]
OFMSW + SS	35	97.7	76	64.6	27	[73]
OFMSW + SS	35	80	215	76.5	-	[76]
OFMSW + SS	35	89.3	265 ÷ 311	-	44 ÷ 71	[77]
OFMSW + FVW + SS	35	80	433.9	80.8	-	[76]

FVW: fruit and vegetable waste, HRT: hydraulic retention time, GP: gas production, M: moisture.

4. Results

Determination of the energetic potential based on the ERWP model was performed for the municipality of Koszalin with statistical data pertaining to the volume of municipal waste amassed in 2018 using the results of research describing waste composition and selected physicochemical parameters. The volume of household and commercial waste that was generated at that time was 30,760 t. The mechanical part of the plant to which the waste is directed comprises recovery of the fraction (Fr. > 200 mm) set for manual separation. The fraction below 200 mm makes up 89.2% and is directed in full to a rotary drum sifter with 90 mm mesh [44]. The average share of the oversized (Fr. > 90 mm) and subscreen (Fr. < 90 mm) fractions was 46.1% and 53.9%, respectively. The percentage shares of separated waste elements in both fractions are shown in Figure 2. Classifying all separated waste elements into four groups, shares in both fractions (Figure 3) were fixed and on this basis average shares of Groups No I–IV in mixed waste were calculated; they were: 44.2, 14.0, 20.7 and 21.1%, respectively.

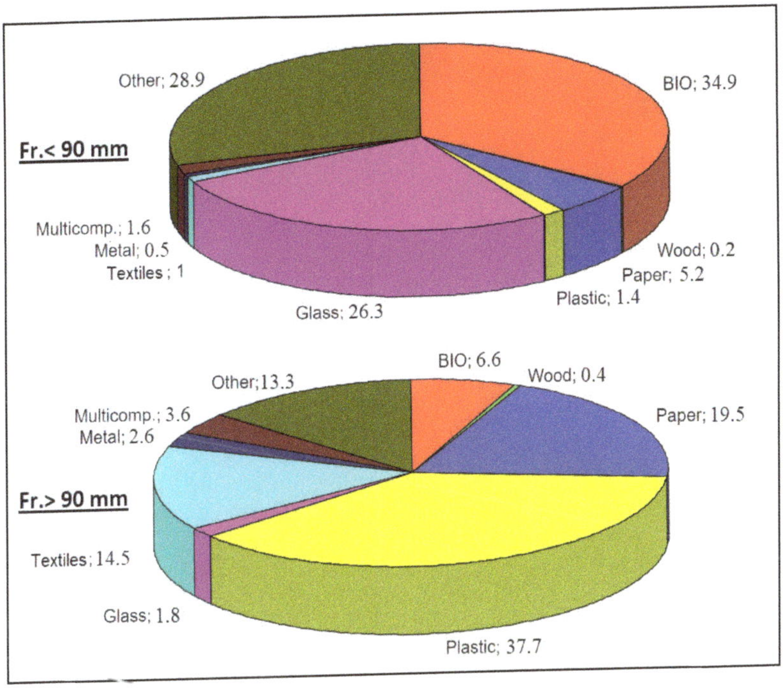

Figure 2. Percentage of waste components in two particle size fractions.

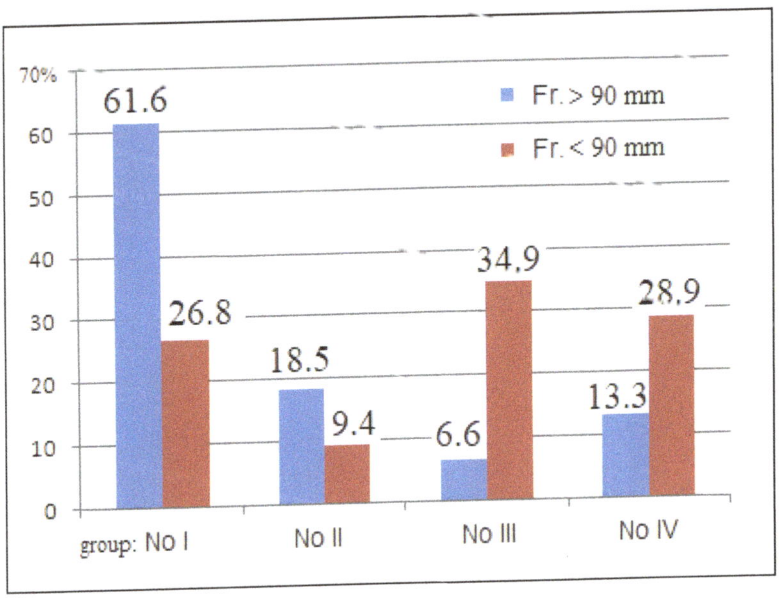

Figure 3. Percentage of four groups (I–IV) in two particle size fractions.

The mass balance was set assuming 280 working days per year, which allowed for fixing the daily volume of mixed waste (M_d) directed to MTP as approx. 110 t. Using Equations (3) and (5), the mass of the waste attributed to Group No II—m_{i2}—directed to CSP and mass of the waste attributed to Group No III—m_{f3}—directed to DFI were calculated. These masses amounted to 13.7 and 20.3 t/d, respectively. The results of physicochemical tests of samples of both waste groups and sewage sludge before and after mechanical dehydrations are shown in Table 6. The content of organic matter found in sewage sludge was in the range of 65–75% *dm*, defined as characteristic [78].

Table 6. Physicochemical test results.

Raw Materials	Dry Mass (*dm*), %	Org. Matter, % *dm*	C_{org}, % *dm*	N, % *dm*
Group No II	78.0	25.0	51.22	0.48
Group No II	72.0	65.0	34.56	1.11
Sewage Sludge (SS)	5.1	74.0	55.0	4.11
Dewatered SS	21.0	72.0	51.12	4.04

5. Discussion

In total, 3.044 t of biodegradable (BIO) waste was stored. In accordance with the earlier description, this waste originated from selective collection performed in individual households, collection from the gastronomic sector and green area cultivation. The quality of BIO waste justifies its delivery to BSAC provided that any wood waste that was not used for the production of wood chips as structure-forming material in any composting process was directed to the CSP [79]. As there is no accurate data to that effect, this stream of waste was omitted by putting 0 in Equation (6). In waste from selective collection, amounting to 100 tons/year, waste window and door woodwork as well as worn furniture make up 25%, which makes stream mass m_{i3} equal to 25 t/a.

In the Koszalin-Jamno wastewater treatment plant featuring throughput of 36,000 m^3/d, approximately 71 tons of mechanically dehydrated sewage sludge (m_{SS}) with 21% *dm* is produced during a day. The predisposed fermentation technique BOF, due to the high concentration of dry matter, is dry fermentation, which proceeds in low-hydration conditions below 85% [80]. This allows for a considerable reduction in reactor volumetric capacity compared to wet fermentation that is performed at a hydration level exceeding 94% [81]. Considering the composition and structure of BOF, this waste, being a coarse-grained raw material, requires preliminary preparation, i.e., comminution. The optimum degree of comminution of the organic fraction for the dry fermentation process should take into account the maximum share of granules with dimensions falling into the 20 to 40 mm interval [82]. The impact of comminution on the increase in fermentation gas yield is not clear-cut. The majority of the available sources of information indicate that the increase in the active surface of organic particles through material comminution facilitates access of microorganisms to nutritive substrates, thus improving the process conditions [83]. The ERWP model is based on the assumption that BOF arrives in its entirety to the DFI module of the GPP. The required content of water for the dry fermentation process, due to the low humidity of BOF at 28%, is secured by supplementation of mechanically dehydrated sewage sludge with 79% water content. To determine the mass of sewage sludge m_{f1} and m_{f2} directed to the methane fermentation process, Equations (1) and (2) were used, respectively. Based on these equations, it was found that 79 t of sewage sludge containing 95% water is directed to the wet fermentation process, whereas 28 tons of sewage sludge containing 81% water mixed with 20.3 tons of the organic fraction separated from mixed waste makes up a mass of daily charge into the dry fermentation process. The decrement of the dry mass of sewage sludge in the methane fermentation process is 31–35%, which causes a decrease in the heat of combustion on average by 22% [69]. The mass of post-fermenter m_{i1} amounting to 43.1 t was calculated using Equation (4).

The volume of gas generated in the GPP in both VFI and DFI modules was calculated using the unit indicators presented in Table 5. In the case of wet fermentation performed in the VFI module, an indicator of 60.6 dm^3/kg *dm* was used as an average value of two empirical indicators given by [72,73]. In both cases, the fermentation process proceeded in mesophilic conditions and a similar period of charge was used in the HRT reactor. The generation of gas in the DFI dry co-fermentation process using a mixture of sewage sludge and organic fraction separated from municipal waste (Group No III) was estimated using an indicator value of GPP = 265.0 dm^3/kg *dm* [76]. The adopted indicator corresponds to the value attained on an industrial scale in a methane fermentation plant based on the LARAN technique employed in the municipality of Tychy (Poland), where the average daily production of gas in mesophilic conditions at HRT = 20 days is 266.3 dm^3/kg *dm* [44]. A slightly lower gas yield obtained from a dry fermentation process amounting to 215.0 dm^3/kg *dm* has been observed in laboratory conditions [73]. For the indicators assumed, the volume of gas generated in WFI and DFI modules is 239.37 and 5280.06 m^3/d, respectively. The energy equivalent to the volume of gas generated in both modules of methane fermentation was calculated by assuming that the combustion heat was 22.5 MJ/m^3 [42] (Table 7).

Table 7. Mass and energy production using ERWP model.

Process	Raw Materials	Daily		
		Mg	GJ	MWh
WFI	m_{f2}	79	5.39	1.49
DFI	$m_{f1} + m_{f3}$	48.3	118.8	33
CSP	m_{i1}	43.5	555.89	154.37
CSP	m_{i2}	13.7	335.58	93.19
CSP	m_{i3}	0.1	11.78	3.26

To CSP generating thermal energy and electricity directed the remains from the dry fermentation process (m_{i1}), Group No II (m_{i2}) and door/window woodwork as well as waste furniture, jointly making a high-calorie raw material SRM (m_{i3}). The content of organic matter in the waste stream m_{i1}, determined as a weighted average in sewage sludge and BOF (Group No II) after the fermentation process, causing partial oxidation of *so*, was 60.56%. To calculate HHV for m_{i1} and m_{i2}, Equations (7) and (8) were used, respectively. Contents of carbon and nitrogen in Group No II determined in the elementary analysis method are similar to the results of research published in numerous papers. For example, the share of carbon determined in the elementary analysis of C, N, O, H, P was 47.81% [48] and 44.72% [63]. HHV values calculated from regression equations or determined in the heat analysis indicated in the papers cited above were 23.19 and 19.50 MJ/kg *dm*, respectively whereas the HHV value found in my own research work, calculated from Equation (7), was 24.5 MJ/kg *dm*. The balance of raw materials mass and energy from their processing using the ERWP model's energy generation processes is presented in Table 7.

6. Conclusions and Recommendation

The ERWP model presented here serves for the planning of a waste management system based on the all known forms of recycling, including recycling of energy. The disposal of municipal waste based on the presented model is, in fact, an implementation of the circular economy strategy, in which waste becomes a raw material. The model, taking into account specific features of the given waste, provides for proper preparation of the raw material constituting a charge to the energy generating process. This pertains mainly to separation, through mechanical separation of mixed waste, of highly energetic RDF having an HHV value close to 24.5 MJ/kg versus 16 MJ/kg of mixed waste directed to combustion in its entirety. The separated organic fraction together with dehydrated sewage sludge is then subjected to methane co-fermentation in low water content conditions featuring a high unit gas yield of 265 dm^3/kg *dm*. The remaining part of the sewage sludge is directed to the wet fermentation process. The application of three energy-generating processes in the model described above provides a source of alternative energy, reducing consumption of fossil fuels such as crude oil, natural gas and coal. In a year, the energetic value of generated biogas and raw materials originating from selective collection as well as separated from the mixed waste stream for a municipality, with a population of almost 106,000, is 79,917.6 MWh, which substitutes approximately 10,000 t of coal.

The described ERWP model can be used in planning and implementation of an effective waste management system of any scale. The effectiveness of the model has been demonstrated on the grounds of a medium-sized city having approximately 110,000 inhabitants. As a rule, bigger cities generate larger volumes of waste, which increases the profitability of the implementation of a system based on the ERWP model. The fundamental advantage of the model is the use of practical processes that make up the system, where products from one site become a raw material for other sites. Thus, potential stakeholders generating or gathering municipal waste are united by one goal—effective waste processing. The unit value of energy generated in the system based on the ERWP model amounting to 754 kWh/inhabitant/a cannot cover all of the energetic demand of the local population. However, the example of Copenhagen shows that the share of energy generated from municipal waste processing may constitute up to 40% of the entire energy demand. Use of the alternative source of energy, which is the municipal waste through the creation of a system based on known processes, is the fundamental advantage of the ERWP model that justifies its application.

Funding: This research received no external funding.

Institutional Review Board Statement: Not applicable.

Informed Consent Statement: Not applicable.

Conflicts of Interest: The author declare no conflict of interest.

References

1. Guar, V.K.; Sharma, P.; Awasthi, M.K.; Dussap, C.G.; Pandey, A. Assessing the impact of industrial waste on environment and mitigation strategies: A comprehensive review. *J. Hazard. Mater.* **2020**, *398*, 123019. [CrossRef]
2. Janowska, B.; Szymański, K.; Sidełko, R.; Walendzik, B.; Siebielska, I. Assessment of mobility and bioavailability of mercury compounds in sewage sludge and composts. *Environ. Res.* **2017**, *156*, 394–403. [CrossRef]
3. Golwala, H.; Zhang, X.; Mdlskander, S.; Smith, A.L. Solid waste: An overlooked source of microplastics to the environment. *Sci. Total Environ.* **2021**, *769*, 144581. [CrossRef]
4. Qugaard, M. *Political Globalization: State, Power, and Social Forces*; Palgrave Macmillan: New York, NY, USA, 2004.
5. Vogli, R.; Kouvonen, A.; Elovainio, M.; Marmot, M.G. Economic globalization, inequality and body mass index: A cross-national analysis of 127 countries. *Crit. Public Health* **2014**, *24*, 7–21. [CrossRef]
6. Schmukler, S.L.; Abraham, F. *Financial Globalization*; World Bank: Washington, DC, USA, 2017.
7. Sánchez-López, C.; Aceytuno, M.T.; De Paz-Bañez, M.A. Inequality and globalisation: Analysis of European countries. *Econ. Sociol.* **2019**, *12*, 84–100. [CrossRef]
8. Shkolnyk, I.; Kozmenko, S.; Kozmenko, O.; Mershchii, B. The impact of economy financialization on the level of economic development of the associate EU member states. *Econ. Sociol.* **2019**, *12*, 43–58. [CrossRef] [PubMed]
9. Balcerzak, A.P. Quality of institutions in the European Union countries. Application of TOPSIS based on entropy measure for objective weighting. *Acta Polytech. Hung.* **2020**, *17*, 101–122. [CrossRef]
10. Halaskova, M.; Gavurova, B.; Korony, S. Change of EU28 countries research and development indicators between 2010 and 2015. *Econ. Sociol.* **2020**, *13*, 230–248. [CrossRef] [PubMed]
11. Škare, M.; Franc-Dąbrowska, J.; Cvek, D. Cointegration analysis and VECM of FDI, employment, export and GDP in Croatia (2002/2017) with particular reference to the global crisis and poor macroeconomic governance. *Equilib. Q. J. Econ. Econ. Policy* **2020**, *15*, 761–783. [CrossRef]
12. Gavurova, B.; Soltes, M.; Kovac, V. Application of cluster analysis in process of competitiveness modelling of Slovak Republic regions. *Transform. Bus. Econ.* **2017**, *16*, 129–147.
13. Simionescu, M.; Balcerzak, A.P.; Bilan, Y.; Kotásková, A. The impact of money on output in Czech Republic and Romania. *J. Bus. Econ. Manag.* **2018**, *19*, 20–41. [CrossRef]
14. Szopik-Depczyńska, K.; Kędzierska-Szczepaniak, A.; Szczepaniak, K.; Cheba, K.; Gajda, W.; Ioppolo, G. Innovation in sustainable development: An investigation of the EU context using 2030 agenda indicators. *Land Use Policy* **2018**, *79*, 251–262. [CrossRef]
15. Kisel'áková, D.; Šofranková, B.; Onuferová, E.; Čabinová, V. The evaluation of competitive position of EU-28 economies with using global multi-criteria indices. Equilibrium. *Q. J. Econ. Econ. Policy* **2019**, *14*, 441–462. [CrossRef]
16. Androniceanu, A.M.; Kinnunen, J.; Georgescu, I.; Androniceanu, A.A. Multidimensional Approach to Competitiveness, Innovation and Well-Being in the EU Using Canonical Correlation Analysis. *J. Compet.* **2020**, *12*, 5–21. [CrossRef]
17. Kuc, M. Is the regional divergence a price for the international convergence? The case of Visegrad group. *J. Compet.* **2017**, *9*, 50–65. [CrossRef]

18. Kijek, A.; Matras-Bolibok, A. Technological convergence across European regions. Equilibrium. *Q. J. Econ. Econ. Policy* **2020**, *15*, 295–313. [CrossRef]
19. Kónya, I.; Ohashi, H. International Consumption Patterns among High income Countries: Evidence from the OECD Data. *Rev. Int. Econ.* **2007**, *15*, 744–757. [CrossRef]
20. Horáková, M. Consumer behavior of college students in the Czech Republic. *J. Compet.* **2015**, *7*, 68–85.
21. Grybaitė, V.; Stankevičienė, J. An empirical analysis of factors affecting sharing economy growth. *Oeconomia Copernic.* **2018**, *9*, 635–654. [CrossRef]
22. Jankiewicz, M.; Pietrzak, M.B. Assessment of trends in the share of expenditure on services and food in the visegrad group member states. *Intern. J. Bus. Soc.* **2020**, *21*, 977–996. [CrossRef]
23. Walczak, D.; Dziawgo, L.; Dziawgo, D.; Buszko, M.; Pawłowski, J.; Żołądkiewicz-Kuzioła, A.; Krupa, D. Attitudes and Behaviors Regarding Environmental Protection in the Financial Decisions of Individual Consumers. *Energies* **2021**, *14*, 1934. [CrossRef]
24. Pietrzak, M.B.; Walczak, D. The Analysis of the Agrarian Structure in Poland with the Special Consideration of the Years 1921 and 2002. *Bulg. J. Agric. Sci.* **2014**, *20*, 1018–1039.
25. Rokicki, T.; Perkowska, A.; Klepacki, B.; Bórawski, P.; Bełdycka-Bórawska, A.; Michalski, K. Changes in Energy Consumption in Agriculture in the EU Countries. *Energies* **2021**, *14*, 1570. [CrossRef]
26. Lee, Y.; Kim, B.; Hwang, H. Which Institutional Conditions Lead to a Successful Local Energy Transition? Applying Fuzzy-Set Qualitative Comparative Analysis to Solar PV Cases in South Korea. *Energies* **2020**, *13*, 3696. [CrossRef]
27. Lin, M.-X.; Liou, H.M.; Chou, K.T. National Energy Transition Framework toward SDG7 with Legal Reforms and Policy Bundles: The Case of Taiwan and Its Comparison with Japan. *Energies* **2020**, *13*, 1387. [CrossRef]
28. Pietrzak, M.B.; Igliński, B.; Kujawski, W.; Iwański, P. Energy Transition in Poland–Assessment of the Renewable Energy Sector. *Energies* **2021**, *14*, 2046. [CrossRef]
29. Singh, J.; Laurenti, R.; Sinha, R.; Frostell, B. Progress and challenges to the global waste management system. *Waste Manag. Res.* **2014**, *32*, 800–812. [CrossRef] [PubMed]
30. Jouhara, H.; Czajczyńska, D.; Ghazal, H.; Krzyżyńska, R.; Anguilano, L.; Reynolds, A.J.; Spencer, A. Municipal waste management systems for domestic use. *Energy* **2017**, *139*, 485–506. [CrossRef]
31. Eurostat Statistics. 2020. Available online: https://ec.europa.eu/eurostat/statisticsexplained (accessed on 1 April 2021).
32. Szymański, K.; Sidełko, R.; Janowska, B.; Siebielska, I. Monitoring of waste landfills. *Zesz. Nauk. Wbiiś Politech. Koszalińskiej* **2007**, *23*, 75–136.
33. Zaman, A.U. A comprehensive study of the environmental and economic benefits of resource recovery from global waste management systems. *J Clean. Prod.* **2016**, *124*, 41–50. [CrossRef]
34. Kaza, S.; Yao, L.; Bhada-Tata, P.; Woerden, F. *What a Waste 2.0. A Global Snapshot of Solid Waste Management to 2050*; Urban Development Series; World Bank: Washington, DC, USA, 2018.
35. Alzamora, B.R.; Baros, R.T. Review of municipal waste management charging methods in different countries. *Waste Manag.* **2020**, *115*, 47–55. [CrossRef]
36. Statistical Yearbook of the Republic of Poland. 2018. Available online: https://stat.gov.pl (accessed on 1 April 2021).
37. Sadeleer, I.; Brattebø, H.; Callewaert, P. Waste prevention, energy recovery or recycling—Directions for household food waste management in light of circular economy policy. *Resour. Conserv. Recycl.* **2020**, *160*, 104908. [CrossRef]
38. Szymanski, K.; Sidelko, R.; Janowska, B.; Siebielska, I. Monitoring of waste landfills. *Book Ser. Sci. Noteb. Wbiis* **2007**, *23*, 75–133.
39. Senetra, A.; Krzywnicka, I.; Tuyet, M.D.T. The Analysis and the Evaluation of Municipal Waste Management in Voivodship Cities in Poland. *Rocz. Ochr. Środowiska* **2019**, *21*, 1076–1098.
40. Nanda, S.; Berruti, F. A technical review of bioenergy and resource recovery from municipal solid waste. *J. Hazard. Mater.* **2021**, *403*, 123970. [CrossRef]
41. Leckner, B. Process aspects in combustion and gasification Waste-to-Energy (WtE) units. *Waste Manag.* **2015**, *37*, 13–25. [CrossRef] [PubMed]
42. Hossein, R.; Izadbakhsh, H.; Shakouri, H. A planning model for using municipal solid waste management technologies-considering Energy, Economic, and Environmental Impacts in Tehran-Iran. *Sustain. Cities Soc.* **2021**, *65*, 102566.
43. CPH 2025 Climate Plan. Available online: http://kk.sites.itera.dk/apps/kk_pub2/pdf/983_jkP0ekKMyD.pdf (accessed on 1 April 2021).
44. Jędrczak, A.; den Boer, E. Final Report of the Third Stage of Expertise Aimed at Conducting Waste Tests in 20 Mechanical-Biological Installations Waste Processing. Zielona Góra 2015. Available online: https://sdr.gdos.gov.pl/Documents/GO/Ekspertyzy/Ekspertyza%20MBP_III%20etap_%2022-06-2015%20+%20za%C5%82aczniki.pdf (accessed on 26 May 2021).
45. Azam, M.; Jahromy, S.S.; Raza, W.; Raza, N.; Lee, S.S.; Kim, K.H.; Winter, F. Status, characterization, and potential utilization of municipal solid waste as renewable energy source: Lahore case study in Pakistan. *Environ. Int.* **2020**, *134*, 105291. [CrossRef] [PubMed]
46. Directive 2008/98/EC of the European Parliament and of the Council of 19 November 2008. *Off. J. EU.* 2008. no. l 312/3. Available online: https://eur-lex.europa.eu/LexUriServ/LexUriServ.do?uri=OJ:L:2008:312:0003:0030:EN:PDF (accessed on 26 May 2021).
47. Melikoglu, M. Vision 2023: Assessing the feasibility of electricity and biogas production from municipal solid waste in Turkey. *Renew. Sustain. Energy Rev.* **2013**, *19*, 52–63. [CrossRef]

48. Shumal, M.; Reza, A.; Jahromi, T.; Ferdowsi, A.; Mohammad, S.; Dehkordi, M.N.; Moloudian, A.; Dehnavi, A. Comprehensive analysis of municipal solid waste rejected fractions as a source of Refused Derived Fuel in developing countries (case study of Isfahan-Iran): Environmental impact and sustainable development. *Renew. Energy* **2020**, *146*, 404–413. [CrossRef]
49. Jedelhauser, M. *The Future of Waste Incineration in a Modern Circular Economy*; Info: Waste Incineration; Naturschutzbund Deutschland (NABU): Berlin, Germany, 2020.
50. Agar, D.A.; Kwapinska, M.; Leahy, J.J. Pyrolysis of waste water sludge and composed organic fines from municipal solid waste: Laboratory reactor characterisation and product distribution. *Environ. Sci. Pollut. Res.* **2018**, *25*, 35874–35882. [CrossRef]
51. Werle, S. Influence of the Gasification Agent Type on Syngas Parameters. *Rocz. Ochr. Środowiska* **2014**, *16*, 440–452.
52. Thome-Kozmiensky, K.J. *Thermische Abfall Behandlung*; EF-Verlag für Energie und Umwelt technik: Berlin, Germany, 1994.
53. Juniper Consultancy Services. *Advanced Conversion Technologies*; Juniper Consultancy Services: London, UK, 2007.
54. Kalinichenko, A.; Havrysh, V. Feasibility study of biogas project development: Technology maturity, feedstock, and utilization pathway. *Arch. Environ. Prot.* **2019**, *45*, 68–83.
55. Nguyen, V.K.; Chaudhary, D.K.; Ram, R.H.; Dahal, H.; Trinth, N.H.; Kim, J.; Chang, W.; Hong, Y.; La, D.D.; Nguyen, X.C.; et al. Review on pretreatment techniques to improve anaerobic digestion of sewage sludge. *Fuel* **2021**, *285*, 119105. [CrossRef]
56. Sosnowski, P.; Wieczorek, A.; Ledakowicz, S. Anaerobic co-digestion of sewage sludge and organic fraction of municipal solid wastes. *Adv. Environ. Res.* **2003**, *7*, 609–616. [CrossRef]
57. Campuzano, R.; Gonzalez-Martinez, S. Start-up of dry semi-continuous OFMSW fermentation for methane production. *Biomass Bioenergy* **2020**, *136*, 105544. [CrossRef]
58. Bachmann, N. *Sustainable Biogas Production in Municipal Waste Water Treatment Plants*; Technical Report; IEA Bioenergy: Paris, France, 2015.
59. Sarra, A.; Mazzocchitti, M.; Nissi, E. A methodological proposal to determine the optimal levels of inter-municipal cooperation in the organization of solid waste management systems. *Waste Manag.* **2020**, *115*, 56–64. [CrossRef] [PubMed]
60. Jaworski, T.J.; Grochowska, G. Circular Economy-the criteria for achieving and the prospect of implementation in Poland. *Arch. Waste Manag. Environ. Prot.* **2017**, *19*, 13–22.
61. Foggia, G.; Beccarello, M. Designing waste management systems to meet circular economy goals: The Italian case. *Sustain. Prod. Consum.* **2021**, *26*, 1074–1083. [CrossRef]
62. Matthews, K. 6 Waste-To-Energy Trends in 2020 That Are Worth Watching. Available online: https://blueandgreentomorrow.com/environment/ (accessed on 1 April 2021).
63. Leckner, B.; Lind, F. Combustion of municipal solid waste in fluidized bed or on grate—A comparison. *Waste Manag.* **2020**, *109*, 94–108. [CrossRef] [PubMed]
64. Ibikunle, R.A.; Titiladunayo, I.F.; Lukman, A.F.; Dahunsi, S.O.; Akeju, E.A. Municipal solid waste sampling, quantification and seasonal characterization for power evaluation: Energy potential and statistical modeling. *Fuel* **2020**, *277*, 118122. [CrossRef]
65. Kuo, W.C.; Lasek, J.; Słowik, J.; Głód, K.; Jagustyn, B.; Li, Y.H.; Cygan, A. Low-temperature pre-treatment of municipal solid waste for efficient application in combustion systems. *Energy Convers. Manag.* **2019**, *196*, 525–535. [CrossRef]
66. Niessen, W.R. *Combustion and Incineration Process—Aplication in Environmental Engineering*; Marcel Dekker: New York, NY, USA, 1995.
67. Channivala, S.A.; Parikh, P.P. A unified correlation for estimating HHV of solid, liquid and gaseous fuel. *Fuel* **2002**, *81*, 1051–1063. [CrossRef]
68. Khan, M.Z.A.; Abu-Gharah, Z.H. New approach for estimating Energy content of municipal solid waste. *J. Environ. Eng.* **1991**, *117*, 376–380. [CrossRef]
69. Szaflik, W.; Iżewska, A.; Dominowska, M. Chemical Energy Balance of Digested Sludge in Sewage Treatment Plant Pomorzany in Szczecin. *Annu. Set Environ. Prot.* **2014**, *16*, 16–33.
70. Tillman, D.A. *Wood as an Energy Resources*; Academic Press: New York, NY, USA, 1978.
71. Khalid, A.; Arshad, M.; Anjum, M.; Mahmood, T.; Dawson, L. The anaerobic digestion of solid organic waste. *Waste Manag.* **2011**, *31*, 1737–1744. [CrossRef]
72. Iżewska, A.; Szaflik, W. Energy processing of sludge come from the Pomorzany Water Treatment Plant in Szczecin. *Ecol. Eng.* **2016**, *48*, 88–95.
73. Sailer, G.; Eichermüller, J.; Poetsch, J.; Paczkowski, S.; Pelz, S.; Oechsner, H.; Müller, J. Datasets on chemical composition and anaerobic digestion of organic fraction of municipal solid waste (OFMSW), digested sewage sludge (inoculum) and ashes from incineratio norgasification. *Data Brief* **2020**, *31*, 105797. [CrossRef]
74. Duan, N.; Dong, B.; Wu, B.; Dai, X. High-solid anaerobic digestion of sewage sludge under mesophilic conditions: Feasibility study. *Bioresour. Technol.* **2012**, *104*, 150–156. [CrossRef]
75. Li, H.; Si, D.; Liu, C.; Feng, K.; Liu, C. Performance of directan aerobic digestion of dewatered sludge in long-term operation. *Bioresour Technol.* **2018**, *250*, 335–364. [CrossRef] [PubMed]
76. Pavi, S.; Kramer, L.E.; Gomes, L.P.; Miranda, L.A.S. Biogas production from co-digestion of organic fraction of municipal solid waste and fruit and vegetable waste. *Bioresour. Technol.* **2017**, *228*, 362–367. [CrossRef] [PubMed]
77. Giordano, A.; Di Capua, F.; Esposito, G.; Pirozzi, F. Long-term biogas desulfurization under different microaerobic conditions in full-scale thermophilic digesters co-digesting high-solid sewage sludge. *Int. Biodeterior. Biodegrad.* **2019**, *142*, 131–136. [CrossRef]

78. Zawieja, I.E. The course of the methane fermentation process of dry ice modifi ed excess sludge. *Arch. Environ. Prot.* **2019**, *45*, 50–58.
79. Sidełko, R.; Janowska, B.; Walendzik, B.; Siebielska, I. Two composting phases running in different process conditions timing relationship. *Bioresour. Technol.* **2010**, *101*, 6692–6698. [CrossRef] [PubMed]
80. Chiumenti, A.; da Borso, F.; Limina, S. Dry anaerobic digestion of cow manure and agricultural products in a full-scale plant: Efficiency and comparison with wet fermentation. *Waste Manag.* **2018**, *71*, 704–710. [CrossRef]
81. Appels, L.; Baeyens, J.; Degreve, J.; Dewil, R. Principles and potential of the anaerobic digestion of waste-activated sludge. *Prog Energy Combust Sci.* **2008**, *34*, 755–781. [CrossRef]
82. Waweru, M.; Herrygers, V.; Van Langenhove, H.; Verstraete, W. Process engineering of biological waste gas purification. In *Biotechnology Set*, 2nd ed.; Wiley: Hoboken, NJ, USA, 2008.
83. Zhang, Y.; Banks, C.J. Impact of different particle size distribution on anaerobic digestion of the organic fraction of municipal solid waste. *Waste Manag.* **2013**, *33*, 297–307. [CrossRef]

energies

MDPI

Article

The Sustainability of Decentralised Renewable Energy Projects in Developing Countries: Learning Lessons from Zambia

Susann Stritzke [1,*] and Prem Jain [2]

1 Smith School of Enterprise and the Environment, University of Oxford, Oxford OX1 3QY, UK
2 Department of Physics, University of Zambia, Lusaka P. O. Box 32379, Zambia; prem.jain@outlook.com
* Correspondence: susann.stritzke@ndl-renewables.com

Abstract: Decentralised renewable energy (RE) systems such as solar PV mini-grids (MG) are considered to be a cornerstone for the strategic achievement of the UN's energy access goals in the developing world. Many of these systems implemented however face substantial technical, financial and social sustainability challenges which are also a recurring theme in the relevant literature. MG analyses however often lack detailed technical or financial data or apply 'silo-approaches' as a comprehensive review of MG case study literature presented in this article reveals. Consequently, this study aims to enhance the understanding of RE MG sustainability in the developing context based on the integrated evaluation of the technical, financial and social dimensions of MG operation through empirical data from community surveys on energy use from Uganda and Zambia and two in-depth MG case studies from Zambia. By presenting detailed technical and financial data in combination with energy consumer perception, the study aims to close existing data gaps on sustainable RE MG operation and offers an approach to evaluate and optimise the operational sustainability of an MG in its individual local context. The article finds that the complex rural community ecosystem is a central, but yet undervalued determinant of MG sustainability in rural developing contexts. The mismatch between energy affordability and MG tariffs threatens MG sustainability and the scaling of energy access projects if not addressed specifically during project development and implementation. Consequently, the article calls for a strategic inclusion of community-ecosystem parameters and MG planning based on realistic energy affordability levels and an added value approach that includes dynamic MG financing mechanisms and targeted measures to generate added value through energy consumption as integral parts of RE MG projects

Keywords: mini-grids; energy access; energy sustainability; SDG 7; energy affordability

Citation: Stritzke, S.; Jain, P. The Sustainability of Decentralised Renewable Energy Projects in Developing Countries: Learning Lessons from Zambia. *Energies* **2021**, *14*, 3757. https://doi.org/10.3390/en14133757

Academic Editors: Michal Bernard Pietrzak and Bartlomiej Iglinski

Received: 28 May 2021
Accepted: 17 June 2021
Published: 23 June 2021

Publisher's Note: MDPI stays neutral with regard to jurisdictional claims in published maps and institutional affiliations.

Copyright: © 2021 by the authors. Licensee MDPI, Basel, Switzerland. This article is an open access article distributed under the terms and conditions of the Creative Commons Attribution (CC BY) license (https://creativecommons.org/licenses/by/4.0/).

1. Introduction

Despite enhanced efforts to increase energy access in sub-Saharan Africa (SSA) and an estimated overall investment of over $1.4 bn for off-grid electrification in the global South [1], the number of people without access to electricity in the region has remained largely unchanged [2]. This means that around 600 million people still lack access to energy in SSA. Over 80% of the people without electricity access living in rural areas [3]. Decentralised renewable energy (RE) systems such as solar PV mini-grids (MGs) are considered to be central solutions to enhance access to clean energy in rural SSA [4] and are estimated to provide connections to 75% of the rural SSA population in a universal energy access scenario by 2030. This scenario translates into roughly 100 to 120 million additional MG household connections, or around 350,000 MGs to be implemented within the next decade [3]. Consequently, questions related to the planning and implementation of RE MGs and their sustainability dimensions have gained substantial public and academic interest [5,6] resulting in two conflicting major overall findings. The first is that RE MGs have the potential to significantly contribute to the achievement of the Sustainable Development Goals (SDGs) in rural SSA. The second is that they face substantial sustainability challenges which hamper their widespread adoption and market-take-off [7] in many rural areas. Hence, the significant capital investment and scaling of RE systems that might cease operation if operation expenses permanently exceed revenues and maintenance costs is a substantial long-term threat to enhance energy access in the Global South.

To assess opportunities to scale the implementation of RE mini-grids to achieve the UN Sustainable Development Goals including clean and reliable energy access for all by 2030, a systematic understanding of the specific sustainability challenges of RE MGs is required to enhance implementation strategies for sustainable off-grid solutions. Previous learning lessons and insights gained during the implementation and operation of RE mini-grids that have been implemented in a developmental context are of pivotal significance in this context but are often performed in silos, either focusing on technical—or financial or socio-cultural aspects of MG implementation or operation.

The goal of this article is to examine factors and their interdependencies that determine the sustainability of decentralised off-grid systems in a developing context from a multi-disciplinary perspective. Based on a comprehensive data analysis, the study aims to extract key learning lessons and strategic recommendations for scaling MG implementation through an interdisciplinary review process that integrates the three interrelated dimensions of MG implementation. These include the technical design, MG finance and economics as well the end-user perspective in the community context.

A systematic analysis of RE mini-grid case studies published between 2010 and 2020, the evaluation of over 1200 community surveys performed in Uganda and Zambia and a longitudinal, in-depth evaluation of two solar PV mini-grid projects in Zambia are the basis of this research. The analysis aims to provide detailed insights on the technical and economic challenges for MG systems and suggests technical, financial and operational solutions to address them in the context of scalability of decentralised energy solutions.

The analysis is guided by three research questions which also structured the research approach: First, how can sustainability of RE MGs be defined and which dimensions need to be taken into consideration when assessing the sustainability of a RE mini-grid? Secondly, what determines or influences the longer-term sustainability of MGs in a rural developmental context and which learning lessons can be derived from previous cases? Thirdly, which strategic implications can be generated from such an analysis for the scaling of RE mini-grid projects? The study aims to derive specific approaches for sustainable MG planning and implementation approaches in developing contexts.

This article is structured into five main sections. The overview of the discussion about the sustainability of RE MGs and the related dimensions in the following section will lead to the presentation of the methods applied to answer the research questions in section three which are a comprehensive literature review on MG sustainability, community surveys to understand the consumer perspective and an in-depth analysis of two case studies from solar PV MGs in Zambia. The findings for each approach, which will be presented in section four will be contextualised and discussed in section five under the aspect of implications for MG project development and operation as well as the scaling of MGs in SSA. The final section six summarizes key conclusions for a way forward to enhance the sustainability of MGs in developing countries.

The study is mainly focused on a rural sub-Saharan African context but includes findings from other regional developmental settings in Asia and Southern America. The data generated from the Zambian case study evaluation and the scale-up implications are mainly focused on solar PV mini-grids. However, the findings also bear implications for other types of RE MGs such as hydro MGs which have a substantial potential to achieve SDG 7 [8–10] and hybrid MGs as well, which however display different system-economics than solar PV MGs [11,12].

2. Background & Hypothesis: The Sustainability of RE Mini-Grids in a Rural African Context

2.1. Renewable Energy (RE) Mini-Grids (MGs) in the Developing Context: Definitions, Trends, Challenges

In the context of this article, a mini- or microgrid is understood as interconnected loads and distributed energy resources that are grouped and controlled as a single entity within a set area and which can operate interconnected to the national electrical grid or as a decentralised island solution [13]. The terms mini- and micro-grid are used interchangeably in this study as both terms are commonly used in the relevant literature [14]. MGs provide connections to a number of customers in villages or towns that are located far from the main electrical grid and which can be owned and operated by utilities, cooperatives, village electrification projects or private sector companies [15]. The size of these systems and number of household connections can range significantly but lies usually between five kWp up to two MWp of installed capacity and can either be a hybrid system combining two or more sources of energy, for example, diesel and solar PV or based on a single energy source such as hydropower or solar PV, normally in combination with a battery system for energy storage [16]. Due to their sizing and capacity, MGs are usually considered to be more suitable for productive uses of energy than solar home systems (SHS) which are usually sized between 10 Wp and 250 Wp installed capacity [17,18].

To date, the overall majority of RE mini-grid projects in developing countries are financed through a mix of private sector investment and donor-funded grants of 50, 70 or in some cases even 100% of the total initial investment [19,20]. This model is designed to mitigate some of the financial risks associated with the implementation and operation of MGs in rural areas and to attract private investment. In theory, the blending of grant and private financing should help deliver quality energy services at an affordable cost for the consumer but the overall question has been raised [21,22], whether this is actually the case and whether MGs co-financed by grants provide 'cheaper electricity', trigger energy consumption in low-income areas and contribute to the overall financial and social sustainability of MGs.

Despite acknowledging the significance of decentralised RE MGs for enhancing access to clean energy in developing regions and the demonstrated positive impacts on rural livelihoods [23] in sub-Saharan Africa (SSA), the sustainable operation of these systems is challenging and largely remains an unresolved issue, especially if the systems are required to operate in a market environment that requires cost-reflective energy tariffs [24]. Despite a number of commonly reported issues such as technical faults [25,26], limited local capacity for maintenance and operation [27], inappropriate financing models or inadequate operational models that lead to limited affordability of energy and payment defaults [7,28–30] the availability of case studies presenting comprehensive financial, technical and operational data and their systematic review focusing on the sustainability of RE MGs is still limited [31]. However, comprehensive case studies of MGs can substantially enhance the understanding of their successful implementation and operation [32]. Hence, the need for qualitative research to understand the drivers of MG sustainability incorporating the demand and supply side have been widely formulated [33,34] and are embedded in the overall discussion of sustainability scenarios and parameters for MGs in rural developmental settings. While specific data revealing details of the technical design, investment including capital expenses (CAPEX), operating expenses (OPEX) and electricity tariff-models can potentially provide essential learning lessons to overcome barriers to the sustainable operation of decentralized renewable energy solutions [34] they are closely connected to the question of how sustainability is defined in that specific context which shall be discussed in the next section.

2.2. Dimensions of RE Mini-Grid Sustainability

The main simple baseline question of this study is, which factors are essential to operate MGs over the desired lifespan, usually twenty to twenty-five years, at optimal capacity without generating significant financial losses which shall be understood as 'operational sustainability' in this context.

While the literature offers various concepts of defining and assessing the sustainability of MGs [9,18,35,36] the overall conclusion is, that only a few MGs currently exist that are operating sustainably and constitute more than 'boutique electrification' [35] establishing a photo opportunity for a foreign donor or a local politician [18] such success-cases, however, are scarcely documented and could not be retrieved for this study.

In many cases, RE MGs in developing countries show deficits with regard to their technical, social and/or financial sustainability [37,38] which shall be discussed in more detail in this section. Hence an evaluation of MG sustainability requires the incorporation of these three interrelated dimensions and to be considered within the local context which has a decisive impact on the operational logic and the sustainability of an MG [18,36].

The aspect of technical sustainability is comprised of four main components which are service reliability, service availability, safety and meeting the demand capacity of the consumers [37]. This means that the technical design and operation are optimised to a degree that outages are limited, efficient maintenance measures are in place to allow high service quality [32,39]. The technical sustainability of MGs can potentially be approached and optimised through comprehensive technical and operational planning processes which can be highly standardised across various regions but is potentially prone to a limited number of unexpected external events such as natural disasters or unexpected system failure which requires contingency plans to limit power outages as much as possible.

The question of social sustainability is slightly more complex than the technical dimension but is a yet under-researched dimension [40]. The social sustainability of MGs can be viewed from two perspectives: the end-user perspective [41] and the wider community perspective which relates to socio-economic development including health and education as well as power—and gender structures. In the context of the baseline question of this paper, the most central elements of the social sustainability dimension are issues regarding affordability and income, the social acceptance of the energy services which is closely related to willingness to pay (WTP) and energy justice [18,42]. These dimensions are contextualised by the local livelihoods, understood as the wider community ecosystem outlined in Figure 1 which establishes specific system requirements, challenges and opportunities such as productive uses to any rural infrastructure solution including energy and water [43–45].

The financial sustainability of an MG in this context can be understood with regard to the initial investment costs, referred here as capital expenses (CAPEX), and the operating expenses (OPEX) of the system over its lifespan. In the ideal case, the minimum financial sustainability of an MG means that the non-grant financed share of the CAPEX and the OPEX over the lifespan of the asset would be retrieved by its revenue.

The financing approach of a system including certain levels of grant funding define to which extent CAPEX needs to be recovered over the lifespan of the system. In the SSA context, three MG financing models are prevalent, auction programs in which developers bid for construction and, most often also the operation of MGs at pre-determined sites, usually coupled with a capital subsidy which ranges between 60–80% of the initial costs versus results-based financing (RBF) in which developers are either paid a subsidy per connection, usually $350–500 [46] or a blend of both.

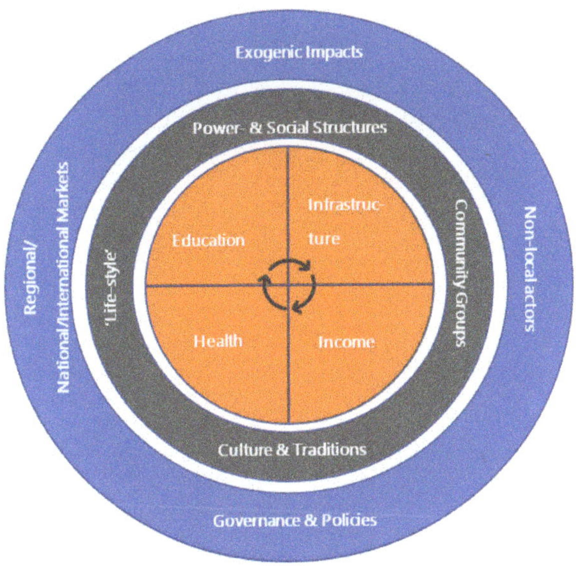

Figure 1. Community Ecosystem: suggested model of interrelated components and meta-level determinants (compiled by the authors).

The tariff schemes applicable to the MGs are usually laid out in the sectoral national regulatory frameworks as illustrated in Table 1. In a Uniform National Tariff, the regulator standardizes national MG tariffs which are equal to the on-grid sector and are usually subject to subsidies. In a 'Bid-Tariff Scheme', which has been piloted for the off-grid sector in Uganda, the MG tariffs are determined through an auction aiming at the lowest price while individual tariff limits are determined by the regulator in an 'Individualized Cost-Based Tariff Scheme' which are cost-reflective. In a 'Willing Buyer/Willing Seller Scheme', tariffs are agreed between the MG developer and the customers and are usually cost-reflective and approved by the regulator [47]. In Zambia for example, the regulator introduced a three-tiered system for off-grid solutions based on the size of the systems and generally require cost-reflective energy tariffs while tariffs in the on-grid sectors remain subsidised.

Table 1. Energy tariff-setting approaches for MGs in selected African countries (compilation based on: (NARUC: National Association of Regulatory Utility Commissioners, 2020).

SSA Country	Uniform Tariff	Bid Tariff	Individualised Cost-Based Tariff	Willing Byer/Willing Seller
Ethiopia			[black]	
Ghana	[light grey]			
Kenya	[light grey]		[black]	
Nigeria			[black]	[black]
Rwanda			[dark grey]	[dark grey]
Tanzania			[black]	[black]
Uganda		[light grey]	[dark grey]	[dark grey]
Zambia			[black]	[black]
[black]	Primary Approach for all projects in a certain category (off-grid)			
[dark grey]	Mixed Approach			
[light grey]	Secondary Approach for specific projects			

The study suggests that the three sustainability dimensions of RE MGs presented in this section are closely interrelated. This means that financial sustainability depends on a well-operated and -maintained, optimised solution that is reliable and meets the demands and affordability levels of its customers who can socioeconomically benefit from the solution and are able and willing to pay the energy tariffs required [48]. Hence, the sustainability assessment and development of MGs need to account for the interdependency of these three sustainability dimensions and in the context of the community ecosystem, but these variables are yet not fully understood due to a lack of longitudinal detailed data and in-depth case studies which account for these interdependencies applying a systemic approach.

3. Materials and Methods

The study aims to provide a better understanding of the specific interrelations between the financial, social and technical sustainability dimensions of decentralised energy systems based on empirical survey data and MG case studies. The novelty of the approach presented lies in its interdisciplinary by evaluating financial—technical—and consumer-based aspects of RE MG operation and revealing their interdependencies.

The country focus of the analysis is Zambia as it shares central features in terms of energy access and socio-economic challenges with other countries in the region and the data generated from Zambia will be contextualised with a regional perspective to discuss its validity for other countries in the region.

Consequently, the research aims to derive strategic implications to enhance the operational sustainability of MGs in a developmental context which are highly relevant for further scaling of MGs in developing countries. The study also aims to shed light on the question, whether substantial grant financing of MG CAPEX in an environment of cost-reflective tariff requirements is the right option to implement MGs that operate sustainably and provide reliable and affordable energy to rural households in the specific context.

Based on the understanding of the functioning of rural infrastructure solutions determined by the community ecosystem, the study approaches these questions through a three-step approach: a comprehensive literature review of MG case studies in a developmental context, the evaluation of empirical data on the socio-economic conditions and energy demands of rural communities in Uganda and Zambia which is mirrored by an in-depth evaluation of OPEX and CAPEX data of two solar PV MGs in Zambia.

The methodological approach is limited by three factors. The first is, that the focus on Zambia as the country context establishes certain specifics, for example with regard to currency fluctuation, energy markets and—supply chain which has specific impacts on energy pricing and which might be different in other developing countries. Secondly, the overall number of respondents who are existing MG customers in Uganda and Zambia is relatively low which however also presents an opportunity for enhancing comparative empirical research on different types of energy consumer perspectives in developing countries. Finally, it has proven to be challenging to gather specific financial data on the MG operation in Zambia including detailed revenues and costs due to the absence of comprehensive and systematic recording of this data by the operators. Consequently, some of the financial calculations presented in this study are based on assumptions that are highlighted and discussed in the results section.

3.1. *Systematic, Integrated Literature Review*

The aim of the integrated literature review [49] is to identify the general availability of case studies and in-depth data including on OPEX—and CAPEX of RE MGs as well as identified specific challenges for the operation of these systems in rural, developmental contexts. The analysis is based on a number of categories such as the type of study and data presented and indications with regard to the system sustainability in relation to the overall research questions and are presented in Table A1.

In total, 26 studies published over the last decade have been evaluated. The studies have been selected on the basis of the technology (RE) and energy system-type (MG) presented, the regional focus (developing country) and the search was based on the keywords "sustainability", "financial sustainability", "technical optimisation" and "consumer perspective". The search was mainly focused on studies published between 2017 to 2021 to capture the most recent findings. The review included academic papers as well as acknowledged reports and academic theses of either specific MG case studies or feasibility studies of RE off-grid systems including solar PV and mini-hydro in a rural developing context mainly in Africa and Asia. The review was focused on the sustainability parameters applied in the studies, identified challenges of the systems, operational data and applied tariff schemes. It was also analysed whether the wider community context has been taken into account in the studies evaluated. The aim of the analysis is to detect MG data gaps and sustainability trends from a cross-regional perspective.

3.2. Empirical Data: Community Surveys, Stakeholder Interviews, Focus Group Discussions

The results from the literature review data inform the evaluation of empirical survey data from Zambia which will be mirrored with survey data from Uganda. This data has been generated through a mixed-method approach of qualitative interviews, focus group discussions, rural household surveys and on-site data collection to gain a comprehensive understanding of rural household energy needs [50] and their socio-economic conditions as essential attributes of the community ecosystem and to account for the specific importance of local end-user perspectives for the sustainability of rural energy solutions [51].

The household surveys were rolled out in two stages. The first round of surveys (N = 1016) was deployed in rural communities in Uganda and Zambia focused on the general socio-economic situation, energy needs and demands which were captured through 106 questions. Based on the findings gathered in this round, the surveys in round two (N = 50) were focused on gaining a deeper understanding of the potential value chain opportunities for the productive use of energy and were rolled out in two rural communities in the Southern Province of Zambia. For the evaluation, the survey data presented in this study has been clustered into various energy-consumer subgroups which are presented in Table 2.

Table 2. Community Surveys Uganda and Zambia—number and groups of respondents.

	Uganda	Zambia	Zambia: SHS-Users	Uganda: SHS-Users	Zambia: MG-Users Sinda	Uganda MG-Users Kalangala	Zambia: No Connection	Uganda: No Connection
# of respondents	441	537	64	164	40	98	433	119

The data of MG users has been collected within the communities of the 30 kWp Sinda solar PV MG in Zambia which is presented as a detailed case study in this analysis and for Ugandan MG users of the 1.6 MWp solar thermal hybrid MG in Kalangala [52].

The survey rounds were complemented by site-visits and focus group discussions with representatives of different community groups such as members of cooperatives and parent-teacher associations (PTAs), local women self-help groups (SHGs), local businesses as well as public representatives such as councillors, health workers and school staff in five rural communities in the Southern Province of Zambia.

The physical data was collected through the lens of a potential RE project development in cooperation with a local energy project developer to gain a better understanding of site selection strategies [53] and included existing local infrastructure, population density, present social or economic groups and environmental conditions including soil erosion and deforestation. The evaluation also included a review of findings from 45 qualitative interviews with off-grid energy companies and public sector stakeholders to embed the community-level perspective into the wider energy policy-making and governance context to derive strategic implications based on previously identified current energy governance structures in Uganda and Zambia [54].

3.3. Mini-Grid Case Studies: Context and Selection

Based on identified data gaps through the integrated literature review, the study incorporates detailed technical, financial and operational data from two existing solar MGs in Zambia located in Mpanta and Sinda. This embedment of in-depth MG case studies aims to support the closing of existing data and knowledge gaps with regard to MG operation and the generation of important strategic learning lessons for rural electrification processes [55]. The MG data has been collected through a number of site visits performed by researchers of the University of Zambia resulting in a technical and financial assessment [56], stakeholder interviews and desk-based research. The technical and financial evaluation of the Mpanta MG complements previous local studies on community-level engagement and energy transition [41,57].

3.4. Country Context: Energy Access and Governance in Zambia

Zambia shares central features in terms of energy access and the socioeconomic structure of rural communities with a number of SSA countries. This includes a low density of the main electrical grid which leaves the majority of rural Zambia unelectrified with off-grid RE systems being the preferred solution to enhance energy access in these areas [58]. The rural electrification rate in Zambia is significantly below ten per cent. Although specific energy access data is currently lacking according to the Rural Electrification Authority' (REA), this rate is well below the average rural electrification rate in SSA, which is estimated at around 22.59% [59,60].

Following the establishment of REA Zambia in 2003, which is mandated to enhance access to electricity in rural areas to 51% by 2030, planning and policy approaches to increase energy access in Zambia have originally been dominated by grid-expansion projects mainly in peri-urban areas and in cooperation with the parastatal national utility ZESCO Ltd. The 'Rural Electrification Masterplan' (REMP) which has been developed with significant support from the Japan International Cooperation Agency (JICA) and which was adopted in 2008 identified 1217 Regional Growth Centres that were clustered in 180 project packages to be electrified by 2030 [61]. According to the 'REMP', a total investment of around $1.1 bn would have been required for the realization of all project packages which amounts to a required annual investment of roughly $50 million per year between 2008 and 2030 to achieve the electrification target [62]. However, due to the slow progress of project implementation, national budget constraints, the high costs of grid extension projects and the emergence of more cost optimised solutions [63], REA began to review the electrification strategy laid out by the REMP [64] and explored a potential shift towards emphasizing decentralized energy solutions like solar PV MGs. Consequently, these off-grid energy solutions have been gaining more and more interest in Zambia since 2011/2012, a process that was increasingly pushed by foreign donor organisations who started to develop and implement off-grid procurement programs, some of these programs in cooperation with REA and other institutional partners in Zambia.

This resulted in the implementation of off-grid electrification projects now being largely 'outsourced' to the private sector and foreign donor organisations [43]. In the course of this strategic shift, RE MGs have become an emerging technology for rural electrification alongside SHS in the country. Over the last seven years, more than 20 MGs have been implemented through partnerships between private sector companies and foreign donor organisations with dozens of more projects in the pipeline of being deployed for example through the EU-funded Increased Access to Electricity and Renewable Energy Production (IAEREP) Project.

According to REA, as of February 2021, around 29 solar PV MGs with installed capacities between ten and 50 kWp and two mini-hydro MGs with an installed capacity of 640 kWp and 750 kWp have been operational in Zambia with two additional PV MGs currently under construction. These MGs have largely been implemented through tender-based procurement processes with CAPEX grant-funding levels of around 70 to 80% supported through various foreign initiatives including the Beyond the Grid Fund for

Zambia (BGFZ), the Mohammed bin Rashid Initiative for Global Prosperity and U.S. African Development Foundation (USADF)/Power Africa.

Most of the Zambian MGs are applying mixed tariff schemes but shifting from subsidised towards 'Individualised Cost-Based Tariff' schemes in alignment with the regulatory framework for MGs introduced in 2018 by the Energy Regulation Board Zambia (ERB). The framework defined certain technical, financial and operational requirements for MGs including the requirement of cost-reflective tariffs while energy tariffs in the on-grid sector remain subsidised in Zambia [65].

Although the challenges of enhancing rural electrification have been acknowledged by leading national and foreign stakeholders, recent research has evidenced that the energy policy framework yet lacks a clear integrated strategy for on- and off-grid electrification and energy access governance structures are still highly top-down oriented and largely lack the inclusion of community-level representatives [54].

4. Results

4.1. The Sustainability of MG Documented in the Literature

The literature review revealed three major types of scientific resources: single case studies, reviews of multiple case studies and feasibility studies including simulations for potential project locations. The analysis shows that, despite a substantial overall number of studies on MG sustainability, the presentation of specific financial data including OPEX, CAPEX, details on revenues and tariff models is yet very limited, even in those studies which focused explicitly on the financial sustainability of MGs.

Secondly, 18 out of 21 case studies reported substantial financial sustainability challenges for all regions due to various combinations of low energy affordability, poor revenue collection and high operation and maintenance costs. The reported financial challenges in the case studies evaluated outweigh other issues such as consumer behaviour and acceptance (14), policy-related or legal challenges (11) or technical problems (8). With regard to MG revenues covering its OPEX, none of the case studies presented specific data for the operational sustainability of the MG. Only six case studies revealed some information regarding the tariff models of the MGs applied which indicates a data gap and evidence to perform a comparative analysis of the connection between cost-reflective vs. subsidised MG tariff models and operational sustainability.

The review also shows that although financial, technical and end-user focused sustainability dimensions have been included to some extent in most of the case studies, the focus on economic community-impact including the productive use of energy in the context of a comprehensive community eco-system approach which includes analysis and understanding of local income patterns and community value-chains is clearly underrepresented. None of the case studies reviewed correlate technical—financial—and community-based approaches or integrate these dimensions but focus mainly on technical and financial analyses instead with the availability of detailed data however being generally limited. The few studies that include the community—or end-user context either lack a specific technical or a financial evaluation.

The reviewed feasibility studies largely follow that pattern of selected modelling usually incorporating technical and financial calculations based on an LCOE approach. They generally demonstrate the tendency of an economic and technical advantage of RE MG systems over grid connections in the rural context and more clearly over fossil fuel alternatives [30,66,67]. The feasibility studies evaluated however usually do not account for potential community eco-system impact factors illustrated in Figure 1 on energy systems including fluctuating income, sensitivity to exogenic effects or existing socio-economic structures. Consequently, the optimised theoretical financial and operational projections of the feasibility studies reviewed are in stark contrast to the actual case studies.

The literature review evaluation a clear gap between theoretical modelling presented in feasibility studies and the operational reality of RE MG operation in a developmental context which substantiates the necessity of the inclusion of the community ecosystems-setting into MG planning and modelling processes. The evaluation also reveals specific MG data gaps and highlights the important links between financial sustainability, technical management and end-user behaviour which informs the requirement of combined research approaches aiming for an understanding of the complex socio-economic community settings for the evaluation of the sustainability of RE MGs in the developing context from which strategic implications can be derived to improve their overall operational long-term sustainability.

4.2. Survey Data: Community Ecosystems—Central Features

While the literature review has pointed toward the necessity of including the community context for an interdisciplinary research and implementation approach of sustainable decentralised energy systems, the following section aims to illustrate how the community context facilitates or limits the operational sustainability of RE MGs.

The empirical survey data as a basis for this evaluation focuses on four main aspects which are relevant with this regard: (1) the income situation of rural households, (2) the current use of energy, (3) the potential creation of added value through energy demands and (4) end-user experience. The data is presented for Uganda and Zambia to increase the applicability and validity of data for generalisable outcomes [68] and has been detailed for the national, regional and community levels in Zambia to account for eventual local and regional variations.

4.2.1. Socio-Economic Characteristics and Rural Income Situation

The financial and hence operational sustainability of MGs is directly interlinked with the income situation of its potential and actual customers as it determines levels of energy affordability, consumption levels and overall MG revenues. Hence, the full utilisation of the energy generated from the energy system and steady cash flow are central to cover at least the operating expenses of the system.

The empirical data presented in Figure 2 indicates, that well over 80% of the respondents in Zambia are self-subsistence farmers with a high dependency on seasonal rainfalls and reported low average household spending levels of around ZMW 288 as of 2020. This amount lies significantly below the reported average rural household income levels of ZMW 810 reported in 2015 [69] and translates into around \$13 per month under currency exchange rates of spring 2021 as Figure 3 illustrates. For Uganda, the data suggests higher average household incomes and monthly spending levels of around \$86 per month.

It must be noted, however, that capturing income data has proven to be challenging due to high variations of income levels stated by the respondents in Zambia for example between ZMW 5000 to ZMW 20 monthly. This is also due to the fact that many respondents only generate an income once or twice annually as they rely on the sales of seasonal farming produce, mainly maize, with no or just very little other income. Hence capturing actual local income levels is challenging and needs local verification. Furthermore, the calculation in \$vs local currency is subject to significant fluctuations. The Zambian Kwacha for example lost over 60% value over the US Dollar between 2019 and 2021.

The high prevalence of over 95% of income generation through some form of self-employment including self-subsistence farming and low levels of wage labour are interrelated with high degrees of income intermittency in Uganda and Zambia. Close to 80% of the respondents in Uganda, where the proportion of income generation through small businesses is slightly higher than in Zambia, and around 90% of respondents in Zambia reported a significant income variation throughout the year. Interestingly, MG or SHS access does not impact or reduce the income variation in both countries. This indicates either a low or no use of productive use appliances or their limited impact to generate stable income throughout the year.

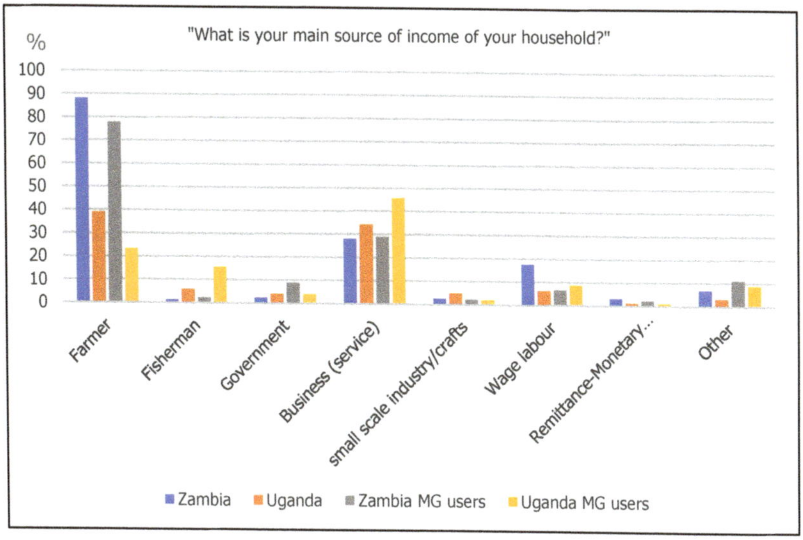

Figure 2. Socio-economic characteristics of rural communities in Uganda and Zambia (N = 1016; Responses in %; Multiple responses possible).

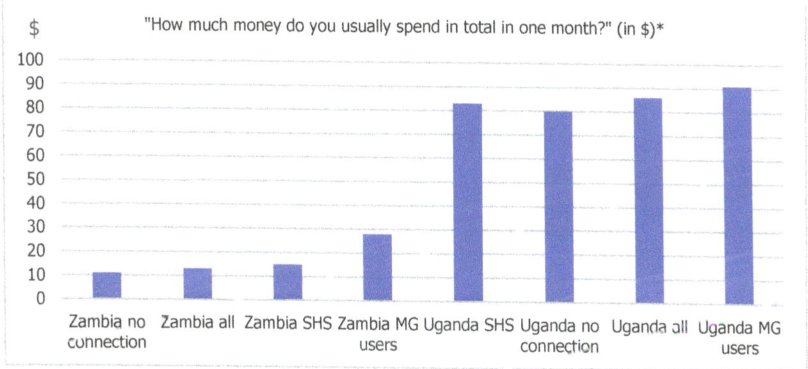

Figure 3. Average monthly spending levels of rural communities in Uganda and Zambia (N = 1016; Responses in %; Multiple responses possible). * Based on currency exchange values of 2020.

The causes of income fluctuation have been evaluated in more detail on community level in Zambia among non-MG users and are mainly related to the direct and indirect dependency on seasonal rainfall patterns due to the absence of irrigation solutions and low degrees of agri-processing which requires small scale industrialisation and affordable energy supply. Climate change-induced droughts as experienced in the rainy season of 2018/2019 in Southern Zambia seriously threaten the socio-economic fragility of these communities and seriously threaten food security in the region.

4.2.2. Energy Access and Usage-Patterns

To gain a better understanding of the end-user adaptation to utilising different energy sources and the conditions of how to facilitate the productive use of energy to generate a stable demand and potentially positive economic impacts in the community, different energy access situations, energy consumption patterns and payment models have been compared. This evaluation is guided by two questions that are central to understand local adaption and the impact of decentralised energy systems. Firstly, what type of energy sources are currently most used by rural communities? Secondly, how does the source of energy influences energy usage patterns?

Figure 4 illustrates respondents with different connection types in Uganda and Zambia and points towards higher access rates to the national grid, Solar Home Systems (SHS) and MGs in Uganda. While this is partially due to the selection process of respondents as for Uganda, also respondents with access to the national grid have been selected during the survey rounds, the data illustrates a clear tendency to higher energy access rates in rural Uganda compared to Zambia which correlates with World Bank data which presents a rural electrification rate of around 10% for Zambia compared to 38% in Uganda [60]. This data, however, is mainly focused on connections to the national grid. Around two-thirds of the respondents in Uganda have access to MGs, the national grid or SHS, in Zambia however, this figure stands at only around 20% with 80% of the Zambian respondents originally claiming to have no electricity access at all. A further survey round that focused on rural community members in Southern Zambia revealed, however, that despite a high share of respondents originally claiming to have no electricity access at all, granular data from two communities showed that over 80% of the respondents use some kind of energy sources like portable solar panels or car batteries.

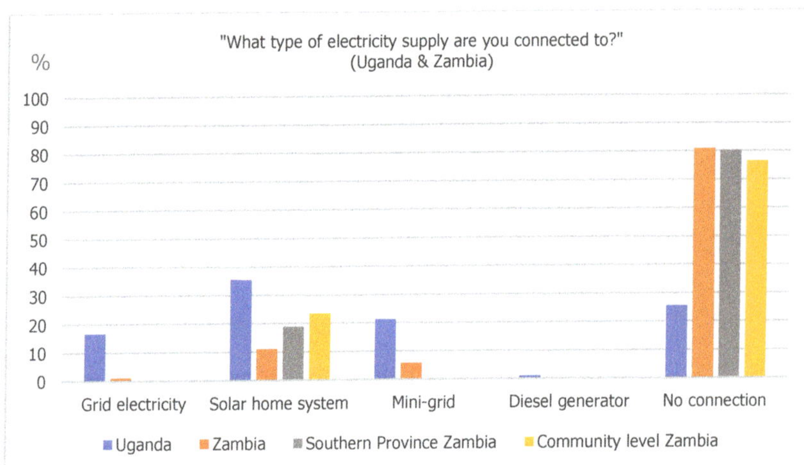

Figure 4. Type of energy supply, Uganda and Zambia (Responses in %; N = 1037).

In both countries, the most prevalent energy source for rural households is SHS with 'Pay-as-you-go' (PAYG) or Rent-to-Own (RTO) purchase models to be seemingly more widespread in Uganda while one-off purchases seem to be more common in rural Zambia which is also due to yet limited availability of these solutions which have been acknowledged as the most favoured type of financing and electricity access model for the Zambian respondents.

With regard to the second question focusing on energy usage pattern it could be hypothesised, that the ownership of appliances and the utilisation of the energy provided, including its productive use, are potentially diversifying and growing in the course of enhanced connectivity. The data presented in Figures 5 and 6 partially supports this hypothesis to some extent for MG and SHS users in both countries but the data also illustrates the yet limited productive use of energy across all connection levels despite the fact that over 80% of the respondents are self-employed entrepreneurs or farmers. The national-level data indicates similar energy usage patterns for Uganda and Zambia with energy being mostly used for lighting, communication including charging devices and entertainment.

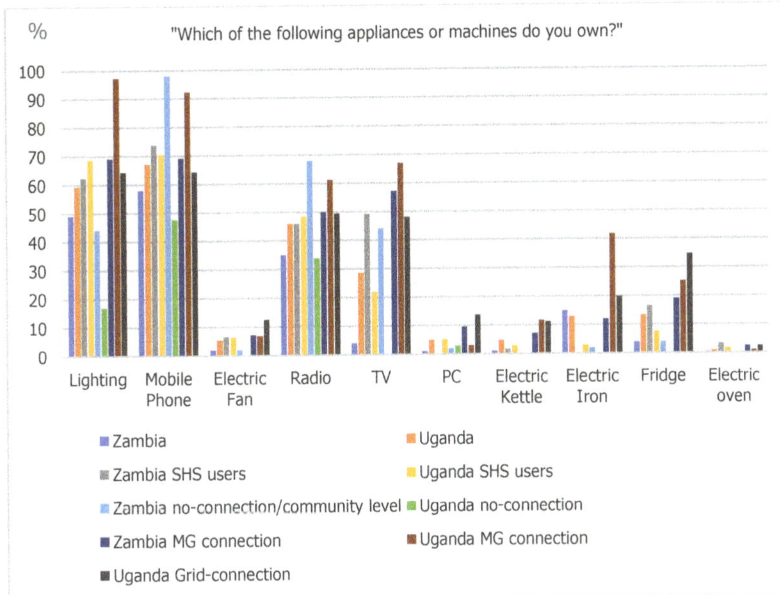

Figure 5. Ownership of appliances, Uganda and Zambia (Responses in %; N = 1037).

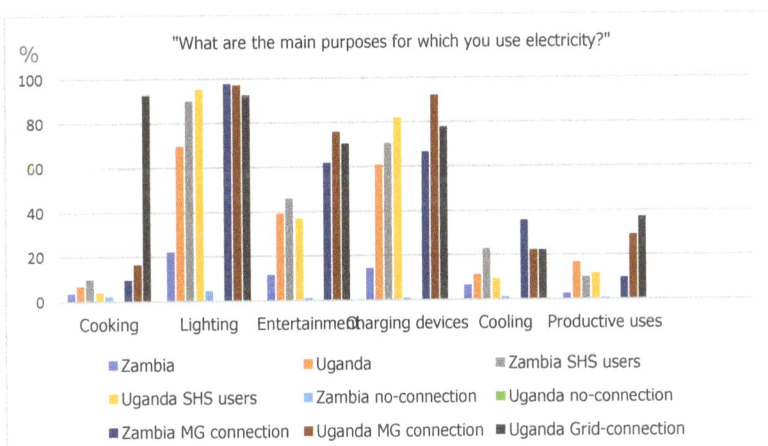

Figure 6. Energy usage patterns; Consumer Groups in Uganda and Zambia (Responses in %; N = 1037).

The availability of MG systems does not automatically trigger the possession of a wider array of appliances among the respondents such as fridges and the productive use of energy but reveals visible national differences instead. While Ugandan MG users have a higher but still limited tendency towards productive energy, this is of less significance to Zambian MG users which could be due to higher income and energy affordability levels as well as better access to appliances in Uganda. Users of the Ugandan Kalangala mini-grid for example pay between 0.18 and 0.22 $ per kWh as of 2019 [70] while Zambian mini-grid users in Sinda pay around 0.26 $ per kWh with income levels in rural Zambia being less than half compared to Uganda as illustrated earlier.

4.2.3. Energy Demands, 'Willingness to Pay' (WTP) and Productive Uses

The majority of the respondents in both countries indicated, that their current energy source is limited and not sufficient to meet their energy needs for example with regard to lighting, productive use or the connection of additional appliances. Consequently, the data presented in Figure 7 suggests a general desire of the respondents to upgrade their current energy supply especially in Zambia, where 100% of respondents indicated their willingness to upgrade while this figure is less pronounced in Uganda, especially among MG users. This might indicate, that the Ugandan MG has the tendency to better fulfil the energy demands of its customers while the Sinda MG in Zambia does not meet the energy- needs of the customers to large extent. Data from Southern Zambia also suggest that consumers would not automatically prefer an MG connection over an SHS as around 55% of the respondents in Southern Zambia stated their preference of an SHS based on a rent-to-own model, independently from their current source of electricity. This dataset also indicated limited popularity of grid-connection provided by the national utility (Zambia Electricity Supply Corporation—ZESCO) as none of the respondents preferred that connection type which might partially be due to the high amount of national power-cuts of up to 12 h daily Zambia experienced in 2018/2019 due to a serious drought in the hydropower-dependent country. The expression of consumer preferences however does not eradicate the question of the impact of national grid-arrival in a market offering subsidised energy tariffs for on-grid connections versus the requirement of cost-reflectiveness in the off-grid sector.

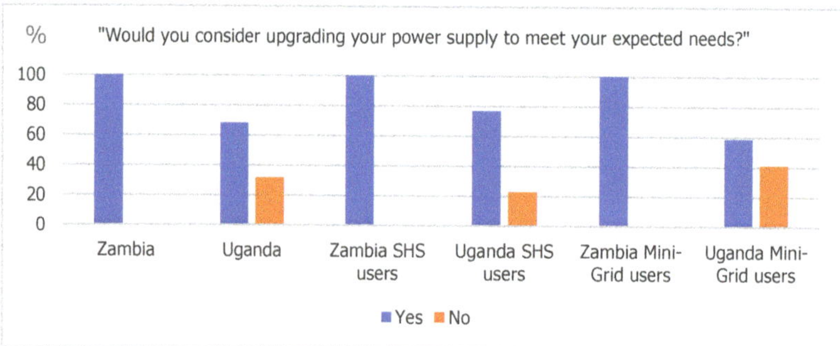

Figure 7. Desire for upgraded electricity supply Consumer Groups in Uganda and Zambia (Responses in %; N = 1037).

The desire for an upgraded energy supply is largely mirrored by the willingness to pay for better energy access which is high in both countries as illustrated in Figure 8. With regard to the intended use of an upgraded energy connection, the responses show a wide array of preferences with significant variations among consumer groups with regard to cooking and cooling. Lighting, charging devices and productive uses and to some extent entertainment, however, are desired by the majority of all respondents in each consumer group as Figure 9 shows which indicates a yet untapped consumer demand in many areas.

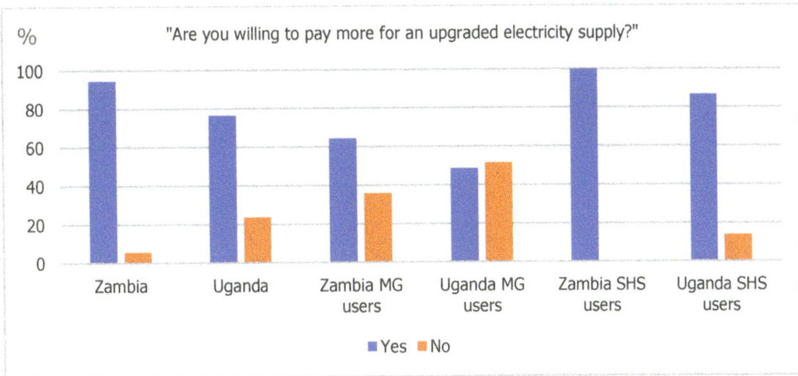

Figure 8. Willingness to pay for an upgraded connection, Energy-use demands, Consumer Groups in Uganda and Zambia (Responses in %; multiple responses; N = 1037).

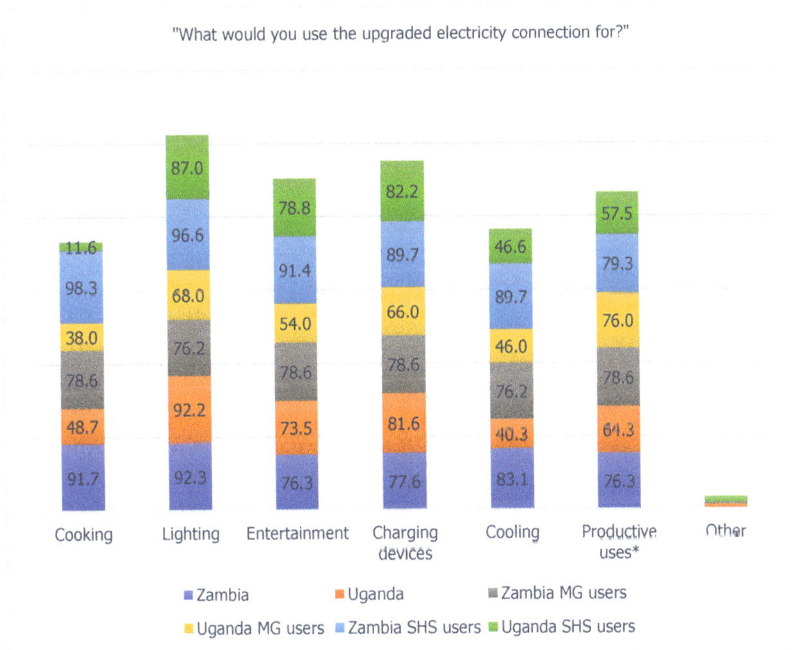

Figure 9. Energy-use demands, Consumer Groups in Uganda and Zambia (Responses in %; multiple responses; N = 1037).

As the productive use of energy is still limited among all consumer groups in both countries, the majority of all respondents confirmed their demand to use energy more productively. Figure 10 illustrates the diversity of the forms of desired productive energy use between various consumer groups and communities. While the data suggests demand-trends for cooling, lights and charging of devices, the demands in other areas are much more diverse between various groups of respondents. Despite these variations, respondents expressed a strong demand for improved access to productive appliances including financing mechanisms that would be based on 'Rent-to-Own' or 'Pay-Go' models which were the preferred solution over one-off purchases which appears yet to be a challenge.

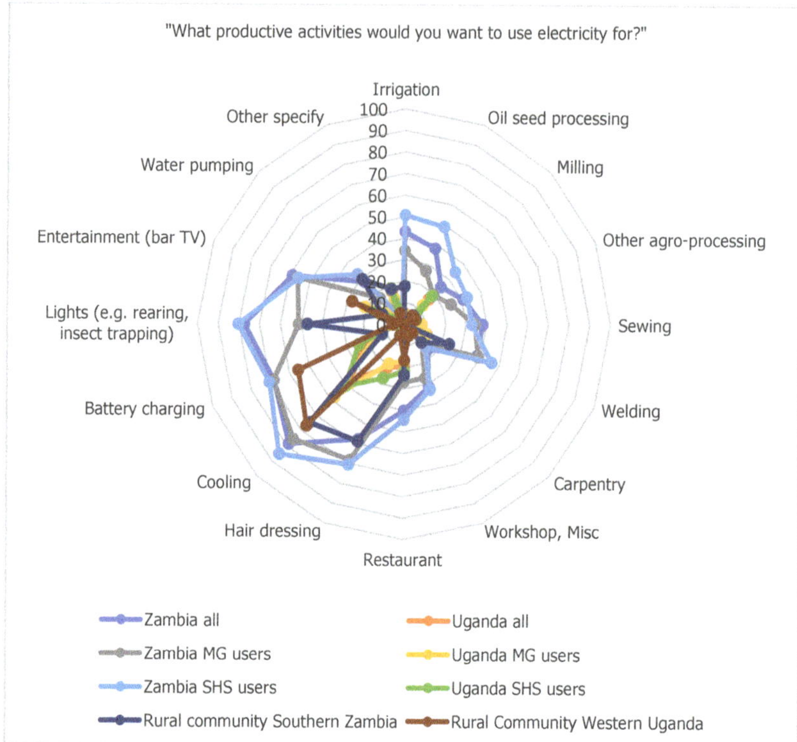

Figure 10. Type of desired productive uses in Uganda and Zambia (Responses in %; multiple responses; N = 1037).

A direct comparison of two rural communities in Western Uganda and Southern Zambia that share similar features like energy access via SHS, low and intermittent income and the absence of wage labour level included presented in Figure 10 reveals that the demands for productive energy use vary despite their similarities. The data suggests that manufacturing and crafts activities are underrepresented, especially in Uganda and the clear trend towards using energy for service provision in both countries. Since most services depend on income patterns of 'their' customers, the data implicate the opportunity of training, capacity building and appliance financing mechanisms coupled with MG development to diversify local business structures to create local added value and positively impact the income situation in local communities.

4.2.4. Impacts of Electrification on Consumer Finance

Since low and intermittent income levels in the rural communities surveyed establish substantial challenges for the financial MG sustainability, the question for impacts of energy availability in the communities arises and more specifically whether energy availability did directly contribute to local income generation to stabilise MG cashflow. Although the overall majority of respondents reported the general improvement of life quality after getting a connection, either through SHS or MG as shown in Figure 11, the data on income change does not draw a conclusive picture with this regard.

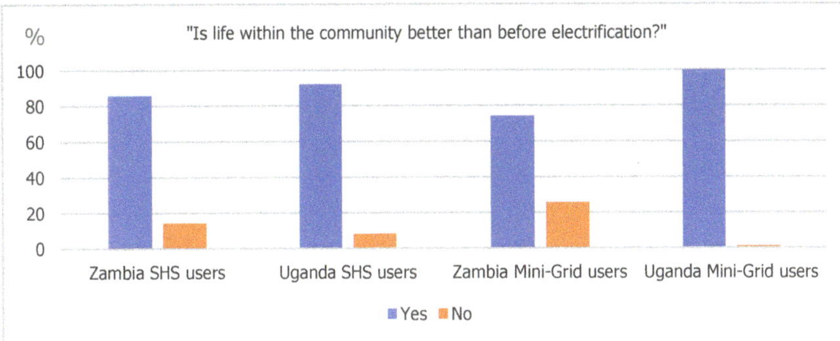

Figure 11. General perceived community impact of electrification on communities in Uganda and Zambia among various subgroups; Responses in %.

As Figure 2 showed earlier for both countries, spending levels between MG users versus respondents without energy connection are generally higher with greater significance in Zambia where spending levels of MG users are almost double as high as those of non-connected respondents. For SHS users this correlation is less visible as SHS users in Uganda reported similar spending levels compared to non-connected respondents in Uganda while the spending levels for SHS users in Zambia are around 40% higher compared to non-connected respondents. Since the data presented in Figure 2 does not provide a blanket indication of energy access automatically generating higher incomes or whether merely consumers who already have a higher income can afford an MG or SHS connection, respondents have been specifically asked for the impact of getting electrified on their income. The responses presented in Figure 12 clearly indicate a positive impact on the income situation for MG and SHS users in Uganda while interestingly about two-thirds of the Sinda MG users in Zambia stated, that their income remained the same or has even decreased after getting electrified. This response pattern seems to replicate for SHS users in Zambia and is largely mirrored by the number of respondents reporting difficulties in energy payment among SHS users and to a larger extend MG users in Zambia as shown in Figure 13.

The impact of electrification on income, however, becomes inconsistent when respondents have been asked about their ability to pay for school fees before and after electrification. While a clear majority of MG users in Uganda and around 53% of SHS users in Zambia confirmed an increased ability to pay school fees after getting electrified, the majority of SHS users in Uganda and MG users in Zambia stated no or negative impact of electrification on the ability to pay for school fees.

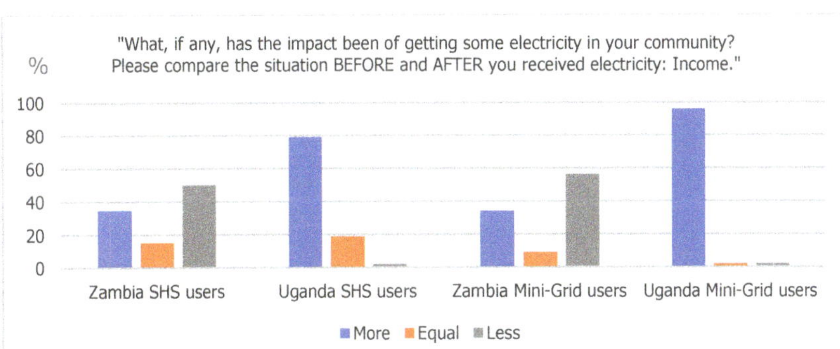

Figure 12. Impact of electrification on income among various subgroups; Responses in %.

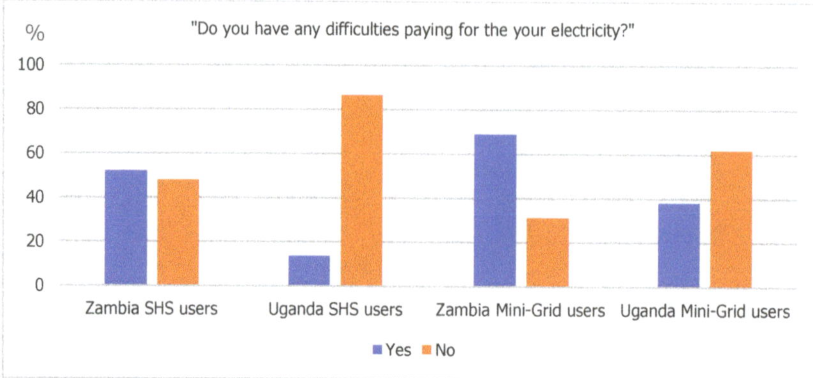

Figure 13. Difficulty in payment for electricity among various consumer subgroups; Responses in %.

Another indicator for the financial community impact of electrification is the development of business activity in the community. The response patterns with this regard reflect positive income development to some extend with the biggest perceived growth of business activity in the community after electrification for MG and SHS users in Uganda, to some extent among SHS users in Zambia but with no perceived positive impact among the respondents from the MG users in Zambia.

A clear majority of respondents across all consumer groups reported increased studying time of children after getting connected which indicates a positive impact on education in both countries.

These variations among the different indicators related to electrification and end-user income illustrate the complexity of this dimension and highlight the importance of a multidimensional approach of measuring the financial impact of energy systems on community level through a systematic evaluation of various indicators as a sole focus on household income when measure impact might not reflect the complex reality of rural communities. For exhaustive coverage, it would also require a measurement system that includes the non-direct monetary value components such as the ability to purchase certain goods, increase of months with income, ability to generate savings etc. which could be a result of changes in end-user income due to electrification.

4.2.5. Experienced Reliability, Technical—And Environmental Issues

As illustrated, the majority of the respondents reported positive impacts of electrification in their communities which is largely mirrored by the question of the negative effects of electrification. Around two-thirds of the respondents using an MG- or SHS-connection in Uganda as well as MG-users in Zambia report no negative effects which are in contrast to 60% of Zambian SHS-users reporting negative effects of electrification in their community. However, experienced energy problems vary among user groups and country as shown in Figure 14. In congruence with the data on consumer finance and energy payment, Zambian SHS users reported problems with regard to affordability and cost as most significant, followed by limited energy supply and reliability issues. These concerns are shared with SHS users in Uganda who, however, see safety issues as their most central problem. Differences can also be observed among MG users in Uganda and Zambia. While respondents connected to the Sinda MG mostly reported issues with regard to reliability and energy cost, MG users in Kalangala overwhelmingly stated certain safety problems such as severe injuries and death due to electrical shocks as well as electric fires as the biggest problems. Possible explanations for these reports from Uganda can be the cases of electricity theft and tampering with the system, as users reported that two or more households are sharing an electric meter which is clearly dedicated to one household only, instead of a general lack of technical quality of the installation. While the environmental impact of the energy connection has been considered low by the respondents in both countries, social problems, especially conflicts between community members with—and without energy access seem to be of some concern to SHS users both in Uganda and Zambia.

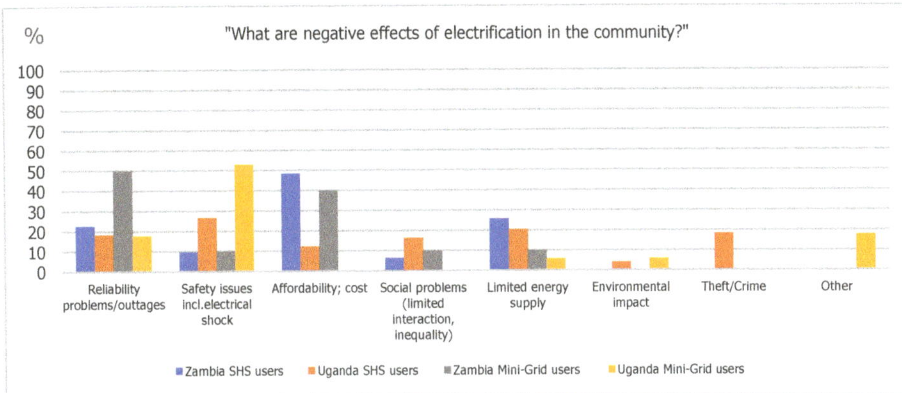

Figure 14. Perceived negative impacts of electrification among various subgroups; Multiple responses; Responses in %.

4.3. Mini-Grid Case Studies Zambia

The literature review presented earlier revealed substantial gaps with regard to the available technical and financial data in MG studies and limited inclusion of community context data while the previous section illustrated end-user perceptions and socio-economic conditions that heavily impact mini-grid operation. Consequently, in the third analytical step of this study, the implementation and operation of two Zambian solar PV MGs located in Mpanta and Sinda are evaluated in more detail to complement and mirror the analysis of community—and consumer perceptions. The goal is to analyse the interdependencies of the technical- financial and socio-economic sustainability components in more detail and illustrate how the community ecosystem impacts MG operation.

4.3.1. Community and System Context: Socioeconomics and Geography

The solar MG in Mpanta developed and commissioned in 2013 by the REA Zambia was the first solar PV MG in Zambia. With an installed capacity of 60 kWp, it was initially designed to provide street lighting in the village as well as basic electricity services to about 450 private households to power small appliances such as radios, lights, TV sets or mobile phone charging, to schools, a rural health centre, small businesses, harbour facilities and churches. The original design of the MG was comprised of 300 × 200 Wp mono-crystalline solar PV modules, 720 × 650 A/2 V deep cycle batteries, 4 × 15 kVA inverters and 4 × 100 A/240 V DC charge regulators. The whole system operates at 240 V DC to AC and electricity is distributed through two 0.4 kV lines within a distance of 1 km. A re-wiring of the whole system, the exchange of all inverters and charge controllers was necessary in 2015 after lightning hit the system a few times causing high voltage surges which led to the failure of the components. During the reconfiguration, the string size has been increased from 10 to 14 panels per string to optimize the voltage requirement of the controllers. The number of panels was reduced from 300 to 224 which resulted in a reduction of energy output by 15.2 kW or 25.3% compared to the former configuration leading to an average energy output of around 209 kWh per day. Accordingly, the number of charge controllers and inverters was also reduced from four to two providing a 45 kVA instead of a 60 kVA power output which was sufficient due to the low energy demand. With the downsizing of the system, the daily power supply is rationed and power is now available only at certain times and for 14 h a day. The estimated annual energy output is 76,500 kWh based on a projected local energy yield of around 1700 kWh per kWp installed [71] excluding transmission and distribution losses per year. The battery system, which is the largest cost position in an off-grid solar system is currently sized at 720 batteries each 2 V and 650 Ah capacity, i.e., total energy storing capacity of 720 × 2 V × 650 Ah = 936 kWh in Mpanta. This configuration if fully operational, would allow two days of battery autonomy so that the users can enjoy uninterrupted power during cloudy days.

The 30 kWp mini-grid in Sinda village, commissioned in 2017, was developed as a pilot project to enhance private sector investment in the RE sector in Zambia and has been developed and owned by the Zambian developer Muhanya Solar Ltd. It is the first MG in Zambia that has been established as a public-private partnership project and is operated by a private company in cooperation with the community. The system is comprised of a 30 kWp solar PV generator, one 20 kW inverter and 140 kWh of battery storage capacity which is connected to four 100 A charge controllers. The 2.5 km overhead distribution network with a voltage of 230 V currently connects around 60 households and 5 businesses.

The implementation of Mpanta in 2013 was part of a wider $4.78 million energy access programme rolled out by REA and funded by the United Nations Industrial Development Organization (UNIDO) with an investment volume for Mpanta MG estimated at around $1.3 m which was 100% grant-funded [72,73]. The CAPEX for the Sinda MG stands at approximately $350,000 based on approximately 70% grant funding from Power Africa Off-Grid Energy Challenge and Musika Development Initiatives which was complemented by 30% equity provided by Muhanya Solar Ltd. [74]. Hence both MGs were implemented with significant grant investment with CAPEX per kWp installed being almost twice as high for Mpanta compared to Sinda. Table 3 presents an overview of the technical specifications of both MGs.

Table 3. Overview technical and financial specifications for Mpanta and Sinda Mini-Grids.

	Mpanta MG	**Sinda MG**
Technology	solar PV	solar PV
Installed capacity	60 kWp (45 kWp)	30 kWp
Inverter capacity	4 × 15 kVA (2 × 15 kVA)	25 kVA
Total CAPEX/Investment Volume (approx.) in $ (and per kWp installed)	1,300,000 (21,667)	350,000 (11,6667)
Current connections (approx.)	190	65 (60 households, 5 businesses)
Funding	100% grant	70% grant, 30% equity
Annual current OPEX (estimation) in $ total/per kWh	43,000/0.63 *	16,178/0.30 **

* See Table 5; [74]. ** See Appendices A and B.

4.3.2. MG Operation and Tariffs

The socioeconomic conditions in both communities pose critical challenges for the operation of the MGs which have been outlined in the previous section. Low energy affordability and high-income fluctuation throughout the year directly impact the revenue and cash flow of both MGs.

The implementation of the energy tariff system in both locations was independent of the Zambian regulatory MG framework which was introduced a few years after the commissioning of the MGs and which now requires cost-reflectiveness in the MG sector. Before the commissioning of Mpanta MG, members of the community discussed the options of a pre-paid metering system and a fixed fee. During the consultation processes, few members raised concerns with regards to trust in a pre-paid system and a vote among community members revealed a decision towards a fixed fee which was introduced, based on the size and type of the connected properties. After an introduction period in Mpanta which allowed free connections, basic fees of around $5 for the connection and $1.5 for wiring were introduced and the management including revenue collection was handed over to the Kafita Cooperative Society (KCS) while REA retains ownership of the plant which operates on an interim generation and distribution license issued by the Energy Regulation Board (ERB). The introduction of the connection fee caused complaints and irritation among the Mpanta community. The perception among the community was that the project was brought by the government as a grant without the requirement for end-users to pay. Additionally, the transition from free to prepaid connections was not well communicated to the community [41].

While Sinda MG also applies a monthly fixed rate, the tariffs are uniform at ZMW 150 ($6.5) per customer per month translating into ZMW 5 ($0.21) per day and are significantly higher than in Mpanta where households pay a monthly fixed fee of between ZMW 30 ($1.3) to ZMW 70 ($3.1). Consequently, the majority of respondents surveyed in Sinda reported difficulties in paying for the electricity and less available income than before getting electrified as Figures 12 and 13 in the previous section illustrated. The fixed load in Sinda per household is limited to 300 W and the average customer consumption in 2018 has been estimated at 55 kWh per month but is probably lower due to outages. The tariff system in Sinda has created complaints especially among low-consuming customers as consumers operating a single electric bulb for example are paying the same price as customers running more appliances including fridges, etc. leaving them with a feeling that clients with low consumption were subsidising those with higher energy consumption. Table 4 provides an overview of the tariff structures in both locations and also reflects the currency fluctuation of the tariffs charged in Zambian Kwacha compared to the US-Dollar.

Table 4. Tariff structures Mpanta and Sinda.

Mpanta MG, 45 kWp, (around 190 Customers)							
	Monthly Fixed Charge (ZMW)	**Value in $ 2016**	**Value in $ 03/2021**	**Commercial Clients/Social Services**	**Monthly Fixed Charge (ZMW)**	**Value in $ 2016**	**Value in $ 03/2021**
Less than 3 roomed house	30	2.6	1.3	1 roomed shop	60	5.4	2.7
4 roomed house	35	3.1	1.6	2 roomed shop	65	5.7	2.9
5 roomed house	60	5.4	2.7	3 roomed shop	70	6.2	3.1
6 roomed house	65	5.7	2.9	Harbour depot	50	4.4	2.2
7 roomed house	70	6.2	3.1	Health Centre	50	4.4	2.2
School staff houses	100	8.8	4.5	Primary schools	50	4.4	2.2
Clinic staff houses	100	8.8	4.5	Churches	50	4.4	2.2
Sinda MG, 30 kWp (around 60 Customers)							
		All customers			150	13.2	9.5

Tariff structure Mpanta MG as of 2016 [41]; Sinda MG (UNZA data 2018/2019).

In Mpanta, 450 users were initially connected for free in 2013. As of 2020, more than 260 of these users had been disconnected as they were unable to pay the monthly fixed fee. In addition to these figures, over 60 users were disconnected for various reasons including due to the collapse of their traditionally built, grass-thatched houses during a heavy rainy season while 4 users were removed for illegal connection. As a consequence, more than 50% of the households in Mpanta still rely on pollutant fuels for lighting and over 95% for cooking. During the operation, community complaints emerged at scale about miscommunication regarding the tariff scheme when households were initially connected or a lack of notification of affected customers about an imminent disconnection when they were unable to pay [41].

While the Sinda MG operates through a pre-paid metered systems which allow for some efficiency of revenue collection, the tariff collection is not efficient in Mpanta as local site visits revealed. As of 2020, the average estimated total monthly collection was only about ZMW 6000 to 8000 which translates into $300 to 400 applying an average exchange rate of $1 = ZMW 20 for 2020 or an annual average total of around ZMW 84,000 (or $4200). These inefficiencies in tariff collection are partially caused by local management problems but are also rooted in high levels of customer dissatisfaction due to reported power outages, restricted system capacity and limited understanding of the operation of the system [41]. The current OPEX for Mpanta MG according to REA is ZMW 108,000 for the two staff salaries. However, there are other costs like office space, materials and equipment, vehicles and travel which are not included here. Hence, the total annual costs are more likely to be around ZMW 300,000 which is calculated without component replacement costs according to a UNZA assessment in 2019. Consequently, the annual OPEX exceeds the collected revenue by around ZMW 24,000 ($1200) in terms of staff costs or ZMW 216,000 or $10,000 based on 2019 currency values. Although these values are only rough estimations as this type of data collection has been proven challenging and may fluctuate over time, the plant is currently generating a significant annual deficit. These costs are currently covered to large extent by the Zambian Rural Electrification Authority (REA) that subsidizes the mini-grid. The OPEX deficit is even higher when replacement costs for system components such as charge controllers, inverters or batteries are becoming part of the equation as they are a substantial component of OPEX calculation. Table 5 presents an approach to calculate the current assumed OPEX for Mpanta MG based on information provided by the operators and market-price estimation for the components installed which however can vary. The cost estimation of the current system is enhanced by an optimisation scenario which shall be discussed in the following section.

Table 5. Current and optimised OPEX for Mpanta MG.

	Current Layout/Estimations	Optimised Model/Assumptions	
Plant capacity in kWp	**45**	**45**	60
annual kWh output *	76,500	76,500	101,400
Minus 10% power losses	73,953	73,953	91,260
kWh per day	203	203	250
OPEX System (components) (pA)	26,343	11,384	15,206
Management costs (pA)	17,000	11,000	11,000
Total annual OPEX in $	43,343	22,384	26,206
Total OPEX per kWh in $	0.59	0.30	0.29
OPEX per kWh incl. Modules & racking system **	0.63	0.34	0.33
Total OPEX per kWh in ZMW ***	14.40	7.88	7.52
* based on estimated local energy yield of around 1700 kWh per kWp installed [71]			
** 0.04 $ per kWh			
*** as of 04/2021/1 $ = 23 ZMW			

Although the OPEX costs for the Mpanta mini-grid are very high and mainly driven by a battery bank that is designed to provide two days of energy back-up supply when fully charged as well as high overhead costs, the analytical approach illustrates the importance of a thorough evaluation during energy project development to determine realistic OPEX costs and energy tariffs within the local socio-economic context which defines the sustainability of an off-grid system. It must be noted at this point that the calculation presented does not contain any cost of capital as the project was 100% grant-funded. The analysis also illustrates the interdependency between technical—financial—and community-context parameters and highlights the necessity of holistic project planning approaches that account for these interdependencies.

In addition to that, the calculation of OPEX, tariff and revenue reveal the impact of the significant currency depreciation of the Zambian Kwacha vs. the US-Dollar over the last years, a problem that most African countries encounter. This poses a serious problem jeopardising the sustainability of future investments in the whole industry as imports for components and spare parts as well as the cost of capital payments are usually made in US-Dollar while local revenues are generated in local currency.

The OPEX for the Sinda MG shown in Table 6 reveals significant differences in the cost per kWh to the Mpanta MG. But like in Mpanta, the analysis discloses a financial sustainability gap of around $0.04 per kWh based on an assumed energy retail price of around $0.26 per kWh and OPEX costs of 0.30 kWh. Taking into account both CAPEX and OPEX, this gap widens to $0.57 per kWh under a scenario that is assuming a price of $0.83 per kWh LCOE based on the grant and debt funding ratio. The annual revenue in year one has been estimated at around $11,000 which results in an annual loss of approximately $5100 with regard to OPEX only. These figures do not include currency depreciation of around 20% per annum between 2018 and 2021.

Table 6. Overview estimated OPEX costs of Sinda Mini-Grid, 2018 [74].

System OPEX Sinda MG	
O&M costs plant ($ p.a.)	5748
O&M costs grid ($ p.a.)	1472
Staff & admin costs ($ p.a.)	8405
Insurance ($ p.a.)	552
Total cost ($ p.a.)	16,178
Approx. annual energy generation in kWh * OPEX—Price per kWh in $ as of 2018	52,410 0.30
No.of customers year 1	65

* 30 kW × 1747 kWh/kW/a = 52,410 kWh.

5. Discussion

The case study analysis presented in this article widely reflects the financial key issues that have been reported in the literature [7,73,74] but extend the existing scope of the studies reviewed by the community perspective. The study illustrates the fundamental impact of the socio-economic community characteristics on off-grid energy systems and vice versa. Although the community- and consumer perspective have been included in RE MG system reviews [7,30,75] or were the sole focus of MG implementation reviews [41], it has been of limited consideration yet, as to what extend the consumer economics influence MG operation and how MG operation can specifically contribute to enhance and stabilise the income situation of MG customers as part of a planning approach.

Secondly, while the relation between technical and financial sustainability has been evaluated in the literature [25,35,74,76], approaches how to implement RE off-grid systems that combine qualitative energy services at affordable tariffs and potential trade-offs, or how to optimise existing systems that are not yet financially sustainable have received less attention with few exceptions [77].

These findings have specific implications for the planning- and development of MGs from the project-level perspective as well as for the scaling of MGs across SSA. Both implications will be discussed in the following subsections.

5.1. The Interdependency of Financial—Technical—and Socio-Economic Sustainability

5.1.1. Towards Break-Even: Optimisation Potential for PV Mini-Grids

As the operation of Mpanta MG is currently seriously financially unsustainable, a projection has been developed to model a potential improvement of the financial and operational sustainability of the system. The following projection, which can be applied to a variety of MG systems, illustrates potential energy availability trade-offs and the conjunction of technical, financial and socio-economic sustainability. Hence, the analysis presents one possible approach shown in this section to counter some of the key sustainability issues of RE MGs detected in developing countries.

Table 5 compares the current system OPEX for Mpanta to an optimized system OPEX for the current 45 kW system configuration and a 60 kW system which would utilize the current installed capacity to the full extent but would face the challenge of local energy demand and affordability to avoid under-usage. The new cost estimates are based on updated price estimations for the system components and a remodelled battery backup system. The figure illustrates that the system OPEX could be significantly reduced by more than a third under a new calculation scheme. In addition to the cost analysis of the system components, the position 'Miscellaneous' as reported by the operator needs a further assessment as it is quite high. We further note that the battery system is the largest cost item. This cost could potentially be reduced through some balancing of cost vs. benefit. The optimisation of the battery size by implementing one-day battery autonomy instead of two days. Secondly, it is assumed that one-third of the total load is being used in the daytime and therefore does not need storage. This could be achieved through an incentivisation of energy use during the day. These two measures reduce the battery size to less than one-third of the current size. Furthermore, the use of modern Li-Ion batteries can potentially be suggested, but the higher cost of this technology needs to be accounted for in relation to the advantages including a much longer life span and higher depth of discharge, which make their lifetime cost often cheaper compared to that of lead-acid batteries.

These measures reduce OPEX significantly as Table 5 suggests. However, this new configuration has some trade-offs as end-users will potentially have limited power supply during cloudy days, especially during the rainy season. These issues would have to be addressed upfront in cooperation with the community to discuss the balance of energy supply, service quality and energy tariffs. In series wiring configurations it is advisable, that each battery should have the same load status which can be achieved through conditioning the batteries. If batteries are not conditioned or at full capacity, the different battery strings do not have the same capacity and can display a significant deviation up to minus 30% or more to the full capacity, especially when they operate under challenging temperature conditions, for example in heat with no cooling system. Conditioning the batteries would require re-wiring the battery system including an option to remove single batteries from the string from time to time to fully charge and recharge that battery, understood as conditioning the battery, and then add it to the string to ensure that all batteries have the same condition. In an optimal case, the remodification of the battery system would also entail introducing temperature balancing and recording of charging/discharging of the batteries.

These optimisation measures could achieve significant OPEX reduction as shown in Table 7. A more in-depth technical and financial analysis for example through a HOMER-simulation [78] or similar software could potentially further refine the optimisation.

Table 7. Annual energy generated and break-even tariff estimation for Mpanta MG.

Item	Value	
Plant capacity	60 kW_p	45 kW_p
Assume transmission loss	10%	10%
Assume average fraction of the power being sold	90%	90%
Annual energy sold	101,400 × 0.9 × 0.9 = 82,134 kWh	76,050 × 0.9 × 0.9 = 61,601 kWh
Average daily energy sold	82,134/365 = 225 kWh	61,601 kWh/365 = 169 kWh
Annual average O&M cost	$26,206	$22,384
Break-even tariff per kWh	$26,206/82,134 = $0.32/kWh	$22,384/61,601 = $0.36/kWh

Although reduction of operating costs as illustrated for the Mpanta case might reduce the price per kWh to be paid in order to cover operation costs, the utilization of the energy produced by the mini-grid is a second key element of the financial sustainability of these systems. A closer look at the socio-economic context in relation to MG financials reveals severe disparities.

5.1.2. Household Income and Energy Affordability vs. Energy Tariffs

Although the equation presented in Table 5 is based on a number of assumptions, it can roughly be estimated, that based on the current number of consumers and a total annual OPEX of $43,343 an average payment of monthly ZMW 437 or $19 per connection would be necessary to cover the costs of operation which is more than half of the average reported household income in Mpanta. Translating this scenario which has been discussed as the 'energy poverty penalty' [79] to Europe in order to illustrate the financial burden for the consumer, the average UK household with an annual income of £29,900 [80] would have to spend around £1245 or $1474 on electricity per month. In a similar translation to illustrate the relation between income and expenditure, an average UK household would have to pay around £31.8 or $37.74 per kWh if affordability and consumption levels of MG users in rural Zambia are translated in a UK scenario.

The average monthly household income in rural areas of Zambia in 2015 was estimated at $77 but appears to be lower according to the survey data presented earlier. This limits the disposable income of private households for electricity to $6–7 US per month when applying an estimation of 10% of potential energy expenditures per household and month. Applying a tariff of around 0.30 $ per kWh that would at least cover the OPEX would allow Sinda customers for example the consumption of approximately 23 kWh which is far below the projected energy consumption per month and rural customer of 49 kWh according to the Rural Electrification Master Plan of 2008 [62] and is unlikely to trigger the productive use of energy. These considerations reveal the extent to which the community context determines the technical and financial parameters of an off-grid system and that the understanding of local energy needs and demands is essential for energy system implementation [81].

A more detailed onsite assessment using household surveys could reveal the actual income situation and the potential consumption levels at the MG sites evaluated in this article as well as other locations. The identification of current barriers for customers of getting and staying connected and their actual energy needs and demands as well as opportunities for productive uses of energy based on local value chains as discussed earlier can be key to enhance the operational sustainability of MGs in three ways: First, they could provide a basis for a tariff scheme that is more adjusted to the consumer needs including options for prepaid-meters based on actual consumption or schemes that incentivize certain private and commercial user profiles in terms of volume and timing. Secondly, evaluating consumer behaviour and satisfaction could also reveal under which conditions an increase in the numbers of connections could be achieved. Thirdly, a focus on aspects of productive use in connection with innovative financing schemes for small business owners or farmers could provide the basis for economic development, income generation and enhance energy consumption in the area.

5.1.3. The End-User Perspective and a Community-Ecosystem Approach

The data presented previously revealed high energy-demand levels and a substantial willingness to pay for upgraded energy services among communities in Uganda and Zambia. The large majority of respondents within all consumer subgroups, including MG users in Sinda, confirmed improved living conditions in the community after electrification. Community surveys in Mpanta also revealed the positive effects of electrification on household and community level [41] as respondents in Mpanta reported positive impacts enhanced security (street lighting), better availability of medical supplies, higher levels of education and growing business opportunities as a result of MG energy access. The number of communal gatherings increased, people feel better connected and informed due to higher accessibility of TV and radio and local women emphasized that household work became much easier due to improved light sources. A significant impact on the use of traditional fuel, however, such as reduced collecting of firewood or use of charcoal has not been observed by the researchers in Mpanta [41] and Sinda.

The response patterns on negative and positive energy impacts presented are only a snapshot of different energy consumer perspectives due to the limited number of respondents in each group but they potentially indicate certain key interdependencies and trends: Low-patterns of productive energy use and energy affordability are closely interrelated but the uptake of energy and its productive use are not necessarily linked to a specific type of connection. The Zambian case illustrates, that connecting users to an MG does not automatically have a greater potential to generate income and business growth compared to SHS. The consumer perspective partially confirms the trend of financial and technical sustainability challenges for MGs as the most dominant problems but their significance varies according to the rural national context. The data also indicates that technical sustainability needs to be considered along with energy generation, transition and distribution of an energy system including the end-user connection.

Reported social tensions and the potential tendency of higher-income households being more likely to obtain a connection as indicated by the data suggests that a further increasing promotion and uptake of decentralised energy systems in rural areas needs to incorporate strategies for inclusive approaches to also reach lower-income groups in the communities to ensure that 'no-one is left behind'.

A lack of community involvement during the planning stage of the mini-grid main probably has contributed to the challenges of the Mpanta mini-grid since the actual energy needs, affordability and social barriers have not been assessed extensively. The planning and implementation process thus did not address sufficiently the potential payment issues or promoted a pre-paid metering system which might have been more suitable for the local economic conditions. In this light, the Mpanta situation the importance of early-stage community involvement for energy project planning which can potentially lead to a system design that meets community energy demands and better matches the income structure. The review of MG case—and feasibility studies has revealed that the socio-economic community context only finds limited consideration and is mirrored in 'top-down' energy access governance approaches and limited strategic community engagement for RE project development [50,54]. Consequently planning and operation of decentralised energy systems such as solar PV MGs must be based on this context and follow an interdisciplinary approach that takes into account financial, socio-economical, cultural, technical and environmental aspects, not only of the mini-grid itself—but the wider community that is expected to use the mini-grid and benefit from the provision of clean energy.

For the operational phase, tracking customer satisfaction, issues with regard to monthly payments and the evaluation of opportunities to introduce other services such as the provision of clean water, irrigation, communication or media appliances—powered by the mini-grid could potentially enhance the utilization of the electricity from the mini-grid, increase revenue and mitigate the current financial losses the system produces.

The low- and seasonal income levels create a volatile financial situation for these households which makes the payment of a monthly fixed fee often challenging. Although around 20 refrigerators are currently in use by businesses and households in Mpanta to cool soft drinks and produce ice blocks to preserve fish, the access to energy has not yet generated an increased productive use that would help to utilise energy generated and stabilise cashflows of the system.

This finding correlates with the survey data shown in the previous section which indicates that electrification, either via SHS or MG does apparently not create an automatism for income increase, particularly in Zambia. This data also indicates the need for cross-national, comparative analyses of the specific socio-economic impact of electrification via SHS and MG on rural communities to filter best practices and strategic cornerstones which can enhance the income situation of rural consumers and stabilise the financial sustainability of off-grid solutions.

The socio-economic environment of rural communities in SSA establish complex demands for energy systems delivery models [41,76,82] but can also create opportunities that are substantial for the longer-term sustainability of the energy system for example with regard to the productive use of energy, modern energy cooking [83] and the local creation of local added value. The project in Mpanta demonstrates the importance of demand utilization and a steady cash flow as conditions for a sustainable mini-grid operation. As most communities in rural Zambia and other African countries face similar challenges like Mpanta village, such as the dependency on seasonal rainfalls, small-scale farming or fishing as well as limited access to productive appliances, solutions are required to strengthen community resilience and overcome seasonality in income. The provision of clean water, internet access, training, innovative and sustainable farming methods coupled with financing schemes for low energy appliances that extend the opportunities to process agricultural products to be explored at the project planning stage. The baseline of this approach is the local value chain and the evaluation of potential added-value creation [84]. These added services can add to an integrated infrastructure service concept that goes beyond the provision of electricity but sees energy [85].

5.2. Implications for the Scalability of Off-Grid Systems in Africa

The literature review illustrated the substantial sustainability challenges that RE MGs face in the developing world. The in-depth evaluation of two Zambia MGs largely confirmed these challenges for the cases evaluated by providing detailed technical and financial as well as consumer-centric data which is largely missing in most case studies. The data on socio-economic community parameters and energy demands in Uganda and Zambia reveals significant challenges for the implementation of off-grid systems and contributes to the debate of the scalability of electrification efforts in SSA [86].

5.2.1. Off-Grid Planning Approaches

Although the market—and socio-economic conditions vary among SSA countries and the development of MGs in Zambia with regard to scale is still at a rather early stage, an energy project that aims to be sustainable requires an individual extensive planning process that closely involves the local communities [25]. These processes are complex and cost- and capacity-intensive which in turn have significant cost implications for the project development and the energy tariffs of the system that have to be charged to achieve financial sustainability which is often unaffordable for the local rural consumers. The tariff overview of Mpanta and Sinda MGs reveals significant differences in the energy tariff structures. While Sinda MG applies a pre-metered flat fee for all customers, Mpanta has a staggered tariff scheme. Despite the average household tariffs being much lower in Mpanta, the reports of payment problems and customers getting disconnected are much higher in Mpanta and in Sinda which might be due to variations in the socio-economic parameters of the community and indicate high in-country discrepancies between rural communities with regard to energy affordability.

This finding generates two important implications. First, energy research and data collection are usually focused on urban versus rural divides in developing contexts. This analysis however shows, that for community energy research, a more granular analysis of inter-community conditions in rural settings can provide important insights for the deployment and function logic of decentralized energy systems as socio-economic rural community structures are not homogenous in a country. Secondly, in the context of energy system development, this data can serve as a baseline for the sizing of the RE systems and the energy tariffs that can be charged in rural contexts. The size of the MG system must be based on the energy-utilisation potential of the energy generated as unsold kWh are revenue losses. The stocking of spare parts instead can reduce downtime and limit revenue losses if maintenance is required and the system can be upsized if energy demand and income situation improve in the future, for example through the productive use of energy [25]. However, the technical and financial optimization potential is limited and even the bundled deployment of MGs which can further reduce maintenance costs to some extent might not achieve substantial kWh retail price reductions to match affordability levels. Interviews among developers revealed that the current average solar PV MG tariff in Zambia charged is between 0.40 to 0.49 $ per kWh. A cost-reflective RE MG tariff for a system of the Mpanta or Sinda size below 0.30 $ per kWh appears to be unrealistic, even with a further decrease in component prices at least for the near future and in similar contexts.

5.2.2. Energy Affordability Benchmarking

A detailed energy-affordability benchmarking on a national or regional level for rural SSA is essential to include in future MG scale-up strategies and can be based on data for national disposable rural household income, which however might vary between different locations. Based on the community survey data presented in this study and regional benchmarking data [87], it can be assumed that the total disposable income for non-housing costs ranges from 1 to 2 $ per day in rural Eastern Africa. In rural Zambia, it is significantly lower. Table 8 provides a scenario for affordable energy tariffs at the Sinda and Mpanta sites based on the current number of connected customers, MG energy output and local household income and spending levels. The calculation which has been made in local currency illustrates the significant gap between affordable and cost-reflective tariffs at both sites. It also indicates that a desired further increase in consumption, for example through the productive use, would require a further tariff reduction or careful calculation of added value and income generation through these activities as an automatic increase in household income through energy access cannot automatically be assumed at all locations as the survey data indicates. The affordability context in this scenario however is purely based on a financial calculation and does not include the dimension of energy justice. If we transfer this scenario to an average UK household connected to Sinda mini-grid, this calculation would reveal an energy tariff of around 9£ per kWh which would be hard to communicate to a UK customer. These scenarios also emphasise that energy affordability does not automatically contribute to energy justice.

Table 8. MG energy-tariff affordability estimation for Mpanta and Sinda MG sites and comparison to UK household scenario based on household income and spending levels.

	Sinda	Mpanta	Comparative Scenario of Average UK Household Connected to Sinda MG in £
Daily output kWh	105	203	203
No. of customers	65	190	190
Total levels levels per month in ZMW	800	550	2000
Disposable spending after fixed-/food costs (30%)	240	165	600
Disposable Income for Energy (50% from disposable income) per month	120	82.5	300
Affordable energy spending levels per day in ZMW	4.00	2.75	10
Required kWh consumption per day based on system output/no of connections *	1.62	1.07	1.07
'Affordable' household tariff per kWh in ZMW with minimal consumption **	2.48	2.57	9.35
Current MG OPEX per kWh in ZMW	9.34	14	0.45

* Percentage based on regional income-benchmarking studies [87]. ** Assuming equal consumption across consumers.

5.2.3. Off-Grid Energy Tariffs

With this regard, it is generally questionable whether MG development at scale is desirable and realistic under the requirement of cost-reflective tariff setting, also in the light of energy justice. In Zambia for example, private consumers in the on-grid sector currently pay between ZMW 0.56 or $0.02 and ZMW 2.31 or $0.10 per kWh after a heavily debated price increase in September 2020 as on-grid energy tariffs remain subsidized in Zambia. These households represent around 31% of the Zambian population of around 5.5 million people and are mainly located in the urban areas with an average monthly income of ZMW 3152 or around $142 [69]. A tariff adjustment towards cost reflectiveness suggested by the Zambian Energy Regulation Board was axed by the Zambian President after serious public outcry [88] but further tariff increases are on the horizon. Over 95% of rural households with an average household income of ZMW 810 and below or $36 do not currently have access to the electrical grid. Interviews with project developers revealed that Zambian mini-grid customers in rural areas are currently paying around ZMW 10 per kWh under the latest MG deployment scheme [89]. Hence, within the further roll-out of MGs in rural areas due to the limited feasibility of scaling on-grid connections exhaustively within the next decade [54], the majority of the rural population in Zambia could be potentially facing energy tariffs that are up to 20 times higher than in grid-connected urban areas despite them having a much lower household income. This is not simply unjust or questionable with regard to economic feasibility, it also bears the potential of future political or civil rifts as the energy sector is a highly debated and critical topic in Zambia [88].

A more just approach could either integrate on- and off-grid energy planning and implementation under a national tariffing scheme which, if scaled, must also address specific requirements in terms of energy distribution and demand-side management which will require significant technology investment in developing countries [90–93]. The management of a potential future integration of MGs in the national electricity grid upon grid arrival should be a component during the MG planning stage and must also be sufficiently addressed by the regulatory frameworks to avoid decommissioning of the MG upon grid-arrival due to limited tariff competitiveness. On national level, strategic energy planning scenarios should account for the projected or potential integration of solar PV MGs and other decentralised energy sources such as wind- or hydropower into the national grid in terms of infrastructure investment—as well as demand-management planning, to avoid future drawbacks on the grid if this is going to be further extended in the future [94,95].

Alternatively and since off-grid electrification is heavily driven by foreign donor engagement [43], funding strategies could shift from a purely OPEX-based approach to a blended model that includes a tariff subsidy component to trigger energy consumption and include longer-term results-based funding (RBF) components [83] that are focused on MG operation and utilisation of electricity generated. This model would then ideally include a scenario of shifting towards cost-reflectiveness after a certain period of time when the targeted measures implemented alongside the MG deployment to increase income through productive use of energy, irrigation or improved farming methods for example triggered the desired income generation and stabilisation. This strategy could also potentially incentivise the implementation of financially and technically more optimised solutions, encourage CAPEX-reduction and incentivise the longer-term sustainable operation of the MG.

6. Conclusions

This study presents an interdisciplinary approach to evaluate the sustainability of RE mini-grids in SSA incorporating the technical, financial and communal dimensions. A comprehensive literature review, community survey data from Uganda and Zambia as well as the in-depth evaluation of two MG case studies from Zambia reveals the interdependency of financial—technical—and social sustainability.

The study has shown that an off-grid system is operationally sustainable if it can provide affordable energy access and deliver the desired outcomes over its estimated lifespan and that these interdependencies, however, exist partially in tension with each other. Social sustainability which includes affordable tariffs, for example, can bear trade-offs for example with financial sustainability in terms of recovering MG development—and operation costs. Consequently, sustainable off-grid electrification in sub-Saharan Africa requires an approach that goes beyond a top-down planning process that is solely focused on the provision of electricity.

Actual community energy needs and demands as well as the socio-economic structure are critical components for the long-term sustainability of energy systems and must be taken into account when an MG is going to be developed and implemented.

The interdisciplinary study allows five general conclusions with regard to the implementation of RE MGs in developing contexts:

Current MG energy tariffs are often yet far beyond local affordability levels which is also related to the requirement of tariff cost-reflectiveness in most SSA countries, high equipment costs and oversized systems. Consequently, local affordability levels should function as a key baseline for MG-planning- as well as wider electrification processes and design the systems based on local income and energy demand levels. This approach also includes close monitoring and benchmarking of system costs (CAPEX), especially if projects are donor-funded, the realistic sizing of the system as well as the acknowledgement of key community challenges that trigger low and intermittent incomes. These considerations also open the general debate whether it is realistic to achieve national energy access goals for low-income, rural areas until 2030 through the application of cost-reflective tariff schemes or whether rural energy consumption can only be triggered through mid- to longer-term energy tariff subsidies.

The complex challenges for rural African communities that are dependent on small-scale farming or fishing undermine the longer-term financial sustainability of MGs and are likely to be exacerbated by the effects of climate change. In addition to that, limited access to productive appliances, a of local know-how, decency on seasonal rainfalls and a lack of local capital often restrict rural communities from the uptake of the energy provided through off-grid systems.

Integrated planning approaches that flank energy system development and which focus on overcoming these specific challenges, for example through providing irrigation, training or access to appliances including financing schemes. Consequently, this paper calls for a change of the energy-access narrative from the focus of systems deployed or connections provided to the consideration of energy as a tool to support rural socio-economic development through the creation of added value and as one infrastructure component to be implemented alongside bundled measures that target key challenges for rural communities. The access to affordable and good quality appliances for private or commercial use in rural areas is still a major challenge and the establishment of affordable supply—and maintenance systems in the appliance sector are thus key to the utilisation of energy.

From a policy level, the findings suggest the promotion of feasibility studies that realistically benchmark local income—and projected energy tariff levels during the project planning and -application stage, support mechanisms for enhanced community engagement beyond the sole local approval of energy projects and the promotion of integrated infrastructure solutions. Projects publicly funded by donors or local government need to be closely evaluated in terms of CAPEX and prospected CAPEX in terms of realistic market pricing.

The scaling of sustainable RE mini-grids requires innovative financing approaches that go beyond pure upfront CAPEX grant-financing. Infrastructure solutions in rural settings of developing countries must be considered in the context of their operational environment, the community ecosystem which determines the sustainability of these solutions. The current trend of implementing largely grant financed, CAPEX intensive solutions with very limited community involvement does neither incentivize the productive use of energy through lower tariffs nor facilitate long-term sustainability and operation of these systems. The clustering of mini-grids can help in lowering down CAPEX and OPEX which can create a positive impact on energy tariffs and consumption. The provision of electricity is only one element of mini-grid planning and need to be aligned with energy uptake and added-value generating productive use of energy. Results-based financing approaches and carbon-credit financing that focus on longer-term mini-grid operation and energy uptake can be a key element of future financing approaches with this regard but require further research in terms of practical applicability.

The scaling of sustainable off-grid energy systems also requires adequate regulatory frameworks that technical quality, frictionless project implementation, strategic energy planning scenarios that integrate the on-and off-grid sectors flanked by investigating future grid-distribution strategies.

The danger of creating 'White Elephants' at scale, grant financed projects that are unsustainable and will be abandoned if operation costs permanently exceed revenues, is very real. The $1.3 million Mpanta mini-grid, which was commissioned in 2013, is currently on the brink of being decommissioned as the national electrical grid is now just about two km away from the village and due to replace the mini-grid very soon. According to REA Zambia, the solar PV mini-grid will be decommissioned. Batteries will be sold to a recycling plant. There are currently no plans available for utilising other assets of the mini-grid.

Author Contributions: S.S.: Conceptualization, methodology, validation, formal analysis, investigation, resources, data curation, writing—original draft preparation, writing—review and editing, visualization, project administration, funding acquisition. P.J.: formal analysis, investigation, data curation, writing—review and editing. All authors have read and agreed to the published version of the manuscript.

Funding: This research was supported by the Global Challenges Research Fund (GCRF) and the Economic and Soucal Research Council (ESRC) and the product of the research projects 'RISE' (https://www.smithschool.ox.ac.uk/research/rise-renewable-energy-innovation-scale/) and Mumuni Singani (https://www.smithschool.ox.ac.uk/research/mumuni-singani/index.html).

Data Availability Statement: Data supporting reported results can be found under https://www.smithschool.ox.ac.uk/research/mumuni-singani/index.html (accessed on 8 June 2021).

Acknowledgments: The authors are grateful for the cooperation and the provision of data by the Department of Physics of the University of Zambia (UNZA), Youth Development Organisation Zambia (YDO), Muhanya Solar Ltd. Zambia, the Rural Electrification Authority Zambia as well as for the technical guidance provided by Juergen Doering (NDL Renewables). The authors gratefully acknowledge the support of Chilangisha Changwe (Department of Physics, UNZA), Tonny Kukeera (University of Oxford) for content support and Aoife B. Haney (University of Oxford) who supported the project as Principal Investigator (https://www.smithschool.ox.ac.uk/research/mumuni-singani/

people.html, accessed on 21 June 2021). The authors also thank the team of project 'RISE' (https://www.smithschool.ox.ac.uk/research/rise-renewable-energy-innovation-scale/, accessed on 8 June 2021) and their partners who were involved in part of the community data collection for Uganda and Zambia.

Conflicts of Interest: The authors declare no conflict of interest.

Abbreviations

Nomenclature	
CAPEX	Capital Expenditures
ERB	Energy Regulation Board
MG	Mini-Grid
OPEX	Operating Expense
RE	Renewable Energy
REA	Rural Electrification Authority
SHS	Solar Home Systems
SSA	sub-Saharan Africa
UNZA	University of Zambia
WTP	'Willingness to pay'
ZESCO	Zambia Electricity Supply Corporation
ZMW	Zambian Kwacha

Appendix A

Table A1. Integrated, systematic literature review: categories and examples.

Source	Documen Type	Content Type	Countries in Scope	# of Cases	Technolog	Approach/Dime	Methodology	Detailed Data (OPEX/CAPEX) Y/N	Content & Findings	Main Challenges	Financing/Tariff Model	Conclusion	Main Sustainability Parameters	MG Financially (OPEX) Sustainable?	Community Context Analysis? (Y/N)
Adenle, 2020 [7]	Academic article	Review of case studies	Ghana, Kenya, South Africa	>30	Mainly SHS	Focus on technology to facilitate SHGs	Literature review; expert interviews	Partially/ limited	Financial problems; low household income; high procurement/operation costs; RE positively impacted most SDGs	Financial constraints (1); (2) technical issues; (3) governance/policy issues	Fully-/partially donor funded; all 4 tariff models	Most energy projects are donor-funded; financial sustainability failures across all countries in scope—especially with regard to maintenance; alternative approaches necessary	Financial; technical	>90% not	Limited extend (financial)
Butchers et.al., 2020 [75]	Academic article	Case study	Nepal	24	MG-hydro	MG operation	Mixed-methods incl.maintenance assessment & interviews	No	Technical issues identified during design, manufacture & installation phases; financial viability aided by charging consumers based on consumption, Community engagement essential but challenging; financial issues & challenges in tariff collection	Financial & technical constraints; community engagement; tariff collection	Mainly donor-funded; Willing Byer/Willing Seller & Individualised Cost-Based Tariff; partially cost-reflective	Community engagement + -context & operation & maintenance planning essential	Financial; operational	Not evaluated in detail	Limited extend (financial)
Ulsrud et.al., 2020 [96]	Academic article	Case study	India, Kenya and Senegal	4	Solar PV MG	Social; financial; access #	Qualitative interviews, quantitative surveys, participation	No	Main question: why is increasing energy access slower than expected? Top-down planning approaches that do not fit community requirements	Lack of affordability mismatch between system-design & local geography; gender inequality	No information	Importance of decentralised solutions; improvement of solar power delivery models needed & to be integrated into energy sector strategies	Social; economical; financial; operational	No information	Yes
Kinhoni, 2019 [22]	Thesis	Case study	Uganda	4	Solar PV MG	Financial, technical, operational	LCOE; empirical MG data	Yes	ROE and ROA were generally negative for the 4 MGs over 6 years; increasing losses; high operation and maintenance costs	Low affordability; low WRP; high operation costs	No information	Operation -/ownership model did not influence financial sustainability; large socioeconomic benefits; longer-term operational support necessary	Financial	No	No

Table A2. Findings literature review.

Source	1. Content & Focus				2. Approach/Dimension						3. Main Sustainability Challenges:					4. Data Available/Presented				
	Type	Region	Tech.	# of Cases	LCOE Optimisation	Technical Optimisation	Financial Evaluation	User-Centric; Community Based Approach	Productive Use/Economic Impact	Other	Financial	Technical	Social Acceptance/ End-User-Behaviour/ Community Engagement	Policy/Legal Issues	Other	Detailed Investment & Operational Data (OPEX/CAPEX)	Technical Data Incl. Generation	Community Ecosystem Approach/ Comprehensive Household Data	Tariff Data/Model	MG Financially Sustainable (OPEX)
Adenle, 2020 [7]	case study	SSA	solar mg	>30	-	-	x	-	-	-	x	-	-	x	-	Partially	-	Partially	Mixed	-
Akinyele, 2018 [97]	feasibility study		hybrid mg	3	-	-	-	-	-	-	n/A	n/A	n/A	n/A	n/A	-	-	-	-	-
Alonso et al., 2020 [76]	case study	SSA	hybrid mg	1	x	-	-	-	-	-	x	-	-	-	-	-	-	-	-	-
Azimoh et al., 2017 [25]	case study	SSA	hybrid mg	1	x	x	x	-	x	-	x	x	x	-	-	x	x	-	Subsidised	-
Bahaj et al., 2019 [23]	case study	Asia	solar mg	1	-	-	-	x	-	-	x	-	-	-	-	-	-	x	-	-
Boliko & Ialnazov, 2019 [37]	feasibility & case study	SSA	hybrid mg	2	-	-	x	x	-	-	x	x	x	x	-	-	-	x	-	-
Brent & Rogers, 2010 [81]	case study	SSA	hybrid mg	1	-	x	x	-	-	Environment policy	x	-	x	x	Environmental	Partially	x	-	Subsidised	-
Butchers et al., 2020 [75]	case study	Asia	mini-hydro mg	24	-	-	-	-	x	-	x	x	x		-	-	-	Partially	-	-
Feron et al., 2017 [77]	case study	South America	solar mg	>20	-	x	x	x	-	Energy governance; environmental	x	x	x	x	Socio-cultural	-	-	-	-	-
Katre et al., 2019 [27]	case study	Asia	solar mg	24	-	x	x	-	-	Environment	x	x	x	x	Operation Maintenance	-	Only generation consumption	x	-	-
Kinhonhi, 2019 [22]	case study	Asia	solar mg	4	x	-	x	-	-	-	x	-	-	x	-	x	-	-	-	-
Korkovelos et al., 2020 [55]	case study	SSA	solar mg/shs	6	-	-	x	-	-	Policy; strategy	x	x	-	x	-	-	-	-	-	-
Muhoza & Johnson, 2018 [41]	case study	SSA	solar mg	1	-	-	x	x	-	-	x	x	x	-	-	-	-	x	Not cost-reflective	-
Ngowi et al., 2019 [33]	case study	Asia	mini-hydro mg	1	-	-	-	-	x	-	-	-	-	-	-	-	-	-	-	OPEX partially/CAPEX not
Nsafon et al., 2020 [98]	feasibility study	SSA	hybrid mg	1	x	x	-	-	-	-	x	-	-	-	-	-	Simulation	-	-	-
Odou et al., 2020a [9]	feasibility study	SSA	hybrid mg	1	x	x	-	-	-	-	n/A	n/A	n/A	n/A	n/A	-	Simulation	-	-	-
Opoku et al., 2020 [99]	case study	SSA	solar mg	3	-	-	Feasibility	Limited	-	-	x	-	-	-	-	-	-	-	Subsidised	-

Table A2. *Cont.*

Source	1. Content & Focus				2. Approach/Dimension						3. Main Sustainability Challenges:					4. Data Available/Presented				
	Type	Region	Tech.	# of Cases	LCOE Optimisation	Technical Optimisation	Financial Evaluation	User-Centric; Community Based Approach	Productive Use/Economic Impact	Other	Financial	Technical	Social Acceptance/ End-User-Behaviour/ Community Engagement	Policy/Legal Issues	Other	Detailed Investment & Operational Data (OPEX/CAPEX)	Technical Data Incl. Generation	Community Ecosystem Approach/ Comprehensive Household Data	Tariff Data/Model	MG Financially Sustainable (OPEX)
Schnitzer et al., 2014 [100]	case study review	Asia, South America	solar mini-hydro, biomass, hybrid	17	-	-	x	x	-	-	x	-	x	-	Maintenance; skilled labour	Partially	Partially	-	Mixed	Partially for subsidised MG
Sharma et al., 2020 [30]	case study	Asia	solar mg	54	x	-	x	x	-	-	-	-	x	-	-	-	Simulation	Partially	-	-
Shrestha et al., 2019 [101]	case study	Asia	mini-hydro mg	6	-	x	x	MG operation	-	-	x	-	x	x	-	Partially	-	-	-	-
Tazvingal et al., 2020 [102]	feasibility study	SSA	hybrid mg	1	-	x	-	-	-	-	n/A	n/A	n/A	n/A	n/A	-	Simulation	-	-	-
Terrapon-Pfaff, 2014 [103]	case study review	Asia, SA, South America	solar mg, hydro, biomass, hybrid	23	-	-	-	x	-	MDG Impact Analysis	-	-	x	x	Ownership models; logistics availability of skills		-	-	-	-
Tsuchiya, et al., 2020 [104]	case study	SSA	solar mg	2	x	x	x	-	-	Operation; maintenance	x	x	-	-	-	Partially (CAPEX)	-	-	-	-
Uamusse et al., 2019 [105]	feasibility study	SSA	mini-hydro mg	2	x	-	Feasibility	x	-	-	x	-	x	x	-	-	-	-	-	-
Ulsrud, 2020 [96]	case study	Asia	solar mg	4	-	-		x	-	-	-	-	x		-	-	-	x	-	-
Ulsrud et al., 2019 [32]	case study	SSA	hybrid mg	6	x	-	x	-	-	Operation; Maintenance	x	-	x	x	-	-	-	Partially	-	-

Table 3. List of RE Mini-grids in Zambia (Source: UNZA research in cooperation with REA).

	Description	Capacity [kW]	District	Province	STATUS
1	Mpanta Solar Mini grid	(60)	Samfya	Luapula	Decommissioning
2	Chitandika Solar Mini grid	28	Chipangali	Eastern	Operational
3	Muhanya Solar	30	Sinda	Eastern	Operational
4	Solera Mini grid–Ken Village	25	Katete	Eastern	Operational
5	Solera Mini grid–Madzi-Atuwa	25	Chipangali	Eastern	Operational
6	Solera Mini grid–Kapasa	25	Chipangali	Eastern	Operational
7	Solera Mini grid–Taferasoni	25	Chadiza	Eastern	Operational
8	Solera Mini grid–Kacholola	25	Nyimba	Eastern	Operational
9	Solera Mini grid–Chanyalubwe	25	Lundazi	Eastern	Operational
10	Solera Mini grid–Chikomeni	25	Lumezhi	Eastern	Operational
11	Solera Mini grid–Luangwa Market	25	Luangwa	Eastern	Operational
12	Solera Mini grid–Mnukwa	25	Chipata	Eastern	Operational
13	Solera Mini grid–Chikalawa	25	Petauke	Eastern	Operational
14	Magodi Solar Mini grid	48	Lundazi	Eastern	Operational
15	Zengamina Mini Hydro	750		North-Western	Operational
16	Kasanjiku Mini hydro	640	Mwinilunga	North-Western	Operational
17	Katamanda Solar mini grid	50	Chipangali	Eastern	Operational
18	Standard Micro grid–Kakolo	nA	Kitwe	Copperbelt	Operational
19	Standard Micro grid–Kamuchanga	nA	Kabwe	Copperbelt	Operational
20	Standard Micro grid–Zambia Compound	nA	Kapiri Mposhi	Central	Operational
21	Standard Micro grid–Ngwerere 1	10	Chongwe	Lusaka	Operational
22	Standard Micro grid–Mugurameno	10	nA	nA	Operational
23	Standard Micro grid–Lower Zambezi	10	nA	nA	Operational
24	Standard Micro grid–Sioma high school	24	Sioma	Western	Operational
25	Standard Micro grid–Undi Village	10	Katete	Eastern	Operational
26	Standard Micro grid–Katente	nA	Nyimba	Eastern	Operational
27	Standard Micro grid–Zambia Compound	nA	Kitwe	Copperbelt	Operational
28	Standard Micro grid–Chapita	nA	Katete	Eastern	Operational
29	Standard Micro grid–Mphila	nA	Katete	Eastern	Operational
30	Chunga Solar mini grid	nA	Mumbwa	Central	Under construction
31	Lunga Solar mini grid	nA	Lunga	Luapula	Under construction
32	Chitokoloki Mission Solar Mini grid	nA	Petauke	Eastern	Operational
33	Shiwang'andu	nA	Shiwang'andu	Muchinga	Operational

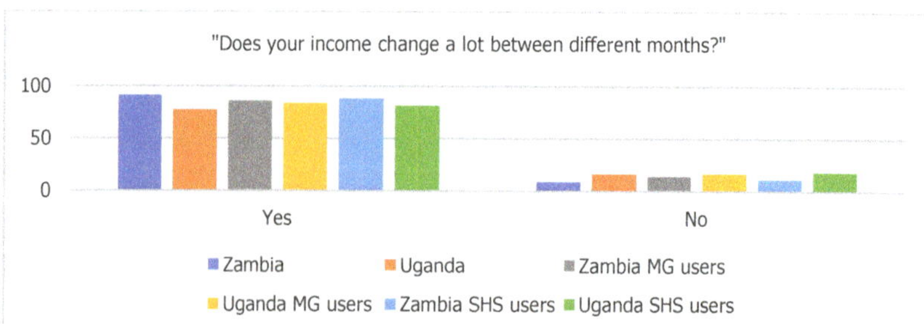

Figure 1. Income intermittency in Uganda and Zambia (N = 1016; Responses in %; Multiple responses possible).

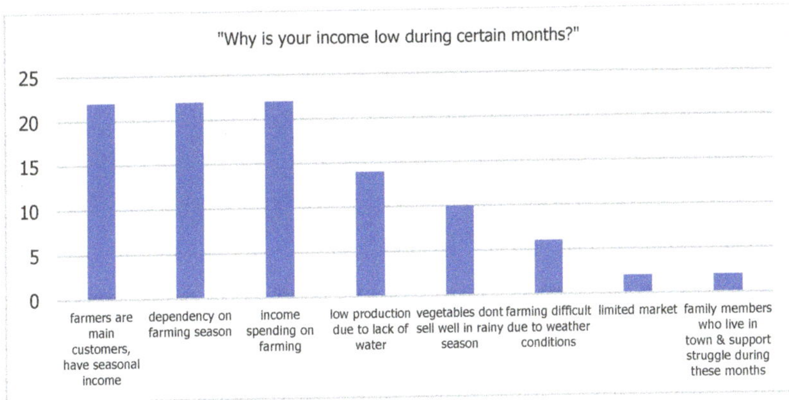

Figure 2. Reasons for low income; Southern Zambia; N = 50; Responses in %; Multiple responses possible.

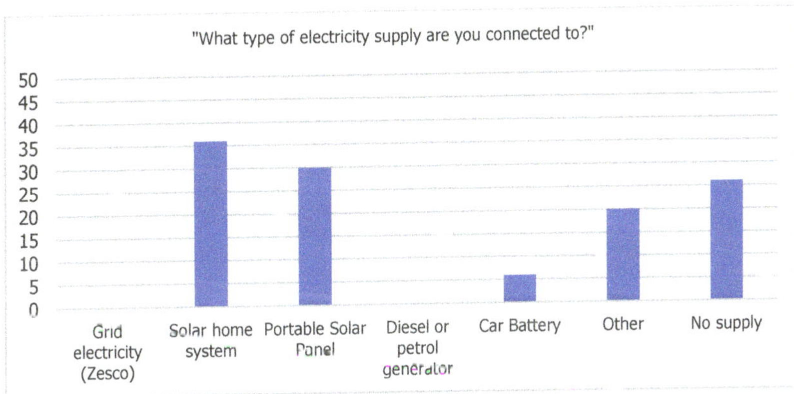

Figure 3. Type of electricity supply, Southern Zambia (Responses in %; Multiple Responses; N = 50).

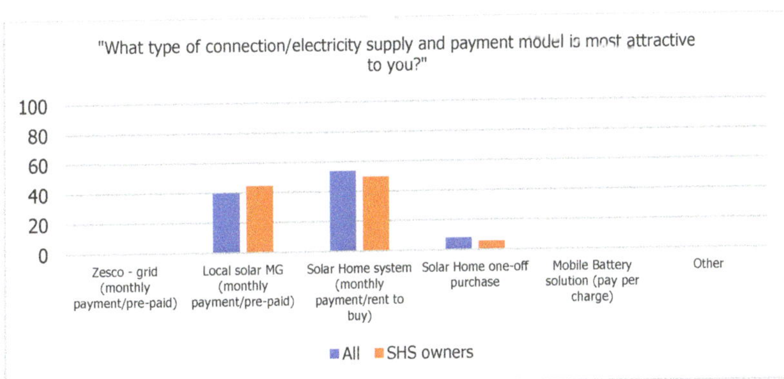

Figure 4. Desired energy connection, Southern Zambia, N = 50, Responses in %.

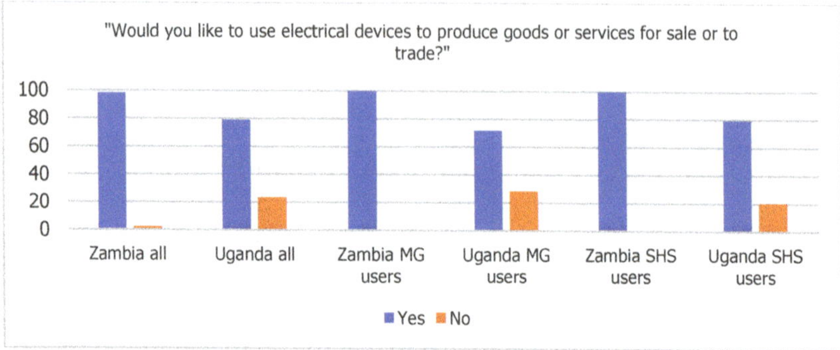

Figure 5. Desire for productive energy use in Uganda and Zambia (Responses in %).

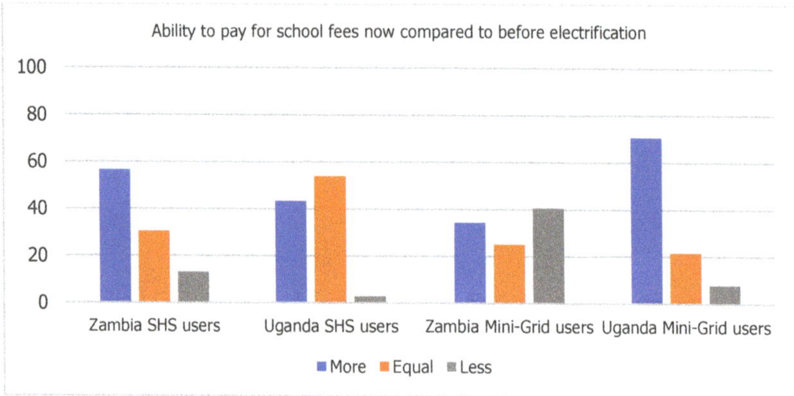

Figure 6. Impact of electrification on ability to pay school fees among various subgroups; Responses in %.

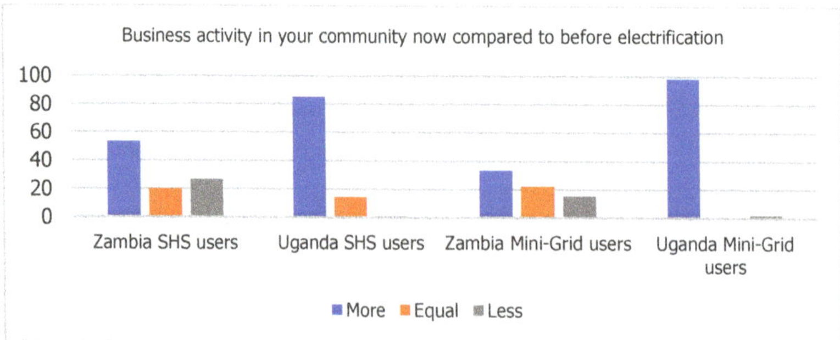

Figure 7. Impact of electrification on business activity among various subgroups; Responses in %.

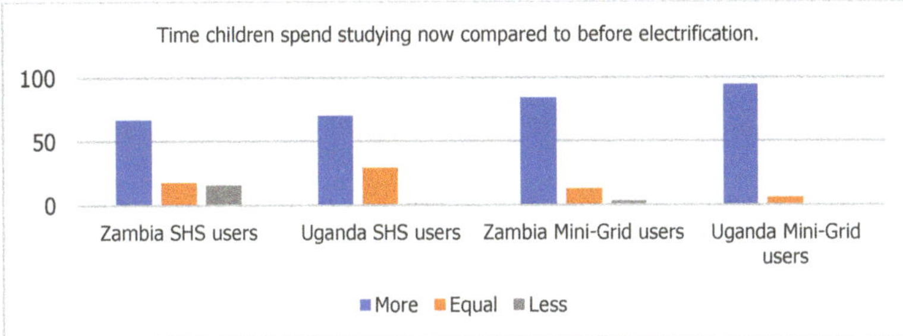

Figure 8. Impact of electrification on education among various subgroups; Responses in %.

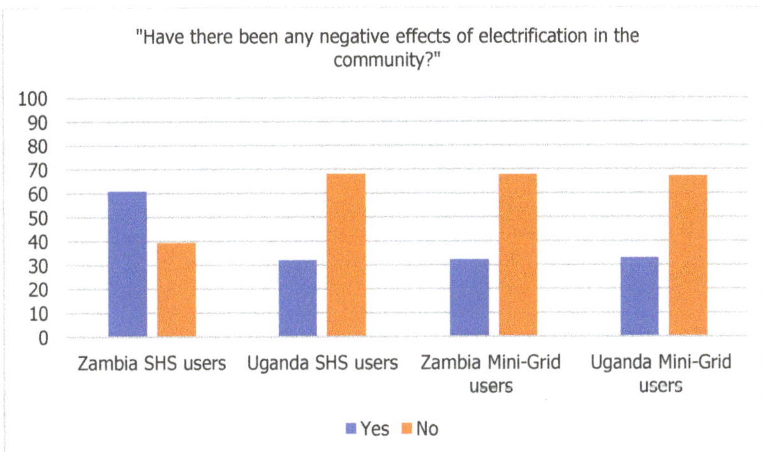

Figure 9. Perceived negative effects of electrification among various subgroups; Responses in %.

Appendix B. Additional Information: Optimisation Approach & Background—60 kWp and 45 kWp Solar Energy Mini-Grid for a Location like Mpanta

Appendix B.1. Estimating PV Power Output

The estimation of solar PV output power in Mpanta, Zambia is calculated through the World Bank Solargis map for Zambia [71]. This map provides a summary of the estimated solar PV power generation potential. It represents the average daily and yearly sum of electricity production from a 1 kWp grid-connected solar PV power plant calculated for the recent 24-year period 1994–2017. The PV system configuration consists of crystalline silicon PV modules mounted at a fixed position and ground-based, free standing structures with an optimal tilt towards the equator. The use of high efficiency inverters is assumed.

Table A4. Annual/Daily PV power output Mpantan MG–modelling approach.

Item	Value	
Plant capacity	60 kW_p	45 kW_p
Average yearly PV power output	1690 Wh/kW_p	1690 kWh/kW_p
Total yearly PV power output	1690 $kWh/kW_p \times 60\ kW_p$ = 101,400 kWh	1690 $kWh/kW_p \times 45\ kW_p$ = 76,050 kWh
Total average daily PV power output	101,400/365 = 278 kWh	76,050/365 = 208 kWh

Appendix B.2. Battery Sizing

The use of high-quality components is important as they allow uninterrupted power-supply and the life-span of the components. Although the initial costs of good quality components are higher, their life-time cost is usually limited due to less failures, a longer life-span and higher efficiencies. Batteries are the most expensive component in a solar PV system with storage. The current trend is to move away from the traditional lead acid batteries to lithium-ion batteries. Li-ion batteries have a life-time of 15 to 20 years with an average warranty of 10 years compared to lead-acid batteries which are supposed to have a life-time of 5 years but often create technical problems and die out before this time. Lithium-ion batteries have a higher 80% depth of discharge (DOD) compared to 50% DOD for lead acid batteries. Lithium-ion batteries have a higher efficiency of over 90% compared to lead acid batteries which are 80% efficient. With these factors, life-time costs of lithium batteries are lower than that of lead acid batteries. Additionally, lithium batteries are much lighter and smaller in size. Consequently, Li-ion are the preferred choice in this new configuration of Mpanta MG.

As a second step, the size of the battery has to be optimised. Keeping in view that Mpanta community's ability to pay is very low, only one day battery autonomy can be provided for cost-reasons. This would mean, that the plant would provide power only for 24 h without the sun. Consequently, residents would have a restricted power supply during cloudy days. Further assumed that on average about one third of the load is used during the day, the need of storage is reduced to two thirds of the daily energy output. Taking 80% as the depth of discharge for Li-ion batteries, the required size of the battery system is outlined in the Table below.

Table A5. Mpanta—Optimised battery-sizing scenario.

MG—system sizing	45 kWp	60 kWp
Daily Energy Demand	208 kWh	278 kWh
Depth of Discharge (Li-ion)	0.8	0.8
Assumed day-time energy demand	1/3 of the daily demand	1/3 of the daily demand
Inverter efficiency	0.9	0.9
Battery size	$208/(0.8 \times 0.9) \times 2/3 = 193$ kWh	$278 \times (0.8 \times 0.9) \times 2/3 = 258$ kWh

Table A6. Current and optimised technical layout and OPEX for Mpanta MG.

Current Layout					Optimised Model with Reduced Battery Back-Up System						
Components	Est. Current Costs ($)	Units	Est. Life Time (Years)	Current Est. OPEX p.a. $	Components	Est. Life Time (Years)	Est. Unit Costs ($)	Annualised Costs			
								45 kW		60 kW	
Batteries ($350 × 320)	112,000	720 ×150 $	5	21,600	Batteries (Li-Ion)	15 **	600	600 $/193 units	7720	600 $/258 units	10,320
Inverter *	8000	2 × 15 kW	7	1143	Inverter ***	7	4000	3×	1714	4×	2286
Charge controllers *	3000	2 × 100 A/240 V	7	600	Charge controllers ***	10	1500	3×	450	4×	600
Misc	-	-	-	3000	Contingency/Misc	-	-	-	1500	-	2000
Total	-	-	-	26,343	Total (in $)	-	-	-	11,384	-	15,206

* The system currently only operates 2.15 kW Inverters and 2 charge controllers which probably decreases the actual kWh output of the 45 kWp solar generator by approximately 10–15%. ** The lifespan of Li-Ion batteries is approx. 3× longer compared to the AGM/LEAD Batteries that are currently used in Mpanta. *** The capacity and the charge controllers has been adjusted in the optimised model to ensure maximum output in kWh in relation to the solar generator.

References

1. Wood Mackenzie Power & Renewables. *Strategic Investments in Off-Grid Energy Access: Scaling the Utility of the Future for the Last Mile*; Wood Mackenzie Power & Renewables, 2019; Energy4Impact London/UK, 28/02/2019; Available online: https://energy4impact.org/sites/default/files/strategic_investments_in_off-grid_energy_access_final.pdf (accessed on 17 June 2020).
2. International Energy Agency (IEA). *Number of People without Access to Electricity in Sub-Saharan Africa in the STEPS*; IEA: Paris, France, 2019; Available online: https://www.iea.org/data-and-statistics/charts/number-of-people-without-access-to-electricity-in-sub-saharan-africa-in-the-steps (accessed on 17 June 2021).
3. International Energy Agency (IEA). *Energy Access Outlook: From Poverty to Prosperity, World Energy Outlook-2017 Special Report*; IEA: Paris, France, 2017.
4. Pedersen, M.B.; Nygaard, I. System building in the Kenyan electrification regime: The case of private solar mini-grid development. *Energy Res. Soc. Sci.* **2018**, *42*, 211–223. [CrossRef]
5. Azimoh, C.L.; Klintenberg, P.; Wallin, F.; Karlsson, B.; Mbohwa, C. Electricity for development: Mini-grid solution for rural electrification in South Africa. *Energy Convers. Manag.* **2016**, *110*, 268–277. [CrossRef]
6. Gollwitzer, L.; Ockwell, D.; Muok, B.; Ely, A.; Ahlborg, H. Rethinking the sustainability and institutional governance of electricity access and mini-grids: Electricity as a common pool resource. *Energy Res. Soc. Sci.* **2018**, *39*, 152–161. [CrossRef]
7. Adenle, A.A. Assessment of solar energy technologies in Africa-opportunities and challenges in meeting the 2030 agenda and sustainable development goals. *Energy Policy* **2020**, *137*, 111180. [CrossRef]
8. Szabó, S.; Bódis, K.; Huld, T.; Moner-Girona, M. Sustainable energy planning: Leapfrogging the energy poverty gap in Africa. *Renew. Sustain. Energy Rev.* **2013**, *28*, 500–509. [CrossRef]
9. Odou, O.D.T.; Bhandari, R.; Adamou, R. Hybrid off-grid renewable power system for sustainable rural electrification in Benin. *Renew. Energy* **2020**, *145*, 1266–1279. [CrossRef]
10. Bhattacharyya, S.C. Financing energy access and off-grid electrification: A review of status, options and challenges. *Renew. Sustain. Energy Rev.* **2013**, *20*, 462–472. [CrossRef]
11. Elkadeem, M.R.; Wang, S.; Azmy, A.M.; Atiya, E.G.; Ullah, Z.; Sharshir, S.W. A systematic decision-making approach for planning and assessment of hybrid renewable energy-based microgrid with techno-economic optimization: A case study on an urban community in Egypt. *Sustain. Cities Soc.* **2020**, *54*, 102013. [CrossRef]
12. Schmidt, T.S. Making electrification models more realistic by incorporating differences in institutional quality and financing cost. *Prog. Energy* **2019**, *2*, 013001. [CrossRef]
13. Hirsch, A.; Parag, Y.; Guerrero, J. Microgrids: A review of technologies, key drivers, and outstanding issues. *Renew. Sustain. Energy Rev.* **2018**, *90*, 402–411. [CrossRef]
14. Bhattacharyya, S.C.; Palit, D. Mini-grid based off-grid electrification to enhance electricity access in developing countries: What policies may be required? *Energy Policy* **2016**, *94*, 166–178. [CrossRef]
15. Hansen, U.E.; Pedersen, M.B.; Nygaard, I. Review of solar PV policies, interventions and diffusion in East Africa. *Renew. Sustain. Energy Rev.* **2015**, *46*, 236–248. [CrossRef]
16. Mandelli, S.; Barbieri, J.; Mereu, R.; Colombo, E. Off-grid systems for rural electrification in developing countries: Definitions, classification and a comprehensive literature review. *Renew. Sustain. Energy Rev.* **2016**, *58*, 1621–1646. [CrossRef]
17. Narayan, N.; Papakosta, T.; Vega-Garita, V.; Qin, Z.; Popovic-Gerber, J.; Bauer, P.; Zeman, M. Estimating battery lifetimes in Solar Home System design using a practical modelling methodology. *Appl. Energy* **2018**, *228*, 1629–1639. [CrossRef]
18. Peters, J.; Sievert, M.; Toman, M.A. Rural electrification through mini-grids: Challenges ahead. *Energy Policy* **2019**, *132*, 27–31. [CrossRef]
19. Maso, M.D.; Olsen, K.H.; Dong, Y.; Pedersen, M.B.; Hauschild, M.Z. Sustainable development impacts of nationally determined contributions: Assessing the case of mini-grids in Kenya. *Clim. Policy* **2019**, *20*, 815–831. [CrossRef]
20. Sustainable Energy for All (SEforALL). *State of the Global Mini-Grids Market Report 2020*; SEforALL: Vienna, Austria, 2020.
21. Derks, M.; Romijn, H. Sustainable performance challenges of rural microgrids: Analysis of incentives and policy framework in Indonesia. *Energy Sustain. Dev.* **2019**, *53*, 57–70. [CrossRef]
22. Kinhonhi, I.V. Financial Sustainability of Mini-Grid Electricity Distribution Companies in Uganda. Master's Thesis, University of Cape Town, Cape Town, South Africa, 2019.
23. Bahaj, A.; Blunden, L.; Kanani, C.; James, P.; Kiva, I.; Matthews, Z.; Price, H.; Essendi, H.; Falkingham, J.; George, G. The Impact of an Electrical Mini-grid on the Development of a Rural Community in Kenya. *Energies* **2019**, *12*, 778. [CrossRef]
24. Pedersen, M.B.; Wehrmeyer, W.; Nygaard, I. Commercial yet social: The practices and logics of bringing mini-grid electricity to rural villages in Kenya. *Energy Res. Soc. Sci.* **2020**, *68*, 101588. [CrossRef]
25. Azimoh, C.L.; Klintenberg, P.; Mbohwa, C.; Wallin, F. Replicability and scalability of mini-grid solution to rural electrification programs in sub-Saharan Africa. *Renew. Energy* **2017**, *106*, 222–231. [CrossRef]
26. Wong, S. Overcoming obstacles against effective solar lighting interventions in South Asia. *Energy Policy* **2012**, *40*, 110–120. [CrossRef]
27. Katre, A.; Tozzi, A.; Bhattacharyya, S. Sustainability of community-owned mini-grids: Evidence from India. *Energy. Sustain. Soc.* **2019**, *9*, 2. [CrossRef]
28. Mulugetta, Y.; Nhete, T.; Jackson, T. Photovoltaics in Zimbabwe: Lessons from the GEF Solar project. *Energy Policy* **2000**, *28*, 1069–1080. [CrossRef]

29. Wamukonya, N. Solar home system electrification as a viable technology option for Africa's development. *Energy Policy* **2007**, *35*, 6–14. [CrossRef]
30. Sharma, A.; Agrawal, S.; Urpelainen, J. The adoption and use of solar mini-grids in grid-electrified Indian villages. *Energy Sustain. Dev.* **2020**, *55*, 139–150. [CrossRef]
31. Nygaard, I.; Hansen, U.E.; Larsen, T.H.; Palit, D.; Muchunku, C. Off-grid Access to Electricity Innovation Challenge. In Accelerating the Clean Energy Revolution—Perspectives on Innovation Challenges. DTU International Energy Report 2018. 2018. Available online: https://data.worldbank.org/indicator/EG.ELC.ACCS.RU.ZS?locations=NP (accessed on 21 June 2021).
32. Ulsrud, K.; Muchunku, C.; Palit, D.; Kirubi, G. *Solar Energy, Mini-Grids and Sustainable Electricity Access: Practical Experiences, Lessons and Solutions from Senegal*; Routledge: New York, NY, USA, 2019.
33. Ngowi, J.M.; Bångens, L.; Ahlgren, E.O. Benefits and challenges to productive use of off-grid rural electrification: The case of mini-hydropower in Bulongwa-Tanzania. *Energy Sustain. Dev.* **2019**, *53*, 97–103. [CrossRef]
34. Moner-Girona, M.; Solano-Peralta, M.; Lazopoulou, M.; Ackom, E.K.; Vallve, X.; Szabó, S. Electrification of Sub-Saharan Africa through PV/hybrid mini-grids: Reducing the gap between current business models and on-site experience. *Renew. Sustain. Energy Rev.* **2018**, *91*, 1148–1161. [CrossRef]
35. Tenenbaum, B.; Greacen, C.; Vaghela, D. *Mini-Grids and Arrival of the Main Grid*; World Bank: Washington, DC, USA, 2018.
36. Goldthau, A.; Sovacool, B.K. The uniqueness of the energy security, justice, and governance problem. *Energy Policy* **2012**, *41*, 232–240. [CrossRef]
37. Boliko, C.M.; Ialnazov, D.S. An assessment of rural electrification projects in Kenya using a sustainability framework. *Energy Policy* **2019**, *133*, 110928. [CrossRef]
38. Bhandari, R.; Saptalena, L.G.; Kusch, W. Sustainability assessment of a micro hydropower plant in Nepal. *Energy. Sustain. Soc.* **2018**, *8*, 3. [CrossRef]
39. Boait, P.J. Technical Aspects of Mini-Grids for Rural Electrification. In *Mini-Grids for Rural Electrification of Developing Countries. Green Energy and Technology*; Bhattacharyya, S., Palit, D., Eds.; Springer: Cham, Switzerland, 2014; pp. 37–61.
40. Eales, A.; Walley, L.; Buckland, H.; Frame, D.; Strachan, S. Social Impacts of Mini-Grids: Towards an Evaluation Methodology. In Proceedings of the 2018 IEEE PES/IAS PowerAfrica, Cape Town, South Africa, 26–29 June 2018; Institute of Electrical and Electronics Engineers: Washington, DC, USA, 2018; pp. 354–359. [CrossRef]
41. Muhoza, C.; Johnson, O.W. Exploring household energy transitions in rural Zambia from the user perspective. *Energy Policy* **2018**, *121*, 25–34. [CrossRef]
42. Fowlie, M.; Khaitan, Y.; Wolfram, C.; Wolfson, D. Solar Microgrids and Remote Energy Access: How Weak Incentives Can Undermine Smart Technology. *Econ. Energy Environ. Policy* **2019**, *8*, 8. [CrossRef]
43. Haney, A.; Stritzke, S.; Trotter, P.; Puranasamriddhi, A.; Madhlopa, A.; Moyo, A.; Batidzirai, B.; Twesigye, P.; Walter, M.; Kovandova, E. *Electricity for Integrated Rural Development. The Role of Businesses, the Public Sector and Communities in Uganda and Zambia. Project RISE Practicioners Report*; University of Oxford: Oxford, UK, 2019.
44. Foster, T.; Hope, R. A multi-decadal and social-ecological systems analysis of community waterpoint payment behaviours in rural Kenya. *J. Rural Stud.* **2016**, *47*, 85–96. [CrossRef]
45. Walther, G.-R. Community and ecosystem responses to recent climate change. *Philos. Trans. R. Soc. B Biol. Sci.* **2010**, *365*, 2019–2024. [CrossRef]
46. Phillips, J.; Attia, B.; Plutshack, V. *Balancing Competition and Subsidy: Assessing Mini-Grid Incentive Programs in Sub-Saharan Africa*; Nicholas Institute for Environmental Policy Solutions Duke Energy Access Project, Policy Brief: Durham, NC, USA, 2020.
47. NARUC: National Association of Regulatory Utility Commissioners. *Exploring Africa's Mini-Grid Tariff Methodology Assessment*; Meister Consultants Group, Inc., a Cadmus Company: Waltham, MA, USA, 2020.
48. Kittner, N.; Gheewala, S.H.; Kammen, D.M. Energy return on investment (EROI) of mini-hydro and solar PV systems designed for a mini-grid. *Renew. Energy* **2016**, *99*, 410–419. [CrossRef]
49. Snyder, H. Literature review as a research methodology: An overview and guidelines. *J. Bus. Res.* **2019**, *104*, 333–339. [CrossRef]
50. Batidzirai, B.; Trotter, P.A.; Brophy, A.; Stritzke, S.; Moyo, A.; Twesigye, P.; Puranasamriddhi, A.; Madhlopa, A. Towards people-private-public partnerships: An integrated community engagement model for capturing energy access needs. *Energy Res. Soc. Sci.* **2021**, *74*, 101975. [CrossRef]
51. Uamusse, M.M.; Tussupova, K.; Persson, K.M.; Berndtsson, R.; Uamusse, M.M.; Tussupova, K. Access to sustainable electrification: Possibilities for rural Mozambique Access to sustainable electrification: Possibilities for rural Mozambique. *Cogent Eng.* **2020**, *7*. [CrossRef]
52. Kalangala Infrastructure Services Ltd. Available online: https://www.kis.co.ug (accessed on 21 June 2021).
53. Kruger, W.; Stritzke, S.; Trotter, P.A. De-risking solar auctions in sub-Saharan Africa—A comparison of site selection strategies in South Africa and Zambia. *Renew. Sustain. Energy Rev.* **2019**, *104*, 429–438. [CrossRef]
54. Stritzke, S.; Trotter, P.A.; Twesigye, P. Towards responsive energy governance: Lessons from a holistic analysis of energy access in Uganda and Zambia. *Energy Policy* **2021**, *148*. [CrossRef]
55. Korkovelos, A.; Zerriffi, H.; Howells, M.; Bazilian, M.; Rogner, H.-H.; Fuso Nerini, F. A Retrospective Analysis of Energy Access with a Focus on the Role of Mini-Grids. *Sustainability* **2020**, *12*, 1793. [CrossRef]
56. The University of Zambia. *Consultancy Services to Assess the Sustainability of Mpanta, Chunga and Lunga Solar Mini-grid Projects and Propose a Sustainable Business Model for the Management of the Plants*; The University of Zambia: Lusaka, Zambia, 2016.

57. Johnson, O.W.; Gerber, V.; Muhoza, C. Gender, culture and energy transitions in rural Africa. *Energy Res. Soc. Sci.* **2019**, *49*, 169–179. [CrossRef]
58. Kemausuor, F.; Sedzro, M.D.; Osei, I. Decentralised Energy Systems in Africa: Coordination and Integration of Off-Grid and Grid Power Systems—Review of Planning Tools to Identify Renewable Energy Deployment Options for Rural Electrification in Africa. *Curr. Sustain. Energy Rep.* **2018**, *5*, 214–223. [CrossRef]
59. Mvula, A.P.; Siwale, L.; Luwaya, E. Design of Circular Arc Blade Hydrokinetic Turbine—A Case of Rural Electrification in Zambia. *J. Power Energy Eng.* **2019**, *7*, 60–69. [CrossRef]
60. The World Bank World Bank Open Data. Available online: https://data.worldbank.org (accessed on 28 May 2021).
61. Madzudzo, L.; Hangooma, G.; Ratner, B. *Strengthening Collective Action to Address Resource Conflict in Lake Kariba, Zambia. Program Report*; Collaborating for Resilience: Lusaka, Zambia, 2014.
62. Japan International Cooperation Agency (JICA). *The Study for Development of the Rural Electrification Master Plan (REMP) in Zambia Final Report Summary Report*; JICA: Lusaka, Zambia, 2008.
63. Bos, K.; Chaplin, D.; Mamun, A. Benefits and challenges of expanding grid electricity in Africa: A review of rigorous evidence on household impacts in developing countries. *Energy Sustain. Dev.* **2018**, *44*, 64–77. [CrossRef]
64. Bowa, K.C.; Mwanza, M.; Sumbwanyambe, M.; Pretorius, J.H. *Solar Photovoltaic Energy Progress in Zambia*; The South African Institute of Electrical Engineers: Johannesburg, South Africa, 2017.
65. Jürisoo, M.; Serenje, N.; Mwila, F.; Lambe, F.; Osborne, M. Old habits die hard: Using the energy cultures framework to understand drivers of household-level energy transitions in urban Zambia. *Energy Res. Soc. Sci.* **2019**, *53*, 59–67. [CrossRef]
66. Ryota Keeley, A.; Managi, S. The importance of financial cost for renewable energy projects: Economic viability assessment of renewable hybrid mini-grid systems in Indonesia. *Green Financ.* **2019**, *1*, 139–155. [CrossRef]
67. Shrestha, P.; Shrestha, A.; Shrestha, N.T.; Papadakis, A.; Maskey, R.K. Assessment on Scaling-Up of Mini-Grid Initiative: Case Study of Mini-Grid in Rural Nepal. *Int. J. Precis. Eng. Manuf. Green Technol.* **2021**, *8*, 217–231. [CrossRef]
68. Eisenhardt, K.M.; Graebner, M.E. Theory Building From Cases: Opportunities And Challenges. *Acad. Manag. J.* **2007**, *50*, 25–32. [CrossRef]
69. Republic of Zambia Central Statistical Office. *Zambia 2015 Living Conditions Monitoring Survey Key Findings*; Government of Zambia: Lusaka, Zambia, 2016.
70. van der Ven, M.J. *An Overview of Recent Developments and the Current State of the Ugandan Energy Sector*; The International Growth Centre (IGC): London, UK, 2020.
71. Solar Resource Maps of Zambia—SolarGIS: Zambia. Available online: https://solargis.com/maps-and-gis-data/download/zambia (accessed on 1 April 2020).
72. Rural Electrification Authority Zambia REA Solar Energy Projects. *Project Brief*; Rural Electrification Authority: Lusaka, Zambia, 2020.
73. Rural and Renewable Energy Division UNIDO. *Renewable Energy-Based Mini-Grids: The UNIDO Experience*; United Nations Industrial Development Organisation: Vienna, Austria, 2017.
74. RECP/EUEI. *Project Case Study—Mini-Grids Zambia' Developed and Issued by RECP/EUEI, 2018*; GET.invest/European Union Energy Initiative Partnership: Brussels, Belgium, 2018.
75. Butchers, J.; Williamson, S.; Booker, J.; Tran, A.; Karki, P.B.; Gautam, B. Understanding sustainable operation of micro-hydropower: A field study in Nepal. *Energy Sustain. Dev.* **2020**, *57*, 12–21. [CrossRef]
76. Alonso, J.B.; Sandwell, P. *Sustainable Mini-Grid Systems in Refugee Camps: A Case Study of RWANDA*; Grantham Institute Briefing Note No 12: London, UK, 2020.
77. Feron, S.; Cordero, R.; Labbe, F. Rural Electrification Efforts Based on Off-Grid Photovoltaic Systems in the Andean Region: Comparative Assessment of Their Sustainability. *Sustainability* **2017**, *9*, 1825. [CrossRef]
78. HOMER Pro®® Microgrid Software by HOMER Energy. Available online: https://www.homerenergy.com/products/grid/index.html (accessed on 1 December 2020).
79. Groh, S. The role of energy in development processes—The energy poverty penalty: Case study of Arequipa (Peru). *Energy Sustain. Dev.* **2014**, *18*, 83–99. [CrossRef]
80. Office for National Statistics Average Household Income, UK: Financial Year 2020. Office for National Statistics, UK. Available online: https://www.ons.gov.uk/peoplepopulationandcommunity/personalandhouseholdfinances/incomeandwealth/bulletins/householddisposableincomeandinequality/financialyear2020 (accessed on 1 January 2020).
81. Brent, A.C.; Rogers, D.E. Renewable rural electrification: Sustainability assessment of mini-hybrid off-grid technological systems in the African context. *Renew. Energy* **2010**, *35*, 257–265. [CrossRef]
82. Katre, A.; Tozzi, A. Assessing the sustainability of decentralized renewable energy systems: A comprehensive framework with analytical methods. *Sustainability* **2018**, *10*, 1058. [CrossRef]
83. Stritzke, S.; Sakyi-Nyarko, C.; Bisaga, I.; Bricknell, M.; Leary, J.; Brown, E. Results Based Financing (RBF) for Modern Energy Cooking Solutions: An effective driver for innovation and scale? *Energies* **2021**, unpublished work.
84. Booth, S.S.; Li, X.; Baring-Gould, I.; Kollanyi, D.; Bharadwaj, A.; Weston, P.; Baring-Gould, E.I. *Productive Use of Energy in African Micro-Grids: Technical and Business Considerations. Productive Use of Energy in African Micro-Grids: Technical and Business Considerations*; Energy4Impact: London, UK, 2018.

85. Stritzke, S.; Kukeera, T. *Project 'Mumuni Singani': A Blueprint for Clean Energy Provision and Sustainable Development in Rural Zambia;* Final Project Report; Smith School of Enterprise and the Environment, University of Oxford: Oxford, UK, 2021.
86. Baptista, I.; Plananska, J. The landscape of energy initiatives in sub-Saharan Africa: Going for systemic change or reinforcing the status quo? *Energy Policy* **2017**, *110*, 1–8. [CrossRef]
87. van de Ven, G.W.J.; de Valença, A.; Marinus, W.; de Jager, I.; Descheemaeker, K.K.E.; Hekman, W.; Mellisse, B.T.; Baijukya, F.; Omari, M.; Giller, K.E. Living income benchmarking of rural households in low-income countries. *Food Secur.* **2020**, *13*, 729. [CrossRef]
88. Bayliss, K.; Pollen, G. The power paradigm in practice: A critical review of developments in the Zambian electricity sector. *World Dev.* **2021**, *140*, 105358. [CrossRef]
89. REEEP Power Africa: Beyond the Grid Fund for Zambia. Available online: https://www.reeep.org/bgfz (accessed on 10 August 2020).
90. Tchao, E.T.; Quansah, D.A.; Klogo, G.S.; Boafo-Effah, F.; Kotei, S.; Nartey, C.; Ofosu, W.K. On cloud-based systems and distributed platforms for smart grid integration: Challenges and prospects for Ghana's Grid Network. *Sci. Afr.* **2021**, *12*, e00796. [CrossRef]
91. Hosseini, S.M.; Carli, R.; Dotoli, M. A Residential Demand-Side Management Strategy under Nonlinear Pricing Based on Robust Model Predictive Control. In Proceedings of the 2019 IEEE International Conference on Systems, Man and Cybernetics (SMC), Bari, Italy, 6–9 October 2019; Institute of Electrical and Electronics Engineers (IEEE): Piscataway, NJ, USA, 2019; pp. 3243–3248. [CrossRef]
92. Karimi, H.; Jadid, S. Optimal energy management for multi-microgrid considering demand response programs: A stochastic multi-objective framework. *Energy* **2020**, *195*, 116992. [CrossRef]
93. Donofrio, S.; Maguire, P.; Zwick, S.; Merry, W.; Wildish, J.; Myers, K. *State of the Voluntary Carbon Markets 2020;* Forest Trends & Ecosystem Marketplace 1203: Washington, DC, USA, 2020.
94. Scarabaggio, P.; Grammatico, S.; Carli, R.; Dotoli, M. Distributed Demand Side Management with Stochastic Wind Power Forecasting. *IEEE Trans. Control Syst. Technol.* **2021**, *2*, 1–16. [CrossRef]
95. Hussain, A.; Arif, S.M.; Aslam, M. Emerging renewable and sustainable energy technologies: State of the art. *Renew. Sustain. Energy Rev.* **2017**, *71*, 12–28. [CrossRef]
96. Ulsrud, K. Access to electricity for all and the role of decentralized solar power in sub-Saharan Africa. *Nor. Geogr. Tidsskr. Nor. J. Geogr.* **2020**, *74*, 54–63. [CrossRef]
97. Akinyele, D. Analysis of photovoltaic mini-grid systems for remote locations: A techno-economic approach. *Int. J. Energy Res.* **2018**, *42*, 1363–1380. [CrossRef]
98. Nsafon, B.E.K.; Butu, H.M.; Owolabi, A.B.; Roh, J.W.; Suh, D.; Huh, J.-S. Integrating multi-criteria analysis with PDCA cycle for sustainable energy planning in Africa: Application to hybrid mini-grid system in Cameroon. *Sustain. Energy Technol. Assess.* **2020**, *37*, 100628. [CrossRef]
99. Opoku, R.; Adjei, E.A.; Obeng, G.Y.; Severi, L.; Bawa, A.-R. Electricity Access, Community Healthcare Service Delivery, and Rural Development Nexus: Analysis of 3 Solar Electrified CHPS in Off-Grid Communities in Ghana. *J. Energy* **2020**, *2020*, 1–10. [CrossRef]
100. Schnitzer, D.; Lounsbury, D.S.; Carvallo, J.P.; Deshmukh, R.; Apt, J.; Kammen, D.M. *Microgrids for Rural Electrification: A critical Review of Best Practices Based on Seven Case Studies;* United Nations Foundation: New York, NY, USA, 2014. [CrossRef]
101. Shrestha, A.; Rana, L.B.; Singh, A.; Phuyal, S.; Ghimire, A.; Giri, R.; Kattel, R.; Karki, K.; Jha, S.K. Assessment of electricity excess in an isolated hybrid energy system: A case study of a Dangiwada village in rural Nepal. *Energy Procedia* **2019**, *160*, 76–83. [CrossRef]
102. TazvingaI, H.; DzoboII, O.; Mapako, M. Towards sustainable energy system options for improving energy access in Southern Africa. *J. Energy S. Afr.* **2020**, *31*. [CrossRef]
103. Terrapon-Pfaff, J.; Dienst, C.; König, J.; Ortiz, W. A cross-sectional review: Impacts and sustainability of small-scale renewable energy projects in developing countries. *Renew. Sustain. Energy Rev.* **2014**, *40*, 1–10. [CrossRef]
104. Tsuchiya, Y.; Swai, T.A.; Goto, F. Energy payback time analysis and return on investment of off-grid photovoltaic systems in rural areas of Tanzania. *Sustain. Energy Technol. Assess.* **2020**, *42*, 100887. [CrossRef]
105. Uamusse, M.; Tussupova, K.; Persson, K.; Berndtsson, R. Mini-Grid Hydropower for Rural Electrification in Mozambique: Meeting Local Needs with Supply in a Nexus Approach. *Water* **2019**, *11*, 305. [CrossRef]

energies

Article

Environmental Production Efficiency in the European Union Countries as a Tool for the Implementation of Goal 7 of the 2030 Agenda

Katarzyna Cheba * and Iwona Bąk

Faculty of Economics, West Pomeranian University of Technology, 70-310 Szczecin, Poland; iwona.bak@zut.edu.pl
* Correspondence: katarzyna.cheba@zut.edu.pl

Abstract: The main purpose of the paper is to present a proposal to measure the relationships between Goal 7 of the 2030 Agenda for Sustainable Development and one of the areas considered in the green growth concept: environmental production efficiency. Both of these areas illustrate the relationship between the natural environment and the economy, emphasizing transformations in the field of energy use. Selected taxonomic methods, TOPSIS, and multicriteria taxonomy, were applied to study the relationships between the two areas. The results of the EU countries classification showed a variety of countries' development pathways within a single economic community. Despite continued attempts to equalize the development levels between European Union countries in many strategic areas, they remain highly diversified. That is also true for the areas analyzed in the paper, which is a disturbing situation, indicating that both strategies might not correlate in all respects. Further research into the relationships linking the remaining dimensions of both strategies is required.

Keywords: green growth; sustainable development; environmental production; relationships; TOPSIS; multicriteria taxonomy

Citation: Cheba, K.; Bąk, I. Environmental Production Efficiency in the European Union Countries as a Tool for the Implementation of Goal 7 of the 2030 Agenda. *Energies* **2021**, *14*, 4593. https://doi.org/10.3390/en14154593

Academic Editors: Bartlomiej Iglinski and Marco Torresi

Received: 20 April 2021
Accepted: 26 July 2021
Published: 29 July 2021

Publisher's Note: MDPI stays neutral with regard to jurisdictional claims in published maps and institutional affiliations.

Copyright: © 2021 by the authors. Licensee MDPI, Basel, Switzerland. This article is an open access article distributed under the terms and conditions of the Creative Commons Attribution (CC BY) license (https://creativecommons.org/licenses/by/4.0/).

1. Introduction

Sustainable development and green growth are ideas for which there has been a renewed increase in interest among politicians, scientists, and the public in recent years. The growing interest in these concepts seems to be directly proportional to the increasingly frequent and more dramatic scientific reports on the progressive degradation and increasing human pressure on the environment. However, treating these concepts as a kind of a "remedy" for the current and future problems of the world is more and more often described as unjustified. Merely investing in increasingly efficient technologies supported by motivating the business sphere so that national economies can grow further and at the same time reduce their unsustainable environmental impact seems insufficient, according to recent studies. It is worth mentioning here, for example, a project of a team led by Monika Dittrich [1], which published its first findings in 2012. It showed that with the economic growth rate of about 2–3% per year, the consumption of, e.g., fish, livestock, the use of forests, metals, and fossil fuels would increase from 70 billion tonnes per year in 2012 to 180 billion tons in 2050.

In contrast, the value that would allow a balance between economic growth and the use of the environment was set at around 50 billion tonnes. As indicated in the report, the value was already exceeded by a civilization in 2000. The Dittrich team also looked at what would happen if each country immediately implemented best practices for the efficient use of natural resources. Consumption would then improve, but it would still be as much as 43 billion tons above the accepted equilibrium level.

Does this mean that we should abandon attempts to implement the assumptions of the concepts for which the overarching goal is to search for the possibility of permanent decoupling of the GDP from the total consumption of the Earth's natural resources? The answer is obvious—no, we should still search for opportunities to implement the assumptions of these concepts effectively. Unfortunately, however, we cannot predict both further obstacles to their implementation and the possibility of evolution of today's technological solutions. Especially that some of these solutions already encounter problems that were difficult to predict earlier.

For example, it is worth mentioning that Norway achieved a record level of electric car sales last year. In 2020, as much as 54% of the cars delivered to this market did not have a combustion engine in any form. However, due to the high repair cost and subsidies, which lower the price of electric cars, otherwise twice as expensive as combustion models, their repair is not profitable. As a result, the number of scrap electric cars is growing exponentially, and at least one in three of them could be repaired. In addition, although the material from scrapped cars is, according to the manufacturers' information, 98% recoverable, only a small percentage of repairs in Norway are made with the use of used car parts [2].

These reports are contradicted by official statistics used to monitor progress in implementing the concept of sustainable development, which show that Scandinavian countries are, in fact, the only countries in Europe that have managed to separate economic growth from negative environmental pressures permanently. What does this mean? Only that the adopted indicators are too general or too unified. Striving to develop similar evaluation standards for all countries, e.g., for the European Union, one forgets about the differences between those countries, which are significant despite relatively similar development conditions.

Hence, there is a need for more detailed analyses and new tools to support the monitoring of progress in implementing the sustainable development strategy, such as green growth. It should be emphasized that this level of detail should not be limited only to creating increasingly advanced indicators. The relationships between individual areas of these concepts are also meaningful, as they will allow for a better understanding of each of them.

The symptoms of similar thinking are visible both in the current and previous concepts of measuring progress in implementing the Sustainable Development Strategy. The strategy, in force until 2018, assumed, for example, various levels of its implementation: from the level of explanatory indicators, through operational indicators, to headline indicators. The main problem of this hierarchical method of presenting indicators was the lack of connections between the successive levels of the pyramid (until 2018, the sustainable development indicators were hierarchical and presented in the form of a pyramid). As it was shown in [3], achieving the level of indicators close to the assumed values at the lower levels of the pyramid did not translate into equally high positions at higher levels. On the other hand, in the current strategy aiming to achieve as many as 17 different goals, it was decided to immediately identify the goals and appropriate indicators that should be related to each other. Moreover, in this case, some of these assumed relationships are not observed in reality, which was confirmed, among others, by the results of the research presented in [4,5], examining in detail the relationship between the various goals of the strategy.

These observations prompted the authors to begin research analyzing the strength and directions of the relationship between the indicators used to measure the latest Sustainable Development Strategy, the 2030 Agenda, and the indicators used to measure the concept of green growth, which is referred to in the literature [6,7] as a tool for sustainable development.

Therefore, the study aims to present a proposal for measuring the relationship between the concepts of sustainable development and green growth. The considerations presented in the paper focus primarily on the relationships between Goal 7 of the 2030 Agenda (SDG 7), i.e., to ensure access to affordable, reliable, sustainable, and modern energy for all and one of the areas considered in the Green Growth Strategy (GGS), i.e., environmental production efficiency. Both of these areas illustrate the relationship between the natural environment and the economy, emphasizing the transformations taking place in energy use. Selected taxonomic methods were used to study the relationships between these areas: TOPSIS [8] and multicriteria taxonomy [9].

The paper, therefore, formulates the following research questions:

1. How are the relationships between sustainable development and green economy shaped in terms of the areas selected for the study related to sustainable energy use (Goal 7 of the 2030 Agenda and one of the areas of Green Growth Strategy)?
2. Does the high level of development in the area of green growth selected for the study impact the achievement of Goal 7 of the 2030 Agenda?

The added value of the study is the research approach used by the authors, in which the values of synthetic measures obtained for both analyzed areas became the basis for grouping the EU countries characterized by similarity and dissimilarity of development. This approach was used for the first time in this study. As a rule, studies of this kind are limited to building rankings of objects in terms of the adopted indicators. The literature on the subject lacks analyses showing the relationship between sustainable development and green growth, which is assumed to be a tool for implementing the sustainable development policy. The research approach used in the paper allows for assessing these relationships from various perspectives, including the similarity and dissimilarity of development. The authors of the article identified a research gap in the literature in this area. There are no studies that show not only the level of development in these two areas but also the way they are related to each other. This approach is demonstrated in this paper.

The layout of this article includes an introduction (Section 1), which presents the main purpose of the work and explains the most important motivations of the authors for researching the relationship between the natural environment and the economy, with particular emphasis on transformations taking place in the field of energy use. Later in the paper, research on the green economy as a sustainable development tool is reviewed (Section 2). Then, the statistical data used in the study are presented, and the research procedure used in the study is described (Section 3). The article ends with presenting the research results (Section 4), discussion, and conclusions resulting from the research (Section 5).

2. Green Growth as a Tool for Sustainable Development-Research Overview

It is assumed that the discussion on the concept of development, taking into account economic and social, and environmental aspects, began in earnest in the 1970s and 1980s [10], although already in the 1960s, the first publications appeared in the literature [11], drawing attention to the negative effects of modern large-scale technologies on the natural environment. Initially, it was rather a political idea aimed mainly at improving the functioning of the economy in terms of its balance with the natural environment, only in subsequent years taking into account also social goals [12]. Since then, this concept has undergone a significant evolution, which was initially aimed at finding the concept's proper definition. As B. Caroll (2002) demonstrated in her work, who analyzed as many as 500 different definitions of sustainable development, most of these proposals did not contribute much to the most frequently cited definition of the Gro Bruntland report [13]. According to this definition, sustainable development is: "development that meets current needs without depriving future generations of the ability to meet their own needs".

Nowadays, more and more often, certain elements (distinctions) are indicated, which are important for a proper understanding of this concept [14]. These are primarily such terms as:

- Development [15–18]—this is usually one of the first terms referred to in constructed definitions;
- Integration [19–21]—also understood as an integration process and balancing most often related to three main orders: social, economic, and environmental ones,
- the quality of life [22–24]—indicated as the overarching goal of sustainable development.

Each of these terms allows for a better understanding of the concept of sustainable development. Their bottom line is to understand this concept as a constantly evaluating process, which means that there is a need for a dynamic approach to this issue. Integration is also essential, understood as a kind of relationality, which means that the three main components of sustainable development form different relationships with each other. In addition, these relationships are formed between them and the environment. Hence the numerous attempts to integrate sustainable development into other areas [25–31]. As a result, there is a growing number of new research areas such as sustainable competitiveness, sustainable agriculture, and sustainable energy. Each of these areas can be treated separately or as a development of particular dimensions of the SD—e.g., sustainable competitiveness or sustainable energy as a development of the economic dimension. The sustainability of the balance between these dimensions, which is now identified as the most important sustainable development goal, is also worth mentioning.

There was also a need to develop concepts that will result in a more accurate picture of what is happening in the economic dimension, which is in a strong relationship with the other dimensions; social and natural. Previous studies by the authors [32,33] show that along with economic development, there is a similar rate of social development, which applies in principle to all European Union countries. On the other hand, in the case of relations between economic and natural measurements, two types of relationships can be distinguished: a positive relationship that means that economic growth also entails an improvement in the environmental dimension and the opposite situation when economic growth comes at the expense of the environment. Therefore, more advanced analyses are needed to show what model or models of development we are currently dealing with in the European Union see: [34,35].

The response to such needs is the so-called concepts supporting research on sustainable development, such as the concept of green growth. This concept was first promoted in 2005 by the UN Economic and Social Commission for Asia-Pacific (UNESCAP), mainly to seek opportunities to introduce a new low-carbon sustainable development model for rapidly developing Asian countries [36]. The term, as in the case of sustainable development, is defined in many different ways. According to the OECD [37] definition, green growth means "fostering economic growth and development while ensuring that natural assets continue to provide the resources and environmental services on which our well-being relies." UNESCAP [38,39] defines this term as: "...growth that emphasizes environmentally sustainable economic progress to foster low-carbon, socially inclusive development." According to World Bank [40], green growth should be described as: "growth that is efficient in its use of natural resources, clean in that it minimizes pollution and environmental impacts, and resilient in that it accounts for natural hazards and the role of environmental management and natural capital in preventing physical disasters." The literature [41,42] also indicates that the concept of green growth does not replace sustainable development but achieving the purposes of sustainable development becomes possible when the economy is functioning right. An essential aspect of this operation is the care of proper efficient, at least diverse, and efficient use of energy sources.

Energy, access, and use are the theme of Goal 7 of the 2030 Agenda: ensure access to affordable, reliable, sustainable, and modern energy for all. As indicated in the explanation to this goal [43]: "Access to affordable, reliable and sustainable energy is crucial to achieving many of the Sustainable Development Goals—from poverty eradication via advancements in health, education, water supply and industrialization to mitigating climate change."

In the case of green growth, the indicators describing energy use were assigned by the OECD to the economy's environmental and resource productivity, together with indicators describing carbon productivity, non-energy material productivity, and environmentally adjusted multifactor productivity. As indicated in the latest OECD report [44,45], "the structure of the country's energy supply and the efficiency of its energy use are key determinants of environmental performance and economic development." This report indicates, for example:

- Increase in energy used most in the services and transport sector;
- Low levels of productivity in many of the major energy-consuming countries despite widespread increases in energy productivity;
- A relatively minor role of renewables in OECD energy mixes.

The results of the research presented in this paper, on the one hand, allow verifying whether these observed trends are still valid. Thanks to the application of selected multivariate methods of statistical analysis, it is also possible to examine the relationship between the degree of achievement of Goal 7 of sustainable development and the results of energy use within the green growth concept. The following part of the paper will examine the relationship between these areas.

An overview of current research directions in the field of green economy and sustainable development can also be found in the works published by other authors. Table 1 contains an overview of selected studies in this field, presenting the main directions of research and analysis in this domain.

Table 1. Research directions in the field of green growth as a tool for sustainable development.

Directions of Research	Papers
exploring the relationship between green growth, green economy, and sustainable development	[46–52]
environmental issues, economic growth, and innovation	[53–56]
low-emission economy, reduction in greenhouse gas emissions	[57–65]
renewable energy, environmental impact and sustainability, barriers, and incentives to the use of renewable energy	[66–77]

3. Data and Methodology

3.1. Statistical Database

According to the adopted assumptions, the analyses covered two groups of data. The first includes the European Commission's indicators to monitor progress in the implementation of Goal 7 of the 2030 Agenda. The second group comprised the indicators describing developments in one of the areas of green growth, i.e., the environmental and resource productivity of the economy, from the OECD database. In the paper, all of the indicators applied by the European Commission (2030 Agenda) and OECD (green growth) were used. The authors decided not to select indicators from these databases with use, e.g., statistical methods. It means that all of the indicators created for this purpose by these international organizations were adopted for the study. This approach will allow for a real comparison of these areas as they were originally designed.

In both cases, because of the comparative nature of the studies conducted, the 2018 data were analyzed. In individual cases, due to the adopted, for example, 5-year period of change monitoring, data from 2015 were included in the analyses, mainly in the case of indicators on the green economy. The study covered the 27 current European Union members and the United Kingdom, which only formally left the European Union in 2020. A comparison of both groups of indicators is presented in Table 2. The symbol *S* next to the indicator (e.g., X_{15S}) means that it is a stimulant—with the increase in the value of this indicator, an improvement is observed in the analyzed area, while the symbol *D* means a destimulant, in this case, the deterioration is observed with the increase of the indicator.

Table 2. Descriptive characteristics of indicators analyzed in the paper, EU countries and the United Kingdom, 2018 [1].

Symbol	Description	$\bar{x}$	V_s (%)	A_s
	Goal 7. Ensure access to affordable, reliable, sustainable and modern energy for all			
$X_{1.1D}$	Primary energy consumption (tonnes of oil equivalent, TOE, per capita)	3.21	258.80	1.66
$X_{1.2D}$	Final energy consumption (tonnes of oil equivalent TOE, per capita)	2.42	210.85	2.79
$X_{1.3D}$	Final energy consumption in households (kg of oil equivalent per capita)	560.14	304.99	0.21
$X_{1.4S}$	Energy productivity (purchasing power standard, PPS per kilogram of oil equivalent)	8.70	318.45	1.62
$X_{1.5S}$	Share of renewable energy in gross final energy consumption (%)	21.09	182.37	1.13
$X_{1.6D}$	Energy import dependency (% of imports in total gross available energy)	56.63	246.50	−0.24
$X_{1.7D}$	Population unable to keep home adequately warm (% of population)	8.61	101.81	1.67
	Green growth: the environmental and resource productivity of the economy			
$X_{2.1S}$	Production-based CO_2 productivity (GDP per unit of energy-related CO_2 emissions)	7.04	37.33	1.13
$X_{2.2D}$	Production-based CO_2 intensity (energy-related CO_2 per capita)	6.36	40.87	1.50
$X_{2.3S}$	Demand-based CO_2 productivity (GDP per unit of energy-related CO_2 emissions)	5.08	20.11	−0.13
$X_{2.4D}$	Demand-based CO_2 intensity (energy-related CO_2 per capita)	7.62	35.61	1.00
$X_{2.5D}$	CO_2 intensity of GDP (CO_2 emissions per unit of GDP)	0.16	38.49	1.29
$X_{2.6S}$	Energy productivity (GDP per unit of TPES, US Dollar, 2015)	13,514.71	37.01	1.87
$X_{2.7S}$	Energy intensity (TPES, tonnes of oil equivalent, TOE per capita)	3.26	37.86	1.03
$X_{2.8S}$	Total primary energy supply (tonnes of oil equivalent, TOE millions per capita)	0.33	37.07	1.04
$X_{2.9S}$	Renewable energy supply (% of total energy supply)	16.94	58.35	1.04
$X_{2.10S}$	Renewable electricity (% of total electricity generation)	36.50	58.08	0.55
$X_{2.11D}$	Energy consumption in agriculture (% of total energy consumption)	2.65	50.98	1.15
$X_{2.12D}$	Energy consumption in industry (% of total energy consumption)	22.93	29.48	0.87
$X_{2.13D}$	Energy consumption in transport (% of total energy consumption)	30.94	26.70	0.90
$X_{2.14S}$	Renewable energy supply, excluding solid biofuels (% of total energy supply)	7.69	54.36	0.88

All the indicators studied have a significant variation level, with coefficients of variation (V_S) for indicators describing changes in Goal 7 of sustainable development strategies are significantly higher than in the case of green growth for the area selected for the study: the economy's environmental and resource productivity. The highest level of coefficient of variation concerns two indicators: X_{13D}—final energy consumption in households (kg of oil equivalent per capita) is 304.99% and $X_{1.4S}$—energy productivity (purchasing power standard, PPS per kg of oil equivalent) is 318.45%. Such a large diversity is influenced, among others, by significant differences between maximum and minimum values. For $X_{1.3D}$, the highest final energy consumption in households per capita, at 1032 kg, was recorded for Finland, and the lowest was recorded at 193 kg for Malta. With respect to the latter, the maximum concerned was Ireland (18.66 PPS per kg of oil equivalent), and the minimum value (4.73) was also Malta. It is also worth noting that most of the indicators adopted for the study were characterized by high (Goal 7 indicators) or moderate (green growth indicators) right-hand asymmetry, which means that for most EU countries, their values were below average. In the case of indicators classified as destimulants, this is a favorable situation for most analyzed countries. Their values are below average.

To describe the second study area of green growth, 14 indicators were selected, 8 of which are stimulants. The coefficients of variation, lower than for Goal 7, range from 20.1% ($X_{2.3S}$) to 58.35% ($X_{2.9S}$). The relatively high variation of the indicator describing renewable energy supply ($X_{2.9S}$, % of total energy supply) is influenced by a significant difference between the maximum value, 40.44% for Latvia, and the minimum value of 5.17% for Malta. It is also worth noting that, compared to the results of the OECD Report [42], there is still an increase in the use of energy in transport in most of the countries under study, the largest for the dynamically developing countries of Eastern Europe: Poland (up 4.60 PPS), Slovak Republic (2.82 PPS), and Hungary (1.45 PPS). There is also a decrease in the use of energy in transport, mainly in the Nordic countries: Norway (down 1.96 PPS), Finland (0.47 PPS), and Sweden (0.55 PPS), but also, e.g., Italy (0.72 PPS) and Latvia (0.50 PPS).

In most countries, the $X_{2.6S}$ energy productivity indicator is also increasing, with the most notable increase observed in Ireland (17.02%), Malta (14.73%), Romania (11.33%), Bulgaria (10.76%), Germany (8.60%), Portugal (8.39%), and Croatia (8.20%). The indicator decreased in the case of Sweden (−4.78%) and Estonia (−0.90%). In many EU countries, there is also an increase in the use of renewable energy ($X_{2.9S}$), even by more than 30%, but at the same time relatively minor role of this energy in energy mixes in the case of, e.g., Malta (up 61.21 PPS, with a share of 5.02%), or Ireland (33.23 p.p. and 8.42%, respectively).

The information provided confirms significant development differences between EU countries, which is important because one of the main objectives of the functioning of such economic organizations is to strive for equal development of all member states. However, it appears that these differences are still significant, and efforts to eliminate them are still needed. The differences between the level of development of EU countries observed at the level of the individual indicators will also be seen in more advanced multidimensional analyses. Therefore, it is essential to check whether the improvement of the analyzed area of green growth: the environmental and resource productivity of the economy is reflected in the implementation of Goal 7, the 2030 Agenda.

3.2. Statistical Methods

A two-stage research procedure was used to study the relationship between indicators describing the degree of implementation of Goal 7 of the 2030 Agenda and Green growth strategy regarding the economy's environmental and resource productivity. At the first stage, synthetic measures were calculated based on the indicators in each area concerned, which allowed ordering the studied countries in terms of their performance. The TOPSIS method was used to determine synthetic measures. A detailed description of this method and examples of its application can also be found in the papers [78–86]. This method is often used in the literature to study customer preferences [78–82]. There are also a growing number of its application in research on the level and directions of regional development [84–86]. Its main advantage is the ability to determine the distance from the so-called pattern and the anti-pattern of development, enabling not only to study the similarity of development in relation to the pattern but also to identify objects similar to the so-called anti-pattern. It appears, which is also confirmed by the results of the research presented in this paper, that a large distance of an object from the pattern (in this case, a country) does not mean a high similarity to the so-called anti-pattern. This observation is important for determining the paths of development of the studied objects within the scope of the studied phenomenon or for comparing objects between one another.

The basis of linear ordering is a synthetic measure whose values are estimated based on observations of diagnostic variables describing the examined objects. TOPSIS is a computational technique that belongs to a group of reference methods for which there are two reference points for objects in multidimensional space, i.e., a pattern and an anti-pattern. The final result of the analysis is a synthetic indicator that creates a ranking of the surveyed objects (in this case, countries). The best object is considered the one with the shortest distance from the pattern and, at the same time, the longest from the anti-pattern.

The determination of synthetic measure in the TOPSIS method is as follows [78,79]:

1. Normalization of variables:

$$z_{ij} = \frac{x_{ij}}{\sqrt{\sum_{i=1}^{n} x_{ij}^2}} \tag{1}$$

2. Determination of the coordinates of the z_{0j}^{+} pattern and the distance of objects from it d_{i0}^{+}:

$$z_{0j}^{+} = \begin{cases} \underbrace{\max}_{i}\{z_{ij}\} \text{ for stimulant variables} \\ \underbrace{\min}_{i}\{z_{ij}\} \text{ for destimulant variables} \end{cases} \quad d_{i0}^{+} = \sqrt{\sum_{j=1}^{m}\left(z_{ij} - z_{ij}^{+}\right)^{2}} \tag{2}$$

3. Determination of the coordinates of the z_{0j}^{-} anti-pattern and the distance of objects from it d_{i0}^{-}:

$$z_{0j}^{-} = \begin{cases} \underbrace{\min}_{i}\{z_{ij}\} \text{ for stimulant variables} \\ \underbrace{\max}_{i}\{z_{ij}\} \text{ for destimulant variables} \end{cases} \quad d_{i0}^{-} = \sqrt{\sum_{j=1}^{m}\left(z_{ij} - z_{ij}^{-}\right)^{2}} \tag{3}$$

4. Determination of the synthetic measure:

$$q_i = \frac{d_{i0}^{-}}{d_{i0}^{+} + d_{i0}^{-}}, \tag{4}$$

with: $q_i \in [0;1]$, $\underbrace{\max}_{i}\{q_i\}$—best object, $\underbrace{\min}_{i}\{q_i\}$—worst object.

5. Division of objects into typological groups according to their distance from the mean value (using the arithmetic mean $\overline{q}$ and standard deviation S_q) as follows:
   - Group I: $q_i \geq \overline{q} + S_q$, containing objects (countries) with the highest values of the synthetic measure;
   - Group II: $\overline{q} + S_q > q_i \geq \overline{q}$;
   - Group III: $\overline{q} > q_i \geq \overline{q} - S_q$;
   - Group IV: $q_i < \overline{q} - S_q$, containing objects (countries) with the lowest values of the synthetic measure.

The second stage was devoted to examining the relationship between the results obtained in the two areas analyzed. The first step is to set linear correlation coefficients: r Pearson (for the value of synthetic measures) and τ Kendall (for the positions occupied by the studied countries), describing the dependencies between the determined measures and the results obtained by individual countries [83–89]. The results allowed examining the relationships between the analyzed areas and, importantly, identifying countries where the assumed objectives of the green economy supporting sustainable development are being achieved.

More advanced statistical analysis methods were used to study the relationships between the analyzed areas. The research used a multicriteria taxonomy described in the literature [90–92]. The mathematical algorithm of this method takes place in several stages. A detailed description of this method can be found in [93,94]. The first step requires a transformation of each indicator utilized in the analyses. The paper proposes that the clustering of countries should be carried out using the distances from the pattern in the TOPSIS method (z_{0j}^{+}) and the anti-pattern (z_{0j}^{-}), which replaces the normalized values of indicators used as a standard for this method. This approach allows countries to be grouped by their similarity in the distance to the pattern (the best object) and the similarity of distance to the anti-pattern (the worst object). In this paper, these distances are defined as baseline distances. The matrices containing information about the baseline distances determined for each indicator analyzed were used to determine two final distance matrices $\mathbf{D^{P}}$ (based on distance from the pattern) and $\mathbf{D^{AP}}$ (based on distance from the anti-pattern). Euclidean distance was used for this purpose. In the next step, based on the values in the distance matrices, a threshold values d^i should be defined. The following formula can be utilized for this purpose:

$$d^i = \min_i \max_j \{d_{ij}\} \tag{5}$$

The transformation of the $\mathbf{D^A}$ and $\mathbf{D^{AP}}$ distance matrices is carried out. For each indocator, the affinity matrix of $(n \times n)$ dimension is defined. The elements of this matrix: $c_{ij}^K (i, j = 1, \ldots, n)$ are equal to:

$$c_{ij}^K = 1 \text{ for } d_{ij} \leq d \tag{6}$$

$$c_{ij}^K = 0 \text{ for } d_{ij} > d \tag{7}$$

If inequality (6) is satisfied, the objects designated as i and j are treated as similar. In opposite, if inequality (7) is satisfied, the analyzed objects are deemed as dissimilar. In the second case, the affinity measure of c_{ij} is equal to zero. Finally, $\mathbf{C^A}_{(n\times n)}$ and $\mathbf{C^{AP}}_{(n\times n)}$ affinity matrices are determined. In this case, the following formula is applied in which c_{ijl} elements of these matrices are equal to the product of relevant elements of $\mathbf{C^A}$ and $\mathbf{C^{AP}}$ matrices for all the analyzed indicators:

$$c_{ijl} = \prod_{K=1}^{r} c_{ij}^K \tag{8}$$

If $c_{ij} = 1(i, j = 1, \ldots, n)$, it means that each of c_{ij}^K elements corresponding to it in $\mathbf{C^K}$ matrices is equal to one, and if $c_{ij} = 0$, if one of the c_{ij}^K elements corresponding to it are equal to zero.

Two analyzed objects (in the paper two countries) are considered to be similar to one another simultaneously on account of all the criteria if they are similar to one another separately on account of those criteria individually. Two objects are treated as dissimilar on account of all the examined criteria if they are not similar to one another, even in terms of one of such criteria. According to this assumption, sometimes it is challenging to find many similar objects in terms of every analyzed indicator.

In the following step, the analyzed objects are divided into typological groups. In the paper, for this purpose, the vector elimination method [87] is used. The procedure in this method, as in the multicriteria taxonomy, involves several stages. In the first step, the final $\mathbf{C}_{(n\times n)}$ affinity matrix is transformed into a $\mathbf{C}^*(_{n\times n})$ dissimilarity matrix. Next, based on $\mathbf{C}^*$ matrix, the c_0 column vector is estimated with n components. In the second step, the row is eliminated from $\mathbf{C}^*$ matrix along with a corresponding column for which thec_0 vector component has a maximum value. If thec_0 vector contains several components whose value reaches the maximum, such a row and column are eliminated. The second step of the procedure is repeated until thec_0 vector components are equal to zero. To the first sub-group, objects corresponding to the rows and columns that were not crossed off and remain in $\mathbf{C}^*$ matrix were assigned.

4. Study Results

4.1. The Results of the TOPSIS Method

The indicators described in the previous parts of the article are now used to construct synthetic measures describing the degree of the implementation of Goal 7, the 2030 Agenda, and the Green Growth Strategy within the framework of environmental production efficiency. The results obtained from the classification and grouping of EU countries due to the examined areas are shown in Table 3.

Table 3. Ranking of European Union countries in 2018 due to Goal 7, 2030 Agenda, and the environmental and resource productivity of the economy of Green Growth strategy.

Country	Goal 7, 2030 Agenda			The Environmental and Resource Productivity of the Economy, Green Growth Strategy		
	q_i	Rank	Group	q_i	Rank	Group
Western Europe						
Austria	0.628	7	II	0.618	2	I
Belgium	0.500	24	IV	0.449	21	III
France	0.599	12	II	0.507	14	II
Germany	0.590	14	II	0.490	17	III
Luxembourg	0.408	28	IV	0.503	15	II
Netherlands	0.554	21	III	0.374	26	IV
Northern Europe						
Denmark	0.695	1	I	0.586	3	I
Estonia	0.630	6	II	0.349	27	IV
Finland	0.533	22	III	0.534	10	II
Ireland	0.615	10	II	0.540	8	II
Latvia	0.674	3	I	0.552	7	II
Lithuania	0.475	25	IV	0.562	5	II
Sweden	0.662	4	I	0.717	1	I
United Kingdom	0.621	8	II	0.515	13	II
Southern Europe						
Croatia	0.654	5	I	0.569	4	II
Cyprus	0.469	27	IV	0.419	23	III
Greece	0.521	23	III	0.459	19	III
Italy	0.561	19	III	0.539	9	II
Malta	0.557	20	III	0.524	11	II
Portugal	0.578	18	III	0.556	6	II
Slovenia	0.619	9	II	0.483	18	III
Spain	0.598	13	II	0.498	16	II
Eastern Europe						
Bulgaria	0.472	26	IV	0.417	24	III
Czechia	0.590	15	II	0.389	25	III
Hungary	0.586	17	II	0.443	22	III
Poland	0.601	11	II	0.327	28	IV
Romania	0.694	2	I	0.520	12	II
Slovakia	0.588	16	II	0.458	20	III

As it has already been mentioned, many of the papers published so far point out that the Scandinavian countries are basically the only countries in Europe that have managed to separate economic growth from the negative environmental pressures permanently. Similar patterns are also visible in the presented list. Scandinavian countries that are members of the EU: Denmark and Sweden, and additionally Croatia, are the only countries in the top five countries with the highest scores in both areas of the survey. For the other countries, two patterns are visible. According to the first one, EU countries ranking high on Sustainable Development Goal 7 have lower rankings on environmental productivity and vice versa. That applies to countries located in different parts of Europe, although more often those located in:

- Northern Europe (Estonia: 6 and 27; Finland: 22 and 10; Lithuania: 25 and 5);
- Southern Europe (Italy: 19 and 9; Malta: 20 and 11; Portugal: 18 and 6; Slovenia: 9 and 18);
- Eastern Europe (Czechia: 15 and 25; Poland: 11 and 28; Romania: 2 and 12).

In Western Europe, this is the case for only one country (Luxembourg: 28 and 15). The results of other countries located in this part of Europe are more often at a similar level in both analyzed areas. This similarity of the results, that is, the second type of regularity, is visible regarding other countries located in different parts of Europe, apart from the already indicated Scandinavian countries and Croatia. The most considerable differences, not exceeding five positions in the rankings, were noted for: Austria (7 and 2), the Netherlands (21 and 26), and the United Kingdom (8 and 13). Several different development models of EU countries (Table 4) can be identified, based on the division into typological groups (I–IV).

Table 4. EU countries' development model in the areas of Goal 7 of SDS and the environmental and resource productivity of the economy of GGS.

Goal 7 of SDS	Environmental and Resource Productivity of the Economy of GGS				Sum
	I	II	III	IV	
I	2	3	-	-	5
II	1	4	5	2	12
III	-	4	1	1	6
IV	-	2	3	-	5
Sum:	3	13	9	3	28

It is clear from the information provided in Table 4 that countries classified in the first typological group in the case of the first ranking (Goal 7, SDS) were also classified in the first two groups in the case of the second ranking (Environmental and resource productivity of the economy of GGS). The first and second typological groups include countries that have above-average values for taxonomic measures of development. However, this division is no longer evident for other typological groups. The countries representing the second typological group in the case of Goal 7 were classified into all groups in the case of the second analyzed area. In the prepared set, however, no country was identified that was classified in group IV (with the lowest scores) in the case of both analyzed areas. These regularities are also confirmed by the assessments of correlation coefficients r Pearson (for taxonomic measures) and τ Kendall (for positions occupied; Table 5).

Table 5. Correlation coefficient matrix r Pearson and τ Kendall, respectively, for the values of the synthetic measures determined and the positions held in the built rankings.

r Pearson	Goal 7	GG
Goal 7	1.0000	0.3205
GG	0.3205	1.0000
τ Kendall	**Goal 7**	**GG**
Goal 7	1.0000	0.2751
GG	0.2751	1.0000

Their analysis indicates a moderate correlation between both the values of the determined synthetic measures and the positions taken by EU countries in the case of Goal 7 and GGS.

Of course, the reasons for the differences in performance between countries vary. As we have already mentioned in the case of the Scandinavian countries, good results in both areas analyzed result from economic development achieved with care for the environment. Countries such as Romania or Croatia owe their high places in the rankings primarily to the lower economic development level compared with other countries. Their GDP per capita (USD 25,805 and USD 26,018, respectively) is well below the EU average (USD 40,192), which results in lower than in other EU countries, environmental interference at this stage of development. The opposite situation can be observed in the case of much more economically developed countries: the United Kingdom (second place in both rankings with a GDP of per capita above average (USD 43,720) or France (USD 42,543). The observed differences in development directions are confirmed by low Pearson's r and Kendall's τ correlation coefficients for the calculated taxonomic measures and GDP per capita not exceeding the level of 0.3, indicating only a moderate correlation.

4.2. The Results of the Multicriteria Taxonomy Method

The wide diversity of EU countries due to the two areas analyzed is one of the many reasons that make it difficult, for example, to develop acceptable by all countries assumptions for various EU policies. Therefore, it is important to check more precisely to what extent (in terms of which indicators) the analyzed countries are similar to each other or which ones make the biggest differences between them. However, the aim is to compare individual indicators from both analyzed areas at the same time. In the literature of the subject [59], the average level of analyzed phenomena is most often used in this case. However, in the approach proposed by the authors, the starting point to more advanced analyses are distance matrices calculated based on the distance between the individual indicators and the adopted pattern (variant V1) and the anti-pattern of development (variant V2). The result is the division of the studied EU countries into groups, as shown in Table 6 and Figure 1a,b.

Table 6. Division of EU countries into typological groups in 2018—variants: V1 and V2.

Group	V1	V2
I	Austria, Belgium, Croatia, Czechia, Denmark, Estonia, France, Germany, Hungary, Italy, Latvia, Poland, Romania, Slovakia, Slovenia, Spain, United Kingdom	Croatia, Cyprus, Czechia, Estonia, Finland, France, Germany, Hungary, Italy, Latvia, Lithuania, Netherlands, Poland, Romania, Slovakia, Slovenia, United Kingdom
II	Bulgaria, Cyprus, Greece, Lithuania, Portugal	Belgium, Denmark, Ireland, Portugal, Spain, Sweden
III	Finland, Ireland, Luxembourg	Bulgaria, Greece
IV	Sweden	Austria
V	Malta	Malta
VI	Netherlands	Luxembourg

(a)

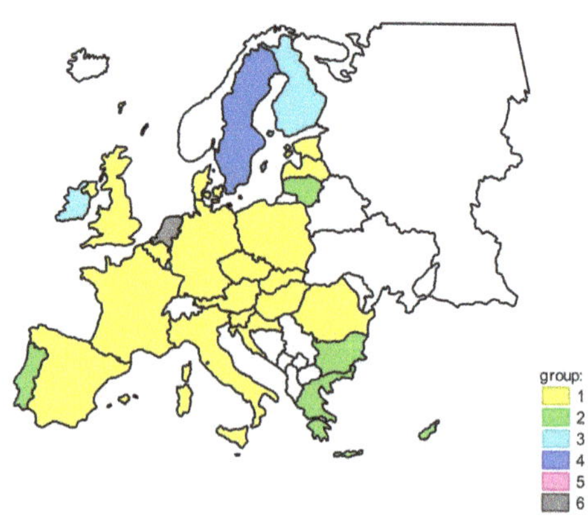

Figure 1. *Cont.*

(b)

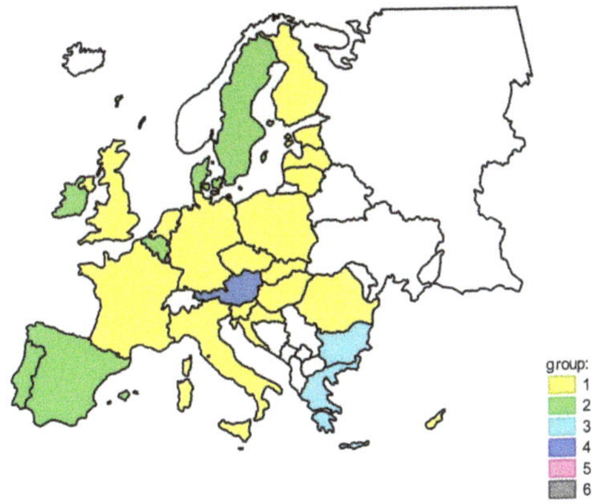

Figure 1. Division of EU countries into typological groups: (**a**) variants V1; (**b**) variants V2.

The compositions of the different groups differ quite significantly, but their comparison allows drawing some interesting conclusions. It is worth noting that the computational algorithm used does not allow to determine the order of groups due to their level of development. The order in which groups are created is conditioned by the number of objects classified into them. The first group is always the most numerous, while single-element groups are distinguished last. Despite the different number of EU countries classified in the first typological group, each of the analyzed cases included countries such as Croatia, Czechia, Estonia, France, Germany, Hungary, Italy, Latvia, Poland, Romania, Slovakia, Slovenia, and the United Kingdom. In their case, there is a substantial similarity of development regarding the achievement of SDG 7. With the exception of Italy, all countries are above the EU average in terms of reaching Sustainable Development Goal 7 (cf. Table 2). The situation in terms of green growth is more diverse, as for six out of thirteen countries (Czechia, Estonia, Germany, Hungary, Poland and Slovakia, and Slovenia), the indicators representing this area often reach unfavorable values compared to other EU countries (below average). However, the fact that these countries are in the same group does not raise any doubts when the distributions of the values of the individual diagnostic characteristics are analyzed in detail. For SDG 7, feature distributions vary distinctively (101.81% to 318.45%), with a strong asymmetry of the vast majority of all indicators. On the other hand, the distributions of individual indicators for environmental production efficiency have slightly different characteristics: variation level from 20.11% to 58.35% and weaker asymmetry. The typological groups presented in Table 4 are the consequence of applying a multicriteria analysis that considers both research areas simultaneously, and the characteristics of the distributions of indicators within each area affect the final result.

Regardless of the adopted variant of grouping, one can clearly see the countries that differ in plus or minus from the rest of European countries, reaching maximum or minimum values. They include:

1. Malta—in the area of achievement of Goal 7 the best in terms of $X_{1.3D}$—final energy consumption in households per capita, kg of oil equivalent, and worst in terms of implementation of indicators: energy productivity, purchasing power standard (PPS) per kilogram of oil equivalent (X_{14S}), energy import dependency, % of imports in total gross available energy ($X_{1.6D}$). In the area of green growth, the country stands out positively in terms of the following indicators: energy productivity, GDP per unit of TPES, US Dollar, 2015 ($X_{2.6S}$), energy consumption in industry, % total energy consumption ($X_{2.12D}$); it stands out negatively for the following: energy intensity, TPES per capita, tonnes of oil equivalent (TOE; $X_{2.7S}$), total primary energy supply, tonnes of oil equivalent (TOE), millions per capita ($X_{2.8S}$), renewable energy supply, % total energy supply ($X_{2.9S}$);

2. Luxembourg—in the area of achieving Goal 7, Luxembourg has achieved unfavorable values in terms of primary energy consumption, tonnes of oil equivalent (TOE) per capita ($X_{1.1D}$), final energy consumption, tonnes of oil equivalent (TOE) per capita ($X_{1.2D}$); whereas in the case of green growth the following can be positively assessed: energy intensity, TPES per capita, tonnes of oil equivalent (TOE; $X_{2.7S}$), total primary energy supply, tonnes of oil equivalent (TOE), millions per capita ($X_{2.8S}$), energy consumption in agriculture, % total energy consumption ($X_{2.11D}$); negatively: production-based CO_2 intensity, energy-related CO_2 per capita ($X_{2.2D}$), demand-based CO_2 intensity, energy-related CO_2 per capita ($X_{2.4D}$), energy consumption in transport, % total energy consumption ($X_{2.13D}$);
3. The Netherlands—this country can be assessed negatively in the case of the indicator on the achievement of Goal 7—share of renewable energy in gross final energy consumption, % ($X_{1.5S}$), and negatively in the forename GG—energy consumption in agriculture, % total energy consumption ($X_{2.11D}$).

It would seem that the grouping of countries in two variants, using distances from the pattern (z_{0j}^{+}) and the anti-pattern (z_{0j}^{-}) determined in the TOPSIS method, should produce similar results, i.e., groups of countries similar to each other due to both their similarity and dissimilarity should be distinguished. This situation occurs only in some cases when most of the characteristics adopted for the study show similar direction and values, which happens especially in the case of single-element groups. It should be noted that most EU countries, due to the characteristics adopted for the study, cannot be unambiguously assigned to the group of those that achieve only desirable or undesirable values. Their situation varies greatly, and this has an impact on the obtained results.

5. Discussion and Conclusions

The results of the EU countries' classification for Goal 7 SD Strategy and Environmental production efficiency of GG Strategy show how different countries' development paths within a single economic community can be. Despite continued attempts to equalize the development levels between European Union countries in many strategic areas, this level remains highly diversified.

Should we expect that the action taken by the European Commission will eliminate these differences in every possible area? In our opinion, this is not possible, especially that this view is also confirmed by the research results presented in this paper. It turns out that even the Scandinavian countries, which in the EU are among the few countries that have managed to separate economic growth from negative environmental impacts, are unable to predict all the pitfalls of their growth and economic development. Problems of this kind are also experienced in the highly developed countries of Northern and Western Europe.

The literature of the subject [88–90] points out that in the initial stage of economic development, environmental pollution increases with economic growth. The higher the level of economic development a society achieves, the more attention it pays to the environment. Less developed countries should therefore exert less pressure on the environment. What should these relationships look like for the areas analyzed in the paper? Given the computational procedure used, where one of the steps is transforming the destimulant into stimulants, we expect that with economic growth, there will be an improvement in the areas of sustainable development and a green economy. The higher the GDP per capita, the higher the results in terms of implementation of Goal 7 SD Strategy and environmental production efficiency under the GG Strategy should be. However, the results of the studies presented in the paper show that these relationships are not so obvious. It was noted that among the countries with the highest GDP values per capita (Austria, Belgium, Denmark, Finland, Germany, Ireland, Luxembourg, the Netherlands, Sweden, and the United Kingdom), there are also those for which the designated values of synthetic measures were lower than the average in the group. Countries such as Belgium, the Netherlands, despite their relatively high GDP per capita, are successful in achieving Goal 7 of the SD Strategy.

Their environmental production efficiency is also lower than that of other less developed EU countries, hence their qualification for Group III or IV.

On the other hand, the opposite was observed for Croatia, Latvia, Romania, and Spain. Despite being lower than the EU average GDP per capita, these countries achieved relatively high results in the analyzed areas, allowing them to qualify for typological groups I or II. These observations are also confirmed by the results of studies by other authors [91–95], which show, among other things, that the relationship between environmental pollution and the wealth of a country has the shape of an inverted letter "U"—with economic growth, the pollution increases, but only to a particular level beyond which environmental pollution decreases with economic growth. In recent years, the above-described curve has flattened, and the peaks at ever-lower levels signify that even poorer countries are beginning to pay more attention to the environment. This observation is fundamental in the analyzed area of environmental production efficiency. It seems that a large group of highly developed countries may not have reached the so-called tipping point yet—hence their worse position than in the case of other countries in terms of indicators showing, for example, CO_2 production [96].

However, it is worth noting that investments, which often rely on costly and durable infrastructure, play an essential role in achieving the objectives of both strategies. Vast amounts of capital are needed to finance infrastructure such as smart grids, renewable energy sources, resource efficiency. The analysis of EU Community Innovation Survey (CIS) data for eco-innovation adoption by EU firms—for energy efficiency and carbon dioxide abatement—suggests that adoption is positively correlated to the emission efficiency of the countries where the companies are based. There are structural differences in this correlation across the EU Member States, with leaders and laggards.

This also means that we need reliable data that will allow us to assess accurately at which development stage the countries currently are and how their development paths may proceed [97]. The choice of computational method is also important. In the literature, there are many different proposals for determining the level of development of the analyzed objects (in this paper, these are the EU countries). These methods focus on determining the average level of development of these objects. In this paper, however, it is proposed to examine the distance of the EU countries in relation to the so-called development pattern and, at the same time, the so-called anti-pattern. On this basis, in the next step, the objects are grouped in order to recognize their current level of development more accurately. It is essential for the evaluation of the studied phenomena because it allows illustrating complex relationships between them. It also enables assessing whether the high level of development of one phenomenon (in this case, within the selected goal of the green economy) influences the development of another one (within the selected goal of sustainable development). The results presented in this paper showed that these relationships are not straightforward. The high level of development in both examined areas concerns only a few countries. Economically less developed countries pollute the environment to a lesser extent, but it can be expected that the environment will be increasingly polluted as the rate and level of economic development increases.

The key question can be formulated as follows: What can be done? What instruments should be applied in order to make this development in the less economically developed EU countries progress in a different way than in the case of the currently most developed EU countries? How to make the transition from one stage of development to another (high economic development and low environmental pressure) as fast as possible?

It is also worth emphasizing that the unique value of the study lies in the research approach that focuses on the relationships between the areas selected for the analysis. In the literature on the subject, there are no works examining the relationships between different areas of development, especially conducted in the manner proposed by the authors (the similarity versus dissimilarity of development). The main concern is to determine the average level of the studied phenomena. The authors of the paper propose a more advanced approach to this issue. The final goal is the degree of correlation between the two areas and the indication of the different development stages currently faced by the EU countries.

In the subsequent research, the authors also plan to extend the range of research methods used to study the relationship between these areas to qualitative and quantitative techniques using, for example, cognitive mapping. The ability to anticipate changes in the relationships that connect the analyzed areas is the advantage of this approach. Such attempts in the study of dependencies can be found in the earlier work of the authors of this article [3–5] and the studies of other authors [98,99]. The authors plan to concentrate on the relationship between the different goals of sustainable development and the green economy in their future studies. In this way, the authors will be able to examine more broadly the relationship between these two areas.

Author Contributions: Conceptualization, K.C. and I.B.; data curation, K.C. and I.B.; formal analysis, K.C. and I.B.; funding acquisition, K.C. and I.B.; investigation, K.C. and I.B.; methodology, K.C. and I.B.; project administration, K.C. and I.B.; resources, K.C. and I.B.; software, K.C. and I.B.; supervision, K.C. and I.B.; validation, K.C. and I.B.; visualization, K.C. and I.B.; writing—original draft, K.C. and I.B.; writing—review and editing, K.C. and I.B. Both authors have read and agreed to the published version of the manuscript.

Funding: The APC was funded by the West Pomeranian University of Technology.

Data Availability Statement: Publicly available datasets were analyzed in this study. This data can be found here: https://ec.europaJru/eurostat/web/energy/data/shares (accessed on 30 January 2021) and https://stats.oecd.org/ (accessed on 30 January 2021).

Conflicts of Interest: The authors declare no conflict of interest.

Abbreviations

Abbreviations	
CIS	The EU Community Innovation Survey
GDP	Gross Domestic Product
GG Strategy	Green Growth Strategy
GGS	Green Growth Strategy
OECD	Organisation for Economic Co-operation and Development
PPS	Purchasing Power Standards SDG–Sustainable Development Goal
SDS	Strategy for Sustainable Development, the 2030 Agenda
TOE	Tonnes of Oil equivalen
TPES	Total primary energy supply
UNESCAP	the United Nations Economic and Social Commission for Asia-Pacific

References

1. Dietrich, M.; Giljum, S.; Lutter, S.; Polzin, C. *Green Economies Around the World? Implications of Resource Use for Development and Environment*; SERI: Vienna, Austria, 2012.
2. Saele, H.; Peterson, I. Electric vehicles in Norway and the potential for demand response. In Proceedings of the 2018 53rd International Universities Power Engineering Conference (UPEC), Glasgow, UK, 4–7 September 2018; pp. 1–6.
3. Szopik-Depczyńska, K.; Cheba, K.; Bąk, I.; Stajniak, M.; Simboli, A.; Ioppolo, G. The study of relationship in a hierarchical structure of EU sustainable development indicators. *Ecol. Indic.* **2018**, *90*, 120–131. [CrossRef]
4. Cheba, K.; Bąk, I. Sustainable development and its relationships with other directions of the development of European Union countries. *Res. Pap. Wroc. Univ. Econ. Bus.* **2020**, *64*, 29–44.
5. Ziolo, M.; Bak, I.; Cheba, K. The role of sustainable finance in achieving Sustainable Development Goals: Does it work? *Technol. Econ. Dev. Econ.* **2021**, *27*, 45–70. [CrossRef]
6. Cheba, K.; Bąk, I. Sustainable Development and Green Economy in the European Union Countries—Statistical Analysis. In *Classification and Data Analysis*; Studies in Classification, Data Analysis, and Knowledge Organization; Jajuga, K., Batóg, J., Walesiak, M., Eds.; Springer: Cham, Switzerland, 2019.
7. Khoshnava, S.M.; Rostmi, R.; Zin, R.M.; Streimikiene, D.; Yousefpour, A.; Strielkowski, W.; Mardani, A. Aligning the Criteria of Green Economy (GE) and Sustainable Development Goals (SDGs) to Implement Sustainable Development. *Sustainability* **2019**, *11*, 4615. [CrossRef]

8. Acuña-Soto, C.M.; Liern, V.; Pérez-Gladish, B. Normalization in TOPSIS-based approaches with data of different nature: Application to the ranking of mathematical videos. *Ann. Oper. Res.* **2018**, *296*, 1–28. [CrossRef]
9. Hwang, C.L.; Yoon, K. Multiple Attribute Decision Making. In *Lecture Notes in Economics and Mathematical Systems*; Springer: Cham, Switzerland, 1981.
10. Meadows, D.H.; Meadows, D.L.; Randers, J.; Behrens, W.W., III. *The Limits to Growth: A Report for the Club of Rome's Project on the Predicament of Mankind*; A Potomac Associates Book: Washington, DC, USA, 1972.
11. Carson, R. *Silent Spring*; Houghton Mifflin Company: Boston, MA, USA, 1962.
12. Forrester, J. *World Dynamics*; Wright-Allen Press: Cambridge, MA, USA, 1971.
13. WCED (World Commission on Environment and Development). *Our Common Future. Un Documents: Gathering a Body of Global Agreements Has Been Compiled by the NGO Committee on Education of the Conference of NGOs from United Nations Web Sites with the Invaluable Help of Information & Communications Technology*; United Nations: New York, NY, USA, 1987.
14. Zhu, J.; Hua, W. Visualizing the knowledge domain of sustainable development research between 1987 and 2015: A bibliometric analysis. *Scientometrics* **2017**, *10*, 893–914. [CrossRef]
15. Lele, S.M. Sustainable development—A critical review. *World Dev.* **1991**, *19*, 607–621. [CrossRef]
16. Ciegis, R.; Ramanauskiene, J.; Martinkus, B. The concept of sustainable development and its use for sustainability scenarios. *Eng. Econ.* **2009**, *62*, 28–37.
17. Glavič, P.; Lukman, R. Review of sustainability terms and their definitions. *J. Clean. Prod.* **2007**, *15*, 1875–1885. [CrossRef]
18. Du Pisani, J.A. Sustainable Development—Historical roots of the concept. *Environ. Sci.* **2006**, *3*, 86–96. [CrossRef]
19. Stafford-Smith, M.; Griggs, D.; Gaffney, O.; Ullah, F.; Reyers, B.; Kanie, N.; Stigson, B.; Shrivastava, P.; Leach, M.; O'Connell, D. Integration: The key to implementing the Sustainable Development Goals. *Sustain. Sci.* **2016**, *12*, 911–919. [CrossRef]
20. Liu, C.-H. Conducting qualitative and quantitative analyses of sustainable behaviour. *J. Retail. Consum. Serv.* **2021**, *60*, 102474. [CrossRef]
21. Ashby, A.; Leat, M.; Hudson-Smith, M. Making connections: A review of supply chain management and sustainability literature. *Supply Chain Manag. Int. J.* **2012**, *17*, 497–516. [CrossRef]
22. Van Kamp, I.; Leidelmeijer, K.; Marsman, G.; de Hollander, A. Urban environmental quality and human well-being: Towards a conceptual framework and demarcation of concepts; a literature study. *Landsc. Urban Plan.* **2003**, *65*, 5–18. [CrossRef]
23. O'Neill, D.W.; Fanning, A.L.; Lamb, W.F.; Steinberger, J.K. A good life for all within planetary boundaries. *Nat. Sustain.* **2018**, *1*, 88–95. [CrossRef]
24. Johnston, P.; Everard, M.; Santillo, D.; Robert, K.H. Reclaiming the definition of sustainability. *Environ. Sci. Pollut. Res.* **2007**, *14*, 60–66.
25. Cheba, K. Sustainable international competitiveness of EU countries. In *Theoretical and Empirical Study*; CeDeWu: Warsaw, Poland, 2019.
26. Cheba, K.; Bąk, I.; Szopik-Depczyńska, K. Sustainable competitiveness as a new economic category—Definition and measurement assessment. *Technol. Econ. Dev. Econ.* **2020**, *26*, 1399–1421. [CrossRef]
27. Kiba-Janiak, M. A Comparative Analysis of Sustainable City Logistics among Capital Cities in the EU. *Appl. Mech. Mater.* **2014**, *708*, 113–118 [CrossRef]
28. Ritchie, J.R.B.; Crouch, G.I. *The Competitive Destination: A Sustainable Tourism Perspective*; CABI Publishing: Cambridge, UK, 2003.
29. Rizos, V.; Behrens, A.; Taranic, I. Measuring progress in eco-innovation. *CEPS Work. Doc.* **2015**, *409*, 1–17.
30. Veleva, V.; Ellenbecker, M. Indicators of sustainable production: Framework and methodology. *J. Clean. Prod.* **2001**, *9*, 519–549. [CrossRef]
31. Vogt, M.; Weber, C. Current challenges to the concept of sustainability. *Glob. Sustain.* **2019**, *2*, 1–6. [CrossRef]
32. Szopik-Depczyńska, K.; Cheba, K.; Bąk, I.; Kędzierska-Szczepaniak, A.; Szczepaniak, K.; Ioppolo, G. Innovation level and local development of EU regions. A new assessment approach. *Land Use Policy* **2020**, *99*, 104837. [CrossRef]
33. Szopik-Depczynska, K.; Kedzierska-Szczepaniak, A.; Szczepaniak, K.; Cheba, K.; Gajda, W.; Ioppolo, G. Innovation in Sustainable Development: An Investigation of the EU Context Using 2030 Agenda. *Land Use Policy* **2018**, *79*, 251–262. [CrossRef]
34. Halaskova, M.; Gavurova, B.; Korony, S. Change of EU28 countries research and development indicators between 2010 and 2015. *Econ. Sociol.* **2020**, *13*, 230–248. [CrossRef] [PubMed]
35. Bilan, Y.; Rzeszów University of Technology; Mishchuk, H.; Roshchyk, I.; Kmecova, I.; National University of Water and Environmental Engineering; Institute of Technology and Business in České Budějovice. An Analysis of Intellecutal Potential and its Impact on the Social and Economic Development of European Countries. *J. Compet.* **2020**, *12*, 22–38. [CrossRef]
36. EAI (European Environment Agency). *Europe's Environment—An Assessment of Assessments*; Publications Office of the European Union: Luxembourg, 2011.
37. OECD. *Green Growth Strategy*; Organization for Economic Cooperation and Development: Paris, France, 2011.
38. UNESCAP. *State of Environment Asia and the Pacific*; UNESCAP: Bangkok, Thailand, 2005.
39. UNESCAP. *Green Growth, Resources and Resilience. Environmental Sustainability in Asia and the Pacific*; UNESCAP: Bangkok, Thailand, 2012.
40. World Bank. *Inclusive Green Growth: The Pathway to Sustainable Development*; The World Bank: Washington, DC, USA, 2012. Available online: http://siteresources.worldbank.org/EXTSDNET/Resources/Inclusive_Green_Growth_May_2012.pdf (accessed on 8 April 2021).

41. Allen, C.; Clouth, S. *A Guidebook to the Green Economy*; UNDESA: New York, NY, USA, 2012.
42. Bąk, I.; Cheba, K. *Green Growth as a Tool of Sustainable Development*; CeDeWu: Warsaw, Poland, 2020.
43. UNEP. Mainstreaming the 2030 Agenda for Sustainable Development Interim Reference Guide to UN Country Teams. Available online: file:///C:/Users/kcheba/AppData/Local/Temp/UNDP-SDG-UNDG-Reference-Guide-UNCTs-2015.pdf (accessed on 8 April 2021).
44. OECD. OECD Work on Green Growth. Available online: https://issuu.com/oecd.publishing/docs/gg_brochure_2019_web (accessed on 10 April 2021).
45. OECD. Green Growth Inicators. Available online: http://www.oecd.org/environment/indicators-modelling-outlooks/Highlights_Green_Growth_Indicators_2017.pdf (accessed on 2 April 2021).
46. Inclusive and Green Growth in Developing Countries. Available online: https://assets.publishing.service.gov.uk/media/5af9702340f0b622dd7aa2c8/Inclusive_green_growth_in_developing_countries.pdf (accessed on 8 April 2021).
47. Meadowcroft, J. Reaching the Limits? Developed Country Engagement with Sustainable Development in a Challenging Conjuncture. *Environ. Plan. C Gov. Policy* **2013**, *31*, 988–1002. [CrossRef]
48. Lorek, S.; Spangenberg, J.H. Sustainable consumption within a sustainable economy—Beyond green growth and green economies. *J. Clean. Prod.* **2014**, *63*, 33–44. [CrossRef]
49. Kasztelan, A. Green Growth, Green Economy and Sustainable Development: Terminological and Relational Discourse. *Prague Econ. Pap.* **2017**, *26*, 487–499. [CrossRef]
50. Hermelin, B.; Andersson, I. How green growth is adopted by local policy–A comparative study of ten second-rank cities in Sweden. *Scott. Geogr. J.* **2018**, *134*, 184–202. [CrossRef]
51. Luigi, A.; Concetto, P.V. Green Economy and Sustainable Development: The Economic Impact of Innovation on Employment. *Sustainability* **2018**, *10*, 3541.
52. Bobylev, S. Sustainable development: Paradigm for the future. *Mirovaya Ekon. I Mezhdunarodnye Otnos.* **2018**, *61*, 1–15. [CrossRef]
53. Davies, A.R.; Mullin, S.J. Greening the economy: Interrogating sustainability innovations beyond the mainstream. *J. Econ. Geogr.* **2010**, *11*, 793–816. [CrossRef]
54. Capasso, M.; Hansen, T.; Heiberg, J.; Klitkou, A.; Steen, M. Green growth—A synthesis of scientific findings. *Technol. Forecast. Soc. Chang.* **2019**, *146*, 390–402. [CrossRef]
55. Klagge, B.; Meister, T. Energy cooperatives in Germany—An example of successful alternative economies? *Int. J. Justice Sustain.* **2018**, *23*, 697–716. [CrossRef]
56. Lee, C.T.; Lim, J.S.; Van Fan, Y.; Liu, X.; Fujiwara, T.; Klemeš, J.J. Enabling low-carbon emissions for sustainable development in Asia and beyond. *J. Clean. Prod.* **2018**, *176*, 726–735. [CrossRef]
57. Yang, C.; Yeh, S.; Zakerinia, S.; Ramea, K.; McCollum, D. Achieving California's 80% greenhouse gas reduction target in 2050: Technology, policy and scenario analysis using CA-TIMES energy economic systems model. *Energy Policy* **2015**, *77*, 129–139. [CrossRef]
58. Delreux, T.; Ohler, F. Climate Policy in European Union Politics. Oxford Research Encyclopedia of Politics. Available online: https://oxfordre.com/politics/view/10.1093/acrefore/9780190228637.001.0001/acrefore-9780190228637-e-1097 (accessed on 28 May 2021).
59. Burns, C.; Tobin, P. The limits of ambitious environmental policy in terms of crisis. In *European Union External Environmental Policy: Rules, Regulation and Governance Beyond Borders*; Adelle, C., Biedenkopf, K., Torney, D., Eds.; Palgrave Macmillan: London, UK, 2018; pp. 319–336.
60. Skovgaard, J. EU climate policy after the crisis. *Environ. Politics* **2013**, *23*, 1–17. [CrossRef]
61. Jankowska, K. Poland's climate change struggle: Greening the East. In *The European Union as a Leader in International Climate Change Politics*; Wurzel, R., Connelly, J., Eds.; Routledge: Abingdon, UK, 2012; pp. 163–178.
62. Tobin, P. Leaders and Laggards: Climate Policy Ambition in Developed States. *Glob. Environ. Politics* **2017**, *17*, 28–47. [CrossRef]
63. Dovgal, O.; Goncharenko, N.; Reshetnyak, O.; Dovgal, G.; Danko, N. Priorities for Greening and the Sustainable Development of OECD Member Countries and Ukraine: A Comparative Analysis. *Comp. Econ. Res.* **2021**, *24*, 45–63. [CrossRef]
64. Vidadili, N.; Suleymanov, E.; Bulut, C.; Mahmudlu, C. Transition to renewable energy and sustainable energy development in Azerbaijan. *Renew. Sustain. Energy Rev.* **2017**, *80*, 1153–1161. [CrossRef]
65. Novacheck, J.; Johnson, J. The environmental and cost implications of solar energy preferences in Renewable Portfolio Standards. *Energy Policy* **2015**, *86*, 250–261. [CrossRef]
66. Elum, Z.A.; Momodu, A.S. Climate change mitigation and renewable energy for sustainable development in Nigeria: A discourse approach. *Renew. Sustain. Energy Rev.* **2017**, *76*, 72–80. [CrossRef]
67. Shahsavari, A.; Akbari, M. Potential of solar energy in developing countries for reducing energy-related emissions. *Renew. Sustain. Energy Rev.* **2018**, *90*, 275–291. [CrossRef]
68. Creutzig, F.; Agoston, P.; Goldschmidt, J.C.; Luderer, G.; Nemet, G.; Pietzcker, R.C. The underestimated potential of solar energy to mitigate climate change. *Nat. Energy* **2017**, *2*. [CrossRef]
69. Yasukawa, K.; Lee, T.J.; Uchida Song, Y. Environmental Barriers to Geothermal Development in Eastern and South-Eastern Asia. In Proceedings of the World Geothermal Congress, Online, 21–26 May 2021.
70. Paulillo, A.; Striolo, A.; Lettieri, P. The environmental impacts and the carbon intensity of geothermal energy: A case study on the Hellisheiði plant. *Environ. Int.* **2019**, *133*, 105226. [CrossRef]

71. Shortall, R.; Davidsdottir, B.; Axelsson, G. Geothermal energy for sustainable development: A review of sustainability impacts and assessment frameworks. *Renew. Sust. Energ. Rev.* **2020**, *44*, 391–406. [CrossRef]
72. Inayat, A.; Nassef, A.M.; Rezk, H.; Sayed, E.T.; Abdelkareem, M.A.; Olabi, A.G. Fuzzy modeling and parameters optimization for the enhancement of biodiesel production from waste frying oil over montmorillonite clay K-30. *Sci. Total Environ.* **2019**, *666*, 821–827. [CrossRef]
73. Wellig, S.D.; Nusslé, S.; Miltner, D.; Kohle, O.; Glaizot, O.; Braunisch, V.; Obrist, M.K.; Arlettaz, R. Mitigating the negative impacts of tall wind turbines on bats: Vertical activity profiles and relationships to wind speed. *PLoS ONE* **2018**, *13*, e0192493. [CrossRef]
74. Adeyeye, K.; Ijumba, N.; Colton, J. Exploring the environmental and economic impacts of wind energy: A cost-benefit perspective. *Int. J. Sustain. Dev. World Ecol.* **2020**, *27*, 718–731. [CrossRef]
75. Hwang, C.-L.; Yoon, K. *Multiple Attribute Decision Making. Methods and Applications A State-of-the-Art Survey*; Springer: Berlin/Heidelberg, Germany, 1981.
76. Zadeh Sarraf, A.; Mohaghar, A.; Bazargani, H. Developing TOPSIS method using statistical normalization for selecting knowledge management strategies. *J. Ind. Eng. Manag.* **2013**, *6*, 860–875. [CrossRef]
77. Shih, H.-S.; Shyur, H.-J.; Lee, E.S. An extension of TOPSIS for group decision making. *Math. Comput. Model.* **2007**, *45*, 801–813. [CrossRef]
78. Soufi, M.D.; Ghobadian, B.; Najafi, G.; Sabzimaleki, M.R.; Yusaf, T. TOPSIS Multi-Criteria Decision Modeling Approach for Biolubricant Selection for Two-Stroke Petrol Engines. *Energies* **2015**, *8*, 13960–13970. [CrossRef]
79. Cao, H.; Jiang, P.; Zeng, M. A Novel Comprehensive Benefit Evaluation of IEGES Based on the TOPSIS Optimized by MEE Method. *Energies* **2021**, *14*, 763. [CrossRef]
80. Tavana, M.; Shaabani, A.; Santos-Arteaga, F.J.; Vanani, I.R. A Review of Uncertain Decision-Making Methods in Energy Management Using Text Mining and Data Analytics. *Energies* **2020**, *13*, 3947. [CrossRef]
81. Androniceanu, A.-M.; Kinnunen, J.; Georgescu, I.; Androniceanu, A. A Multidimensional Approach to Competitiveness, Innovation and Well-Being in the EU Using Canonical Correlation Analysis. *J. Compet.* **2020**, *12*, 5–21. [CrossRef]
82. Tatarczak, A.; Boichuk, O. The multivariate techniques in evaluation of unemployment analysis of Polish regions. *Oeconomia Copernic.* **2018**, *9*, 361–380. [CrossRef]
83. Miola, A.; Schiltz, F. Measuring sustainable development goals performance: How to monitor policy actionin the 2030 Agenda implementation? *Ecol. Econ.* **2019**, *164*, 10637. [CrossRef] [PubMed]
84. Kendall, M.G. A new measure of rank correlation. *Biometrika* **1938**, *30*, 81–93. [CrossRef]
85. Lapata, M. Automatic Evaluation of Information Ordering: Kendall's Tau. *Comput. Linguist.* **2006**, *32*, 471–484. [CrossRef]
86. Wawrzyniak, K. The similarity of powiats in Zachodniopomorskie voivodship according to the educational level of unemployed persons in Folia Pomeranae Universistatis Technologiae Stetinensis. *Oeconomica* **2012**, *297*, 95–106.
87. Kisel'áková, D.; Šofranková, B.; Onuferová, E.; Čabinová, V. The evaluation of competitive position of EU-28 economies with using global multi-criteria indices. *Equilib. Q. J. Econ. Econ. Policy* **2019**, *14*, 441–462. [CrossRef]
88. Piersiala, L. The usage pattern of development method to assess the functioning of special economic zones: The case of Poland. *Equilib. Q. J. Econ. Econ. Policy* **2019**, *14*, 167–181. [CrossRef]
89. Rogalska, E. Multiple-criteria analysis of regional entrepreneurship conditions in Poland. *Equilib. Q. J. Econ. Econ. Policy* **2018**, *13*, 707–723. [CrossRef]
90. Ginevičius, R. Multi-criteria assessment of socioeconomic systems' conditions based on hierarchically structured indicator systems. *Econ. Sociol.* **2020**, *13*, 256–266. [CrossRef]
91. Pietrzak, M.; Igliński, B.; Kujawski, W.; Iwański, P. Energy Transition in Poland—Assessment of the Renewable Energy Sector. *Energies* **2021**, *14*, 2046. [CrossRef]
92. Flint, R.; Warren, R. Basics of Sustainable Development. *Pract. Sustain. Community Dev.* **2012**, *2*, 25–54.
93. Ziolo, M.; Filipiak, B.Z.; Bąk, I.; Cheba, K. How to Design More Sustainable Financial Systems: The Roles of Environmental, Social, and Governance Factors in the Decision-Making Process. *Sustainability* **2019**, *11*, 5604. [CrossRef]
94. Andersen, M.M. Eco-innovation Dynamics–Creative Destruction and Creative Accumulation in Green Economic Evolution. In Proceedings of the Summer Conference 2010 on "Opening up Innovation: Strategy, Organization and Technology", London, UK, 16–18 June 2010.
95. Barbier, E.B. *Global Green New Deal, Economics Discussion Papers 2009–2038*; Kiel Institute for the World Economy (IfW): Kiel, Germany, 2009.
96. Susta, G. *European Union Policies for a Green Economy: Analysis and Objectives towards 2020*; Institute of European Democrats: Brussels, Belgium, 2011.
97. Li, Q.; Cherian, J.; Shabbir, M.S.; Sial, M.S.; Li, J.; Mester, I.; Badulescu, A. Exploring the Relationship between Renewable Energy Sources and Economic Growth. The Case of SAARC Countries. *Energies* **2021**, *14*, 520. [CrossRef]
98. Pearce, D.; Markandya, A.; Barbier, E. *Blueprint for a Green Economy*; Earthscan: London, UK, 1989.
99. Malkowska, A.; Malkowski, A. International Trade in Transport Services between Poland and the European Union. *Sustainability* **2021**, *13*, 424. [CrossRef]

energies

MDPI

Article

The Consumption of Renewable Energy Sources (RES) by the European Union Households between 2004 and 2019

Marlena Piekut

College of Economics and Social Sciences, Warsaw University of Technology, Plac Politechniki 1, 00-661 Warsaw, Poland; marlena.piekut@pw.edu.pl

Abstract: The paper provides the analysis of fuel and energy transition in households sector and its sustainable development in the period 2004–2019. The main purpose of the paper is to determine the development trends in the use of renewable energy sources (RES) in the EU countries household sector in 2004–2019, to recognize the state of development and functioning of the studied area as well as to indicate their successes and shortcomings in observed reality. The article employs the results of Energy balance sheets from Eurostat. The research entity were households from 28 European Union countries, with particular emphasis on households from Poland and selected neighboring countries. The research subjects there were different sources of renewable energy used by households, i.e., solar thermal system, geothermal technologies, primary solid biofuels, charcoal, biogases, blended biogasoline, blended biodiesels, ambient heat (heat pumps). To achieve the research objective a number of statistical measures ands methods, including cluster analysis and linear trend indicator applied. In the analyzed 16 years, an absolute and relative increase in the use of RES in the household sector was noticed. Taking into account the specificity of using RES in households, 6 clusters of countries were distinguished. In Poland, it was noted that there was a significant increase in the use of RES in households, with stagnation in the use of non-renewable energy sources, such as, for example, hard coal.

Keywords: renewable energy sources; household; primary solid biofuels; solar thermal system; ambient pumps

Citation: Piekut, M. The Consumption of Renewable Energy Sources (RES) by the European Union Households between 2004 and 2019. *Energies* **2021**, *14*, 5560. https://doi.org/10.3390/en14175560

Academic Editor: Adélio Rodrigues Gaspar

Received: 3 August 2021
Accepted: 1 September 2021
Published: 6 September 2021

Publisher's Note: MDPI stays neutral with regard to jurisdictional claims in published maps and institutional affiliations.

Copyright: © 2021 by the author. Licensee MDPI, Basel, Switzerland. This article is an open access article distributed under the terms and conditions of the Creative Commons Attribution (CC BY) license (https://creativecommons.org/licenses/by/4.0/).

1. Introduction

For several decades, strong processes of globalization of economies, intensive interconnections between countries have influenced the interaction between consumption patterns [1,2]. It seems clear that as fossil fuel resources are depleted, greater use of renewable energy sources (RES) will become necessary. [3,4]. In the last decades of the 20th century, the cost of household investment in renewable energy sources was extremely high and, thus, out of reach for many consumers. The progress in research and development activities resulting in creation numerous innovative solutions [5,6], as well as financial support from EU funds led to the situation that RES began to play more and more increasing role as energy carriers in households.

Renewable energy sources are the basis for sustainable economic development. Between 2004 and 2019, in the EU-28 countries, the usage of renewable energy sources by all sectors of the economy in final energy consumption increased from 5.1% in 2004 to 10.2% in 2019, while in the household sector from 10.8% to 18.2% [7].

Sustainable development is a basic and main objective of the European Union. The EU sustainable development strategy aims for the continuous improvement of quality of life for society [8]. Sustainable production and consumption patterns are important elements in tackling climate change. Reduction in energy consumption and changes in the fuel mix, by switching to less carbon-intensive energy sources, is linked to lower CO_2 emissions. Transition to a low carbon economy would be an important stage towards meeting this demand for climate stability.

The use of energy from renewables brings numerous advantages: decrease in the cost of energy supply, reducing both environmental pollution and the pressure on fossil fuel energy production, and also more convenient form of energy production and supply [9,10]. It can also contribute to energy poverty alleviation [11,12]. Replacing fossil fuels with renewable energy sources also reduces particular countries need for energy imports and their dependence on countries exporting non-renewable fuels [13].

The paper covers the issue of renewable energy sources (RES) consumed by European Union households. It considers the use of RES as a whole and also its different type categories in the household sector from 2004 to 2019, and (in some parts of the analysis) the time range also includes 1990.

The main purpose of the paper is to determine the development trends in the use of RES in the EU countries household sector in 2004–2019, to recognize the state of development and functioning of the studied area as well as to indicate their successes and shortcomings in observed reality.

In view of the research goals the following research questions were formulated:

Research Question 1. What changes have occurred in the use of renewable energy between 2004 and 2019?

Research Question 2. What groups of countries can be distinguished according to the category of renewable energy sources used in the household sector?

Research Question 3. How is the use of renewable and non-renewable energy sources changing in the household sector from selected EU-28 countries?

The undertaken research topic is important for several reasons. Households are one of the important sectors consuming energy. In 2019, households accounted for 26.9% of final energy consumption [7] in the EU-28 countries. Thus, from a practical point of view, involving them in the process of improving environmental quality can contribute towards climate improvement by reducing emissions of harmful substances into the atmosphere. Observing the changes in the household sector in terms of RES consumption gives an overview of the situation of individual EU members and can be a wake-up call for those who have accomplished little in the period under study to involve citizens in initiatives to install new technological solutions based on cleaner energy. From a theoretical point of view, the diagnosis of the situation of RES use in households is cognitively interesting, since analyses of RES use, as a rule, appear only for their overall use in the economy or in relation to transport. Moreover, the use of energy commodities in households can also be associated with responsible consumption. The problem addressed is also important from the point of view of the 17 Goals of Sustainable Development [14,15] formulated by the United Nations.

The originality of the presented article in comparison to the existing literature data lies in the emphasis on the important role of households in energy consumption and related problems. The novelty is also the presentation of the author's types of households observed in the EU-28 countries, according to the patterns of RES categories.

2. Literature Review

2.1. EU Policy for RES

The European Union, by developing numerous binding policies, is the biggest promoter of renewable energy investments. Within the European Union policies, RES and its support are strongly anchored in the form of strategies, development goals, priorities and current legislation [16]. The principal initiative to promote renewable energy at the European Union level was launched in 1997, when the European Council and the European Parliament adopted the "White Paper for a Community Strategy and Action Plan", aiming to increase the share of renewable energy, which at that time accounted only 6% of gross energy consumption [17].

EU policy standards for the use of RES have been acquired especially in the last two decades. Firstly, the EU Sustainable Development Strategy [18] stipulated that 12% of energy consumption and 21% of electricity consumption should be covered by RES by

2010, with an increase to 15% by 2015. The Europe 2020 Strategy [19], on the other hand, has indicated policy standards towards increase of the RES usage and to promote energy efficiency and greater energy security.

Directive 2009/28/EC of the European Parliament and of the Council (2009) [20] set a mandatory target by 2020 at the level of a 20% RES energy share. The Directive also sets out various mechanisms that Member States can implement to achieve their goals (joint projects, support schemes, cooperation between Member States and other countries). Moreover, it also defined national renewable energy targets for each country, taking into account its starting point and overall renewable energy potential. These goals range from a low of 10% in Malta to a high of 49% in Sweden. The EU countries define how they plan to meet these objectives and all their renewable energy policies in their National Plans. The results in achieving the national goals are measured every two years when EU countries publish national renewable energy progress reports [21].

The Article 3 of Directive 2018/2001 (2018) [22] on the promotion of the use of energy from renewable sources established a new and binding overall EU target for the total share of energy from renewable sources in the Union's gross final energy consumption for 2030 of at least 32%.

The Green Deal for Europe, proposed at the end of 2019, creates new conditions for a very ambitious climate protection agenda. The vision is for Europe to become the world's first climate-neutral continent by 2050. The package of measures in the European Green Deal should enable European citizens and businesses to reap the benefits of a sustainable green transition. The measures, together with an initial 'roadmap' of key policies, cover tasks such as reducing emissions, investing in cutting-edge research and innovation, together with protecting Europe's environment [23]. The implementation of the Green Deal for Europe idea will let EU countries achieve energy independence, which may have a significant impact on the energy market as well as on their regional policies. It is also assumed that the achieved climate neutrality will contribute to the dynamic development of the economy and improve its competitiveness [24].

The support accompanying implementation the abovementioned policies in EU countries resulted in increased the renewable energy use throughout the Community. In the EU-28 final energy consumption, renewable energy sources accounted for 3.8% in 1990, 5.1%, while in 2019 this percentage increased by a further 5.1 p.p. to 10.2%. A relatively high share of RES consumption is characterized by the household sector, where the share of RES in household final energy consumption in 1990 was 8.5%, in 2004—10.8%, and 16 years later by 7.4 p.p. more. [7].

The construction sector in Europe accounts for 40% of energy consumption and 36% of CO_2 emissions [25]. Due to the estimated, high energy saving potential of the housing construction sector, the European Union has established a policy framework focused on reducing energy consumption in buildings, consisting of different policy actions, i.e., Energy Performance of Buildings Directive (EPBD) [26], Energy Efficiency Directive (EED) [27], Ecodesign Directive [28], Energy Labelling Regulation [29] and the mentioned Renewable Energy Directive (RED) [20].

In summary, the EU countries' policies, since the 1990s, have systematically increased the importance and role of RES in the energy sector structure s [30]. All plans for the presumed targets are presented in the National renewable energy action plans 2020 [31]. National RES energy targets for individual countries are established at various levels and the process of reaching them is different depending on the country [24], which also affects the use of RES in the household sector.

2.2. The Role of Authorities in Promoting New Energy Solutions

The EU countries are trying to influence and propose different solutions to improve the situation regarding energy consumption. One of the more effective ways is to exploit renewable energy obtained from natural resources such as wind, sunlight, geothermal heat, etc. instead of non-renewable energy [32].

The consumption of renewable energy is growing in importance and there is an increasing need to encourage households to cooperate. For the development of renewable energy, it is important to ensure an active role of government and other authorities at all levels of general interest. It is even suggested [33] that public institutions, both state and municipal, should be legally obliged to install solar systems on the roofs of the buildings where they hold office. Also, electricity buyers can have an important influence on the way energy is generated in public procurement, where RES energy should be given priority.

However, not all countries have a good appreciation of government policies to promote RES. For example, in Croatia it is found that citizens are not actively encouraged to participate in investments that would largely benefit the environment [33]. A study in Malta [34] concluded that future programs promoted by the government should consider the role of pro-(and anti-)government sentiment in predicting their adoption in the initial stages. Strong pro-government sentiment can strengthen citizen initiatives to install new RES technologies. Delegating RES promotion to municipalities or even commercial entities may also result in increased citizen interest in such installations. The financial constraints faced by low-income households are also pointed out [34]. RES technologies are often expensive and the way in which support schemes are implemented require upfront investment, leaving households unable to afford to pay (in cash) for any investments. It is suggested that subsidies and programs offered by the government, should offer staggered payments for the initial investment in order to provide an incentive. Furthermore, the experience in Malta [34] emerged that requiring consumers to pay for net rather than gross value would enable more households to benefit from the program. Support schemes should also pay attention to vulnerable groups of society. Helping elderly households (for example, through preferential feed-in tariffs or targeted communication) could unlock further potential by encouraging older people to engage in investment. Similarly, significant scope appears to encourage investment in rental housing. Some programs can be designed to promote investment agreements between landowners and tenants [34].

Local authorities should assume a central role and responsibility in the task of solarizing their territories. They have the autonomy to regulate the situation on-the-spot, in particular through their well-known water and wastewater utilities, which can also provide other RES for local energy production (biogas from bio-waste and sewage sludge, energy stored in water). Such companies would therefore integrate power generation into their regular activities and could provide installation and maintenance services for power generation systems installed in their area of competence [33].

2.3. *Willingness to Involving in the RES Use by the Household Sector*

Energy resources have always played a key role in human life. Sufficient energy resources influence economic and social development. A kind of interdependence is observed between technological development, energy consumption and world population growth [35]. Providing adequate energy resources for entities such as households is also about meeting basic social needs. Energy is one of the main categories of consumer expenditure in households [36]. Maintaining adequate thermal comfort affects the consumer life quality [37,38]. On the other hand, improving this kind of comfort is associated with an increase in the consumption of fossil energy carriers, which in turn is associated with an increase in environmental pollution [39,40]. Thus, in order to reconcile social and environmental objectives, it is important to widely involve households in initiatives for the use of energy from renewable energy sources.

An important issue is the problem of consumers' attitude and propensity to make decisions on the use of renewable energy technologies. As indicated by Ropuszyńska-Surma and Węglarz [41], social acceptance of RES technologies is important for their development and should be taken into account when shaping policies for sustainable development in the region.

The results of research by A. Jacksohn, P. Grösche, K. Rehdanz and C. Schröder [42] suggest that households tend to act fairly rationally in the sense that investors consider the costs and benefits of their decision. Since economic factors influence the decision to invest in a renewable energy system, policy makers can provide reasonable financial incentives to steer households in the desired direction.

A number of studies show that financial incentives become a strong motivation for households to switch to renewables. A study of Italian and Austrian households investing in solar PV found that higher financial support was more likely to attract younger and less educated people, as well as those with an anthropocentric attitude towards nature [43].

In contrast, Wasi and Carson [44] investigated how households' decisions to switch to more environmentally friendly water heaters changed with the introduction of a rebate scheme for hot water systems. They concluded that the likelihood of households choosing to use the aforementioned renewable system for hot water, or a heat pump, increased significantly after the introduction of a scheme to financially support these initiatives. Furthermore, the impact of this rebate policy varied with household income, education, access to the gas grid, hot water consumption and expectations of future electricity prices.

In Germany, research on household investment in RES found that the propensity to adopt them was influenced by housing characteristics, household energy consumption and geographical factors, while most socio-demographic variables were found to be insignificant [45]. Other studies [46,47] have considered the motives for adopting an innovative residential heating system based on renewable energy. It was revealed [46] that the influence of socio-demographic, housing and spatial characteristics was more significant for households replacing a heating system in an existing house than for households choosing a heating system for a newly built house.

The likelihood of investing in RES also increases with environmental concern, income, number of children and solar radiation intensity [48]. Men and well-educated people were more willing to engage in RES installations than older people.

In Poland, social acceptance of RES varies, and depends on age, gender, education, income and type of building inhabited. The groups that showed the highest acceptance for RES installation were men, people aged 30–49 years, having secondary technical education, low income and people living in a single-family house. The rationale for installing RES was the expected long-term savings, while the biggest barrier was the lack of financial resources. The financial aspect is crucial for the installation of RES in Polish households. In the case of prosumers, besides the financial aspect, were also pointed out technical possibilities, unclear regulations, complicated grid connection process and lack of knowledge. Thus, besides financial support, additional support in the form of consumer education, promotion of RES development, technical and legal support of potential prosumers seems to be necessary [49].

D. Štreimikienė and A. Baležentis [50] studying households in Lithuania paid attention to employment status and income level that have a significant impact on the willingness to purchase renewable energy sources in households of this country. The self-employed showed the highest willingness to purchase RES. Private sector employees and, surprisingly, pensioners also showed more willingness to invest in RES compared to other social groups. In terms of education level, only respondents with higher education showed a higher willingness to buy RES than the rest. Thus, the willingness to pay for RES in Lithuanian households was determined by factors such as awareness of their existence, education level and income. Other studies showed that the willingness to pay for RES was influenced by age, gender, education, income, price, geographical place of residence. In contrast, membership in environmental organizations, race, political views and perceived health effects had less influence on household usage of renewable energy technologies [51].

A study conducted in Malta [32] revealed that factors associated with the use of RES energy (in this case, energy from photovoltaic devices) were the age of those forming households (the younger the individuals, the higher the involvement in RES) and unemployment (if present, there were fewer opportunities to invest in RES), both of which suggest that financial motives and constraints are crucial to household uptake of RES initiatives.

In contrast, a study by Luttenberger [33] for Croatia, showed that people support for RES projects increased together with knowledge about them. Positive attitudes towards new clean technologies prevail in this country. However, there is a general lack of solid information and understanding of the concepts that are necessary for the possible benefits for end-users in a household. There is also a lack of professional and trained human resources for renewable energy issues, there are no relevant courses at universities and colleges, no systematic research, a lack of experience of local companies in organizing projects and the volume of theoretical knowledge about RES and practical capabilities involved is limited [52]. Even though household purchasing power is modest, their owners are at least declaratively putting aside more money for RES, but only after sufficient additional information has been provided. Croatians also mention obstacles such as national solar quotas, administrative barriers and complexity of the procedure [53].

Summarizing the abovementioned studies, it can be pointed out that the financial motive is an essential motive for household members to undertake RES installations. On the other hand, educational level and age are indicated among the important socio-economic characteristics of those willing to accept RES. Providing consumers with extensive information and educating them on new technologies for the overall social and environmental good is also an important determinant of consumers' commitment to RES application. So, the impacts of socioeconomic factors provide substantial policy implications for the design of green electricity programs [54].

2.4. Application of Cluster Analysis in Comparative Research on RES Use in Different Countries

Cluster analysis was first introduced in the work of R.C. Tryon [55]. It is a useful tool for exploratory data analysis that aims to arrange individual objects into groups so as to acquire objects within the same group most similar to each other and the objects between other groups most dissimilar to each other [56,57]. This analysis, using several different classification algorithms, detects the data structures without explaining why they occur.

European Union member states vary in terms the exploited both of total energy sources [58] and also of RES. The cluster analysis method may be useful for searching the similarities between individual member states. This method is often represented by the simple hierarchical method. The common feature of the stepwise algorithms used in this method is the clustering by combining smaller clusters, created in the previous steps of the algorithm. The basis of all algorithms of this method is the appropriate determination of the measure for object dissimilarity [59].

Below are presented a few selected studies reporting on the results produced by cluster analysis performed for country classifications regarding RES market. Therefore, for example, the study by Bluszcz, Manowska [58] is applying the agglomeration procedure, which results in the division of European Union member states into clusters according to their similarity, with regard their energy markets. The research results constitute an interesting study that could potentially provide a model for the creation of so-called regional energy markets in the transitional integration phase. The above-mentioned authors chose the following diagnostic variables for the analysis: consumption of electric energy which is generated from renewables per capita (TWH per person), consumption of hard coal (million ton per person), emissions of greenhouse gas per capita, available for final consumption (Gigawatt-hour per person), final energy consumption (thousand ton of oil equivalent (TOE) per person), petroleum available for final consumption (Gigawatt-hour), natural gas (Terajoule gross calorific value—GCV) per person, energy intensity of GDP (kilograms of oil equivalent (KGOE) per thousand euro), import dependency (%). The paper distinguishes six clusters, consisting of countries with similar levels of energy system development. The cluster formed by Finland and Sweden presented renewable energy production at the highest level in the EU. Finland had slightly higher greenhouse gas emissions per capita in the energy mix and a higher energy consumption factor than Sweden, as solid fuel use accounted for 9%. Luxembourg formed a one-element cluster. This country had the highest level of electricity consumption per capita compared to other UE member states. Greenhouse gas (GHG) emissions per capita and energy dependency levels were also the highest there. Another cluster, comprising France, Slovenia, United

Kingdom, Latvia and Romania had levels of electricity consumption per capita below the EU average. Energy dependency levels were relatively low in this cluster, ranging from 50% in Slovenia to 24% in Romania. The countries grouped in the cluster comprising Greece, Lithuania, Croatia, Hungary, Spain, Portugal and Italy were characterized by high levels of energy dependency, ranging from 52% for Croatia and 58% for Hungary to over 70% for all other countries in this cluster. In these countries, the contribution of solid fuels to electricity production was significantly higher than in the first and second clusters. The cluster including countries such as the Czech Republic, Slovakia, Poland, Bulgaria and Estonia was characterized by a level of greenhouse gas emissions per capita close to the average and thus exceeded the desired GHG emission factors. For this cluster, renewable energy consumption per capita was one of the lowest in the EU. The level of energy dependency was low, less than 1% in Estonia, 36% in Bulgaria and Czech Republic and 44% in Poland (with the exception of Slovenia, at 63%, the highest in this cluster). The last cluster containing the following countries: Netherlands, Germany, Belgium, Ireland, Denmark and Austria had the lowest level of energy intensity, while the consumption of energy from renewable sources was high [58].

Other researchers [60] analyzed in their paper the renewable energy sector in European Union countries. The k-means clustering method was used for grouping of countries. This method is widely used in various areas of science for the data analysis. The advantage of this method is the intuitiveness and simplicity of the basic calculation idea. In the k-means method, the distances between objects are determined by the Euclidean distance or its square (the peculiarity of the algorithm makes the results in both cases the same). The k-means algorithm can be described in three points: 1. The starting point is the division of a given set of objects into k subsets (usually generated by assigning each element to the "closest" preselected representative of the k groups). 2. For each group, the centers of gravity in the space of diagnostic variables are determined. 3. Each element is assigned to the nearest center of gravity, and then it is necessary to return to step 2, if at least one element has been moved to another group [61]. The algorithm of the k-means method can be regarded as a kind of "inverse" of the analysis of variance. It is helpful in finding a division of the studied community into k groups, so as to maximize the intergroup variance and, consequently, the F-statistic [61,62].

In the abovementioned paper, Parobek and colleagues [60] created nine clusters and the following diagnostic variables were selected forest cover, roundwood production, primary energy consumption, primary production of energy from renewable resources, share of renewable in gross final energy consumption, greenhouse gas emissions, employment, gross value added, GDP growth rate, expenditures on R&D, price of electricity, energy dependence. The 1st cluster was formed by the leaders in the use of renewable energy sources and the average production of primary energy from biomass, i.e., Austria, Portugal, Sweden, Finland. However, these countries were below the EU average in terms of employment rate and energy dependence. The 2nd cluster, consisting of Hungary and Belgium, had the use of renewable energy sources below the average. The 3rd cluster that contained Greece and Romania had a utilisation of renewable energy sources slightly above the EU average. These countries had significant production of wind, solar and hydropower energy. The 4th cluster, consisting of Bulgaria, Lithuania, Czech Rep., Denmark, Ireland and Slovakia had a medium level of renewable energy use. The 5th cluster was formed by Cyprus, Estonia, Latvia, Slovenia, Luxembourg and Malta. In these countries the share of renewable energy sources remained above average, but on the other hand they their production of primary energy from renewable resources was insignificant. Germany formed a one-element cluster (the 6th cluster) with the highest production of primary energy from biomass and RES consumption below average. It was also noted that this country is one of the largest producers and users of wood. Another one-piece cluster formed France (the 8th cluster), which (just as Germany) was a leader in the production of energy from wood and had a relatively low level of greenhouse gas emissions compared to the countries in the other clusters. The 7th cluster included Netherlands, Poland and United Kingdom, i.e., the countries having lower primary energy production from renewable resources and the lowest share of use of renewable resources. Italy and Spain formed the last, the 9th cluster. Use of renewable energy resources in these countries was below the average, but primary energy production from renewable resources—above the average [60].

In turn, K. Chudy-Laskowska and co-authors [63] in their article distinguished clusters on the basis of similarity such EU countries regarding the current level of the wind energy development. The research applied Ward's analysis and Wroclaw taxonomic methods. The Wroclaw method was described and employed in the mentioned article and in other studies [64,65]. Seven clusters containing from 1 to 7 countries were obtained. Denmark, forming one-element cluster, turned out to be the wind energy leader in the EU. This country had not only very profitable use of wind energy but also the highest rates of wind turbine electricity generation, and also the share of wind energy in gross inland energy consumption. The following diagnostic variables were adopted: wind farms per 100 thous. People, number of turbines per one wind farm, renewable (wind offshore) electricity capacity (MW) per thous. people, renewable (wind onshore) electricity capacity (MW) per thous. people, wind cumulative capacity growth rate, renewable (wind) electricity generation (GWh) per thous. people, share of renewables in gross inland energy consumption of which: wind power. The cluster comprising Spain, Portugal, Ireland, Sweden and Germany also had excellent wind energy technologies. These countries have been leaders in the introduction and development of wind energy for many years and have had a significant number of wind farms (above the European average) and these farms were quite large. Finland is another one-element cluster. This country invests and develops the mentioned branch of renewable energy. The index on the number of wind farms was above the world average, so there were more these farms than the EU average, but they were not large. Three countries: the United Kingdom, the Netherlands and Belgium, forming another cluster, were relatively strong in terms of both wind energy potential and the level of its development. There was also an upward trend in wind energy resources in this group. In the countries forming cluster that included Cyprus, Romania, Greece, Italy and Estonia there were not many wind farms, but the existing were relatively large. In addition, all wind energy indices were below the EU average. Another cluster includes six countries: Austria, Luxembourg, Lithuania, Croatia, Poland and France. In these countries there was a chance to change their position for the better by possible investing in wind energy. However, during the analyzed period, the level of wind energy development was still below the world average there. The last and worst group in terms of both wind energy potential and development level was a cluster including seven countries: Slovakia, Slovenia, Malta, Czech Republic, Hungary, Latvia and Bulgaria. This cluster also included Malta, which did not have a single wind farm, and the other countries in this group had almost no wind farms. These countries invested in other renewable energy sources. Slovakia, Latvia, Hungary and Bulgaria invested mainly in hydropower energy, while Hungary and Malta in solar energy [63].

Overall, the methods k-means and Ward were the most frequently applied clustering techniques. Other studies using cluster analysis in research on the use of renewable energy are [66–69].

3. Sources and Research Methods

3.1. Data Sources

The study was focused on European Union households. The research material was from Eurostat "Energy balance sheets" [7].

The energy balance is the most complete statistical data of energy products and their flow in the economy. The energy balance offers a complete view on the energy situation of a country and of individual sectors (e.g., households). The energy balance is a multi-purpose tool, is the natural starting point to study the energy sector [70].

The research entities were the countries of the European Union. The subject of the research were various sources of renewable energy used by households, i.e.,

- Solar thermal system,
- Geothermal technologies,
- Primary solid biofuels,
- Charcoal,
- Biogases,
- Blended biogasoline,
- Blended biodiesels,
- Ambient heat (heat pumps).

The data for the different products are expressed in a common energy unit: thousands of tons of oil equivalent (ktoe) [70].

3.2. Study Design

To achieve the research objective and carry out the tasks presented in this article, the researcher had to apply a number of statistical measures and methods. Subsequent research tasks were subordinated to research questions.

Step 1. Monitoring the evolution of RES consumption in European households required:

- aggregation of data from the Eurostat Energy balance sheets database for individual countries;
- calculating the structure of the RES energy carriers used in each country;
- generation of a ranking of countries in terms of the RES amount and its share in total household energy consumption.

Step 2. Identification of European household types in terms of RES consumption. This task required:

- data standardization;
- Ward clustering method and establishing the optimal number of clusters;
- carrying out clustering with the k-means method;
- description and labelling of clusters.

The use of Ward's method cluster analysis provided an answer to the question about the optimal number of country (state)groups [71,72]. A plot of clustering distances by clustering step indicated that the first clear spike was at the level 23.45. The dendrogram was cut at this level, yielding six clusters of households (Figures 1 and 2). Ward's method for determining the optimal number of clusters is also used by other researchers [73].

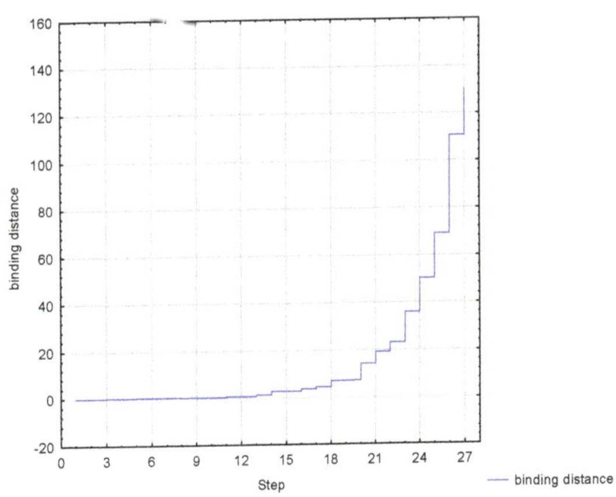

Figure 1. Results binding distance according to binding steps. Source: own elaboration based on data from [7].

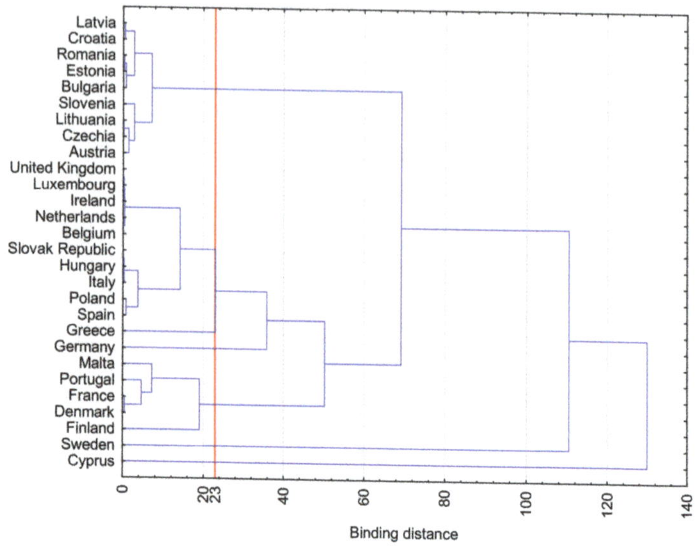

Figure 2. Results of the hierarchical grouping of similarities between the EU countries in final energy consumption from renewable energy sources in households' sector in 2019 using the Ward's method. Source: own elaboration based on data from [7].

Then, the k-means method was used to group countries. K-means cluster is widely described in the literature on the subject [74–77]. The algorithm for the k-means method is presented in Theorem 1.

Theorem 1. *The algorithm for the k-means method*

$$J = \sum_{i=1}^{k} \sum_{dt \epsilon Di}^{k} sim(c_i d_t) \quad (1)$$

The research algorithm of the *k*-means method consists of several stages, presented in the publication [78]. More about the *k*-means algorithm on the example of country grouping in the article [78,79].

Step 3. Examining trends of changes in RES use among households in Poland and neighboring countries. This task required the use of a directional trend indicator. A linear trend is a special case of linear regression, where the explanatory variable X is the time variable t [80]. A trend model belongs to a special class of econometric models in which the variability of the explained variable is described by a specific explanatory variable, namely time. In general, these models do not explain the mechanism of development of the considered explanatory variable but illustrate the development of this variable over time. In this case, therefore, a time series is considered, that is, data that are time-stratified.

Theorem 2. *A formula of linear trend function.*

$$Y = a \cdot t + b \quad (2)$$

where: a—trend slope:

$$a = \frac{\Sigma(t_i - \overline{X}) * (Y_i - \overline{Y})}{\Sigma(t_i - \overline{t})^2}$$

b—trend intercept:

$$b = \overline{Y} - a * \overline{t}$$

t_i, Y_i values of the variables t and Y.

$\overline{t}$, $\overline{Y}$—means of variables t and Y.

When $a > 0$ we are dealing with a growing trend. The greater the a, the faster the value of Y increases over time.

When $a < 0$, there is a downward trend. The smaller the a, the faster the Y value decreases over time.

4. Results

4.1. Changes in the Use of RES in European Households

The aim of this section is to analyze the changes in the use of renewable energy sources in EU-28 households between 2004 and 2019.

In absolute terms, the leaders in household sector RES consumption are the French, Germans and Italians. In 2019, RES energy in these households represented between 6508.6 thousands of tons of oil equivalent (ktoe) in Italy, 8293.3 ktoe in Germany and 9094.3 ktoe in France. The least RES energy consumed households in Malta (14.0 ktoe) and in Luxembourg (19.3 ktoe) (Figure 3). Between 2004 and 2019, the highest increase in energy in absolute units was observed precisely in the mentioned leader countries. In Italy the amount of energy used in the household sector increased by 4247.9 ktoe, in Germany by 3348.3 ktoe and in France by 2123.4 ktoe. Relatively large increases in RES consumption over the 16-year period were also observed in the UK (by 1897.8 ktoe) and the Czech Republic (by 1027.4 ktoe). Even though in the majority of EU countries an increase in household RES consumption was observed, there were also decreases, such as in Latvia (by 255.6 ktoe), Portugal (by 110.4 ktoe), Croatia (by 96.0 ktoe), Lithuania (by 57.1 ktoe) and Slovenia (by 2.1 ktoe).

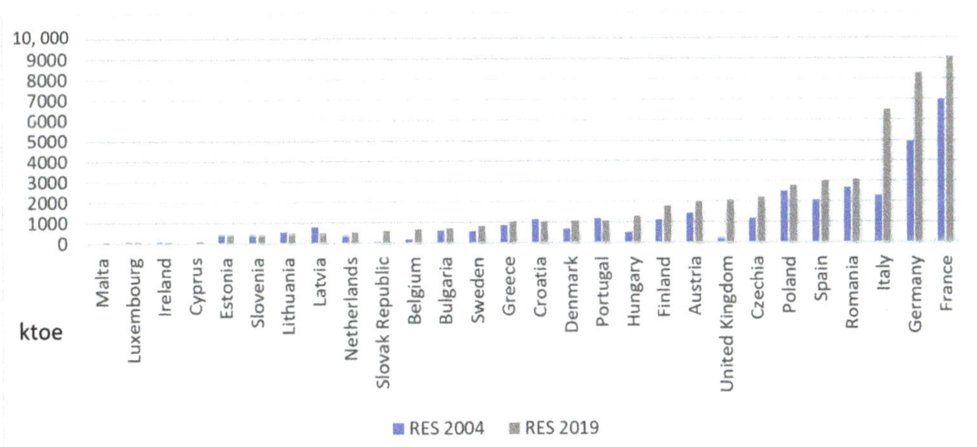

Figure 3. Energy from RES in EU households in 2004 and 2019, in thousands of tons of oil equivalent. Source: own elaboration based on data from [7].

In absolute terms, the consumption of RES by the household sector depends, among others, on the country size and the number of households in it. In the further section of this paper, the RES consumption in relative terms and the type structure of the energy used will be analyzed. Thus, in 2019—in relative terms—the first place in the ranking of renewables use by EU-28 households had Croatia, where RES accounted for more than

46% of household energy exploited. A high share of RES in the energy carriers used in households was also recorded in Slovenia and Latvia—more than 43% of final energy consumption each. In turn, Estonian, Romanian and Portuguese households had more than 36% share of RES in energy carriers used. At the other end of the scale were households of Ireland, Iceland, Luxembourg, the UK, the Netherlands and Belgium, where the share of RES in household final energy consumption did not exceed 9% (Figure 4).

The greatest progress—in relative terms—in the use of RES in EU-28 households was observed in Cyprus. Over a period of 16 years, the increase in the percentage of RES used by these households was 27.3 p.p. However, it is worth mentioning that the first data for this country appeared in 2005, where 11.1 p.p. of RES in household final energy consumption was recorded and all of this energy came from solar thermal system. The next highest increase in RES use by households was in the Slovak Republic (increase by 22.1 p.p.). A relatively high progression was observed in the case of RES usage between 2004 and 2019 by households of Hungary, the Czech Republic, Malta, Italy, i.e., by about 14 p.p. Here it should also be noted that in the case of Malta—as well as the aforementioned Cyprus—no RES usage was recorded in households in 2004 at all. In 2010 Maltese households started using RES (6.2%), which came mainly from solar thermal technology (close to 83%), as well as from primary solid biofuels.

The largest decrease in RES consumption in the household sector was recorded Between 2004 and 2019 in Latvian and Lithuanian households, with a 9.0 p.p. and by 3.1 p.p., respectively, decrease in the share of energy used from RES by these households.

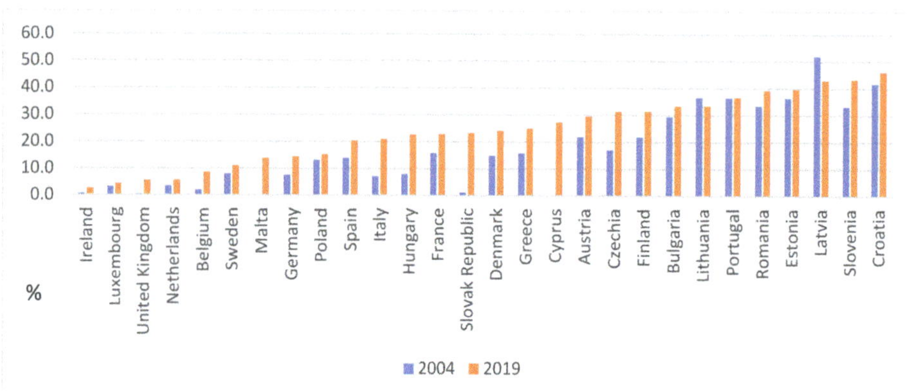

Figure 4. The share of energy from RES in the final energy consumption in European Union households in 2004 and 2019. Source: own elaboration based on data from [7].

The main renewable energy sources used in the EU-28 households are primary solid biofuels. In 2019, the share of primary solid biofuels in the total consumption of renewable energy sources in households was 83%, while in 2004—96.6%. This was followed by ambient heat at 11.6% (0.9% in 2004) and solar thermal technology at 4.0% (1.3%). Charcoal and biogases each accounted for 0.6% and geothermal heat for 0.1% (Figure 5).

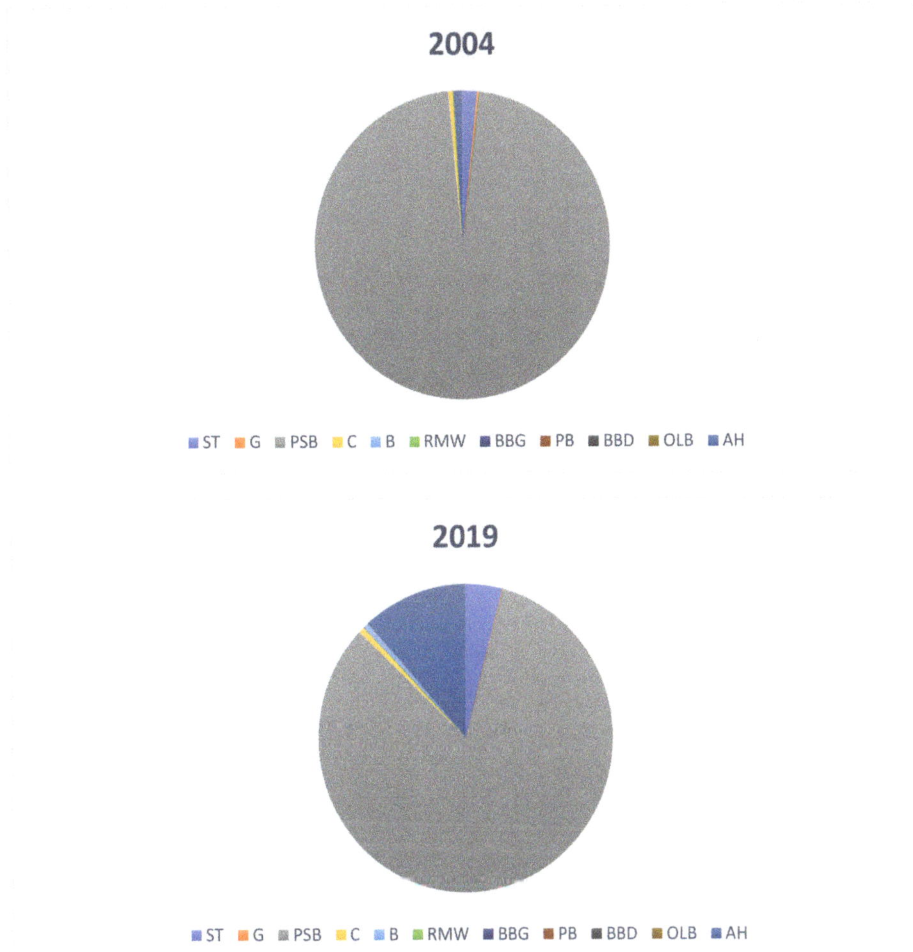

Figure 5. Structure of RES used by the household sector in 2004 and 2019, percentage. Comment. ST—Solar thermal system, G—Geothermal system, PSB—Primary solid biofuels, C—Charcoal, B—Biogases, RMW—Renewable municipal waste, BBG—Blended biogasoline, PB—Pure biodiesels, BBD—Blended biodiesels, OLB—Other liquid biofuels, AH—Ambient heat. Source: own elaboration based on data from [7].

4.2. Identification of Household Types in Terms of RES Consumption

The aim of this section is to identify the types of European households showing similarities in the structure of use of the different renewable energy sources. In the first step of the analysis, the EU countries were divided into similar groups, according to the total amount of energy used from RES in households and the structure of particular sources of this energy in 2019. The EU countries were divided into six clusters (k-means method), based on the assumed optimal number of groups (Ward's method). The adopted division was made on the basis of the first significant jump in the bonding distance related to the bonding stages (Figures 1 and 2, Table 1).

Table 1. Elements of clusters with distances from centers.

Cluster 1	Distances from Centre of Cluster 1	Cluster 2	Distances from Centre of Cluster 2	Cluster 3	Distances from Centre of Cluster 3	Cluster 4	Distances from Centre of Cluster 4	Cluster 5	Distances from Centre of Cluster 5	Cluster 6	Distances from Centre of Cluster 6
Sweden	0.00	Slovak Repub-lic	0.373972	Cyprus	0.00	United King-dom	0.114460	Latvia	0.242084	Malta	0.635871
		Italy	0.321990			Poland	0.455330	Slovenia	0.375570	Portugal	0.484746
		Hungary	0.413577			Ireland	0.185776	Romania	0.224730	France	0.279430
		Spain	0.345554			Netherlands	0.103627	Lithuania	0.217831	Denmark	0.419897
		Greece	0.943778			Luxembourg	0.139628	Croatia	0.393219	Finland	0.919940
		Germany	1.259942			Belgium	0.103894	Estonia	0.163824		
		Austria	0.336421					Czechia	0.336215		
								Bulgaria	0.231347		

Source: own elaboration based on data from [7].

In accordance to the analyses performed, measures of intra- and inter-cluster variation and degrees of freedom (df) were determined. The obtained values of the F-statistic, which is the ratio of the variation between clusters to the variation within clusters, made possible to identify the most important clustering variables in terms of their discriminatory power. This means that the higher value of the F-statistic for a given variable, the more important the assignment of given countries to particular clusters.

The analysis of variance showed that blended biodiesel use played the greatest role in the assignment of EU countries to particular clusters. The value of the F statistic for this variable was the highest at 2995.72, followed by geothermal energy (47.05). Biogases, for which the value of the F statistic was 3.07, were the least significant factor in the assignment of EU countries according to the criterion adopted. However, it is worth remembering that each diagnostic variable, i.e., each renewable energy source, ultimately influenced the grouping of EU countries into homogeneous clusters both in terms of structure and volume of energy exploited from these sources.

As blended biodiesels and geothermal heat were important factors in assigning EU countries to particular clusters, countries using these sources formed single-element clusters. The compositions of the formed clusters and the distance from their centers (cluster centers) are presented in Table 2. The greater the distance of a given EU country from the center of the cluster in which this country was located, the greater its variability in relation to the countries whose distance from the cluster center was smaller.

Table 2. Analysis of variance for renewable energy consumption in households' sector.

Specification	Between SS	Specification	Between SS	Specification	Between SS	Specification
Renewables and biofuels	22.47	5	4.53	22	21.85	0.0000
Solar thermal	22.79	5	4.21	22	23.84	0.0000
Geothermal	24.69	5	2.31	22	47.05	0.0000
Primary solid biofuels	22.47	5	4.53	22	21.84	0.0000
Charcoal	18.99	5	8.00	22	10.44	0.0000
Biogases	11.10	5	15.90	22	3.07	0.0298
Blended biogasoline	19.15	5	7.85	22	10.73	0.0000
Blended biodiesels	26.96	5	0.04	22	2995.72	0.0000
Ambient heat (heat pumps)	22.10	5	4.90	22	19.86	0.0000

Source: own elaboration based on data from [7].

The results indicate the homogeneity of two clusters (the1st and the 3rd clusters). Sweden (the 1st cluster—Specific A), as the only country using blended biogasoline and blended biodiesels, is located in the 1st cluster. No other country has reported the use of this renewable source in households. The Swedes were also ones of the few to use biogases (3.2%), while primary solid biofuels consumption accounted for nearly 94% (Figure 6, Table 3).

Cyprus (the 3rd cluster—Specific B), on the other hand, was one of the few countries with a relatively high geothermal energy consumption in households (1.6%). Cypriot households are also leaders in the use of solar thermal energy (62.5%) and also a relatively low use of primary solid biofuels compared to other countries (18.5% of total household RES consumption) is reported there. In addition, ambient heat (11.4%) and charcoal (6.0%) and geothermal heat (1.6%) appeared among the RES sources in households of this country. The overall use of RES by Cypriot households in final consumption energy was 27.3%.

The 2nd cluster (Follower) comprised seven countries with total RES use ranging from over 14% in Germany to 30% in Austria. These countries were characterized by relatively high use of primary solid biofuels, i.e., from nearly 62% in Greece to over 98% in Hungary. In almost all countries the RES categories observed in the renewable energy mix were solar thermal energy (from 1.0% in Hungary to 26.7% in Greece), charcoal (from 0.1% in Slovakia to 4.4% in Greece, in Hungary this category was generally absent) and ambient heat (from 0.7% in Hungary to 13.9% in Germany).

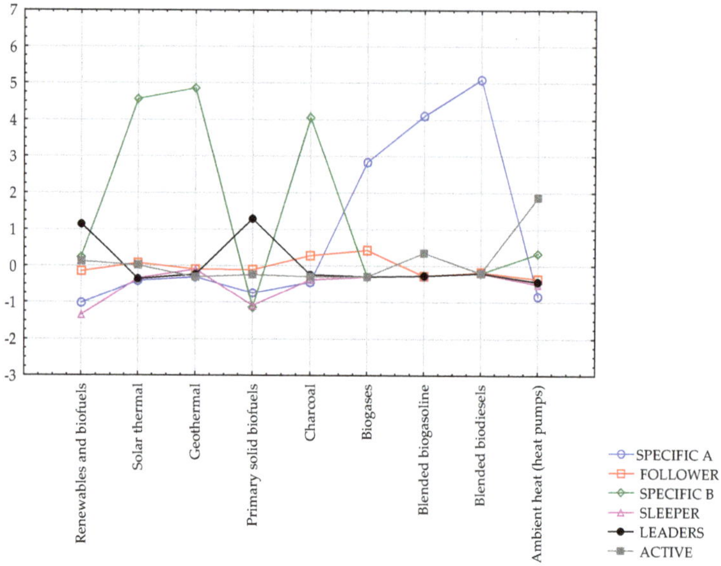

Figure 6. Mean values of energy consumption from renewable energy sources for individual clusters (results of the non-hierarchical grouping of similarities between the EU countries in final energy consumption from renewable energy sources in households' sector in 2019 using the k-mean's method). Source: own elaboration based on data from [7].

The obtained results revealed that the six countries in the 4th cluster (Sleeper) acquired the lowest average value of energy obtained from RES, ranging from 2.5% in Ireland to over 15% in Poland. Primary solid biofuels in this cluster were the relevant source of RES for households ranging from nearly 35% in Ireland to over 93% in the UK. The second place was for heat pumps, i.e., its total household RES use was from 5% in the UK to 47% in Ireland. Luxembourg and Ireland had also relatively high use of solar thermal energy, i.e., 13% and 19% of household RES consumption, respectively.

The most numerous cluster was formed by 8 countries (the 5th cluster—Leaders). This cluster had the highest value of renewable energy used by households. It was found that in 2019 the use of renewable energy sources in final energy household consumption ranged from just over 31% in the Czech Republic to over 46% in Croatia. These households relied mainly on primary solid biofuels, ranging from over 89% in Slovenia to 100% in Estonia.

In the 6th cluster (Active), which includes five countries, there are countries with rather low total RES use, i.e., from 14% in Malta to 37% in Portugal. In these countries there was relatively less use of primary solid biofuels in the household sector (in total RES use ranging from 12% in Malta to 80% in Denmark) and more use of ambient heat (ranging from 19% in Denmark to 52% in Malta). In Malta, solar thermal was a relatively popular source of RES, with more than 36% of RES coming from this source. An overview of the average RES energy shares for each country, cluster by cluster, is presented in Table 3.

Between 2004 and 2019 a diversification of RES sources is observed. Even though primary solid biofuels still remain the dominant source of RES in households, other sources are emerging, notably ambient heat and solar thermal system. In 2004, a sector of households in nine EU-28 countries used ambient heat, while 16 years later this source appeared in 23 countries. In 2004, households in 16 countries used solar thermal energy, while in 2019—in 24 countries. Another RES source, charcoal appeared in households of six EU-28 countries in 2004, while in 2019—it extended to 12 countries.

Table 3. Groups of EU-28 countries due to the use of RES in the household sector in 2019, in percentages.

	Affiliation to Clusters	Renewables and Biofuels	Solar Ther-mal	Geothermal	Primary Solid Biofuels	Charcoal	Biogases	Blended Biogasoline	Blended Biodiesels	Ambient Heat (Heat Pumps)
1	2	3	4	5	6	7	8	9	10	11
						% renewables and biofuels (100% = 3th column)				
SPECIFIC A										
Sweden	1	11.1	1.3	0.0	93.7	0.0	3.2	0.6	1.0	0.0
FOLLOWER										
Slovak Republic	3	23.3	1.1	0.0	94.2	0.1	0.0	0.0	0.0	4.6
Italy	3	20.9	2.6	0.0	95.2	0.7	0.0	0.0	0.0	1.5
Hungary	3	22.6	1.0	0.0	98.3	0.0	0.0	0.0	0.0	0.7
Spain	3	20.3	9.2	0.4	84.3	0.9	0.0	0.0	0.0	5.3
Greece	3	24.9	26.7	0.0	61.8	4.4	0.0	0.0	0.0	7.1
Germany	3	14.4	8.4	0.3	72.0	2.0	3.4	0.0	0.0	13.9
Austria	3	29.5	5.7	0.0	84.9	0.3	0.3	0.0	0.0	8.7
SPECIFIC B										
Cyprus	2	27.3	62.5	1.6	18.5	6.0	0.0	0.0	0.0	11.4
SLEEPER										
United Kingdom	4	5.5	1.7	0.0	93.4	0.0	0.0	0.0	0.0	4.8
Poland	4	15.2	2.4	0.7	88.4	0.0	0.0	0.0	0.0	8.5
Ireland	4	2.5	18.6	0.0	34.6	0.0	0.0	0.0	0.0	46.8
Netherlands	4	5.6	4.4	0.0	73.6	1.2	0.0	0.0	0.0	20.8
Luxembourg	4	4.2	12.7	0.0	80.1	0.0	0.0	0.0	0.0	7.3
Belgium	4	8.6	3.8	0.0	82.8	0.8	0.0	0.0	0.0	12.5
LEADERS										
Latvia	5	43.0	0.0	0.0	99.5	0.4	0.0	0.0	0.0	0.0
Slovenia	5	43.4	2.3	0.0	89.1	0.0	0.0	0.0	0.0	8.6

Table 3. *Cont.*

	Affiliation to Clusters	Renewables and Biofuels	Solar Thermal	Geothermal	Primary Solid Biofuels	Charcoal	Biogases	Blended Biogasoline	Blended Biodiesels	Ambient Heat (Heat Pumps)
Romania	5	39.3	0.0	0.1	99.9	0.0	0.0	0.0	0.0	0.0
Lithuania	5	33.5	0.0	0.0	95.2	0.0	0.0	0.0	0.0	4.8
Croatia	5	46.1	1.0	0.0	96.7	0.8	0.0	0.0	0.0	1.4
Estonia	5	39.8	0.0	0.0	100.0	0.0	0.0	0.0	0.0	0.0
Czechia	5	31.3	0.7	0.0	92.0	0.0	0.0	0.0	0.0	7.4
Bulgaria	5	33.4	1.5	0.0	98.5	0.0	0.0	0.0	0.0	0.0
ACTIVE										
Malta	6	13.8	36.3	0.0	12.0	0.0	0.0	0.0	0.0	51.7
Portugal	6	36.7	5.4	0.0	71.1	0.7	0.0	0.0	0.0	22.8
France	6	22.8	1.9	0.0	70.9	0.0	0.0	0.0	0.0	27.2
Denmark	6	24.1	1.2	0.0	80.1	0.0	0.0	0.0	0.0	18.7
Finland	6	31.4	0.1	0.0	69.5	0.0	0.0	0.2	0.0	30.3

Comment: the percentages for individual clusters are presented horizontally, i.e., the summed results in the rows should give 100%. Source: own calculation based on data from [7].

4.3. Changes in the Use of RES in Households from Poland, Slovakia and Germany

The aim of this subsection is to analyze changes in the use of energy carriers in Polish households and selected neighboring countries, i.e., Slovakia and Germany. The case of Poland was chosen due to the fact of high use of hard coal in this country. Poland is the largest hard coal producer and the second largest brown coal producer in the EU, generating about 80% of electricity from coal. Resistance to limiting coal mining and consumption comes from various sides, namely coal corporations, trade unions, parts of civil society and the government—and their coalition partners. Their objection centers around the prospect of lost business, previous negative experiences of structural change, fears of rising energy prices and concerns about energy security, as well as potential unemployment in regions almost entirely dependent on coal [81].

Germany and Slovakia were chosen for comparison for the following reasons. Firstly, they are neighboring countries. The winter climate in these countries is quite cold, which makes it necessary to use heat energy in every sector, especially in households. Secondly, Germany is a country from the so-called 'richer' west of Europe, while Slovakia, similar to Poland, is a central and eastern European country with a similar history and socio-economic development. There are more than 20 countries in the world whose share of renewable energy sources in total energy consumption exceeds 20%, and Germany is among these countries. By 2050, Germany plans to achieve a 60% share of renewable energy in the country's total energy balance and 80% in electricity production [82].

Raising the level of economic and human development has increased the demand for fossil fuels. Currently, conventional energy sources dominate in terms of resources used by economies, including some European households. Many countries in the world, facing the problem of energy and national security, have intensified their efforts to transition from conventional energy sources (primarily fossil fuels) to alternative energy sources [83]. However, these transitions differ from country to country.

Polish households are a kind of "coal island" on the map of Europe [84]. In 1990, the use of hard coal in total energy consumption in Polish households accounted for 38.2%, and for nearly three decades this percentage has been reduced by 10.8 p.p. In the same period the use of hard coal in German households decreased from 15.3% to 0.6% and in Slovak households from 19.4% to 1.1% (Figure 7). Thus, the progress in this respect—beneficial from the environmental point of view—was definitely better in the countries neighboring Poland.

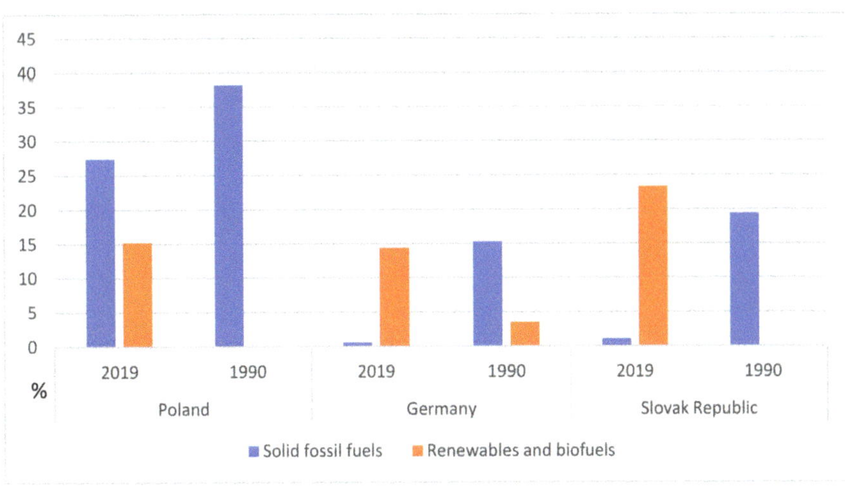

Figure 7. Use of solid fuels and renewable energy sources in the Polish, German and Slovak household sector in 2019. Scheme 7.

Between 1990 and 2019, the use of RES in households in relative terms increased, most notably in Slovak households—by 23.3 p.p., while German (10.9 p.p.) and Polish (15.2 p.p.) households had smaller achievements in this respect. Even though looking at this aspect in absolute terms, between 1990 and 2019 the use of RES by the household sector in Poland increased by 1662.7 ktoe, in Germany by 5308.8 ktoe and in Slovakia by 615.5 ktoe. However, it is worth noting that in 1990 in Slovakia there was no RES consumption in the household sector at all.

In Poland, residential heating is mainly provided by hard coal. In the total consumption of energy carriers in Polish households, hard coal and other bituminous coal accounted for 54.5% in 2019. In the 16-year period, since 2004, coal consumption in the household sector has decreased slightly, i.e., by 2.6 p.p. [7]. The high share of coal in the total consumption of energy carriers is the cause of air pollution, among others in sulphur dioxide, nitrogen oxides and dusts. The household sector is responsible for total atmospheric emissions of sulphur dioxide in nearly 23%, nitrogen oxides in about 8% and dusts in nearly 36% [85].

Based on the directional coefficient of the trend, it can be concluded that the consumption of renewables in Polish households has statistically significantly increased between 2004 and 2019, on average by 31.1 thousands of tons of oil equivalent (ktoe). Similar to other European households, among renewable energy sources, primary solid biofuels are here in the lead, accounting for 88.4% of total RES consumption in these households. Ambient heat accounts for 8.5%, solar thermal system for 2.4% and geothermal energy for 0.7%. Thus, the hierarchy of RES consumption in Polish households is consistent with that observed in European ones. Regarding the different types of RES, a significant increase was recorded for solar thermal (by 4.47 ktoe on average per year), geothermal power (0.77 ktoe) and ambient heat (15.5 ktoe). In addition to the increase in renewable energy sources, Polish households enhanced the natural gas consumption, on average by 28.4 ktoe per year. Other energy sources showed a decrease (manufactured gases, gasworks gas, oil and petroleum gases, gas oil and diesel oil) or stabilization (solid fossil fuels, other bituminous coal, lignite, coke oven coke) (Table 4).

Table 4. Trends in the use of energy sources in Polish households in 2004–2019.

Specification	The Trend Slope Factor	p Value	R^2
Total	−33.10	0.5164	0.0307
Solid fossil fuels	−23.10	0.5484	0.0263
Other bituminous coal	−23.09	0.5314	0.0285
Lignite	0.96	0.3427	0.0644
Coke oven coke	−0.96	0.6923	0.0115
Manufactured gases	−0.19	0.0000	0.9550
Gas works gas	−0.19	0.0000	0.9518
Oil and petroleum products	−34.35	0.0000	0.7717
Liquefied petroleum gases	−3.88	0.0245	0.2626
Gas oil and diesel oil	−30.47	0.0000	0.7577
Natural gas	28.4	0.0028	0.4453
Renewables and biofuels	31.1	0.0001	0.6327
Solar thermal	4.47	0.0000	0.8870
Geothermal	0.77	0.0000	0.9262
Primary solid biofuels	10.4	0.1531	0.0787
Ambient heat	15.5	0.0000	0.9124

Comment: Tables 3–5 include the categories of energy sources that are used in a given country. The list of energy source categories in Tables 3–5 differs due to the differences (the absence of certain categories) in their use between countries. Source: own calculations based on data from [7].

Slovak households in 2019 were dominated by two energy carriers, natural gas and RES, which accounted for 42.4% and 23.3% of the total consumption of energy carriers, respectively. Since 2004, natural gas consumption in the household sector has decreased by 13.4 p.p., instead of renewables and biofuels, which have increased by 22.1 p.p. In particular, between 2018 and 2019 consumption of renewables and biofuels in Slovak households increased markedly, due to the increase of primary solid biofuels in total consumption. The share of hard coal in the total consumption of energy carriers in Slovak households was 1.1% in 2019, while 16 years earlier it was 3.9% [7].

The directional coefficient of the trend demonstrates that the changes in consumption of renewable energy sources in Slovak households in 2004–2019 were not statistically significant. However, the analysis of particular types of renewable energy sources shows that the significant increase was recorded only in relation to solar thermal energy (by 0.52 ktoe on average per year) as well as other bituminous coal (average annual by 0.95 ktoe) and brown coal briquettes (average annual by 0.09 ktoe) (Table 5).

Table 5. Trends in the use of energy sources in Slovak households in 2004–2019.

Specification	The Trend Slope Factor	p Value	R^2
Total	−17.38	0.1398	0.0723
Solid fossil fuels	−3.205	0.0011	0.5107
Other bituminous coal	0.95	0.0000	0.7285
Patent fuel	0.0197	0.1148	0.1086
Lignite	−4.073	0.0001	0.6590
Coke oven coke	−0.1950	0.0024	0.4564
Brown coal briquettes	0.089	0.0189	0.2870
Oil and petroleum products	−0.3894	0.1060	0.1169
Liquefied petroleum gases	−0.3894	0.1059	0.1169
Natural gas	−20.09	0.0014	0.4973
Renewables and biofuels	13.2	0.0909	0.1327
Solar thermal	0.52	0.0000	0.8439
Primary solid biofuels	12.0	0.1060	0.1168
Characoal	0.098	0.0929	0.1305
Ambient heat	0.62	0.1052	0.1765
Heat	−10.82	0.0002	0.5973

The same comment as in Table 3. Source: own calculations based on data from [7].

Regarding German households, three energy carriers dominate the overall consumption of energy carriers, i.e., oil and petroleum products, gas oil and diesel oil, renewables and biofuels, which accounted for 20.6%, 19.0% and 14.4%, respectively, in 2019. Since 2004, consumption of oil and petroleum products in the household sector has decreased by 6.7 p.p., gas oil and diesel oil by 6.6 p.p. and renewables and biofuels have increased by 6.7 p.p. The share of hard coal in the total consumption of energy carriers in German households in 2019 was—0.6% [7].

From the directional coefficient of the trend, it can be concluded that the consumption of renewable energy sources in German households increased statistically significantly between 2004 and 2019, on average by 198 ktoe. As for the individual types of renewable energy sources, significant increases were recorded for solar thermal technology (average annual increase of 34.7 ktoe), geothermal heat (1.81 ktoe), biogases (22.7 ktoe) and ambient heat (68 ktoe). In German households, apart from the increase in renewable energy sources, there was no significant increase in the consumption of other energy carriers (Table 6).

Table 6. Trends in the use of energy sources in German households in 2004–2019.

Specification	The Trend Slope Factor	p Value	R^2
Total	−543	0.0045	0.4104
Solid fossil fuels	−15.70	0.0533	0.0487
Anthracite	−2.407	0.0340	0.2313
Other bituminous coal	−5.99	0.0272	0.2530
Patent fuel	4.04	0.2815	0.0167
Coke oven coke	−2.590	0.0002	0.5989
Brown coal briquettes	−8.76	0.0343	0.2310
Oil and petroleum products	−432.7	0.0002	0.6243
Liquefied petroleum gases	−5.86	0.3362	0.0661
Motor gasoline (excluding biofuel portion)	−1.085	0.0000	0.7845
Other kerosene	−0.007	0.9442	0.0004
Gas oil and diesel oil (excluding biofuel portion)	−425.7	0.0001	0.6366
Natural gas	−115.2	0.11727	0.0663
Renewables and biofuels	198	0.0004	0.5791
Solar thermal	34.7	0.0000	0.9714
Geothermal	1.81	0.0000	0.8675
Primary solid biofuels	67	0.1296	0.0960
Characoal	3.33	0.0011	0.5137
Biogases	22.7	0.0000	0.9422
Blended biodiesels	0.059	0.1057	0.1176
Other liquid biofuels	0.0021	0.9771	0.0000
Ambient heat (heat pumps)	68	0.0000	0.9926
Heat	−79.6	0.0044	0.4122

The same comment as in Table 3. Source: own calculations based on data from [7].

5. Discussion of the Findings

Investment in renewable energy technologies is essential to cut greenhouse gas emissions [86]. However, global CO_2 emissions are increasing, between 2004 and 2019 the increase was 28.6% [87]. In 2020, the COVID-19 pandemic reduced energy demand and reduced global CO_2 emissions by 8% compared to the previous year. This reduction resulted in a return to the CO_2 emissions of a decade ago [88]. It is conjectured [88] that, as after previous crises, the rebound in emissions may nevertheless be greater than the decline, unless the wave of investment to restart the economy will be allocated more widely to cleaner energy infrastructure. Nevertheless, there is an urgent need for policy action to curb the upward trend in CO_2 emissions. On the positive side, there has been an increase in the use of energy from renewable sources—the only one at the time of the COVID-19 pandemic. It should also be noted that, despite the global increase in CO_2 emissions, Europe has seen a fall in these emissions, thanks precisely to investment in solutions that promote renewable energy sources.

The transition to low-emission energy sources is occurring at different speeds in individual Member States of the Community. This is due to the divergent energy security interests of these countries and leads to dissonance in the energy union [89]. Achieving the objectives of the new EU energy strategy setting the goal of reaching 32% of the energy balance from RES in 2030 [90] requires reducing dependence on fossil fuels. The unbalanced perception of this issue and the different security priorities among the EU Member States result in a new west-east division of the Community, thus perpetuating the division that has existed since the former Union of Soviet Socialist Republics (USSR) satellite countries joined the EU organization [91]. The own study shows that with regard to the household sector, a higher share of RES energy—in relative terms—in final energy consumption occurs in the so-called "poorer" part of Europe, while in countries with higher living standards [92] (Ireland, Luxembourg, UK, the Netherlands, Belgium, Sweden, Germany) RES use remains at a lower level. This may be due to the fact that improvements in the quality of life manifested by an increase in the area and standard of furnishings of dwellings and the possibility of obtaining an optimal indoor microclimate [93,94] result in significant energy consumption, which consequently diminishes the relative increase of RES in final energy consumption. In EU countries, the share of expenditure on energy consumed by households has an increasing trend [95]. Income growth is a strong determinant of increase in household energy expenditure [37,96,97]. Income is also a key determinant of investment spending on RES energy extraction. Thus, when considered as a whole, the reason for the smaller improvement in overall RES use is the higher living standards and changing lifestyles (more household appliances and larger dwellings) of today's consumers. Even though the share of RES energy is increasing, the overall increase in consumer energy use diminishes these achievements.

It is worth mentioning that the leaders in absolute terms in the use of RES by sector households are the French, Germans and Italians. According to own research, these countries also saw the highest growth in RES use by sector households between 2004 and 2019. It is noted that Western European countries, but also Scandinavian countries, have for many years been taking and supporting measures to increase the share of RES in total energy production. This includes tax incentives and educational measures as well as public support programs in subsidies for investments in renewable energy [98–100]. At the same time, the energy transition is more manageable in these countries because these economies are not based on fossil fuels (coal) [101]. The leadership role in the use of RES by Western European and Nordic countries is emphasized in many publications [60]. It is worth mentioning that Germany is the largest energy producer in the EU, i.e., 19.7% in total energy production in the whole European Union. Germany is followed by France, with an energy production of 17.6% and the United Kingdom with 10.6%. Italy 8.8%, Spain 8.7%, Poland 5.0% and Sweden 4.8% are also responsible for $\frac{3}{4}$ of total energy production in the EU [102].

The self-analysis grouped the EU-28 countries into 6 groups according to the size and structure of household RES used. This classification indicates that the countries with the highest use of RES in households are the countries of Central, Eastern and Southern Europe. In these countries the dominant RES category is primary solid biofuels. It can be expected that in these countries, due to lower living standards [92], citizens cannot afford to invest in more expensive RES technologies. Deciding to use primary solid biofuels is a cheaper RES alternative than, for example, heat pump installations. Primary solid biofuels include, among others, wood pellets. In Poland, for example, the introduction of the "Clean Air" program (EU support) has widely promoted the replacement of old boilers using non-renewable solid fuels with boilers using biomass. Of all heat source applications, 1/5 requested subsidies for biomass boilers [103]. Some countries have even committed to increasing the use of wood pellets. Croatia, Slovenia and Slovakia by signing the Kyoto Protocol (The Kyoto Protocol) have committed to reduce GHG emissions precisely by promoting the use of wood biomass, primarily wood pellets [104].

In countries in the richer part of Europe, the use of primary solid biofuels in the RES use structure is usually at a lower level. In Western and Northern Europe (e.g., Ireland, The Netherlands, Finland, France) a relatively high use of heat pumps in households is observed. Heat pump technology is regarded as one of the environmentally friendly solutions for increasing energy efficiency and reducing harmful gas emissions into the atmosphere. As indicated, the use of heat pumps is economically beneficial in the Baltic region and the market share of these systems is increasing [105].

Solar thermal technology is also popular in richer European countries such as Ireland and Luxembourg. It is submitted that the interest in solar thermal energy may continue to grow due to the COVID-19 epidemic, which has changed the balance of energy consumption across countries. In particular, household energy consumption increased as people were encouraged to stay indoors [106]. This fact may lead to an increase in solar PV investments in residential buildings [107].

For climatic reasons, it is obvious that solar thermal technologies are most popular in households in the south of Europe (Cyprus, Malta, Greece). Cyprus, which formed a single-element cluster, records the highest use of RES in households. Cyprus is the world champion in terms of solar energy applications [108]. Already in the first decade of the 21st century, it was estimated that about 90% of residences in this country had a solar water heater installed. This record in thermal applications was mainly attributed to favorable weather conditions, a pioneering solar energy industry and a strong coordinated effort by all the concerned [109]. In Malta, on the other hand, the use of solar thermal energy, although dominant among household sources used, is generally in a smaller proportion of households. Solar water heaters in the first decade of the 21st century were present in several percent of all dwellings. It is surprising as the Maltese islands have the highest insolation in Europe [110]. The wider uptake of RES in Maltese households occurred in the second decade of the 21st century. As indicated [111], driven by the need to meet mandatory European Union (EU) renewable energy targets and facing the constraints of a limited territory, Malta was one of the first countries to rely almost exclusively on households to meet its clean energy targets. In 2009, a subsidy scheme was launched to encourage households to install photovoltaic systems on their own properties to feed into the energy network [111].

The results of our own research indicated that the leaders in EU-households respecting the RES share in final energy consumption (in relative terms) are Croatia, Slovenia and Latvia, among others. Croatia has the technical and economic potential of renewable energy sources necessary to achieve 100% RES in energy consumption [33]. The solar energy potential in Croatia far exceeds both existing and future, energy needs. The sunniest parts of Croatia, receive about 40% more solar energy than Central Europe and 60% more than Northern Europe. In winter, the continental part of Croatia receives twice as much solar energy as Northern Europe, with the central and southern coastal parts receiving 3–5 times more than Northern Europe, or twice as much as Central Europe [33]. Croatia is also rich in biomass and waste, hence the high use of solid biofuels. It is worth noting that in the late 1970s this country was one of the few regions in the world to initiate a solar energy program in response to the 1973 oil crisis [34,112].

From our own research we found that in Polish households, although there is a significant statistical increase in RES, the share of solid fossil fuels still remains at a relatively high level. Between 2004 and 2019, the increase in the amount of renewable energy used was at the level of the EU-28 average. As indicated [113], there is potential for development of RES technologies in Poland. Solar energy, wind power and solid biomass processing have the greatest chances for development.

Yang and Zhao [114] analyzing the financial aspect of fossil resource use for shorter periods found that fossil energy generally has lower financial costs compared to renewable energy but based on the conditions of sustainable development and then long-term projections, renewable energy is the only way to achieve sustainable living in the world. The energy system must provide energy services that are socially acceptable, economically sustainable and environmentally friendly [114].

In comparison, in countries neighboring Poland, i.e., Germany and Slovakia, the process of transition of households from fossil energy sources to renewable energy sources has been more favorable. Currently, household sector uses only a small percentage of fossil fuels in these countries. In general, solid fuels in Poland constitute the largest part of the energy mix on the EU scale, which significantly complicates the achievement of environmental goals, especially with the emphasis on decarbonization [84,115]. The transformation of the energy market in Poland depends on the financial situation of all energy producers (mainly coal companies in Poland) and power generators, whose activity is exposed to high financial risks [116,117]. The evolution of the Polish energy system is mainly influenced by the necessity to integrate energy markets in the EU. Despite the difficult conditions of the energy system in Poland, structural changes towards meeting the adopted environmental requirements are taking place. Therefore, the energy policy of the Polish country focuses on the energy supply security. In addition, competitive costs, minimal environmental impact and increase in energy efficiency are taken into account [58].

In conclusion, it can be stated, following Bąk and co-authors [101], that there are many factors influencing the disproportions between EU countries in the use of RES. Each EU Member State should look for the reasons why these disproportions become in order to answer the question how to improve its position in terms of RES use and how to change its policy to be more effective [101]. However, it is indispensable to include the household sector in the transition process from non-renewable to renewable energy sources.

It is impossible to compare the results presented in this paper with the results of similar studies because they have not been conducted so far. To date, taxonomic studies have been carried out on renewable energy in a broad sense, but not only on household consumption of renewable energy. The country classifications obtained in the studies quoted in the second part of the article do not coincide with the classification presented in this paper. It is not surprising that diverse cluster analysis results have been found, since they refer to a variety of aspects in the renewable energy sector and take into account multiple variables for the study. For this reason, the results presented in the paper contributes further to the issue of renewable energy sector in EU countries.

6. Conclusions

The household sector is an important contributor to overall energy consumption in the economy and should therefore be actively involved in measures to improve environmental quality. For a common future, it is necessary to include renewable energy in the long-term planning process of the energy sector. The use of renewable energy has numerous benefits, but investment in new RES technologies often proves costly, so many countries rely on the cheapest solutions using, for example, primary solid biofuels. Nevertheless, every year there is an increase in the use of RES in EU households, with diversification of the sources of this energy

As in other aspects of life, there is a certain polarization among EU countries regarding the use of RES in the household sector. In central and eastern European countries there is a greater use of energy sources such as primary solid biofuels, while households in western European countries are more likely to install ambient heat, solar thermal systems or use biogases.

Many factors influence the disproportion between EU countries in the use of RES. Each EU Member State should look for optimal and efficient solutions to develop RES in the household sector in order to improve its position in terms of their use. The literature review shows that an important factor is an active State policy and extensive education of citizens on RES. Raising citizens' awareness of the opportunities and benefits of installing RES-based solutions with parallel taking care of the energy efficiency in residential building should bring tangible benefits in the long term.

The conducted analyzes gave answers to the research questions posed in the introduction to the study. Based on the analyses carried out, the following conclusions can be drawn:

- RES consumption in households is increasing, with the diversification of renewable energy sources. In some countries the growth of RES use by individual consumers is accelerating, while in other countries there is a kind of stabilization or even regression, which should prompt public authorities to become more involved in promoting the use of RES in households, which is the answer to 1 research question;

- The leaders in terms of absolute use of RES by the household sector are France, Germany and Italy. These countries also have the highest growth in household RES consumption. On the other hand, in relative terms, the CEE and SEE household sectors perform better than the households of the other EU members. The share of RES in final energy consumption in the CEE and SEE countries is the highest. It can be assumed that there is potential for development of RES technologies in the EU countries. Properly created mechanisms can further push forward investments from national budgets to develop renewable energy sectors, which is the answer to 1 and 3 research questions;
- EU-28 countries can be divided into six groups according to household RES use, with—in relative terms—poorer European countries (e.g., Croatia, Romania, Bulgaria) having higher household RES use than richer Western countries (e.g., Luxembourg, Belgium, the Netherlands). However, these poorer countries rely primarily on primary solid biofuels, while in the richer parts of Europe there is a wider use of ambient heat, solar thermal technology and even biogases. The use of primary solid biofuels is a cheaper solution in RES use and this enables the poorer countries to meet the energy policy targets set at EU level, which is the answer to 2 research question;
- In Poland the use of RES by the household sector is increasing year by year, in particular the use of solar thermal or geothermal technology, and ambient heat. However, the RES consumption is still at a relatively low level, which demonstrates the need to make household members aware of the advantages of investing in modern RES technologies for households. At present, Poland dependence on fossil fuels cannot be denied; on the other hand, the development of RES in Poland should be looked upon positively, which is the answer to 3 research question.

Some limitations in this research should be considered. Identifying the situation of RES use by the household sector at the country level gives only a general issues characterizing the discussed subject understanding of the topic under study is a considerable limitation for this study. The household sector is diversified in terms of demographic and socio-economic characteristics. Thus, a number of studies could be usefully developed basing on these results. Subsequent research should take into account the RES use by specific types of households, for example, by demographic, social, economic and culture characteristics.

The research could be extended with the analysis of specific categories of renewable energy sources and could also include other countries, outside Europe. This can be helpful in an assessment of the influence of various factors (e.g., social, economic, legislative, environmental, political, etc.) on the cluster structure. Similar future studies will also give a guidance to determine if there are any changes in the structures of separate clusters.

For energy policy makers and managers offering modern RES technologies, the information that the country remains low in RES consumption by household sector is a signal to deepen the work to raise public awareness about the advantages of introduction this modern and environmentally friendly technology into households.

Funding: This research was funded by Warsaw University of Technology" and "The APC was funded by Warsaw University of Technology.

Institutional Review Board Statement: Not applicable.

Informed Consent Statement: Not applicable.

Data Availability Statement: Data available on request due to restrictions. The data presented in this study are available on request from the corresponding author.

Conflicts of Interest: The authors declare no conflict of interest.

References

1. Sheehan, B. *The Economics of Abundance: Affluent Consumption and the Global Economy*; Edward Elgar Publishing: Cheltenham, UK, 2010.
2. Nützenadel, A.; Trentmann, F. (Eds.) *Food and Globalization: Consumption, Markets and Politics in the Modern World*; Berg: Oxford, UK; New York, NY, USA, 2008.
3. Petrović-Ranđelović, M.; Kocić, N.; Stojanović-Ranđelović, B. The importance of renewable energy sources for sustainable development. *Econ. Sustain. Dev.* **2020**, *4*, 15–24. [CrossRef]
4. Abbasi, K.; Jiao, Z.; Shahbaz, M.; Khan, A. Asymmetric impact of renewable and non-renewable energy on economic growth in Pakistan: New evidence from a nonlinear analysis. *Energy Explor. Exploit.* **2020**, *38*, 1946–1967. [CrossRef]
5. Purwanto, S.K.; Sinaga, O.; Sidik, M.H.J. Ensuring Renewable Energy Consumption through Innovation, R&D and Energy Import in Indonesia: A Time series Analysis. *Int. J. Energy Econ. Policy* **2021**, *11*, 577.
6. Mengova, E. What Determines Investment in Renewable Energy? *J. Strateg. Innov. Sustain.* **2020**, *15*, 22–38.
7. Energy Balance Sheets. Eurostat. Available online: https://ec.europa.eu/eurostat/web/energy/data/database (accessed on 14 April 2021).
8. *Sustainable Development in the European Union: 2011 Monitoring Report of the EU Sustainable Development Strategy*, 2011th ed.; Eurostat Statistical Yearbook; Eurostat: Luxembourg, 2011.
9. Afsharzade, N.; Papzan, A.; Ashjaee, M.; Delangizan, S.; Van Passel, S.; Azadi, H. Renewable energy development in rural areas of Iran. *Renew. Sustain. Energy Rev.* **2016**, *65*, 743–755. [CrossRef]
10. Rezaei, R.; Ghofranfarid, M. Rural households' renewable energy usage intention in Iran: Extending the unified theory of acceptance and use of technology. *Renew. Energy* **2018**, *122*, 382–391. [CrossRef]
11. Lee, J.; Shepley, M.M. Benefits of solar photovoltaic systems for low-income families in social housing of Korea: Renewable energy applications as solutions to energy poverty. *J. Build. Eng.* **2020**, *28*, 101016. [CrossRef]
12. Adom, P.K.; Amuakwa-Mensah, F.; Agradi, M.P.; Nsabimana, A. Energy poverty, development outcomes, and transition to green energy. *Renew. Energy* **2021**, *178*, 1337–1352. [CrossRef]
13. Gökgöz, F.; Güvercin, M.T. Energy security and renewable energy efficiency in EU. *Renew. Sustain. Energy Rev.* **2018**, *96*, 226–239. [CrossRef]
14. General Assembly. *Transforming Our World: The 2030 Agenda for Sustainable Development.*; General Assembly of the United Nations: New York, NY, USA, 2015.
15. Staníčková, M. Can the implementation of the Europe 2020 Strategy goals be efficient? The challenge for achieving social equality in the European Union. Equilibrium. *Q. J. Econ. Econ. Policy* **2017**, *12*, 383–398.
16. Tanil, G.; Jurek, P. Policies on renewable energy at the European and national level of governance: Assessing policy adaptation in the Czech Republic. *Energy Reports* **2020**, *6*, 548–553. [CrossRef]
17. European Comimission. Energy for the Future: Renewable Sources of Energy; White Paper for A Community Strategy and Action Plan COM(97)599 Final; Commission of the European Communites. 1997. Available online: https://europa.eu/documents/comm/white_papers/pdf/com97_599_en.pdf (accessed on 4 September 2021).
18. Renewed EU Sustainable Development Strategy. Brussels: Council of the European Union. 2006. Available online: www.eea.europa.eu/policy-documents/renewed-eu-strategy-for-sustainable-development (accessed on 10 July 2021).
19. Europe 2020 Strategy. Brussels: European Commission 2010. Available online: https://ec.europa.eu/eu2020/pdf/COMPLET%20EN%20BARROSO%20%20%20007%20-%20Europe%202020%20-%20EN%20version.pdf (accessed on 10 July 2021).
20. Directive 2009/28/Ec of the European Parliament and of the Council of 23 April 2009 on the Promotion of the Use of Energy from Renewable Sources and Amending and Subsequently Repealing Directives 2001/77/EC and 2003/30/EC. Available online: https://eur-lex.europa.eu/legal-content/EN/ALL/?uri=CELEX%3A32009L0028 (accessed on 10 July 2021).
21. Gouarderes, F.; Beltrame, F. *Energy Policy: General Principles*; European Parliament: Brussels, Belgium, 2019.
22. Directive 2018/2001. Directive of the European Parliament and of the Council of 11 December 2018 on the Promotion of the Use of Energy from Renewable Sources. Available online: https://eur-lex.europa.eu/legal-content/EN/TXT/?uri=CELEX%3A32018L2001 (accessed on 10 July 2021).
23. Montanarella, L.; Panagos, P. The relevance of sustainable soil management within the European Green Deal. *Land Use Policy* **2021**, *100*, 104950. [CrossRef]
24. Brodny, J.; Tutak, M. Analyzing similarities between the European Union countries in terms of the structure and volume of energy production from renewable energy sources. *Energies* **2020**, *13*, 913. [CrossRef]
25. Prek, M. Primary Energy Factor for Electricity Mix: The Case of Slovenia. In *Energy Policy*; IntechOpen: London, UK, 2019. Available online: https://www.intechopen.com/books/energy-policy/primary-energy-factor-for-electricity-mix-the-case-of-slovenia (accessed on 15 July 2021). [CrossRef]
26. Directive 2010/31/Eu of the European Parliament and of the Council of 19 May 2010 on the Energy Performance of Buildings. Available online: https://eur-lex.europa.eu/legal-content/en/TXT/?uri=celex%3A32010L0031 (accessed on 10 July 2021).
27. Directive 2012/27/Eu of the European Parliament and of the Council of 25 October 2012 on Energy Efficiency, Amending Directives 2009/125/EC and 2010/30/EU and Repealing Directives 2004/8/EC and 2006/32/EC. Available online: https://eur-lex.europa.eu/legal-content/EN/TXT/?uri=CELEX:32012L0027 (accessed on 10 July 2021).

28. Directive 2009/125/Ec of the European Parliament and of the Council of 21 October 2009 Establishing a Framework for the Setting of Ecodesign Requirements for Energy-Related Products (Recast). Available online: https://eur-lex.europa.eu/legal-content/EN/ALL/?uri=celex%3A32009L0125 (accessed on 10 July 2021).
29. Regulation (Eu) 2017/1369 of the European Parliament and of the Council of 4 July 2017 Setting a Framework for Energy Labelling and Repealing Directive 2010/30/EU. Available online: https://eur-lex.europa.eu/eli/reg/2017/1369/oj (accessed on 10 July 2021).
30. Share of Energy from Renewable Sources. Eurostat Statistics | Eurostat (europa.eu). Available online: https://ec.europa.eu/eurostat/statistics-explained/index.php?title=Renewable_energy_statistics#:~{}:text=1%20Share%20of%20renewable%20energy%20more%20than%20doubled,renewable%20energy%20used%20in%20transport%20activities%20in%202019 (accessed on 10 July 2021).
31. European Commission, National Renewable Energy Action Plans 2020 | Energy (europa.eu). Available online: https://ec.europa.eu/energy/topics/renewable-energy/directive-targets-and-rules/national-renewable-energy-action-plans-2020_en (accessed on 10 July 2021).
32. Slavnikova, A. *Renewable Energy Systems for Zero Energy Buildings in Finland*; HAMK HAME University of Applied Sciences: Hämeenlinna, Finland, 2019.
33. Luttenberger, L.R. The barriers to renewable energy use in Croatia. *Renew. Sustain. Energy Rev.* **2015**, *49*, 646–654. [CrossRef]
34. Briguglio, M.; Formosa, G. When households go solar: Determinants of uptake of a Photovoltaic Scheme and policy insights. *Energy Policy* **2017**, *108*, 154–162. [CrossRef]
35. Afgan, N.H. *Sustainable Development of Energy, Water and Environment Systems*; Afgan, B.D., Ed.; CRC Pier: London, UK, 2004.
36. Afgan, N.H. *Sustainability Concept for Energy, Water and Environment Systems*; UNESCO Chair Holder; Instituto Superior Tecnico: Lisbon, Portugal, 2004.
37. Piekut, M. Patterns of energy consumption in Polish one-person households. *Energies* **2020**, *13*, 5699. [CrossRef]
38. Santamouris, M.; Kolokotsa, D. On the impact of urban overheating and extreme climatic conditions on housing, energy, comfort and environmental quality of vulnerable population in Europe. *Energy Build.* **2015**, *98*, 125–133. [CrossRef]
39. Mendes, A.; Papoila, A.L.; Carreiro-Martins, P.; Aguiar, L.; Bonassi, S.; Caires, I.; Teixeira, J.P. The influence of thermal comfort on the quality of life of nursing home residents. *J. Toxicol. Environ. Health Part A* **2017**, *80*, 729–739. [CrossRef] [PubMed]
40. Schieweck, A.; Uhde, E.; Salthammer, T.; Salthammer, L.C.; Morawska, L.; Mazaheri, M.; Kumar, P. Smart homes and the control of indoor air quality. *Renew. Sustain. Energy Rev.* **2018**, *94*, 705–718. [CrossRef]
41. Maina, Y.; Kyari, B.; Jimme, M. Impact of household fuel expenditure on the environment: The quest for sustainable energy in Nigeria. *Cent. Asian J. Environ. Sci. Technol. Innov.* **2020**, *1*, 109–118.
42. Ropuszyńska-Surma, E.; Węglarz, M. Skłonność gospodarstw domowych do instalowania RES–wyniki badań na Dolnym Śląsku. *Studia I Pr. WNEiZ US* **2017**, *49*, 305–316. [CrossRef]
43. Jacksohn, A.; Grösche, P.; Rehdanz, K.; Schröder, C. Drivers of renewable technology adoption in the household sector. *Energy Econ.* **2019**, *81*, 216–226. [CrossRef]
44. Braito, M.; Flint, C.; Muhar, A.; Penker, M.; Vogel, S. Individual and collective socio-psychological patterns of photovoltaic investment under diverging policy regimes in Austria and Italy. *Energy Policy* **2017**, *109*, 141–153. [CrossRef]
45. Wasi, N.; Carson, R.T. The influence of rebate programmes on the demand for water heaters: The case of New South Wales. *Energy Econ.* **2013**, *40*, 645–656. [CrossRef]
46. Mills, B.F.; Schleich, J. Profits or preferences? Assessing the adoption of residential solar thermal technologies. *Energy Policy* **2009**, *37*, 4145–4154. [CrossRef]
47. Michelsen, C.C.; Madlener, R. Homeowners' preferences for adopting innovative residential heating systems: A discrete choice analysis for Germany. *Energy Econ.* **2012**, *34*, 1271–1283. [CrossRef]
48. Michelsen, C.C.; Madlener, R. Motivational factors influencing the homeowners' decisions between residential heatingsystems: An empirical analysis for Germany. *Energy Policy* **2013**, *57*, 221–233. [CrossRef]
49. Busic-Sontic, A.; Fuerst, F. *The Personality Profiles of Early Adopters of Energy-Efficient Technology*; SOEPpapers on Multidisciplinary Panel Data Research at DIW Berlin No. 924; SOEP: Berlin, Germany, 2017.
50. Ropuszyńska-Surma, E.; Węglarz, M. Społeczna akceptacja dla RES–perspektywa odbiorców (prosumentów). *Referat Zgłoszony na Konferencję Rynek Energii Elektrycznej REE* **2017**, *17*, 25–27. (In Polish)
51. Štreimikienė, D.; Baležentis, A. Assessment of willingness to pay for renewables in Lithuanian households. *Clean Technol. Environ. Policy* **2015**, *17*, 515–531. [CrossRef]
52. Streimikiene, D.; Balezentis, T.; Alisauskaite-Seskiene, I.; Stankuniene, G.; Simanaviciene, Z. A review of willingness to pay studies for climate change mitigation in the energy sector. *Energies* **2019**, *12*, 1481. [CrossRef]
53. Šilić, Đ. *Mogućnost Domaće Industrije u Proizvodnji Uređaja i Korištenju Obnovljivih Izvora Energije*; Brodarski Institut: Zagreb, Croatia, 2004.
54. Licul, M. Analysis of the renewable energy sources perspectives in the state of Croatia. In *Università Degli Studi Di Trieste Xxv Ciclo Del Dottorato Di Ricerca in Transborder Policies For Daily Life*; Università Degli Studi Di Trieste: Friuli Venezia, Italy, 2014.
55. Zheng, X.; Li, C.; Fang, X.; Zhang, N. Price sensitivity and consumers' support for renewable energy in China. *Energy* **2021**, *222*, 119862. [CrossRef]

56. Tryon, R.C. *Cluster Analysis; Correlation Profile and Orthometric (Factor) Analysis for the Isolation of Unities in Mind and Personality*; Edwards Brother, Inc., Lithoprinters and Publishers: Ann Arbor, MI, USA, 1939.
57. Caliński, T.; Harabasz, J. A dendrite method for cluster analysis. *Commun. Stat. Theory Methods* **1974**, *3*, 1–27. [CrossRef]
58. Sharma, S.; Batra, N. Comparative Study of Single Linkage, Complete Linkage, and Ward Method of Agglomerative Clustering. In Proceedings of the 2019 International Conference on Machine Learning, Big Data, Cloud and Parallel Computing (COMITCon), Faridabad, India, 14 February 2019; pp. 568–573.
59. Bluszcz, A.; Manowska, A. The Use of Hierarchical Agglomeration Methods in Assessing the Polish Energy Market. *Energies* **2021**, *14*, 3958. [CrossRef]
60. Majerova, I.; Nevima, J. The measurement of human development using the Ward method of cluster analysis. *J. Int. Stud.* **2017**, *10*, 239–257. [CrossRef]
61. Parobek, J.; Paluš, H.; Kalamárová, M.; Loučanová, E.; Šupín, M.; Križanová, A.; Štofková, K.R. Energy Utilization of Renewable Resources in the European Union—Cluster Analysis Approach. *Bioresources* **2015**, *11*, 984–995. [CrossRef]
62. Sobolewski, M.; Sokołowski, A. Grupowanie metodą k-średnich z warunkiem spójności. *Res. Pap. Wroc. Univ. Econ./Pr. Nauk. Uniw. Ekon. We Wroc.* **2017**, *468*, 215–221. (in Polish).
63. Na, S.; Xumin, L.; Yong, G. Research on k-means clustering algorithm: An improved k-means clustering algorithm. In Proceedings of the 2010 Third International Symposium on Intelligent Information Technology and Security Informatics, Jinggangshan, China, 2–4 April 2010; pp. 63–67.
64. Chudy-Laskowska, K.; Pisula, T.; Liana, M.; Vasa, L. Taxonomic Analysis of the Diversity in the Level of Wind Energy Development in European Union Countries. *Energies* **2020**, *13*, 4371. [CrossRef]
65. Senetra, A.; Szarek-Iwaniuk, P. The implementation of the Wroclaw taxonomic method for the identification and evaluation of problem areas in the Warmia and Mazury Region in Poland—A case study. *Socio-Econ. Plan. Sci.* **2019**, *67*, 43–57. [CrossRef]
66. Gawroński, K.; Król, K.; Gawrońska, G.; Leśniara, N. Spatial diversity of tourism attractiveness of the Nowy Sącz district, using the Wrocław taxonomic method. *Geomat. Landmanagement Landsc.* **2019**, *2*, 37–54. [CrossRef]
67. Simionescu, M.; Strielkowski, W.; Tvaronavičienė, M. Renewable energy in final energy consumption and income in the EU-28 countries. *Energies* **2020**, *13*, 2280. [CrossRef]
68. Ntanos, S.; Skordoulis, M.; Kyriakopoulos, G.; Arabatzis, G.; Chalikias, M.; Galatsidas, S.; Katsarou, A. Renewable energy and economic growth: Evidence from European countries. *Sustainability* **2018**, *10*, 2626. [CrossRef]
69. Bórawski, P.; Bełdycka-Bórawska, A.; Szymańska, E.J.; Jankowski, K.J.; Dubis, B.; Dunn, J.W. Development of renewable energy sources market and biofuels in The European Union. *J. Clean. Prod.* **2019**, *228*, 467–484. [CrossRef]
70. Pacesila, M.; Burcea, S.G.; Colesca, S.E. Analysis of renewable energies in European Union. *Renew. Sustain. Energy Rev.* **2016**, *56*, 156–170. [CrossRef]
71. Eurostat. *Statistical Books. Energy Data*, 2020th ed.; Publications Office of the European Union: Luxembourg, 2020.
72. Reiff, M.; Surmanová, K.; Balcerzak, A.P.; Pietrzak, M.B. Multiple criteria analysis of European Union agriculture. *J. Int. Stud.* **2016**, *9*, 62–74. [CrossRef]
73. Nowak, P. Regional variety in quality of life in Poland. *Oeconomia Copernic.* **2018**, *9*, 381–401. [CrossRef]
74. Markowska, M.; Sobolewski, M. Wrażliwość regionalnych rynków pracy Unii Europejskiej na kryzys ekonomiczny. Klasyfikacja metodą Warda z warunkiem spójności. *Acta Univ. Lodziensis. Folia Oeconomica* **2014**, *6*, 308. (In Polish)
75. Reddy, C.K. *Data Clustering: Algorithms and Applications*; Chapman and Hall/CRC: London, UK, 2018.
76. Aggarwal, C.C.; Reddy, C.K. *Data Clustering. Algorithms and Applications*; Chapman&Hall/CRC Data mining and Knowledge Discovery Series; Chapman&Hall/CRC: London, UK, 2014.
77. Rollnik-Sadowska, E.; Dąbrowska, E. Cluster analysis of effectiveness of labour market policy in the European Union. *Oeconomia Copernic.* **2018**, *9*, 143–158. [CrossRef]
78. Charu, C.A.; Chandan, K.R. *Data Clustering. Algorithms and Applications*; CRC Press, Taylory & Francis Group: Boca Raton, FL, USA; London, UK; New York, NY, USA, 2014.
79. Mardia, K.V.; Kent, J.T.; Bibby, J.M. Multivariate Analysis. In *Probability and Mathematical Statistics*; Academic Press: London, UK, 1979.
80. Li, Y.G. A clustering method based on K-means algorithm. In *Applied Mechanics and Materials*; Trans Tech Publications Ltd.: Bäch, Switzerland, 2013.
81. Zaręba, L. Porównanie dynamiki zmian wymogów kapitałowych z tytułu ryzyka operacyjnego w latach 2010–2014 na przykładzie trzech sektorów bankowych w Polsce. *Metod. Ilościowe W Bad. Ekon./Szkoła Główna Gospod. Wiej.* **2017**, *18*, 361–373. (In Polish) [CrossRef]
82. Brauers, H.; Oei, P.Y. The political economy of coal in Poland: Drivers and barriers for a shift away from fossil fuels. *Energy Policy* **2020**, *144*, 111621. [CrossRef]
83. Bórawski, P.; Yashchenko, T.; Sviderskyi, A.; Dunn, J.W. Development of renewable energy market in the EU with particular regard to solar energy. In Proceedings of the Conference Proceedings Determinants of Regional Development, Pila, Poland, 9 January 2019.
84. Yang, Z.; Zhao, Y. Energy consumption, carbon emissions, and economic growth in India: Evidence from directed acyclic graphs. *Econ. Model.* **2014**, *38*, 533–540. [CrossRef]

85. Piekut, M.; Piekut, K. Źródła Energii W Gospodarstwach Domowych Z Polski I Ukrainy Na Tle Innych Krajów Europejskich. In *Selected Issues of Socio-Economic Development in Poland and Ukraine*; Piekut, M., Smetyna, N., Eds.; WUT: Plock, Poland, 2020. (In Polish)
86. CS 2019. Environment 2019. Statistical Analyses. Central Statistics: Warsaw, Poland, 2019.
87. Ameli, N.; Brandt, N. *What Impedes Household Investment in Energy Efficiency and Renewable Energy?* OECD: Paris, France, 2015.
88. IEA. Global Energy&CO2 Status Report. The Latest Trends in Energy and Emissions in 2018. (windows.net). 2019. Available online: https://www.iea.org/reports/global-energy-co2-status-report-2019 (accessed on 17 July 2021).
89. IEA. *Global Energy Review 2020 The Impacts of the Covid-19 Crisis on Global Energy Demand and CO2 Emissions. (windows.net)*; OECD: Paris, France, 2020; Available online: https://www.ieaghg.org/files/2020-IP08_on_IEA_Covid-19_impact.pdf (accessed on 17 July 2021).
90. Zhou, S.; Matisoff, D.C.; Kingsley, A.; Brown, M. Understanding Renewable Energy Policy Adoption and Evolution in Europe: The Impact of Coercion, Normative Emulation, Competition and Learning. *Energy Res. Soc. Sci.* **2019**, *51*, 2. [CrossRef]
91. Council of the European Union. Renewable Energy: Council Confirms Deal Reached with the European Parliament 27 June 2018. Available online: https://www.consilium.europa.eu (accessed on 9 April 2021).
92. Kochanek, E. The energy transition in the Visegrad group countries. *Energies* **2021**, *14*, 2212. [CrossRef]
93. Piekut, M. The rich north-west, the poor middle-east: Consumption in EU households. *Comp. Econ. Res.* **2015**, *18*, 43–63. [CrossRef]
94. Xia, D.; Lou, S.; Huang, Y.; Zhao, Y.; Li, D.H.; Zhou, X. A study on occupant behaviour related to air-conditioning usage in residential buildings. *Energy Build.* **2019**, *203*, 109446. [CrossRef]
95. Yoshida, A.; Manomivibool, P.; Tasaki, T.; Unroj, P. Qualitative Study on Electricity Consumption of Urban and Rural Households in Chiang Rai, Thailand, with a Focus on Ownership and Use of Air Conditioners. *Sustainability* **2020**, *12*, 5796. [CrossRef]
96. Final Consumption Expenditure of Households by Consumption Purpose (COICOP 3 Digit) [NAMA_10_CO3_P3__custom_1149825] Eurostat Database. Available online: https://ec.europa.eu/eurostat/web/products-datasets/-/NAMA_CO3_P3 (accessed on 15 July 2021).
97. De Lauretis, S.; Ghersi, F.; Cayla, J.M. Energy consumption and activity patterns: An analysis extended to total time and energy use for French households. *Appl. Energy* **2017**, *206*, 634–648. [CrossRef]
98. Kotsila, D.; Polychronidou, P. Determinants of household electricity consumption in Greece: A statistical analysis. *J. Innov. Entrep.* **2021**, *10*, 1–20. [CrossRef]
99. Ragwitz, M.; Huber, C. *Feed-In Systems in Germany and Spain and A Comparison*; Report of the Fraunhofer Institut für Systemtechnik und Innovationsforschung (ISI); ISI: Karlsruhe, Germany, 2005.
100. Karakaya, E.; Sriwannawit, P. Barriers to the adoption of photovoltaic systems: The state of the art. *Renew. Sustain. Energy Rev.* **2015**, *49*, 60–66. [CrossRef]
101. Antonelli, M.; Desideri, U. The doping effect of Italian feed-in tariffs on the PV market. *Energy Policy* **2014**, *67*, 583–594. [CrossRef]
102. Bąk, I.; Spoz, A.; Zioło, M.; Dylewski, M. Dynamic Analysis of the Similarity of Objects in Research on the Use of Renewable Energy Resources in European Union Countries. *Energies* **2021**, *14*, 3952. [CrossRef]
103. Bluszcz, A.; Manowska, A. Differentiation of the level of sustainable development of energy markets in the European Union countries. *Energies* **2020**, *13*, 4882. [CrossRef]
104. Program Czyste Powietrze. Available online: https://czystepowietrze.gov.pl/ (accessed on 17 July 2021).
105. Barčić, A.P.; Grošelj, P.; Oblak, L.; Motik, D.; Kaputa, V.; Glavonjić, B.; Perić, I. Possibilities of Increasing Renewable Energy in Croatia, Slovenia and Slovakia-Wood Pellets. *Wood Ind./Drv. Ind.* **2020**, *71*, 395–402.
106. Valancius, R.; Singh, R.M.; Jurelionis, A.; Vaiciunas, J. A review of heat pump systems and applications in cold climates: Evidence from Lithuania. *Energies* **2019**, *12*, 4331. [CrossRef]
107. Graff, M.; Carley, S. COVID-19 assistance needs to target energy insecurity. *Nat. Energy* **2020**, *5*, 352–354. [CrossRef]
108. IRENA. The Post-COVID Recovery: An Agenda for Resilience, Development and Equality. 2020. Available online: https://www.irena.org/-/media/Files/IRENA/Agency/Publication/2020/Jun/IRENA_Post-COVID_Recovery_2020.pdf (accessed on 14 July 2020).
109. Maxoulis, C.N.; Kalogirou, S.A. Cyprus energy policy: The road to the 2006 world renewable energy congress trophy. *Renew. Energy* **2008**, *33*, 355–365. [CrossRef]
110. Panayiotou, G.P.; Kalogirou, S.A.; Florides, G.A.; Maxoulis, C.N.; Papadopoulos, A.M.; Neophytou, M.; Georgakis, G. The characteristics and the energy behaviour of the residential building stock of Cyprus in view of Directive 2002/91/EC. *Energy Build.* **2010**, *42*, 2083–2089. [CrossRef]
111. Grech, A.; Yousif, C. *Lifestyle Trends for Heating and Cooling in Maltese Households, Occasional Papers on Islands and Small States*; No: 5/2013; Islands and Small States Institute: Msida, Malta, 2013.
112. Heng, Y.; Lu, C.L.; Yu, L.; Gao, Z. The heterogeneous preferences for solar energy policies among US households. *Energy Policy* **2020**, *137*, 111187. [CrossRef]
113. Franković, B.; Trp, A.; Jedriško, C. Sunen-National Solar Energy Programme: A Review. 2012. Available online: https://www.bib.irb.hr/16062 (accessed on 12 July 2021).
114. Pietrzak, M.B.; Igliński, B.; Kujawski, W.; Iwański, P. Energy transition in Poland—Assessment of the renewable energy sector. *Energies* **2021**, *14*, 2046. [CrossRef]

115. Ahmad, A.; Zhao, Y.; Shahbaz, M.; Bano, S.; Zhang, Z.; Wang, S.; Liu, Y. Carbon emissions, energy consumption and economic growth: An aggregate and disaggregate analysis of the Indian economy. *Energy Policy* **2016**, *96*, 131–143. [CrossRef]
116. Ranosz, R. Analysis of the structure and cost of capital in mining enterprises. *Min. Res. Manag.* **2017**, *33*, 77–91. [CrossRef]
117. Jonek-Kowalska, I. Efficiency of Enterprise Risk Management (ERM) systems. Comparative analysis in the fuel sector and energy sector on the basis of Central-European companies listed on the Warsaw Stock Exchange. *Resour. Policy* **2019**, *62*, 405–415. [CrossRef]

energies

Article

Corporate Social Responsibility and Financial Performance among Energy Sector Companies

Magdalena Kludacz-Alessandri [1,*] and Małgorzata Cygańska [2]

[1] College of Economics and Social Sciences, Warsaw University of Technology, 09-400 Plock, Poland
[2] Faculty of Economics, University of Warmia and Mazury in Olsztyn, 10-719 Olsztyn, Poland; m.cyganska@uwm.edu.pl
* Correspondence: magdalena.kludacz@pw.edu.pl

Abstract: Corporate social responsibility (CSR) is one of the main drivers of corporate reputation. Many studies show that CSR can positively affect financial performance (FP) and vice versa. However, the relationship between FP and CSR depends on the type of industry in which the company operates, and there is little research regarding the energy sector in this area. The basis of empirical research in this study is slack resource theory which argues that financial performance is the cause of corporate social performance. This paper aims to analyze if financial performance affects corporate social responsibility adoption in energy sector companies. In order to achieve this goal, the study specifically examines the relationship between selected financial performance indicators and CSR adoption. Analyzing an international sample of 219 companies from thirty-two countries for 2020, we observed the statistically significant relations between financial performance and the implementing of the CSR strategy of the energy industry companies. The Return on Assets measure (ROA) and the Earnings Before Interest and Taxes measure (EBIT) were significantly higher among companies implementing the CSR strategy. The Enterprise Value to earnings before interest, taxes, depreciation, and amortization ratio (EV EBITDA) was lower among companies that adopted CSR. We did not confirm that the Return on Equity measure (ROE), Beta coefficient, and EBITDA per Share correlated with CSR adoption. Our research had implications for firms' investment policies in social initiatives and highlighted the relation between the financial performance and CSR initiatives of the energy sector companies.

Keywords: CSR strategy; financial performance; energy sector

Citation: Kludacz-Alessandri, M.; Cygańska, M. Corporate Social Responsibility and Financial Performance among Energy Sector Companies. *Energies* **2021**, *14*, 6068. https://doi.org/10.3390/en14196068

Academic Editors: Wen-Hsien Tsai, Bartlomiej Iglinski and Michal Bernard Pietrzak

Received: 28 July 2021
Accepted: 21 September 2021
Published: 23 September 2021

Publisher's Note: MDPI stays neutral with regard to jurisdictional claims in published maps and institutional affiliations.

Copyright: © 2021 by the authors. Licensee MDPI, Basel, Switzerland. This article is an open access article distributed under the terms and conditions of the Creative Commons Attribution (CC BY) license (https://creativecommons.org/licenses/by/4.0/).

1. Introduction

Corporate Social Responsibility (CSR) is a concept that is often defined in scientific literature. Most definitions of corporate social responsibility describe CSR as a concept whereby companies voluntarily incorporate social and environmental issues into their economic activities and have interactions with stakeholders [1]. At an early stage, it was thought that CSR could only be satisfied by fulfilling responsibilities to individuals and not to society as a whole [2]. However, later, other researchers started to define CSR as a more integral concept relating to the full range of business obligations to society, including legal, economic, ethical, and other optional categories of business activity [3]. For example, CSR is defined as activities that seem to serve a particular social good, going beyond the company's interests and what is required by law [4]. They can include supporting local businesses or charities, developing recycling programs, promoting minority employment [5], adopting advanced human resource management programs, and producing products which integrate social attributes [6]. Therefore, the basic concepts of corporate social responsibility are to reflect the company's total commitment to its internal stakeholders, such as employees, shareholders and external stakeholders, including suppliers, customers and the community. Thus, CSR combines economic, public, and social responsibilities. The content and goals of CSR may differ depending on the country of origin of the company.

The implementation of CSR may depend on the macroeconomic conditions in the country [7] and the existing differences in regional economic development [8], special economic zones [9], differences in productivity, regional innovation performance, the effectiveness of labor market policy [10], and productivity convergence in the regions from which the researched companies come from [11,12].

The result of CSR is corporate social performance (CSP), defined in the literature as a configuration of CSR principles, social-response processes, programs, policies, and observable results related to the company's social relations [13]. Undoubtedly, activities in the field of CSR bring about social benefits. Still, in the literature, there is a discussion of whether they improve the financial performance of the companies and if the CSR depends on the financial performance, and empirical research in this area has not yet reached a consensus.

A review of empirical studies on the relationship between CSR and financial performance shows that most studies treat social performance as an independent variable used to predict financial performance [14]. Approximately 50% of the studies found a positive relationship between CSR and FP [15]. The fact that CSR gives a competitive advantage translates into better financial performance is justified by the theory of instrumental stakeholders and the hypothesis of social impact. The instrumental stakeholder theory framework assumes that if the interests of multiple stakeholders of an organization are taken into account, they can improve the company's image and status. The focus on such aspects can positively affect the company's productivity, financial performance and value creation [16,17]. Inspired by this theory, the hypothesis of social impact suggests that good (bad) social outcomes generate good (bad) financial results [18,19]. It postulates that if a company satisfies its stakeholders, e.g., by implementing social projects, it will improve its image and reputation, and thus its financial results. On the other hand, if a company fails to achieve positive social impact, it will create image concerns among stakeholders, increasing costs and reducing profits [20].

Good social performance is also associated with managerial competencies and good management practices, leading to good financial results. For example, large companies benefit from favorable long-term stock performance [21], and companies with substantial shareholder rights tend to have a lower cost of equity capital than their competitors [22]. Good stakeholder relations and acceptance by the community have a positive effect on the financial outcomes of the companies in the long term. For instance, building a new plant in such a case is more accessible because of lower costs through government regulation; this can also obtain government tax breaks [23]. Moreover, CSR can stimulate human capital accumulation. A firm that adopted CSR on a high level is usually more attractive to employees and has a low turnover, which reduces the costs of recruiting and training employees [24]. The literature describes many other different ways in which CSR adoption can affect a company's FP, e.g., CSR can positively influence a company's resources and capabilities. It can positively impact reputation, which can lower operating costs in terms of reducing waste and risks, or can positively impact employee engagement and productivity [25]. The four benefits of a commitment to CSR are cost reduction, competitive advantages, reputation and legitimacy building, and the search for win–win outcomes. These benefits create a solid resource base and lead to excellent financial results [26].

Other researchers (around 5%) believe that CSR adoption has a negative impact on FP [15,27]. According to them, investments in any CSR activities increase costs due to inefficient resource allocation [28,29], create conflicts of interest between stakeholders [6], thus creating unfavorable competition conditions for firms in a competitive market, and ultimately harm the company's performance. Therefore, firms which have adopted CSR bear higher expenditures and have a lower competitive advantage than companies without CSR [30]. The spending on CSR activities may therefore not be covered by the generated profits. Consequently, CSR activities have a negative effect on the company's FP [31,32]. The negative relationship of CSR–FP is theorized within the compromise hypothesis [33]. This theory is that the company must meet its different needs within a limited resource base. Directing resources towards CSR can consume vital resources that could be used for more productive purposes. The negative relationship is also explained by the hypothesis of managerial opportunism, which assumes that the goal of managers and stakeholders may be contradictory, and in such a case, managers can only support their own interests [33].

Only a few studies regard the inverse relationship between CSR and FP, implying that FP precedes CSR and treats enterprises' CSR as a dependent variable [14]. Empirical evidence showing that better FP affects good CSR confirms the slack resource theory, according to which a financially successful company is better positioned to invest in CSR. This theory states that companies only engage in CSR when the company brings financial benefits. In this case, the firm has enough financial resources to invest in social projects. Financial success is, therefore, the main driver of CSR. The first empirical study to support this theory found that a company's social performance was positively related to the company's previous and future FP [20]. In most other studies, the relationship between social and financial performance was positive in many different contexts and sectors because the companies with better financial results spent more resources on social activities.

In turn, the negative relations between CSR and FP are justified by the hypothesis of managerial opportunism. According to this hypothesis, the more an enterprise is financially effective, the less it will be socially effective. This is explained by the fact that managers who do not achieve good financial results invest in social activities to justify their poor performance. However, when FP is high, they avoid investing in social activities to increase their private profit in the short term [33].

In conclusion, the cause-and-effect relationship between CSR and FP can veer in both directions. Most authors consider the possibility of a "virtuous circle" created by simultaneous and interactive interaction as increased CSR leads to better financial performance and vice versa [17,20].

According to some researchers, applying a corporate social responsibility strategy depends on the industry's sensitivity to the environment. Companies with production processes that harmfully affect the environment need more information than companies from other industries. Such companies include, among others, companies from the energy sector [34]. Energy companies are increasingly forced to take on greater social responsibilities, including labor rights, stakeholder engagement, environmental performance, human rights, and social impact [35]. This is mainly because the energy sector, responsible for the vast majority of emissions, requires optimization measures to reduce emissions. These activities can have different costs, application difficulties, environmental and social impacts [36]. By its nature, the energy sector plays a crucial role in sustainable development and is also a forerunner in CSR issues [37]. However, in this sector there are many varied challenges related to the management and implementation of CSR. These include high costs, a lack of information and awareness, insufficient human resources, poor cooperation with stakeholders, a lack of beneficiary involvement and the integration of CSR initiatives into more extensive development plans, an excessive focus on technical and management solutions [38,39].

Our study considered energy companies that are recently being forced to address a broader set of CSR and sustainability-related efforts and activities. Due to increasing demands from stakeholders related to CSR and sustainability issues, energy companies are under pressure to respond adequately to these needs and expectations, comply with national and international laws and regulations; and follow global initiatives and practices to improve sustainability performance. As a result of the environmental and social issues caused by business organizations operating in environmentally sensitive industries, the importance of CSR and sustainability-related reporting practices based on globally recognized reporting guidelines, has increased.

There is a wide range of research focusing on CSR in the context of energy. However, studies examining the relationship between CSR adoption and FP are scarce. Most of the research on the energy sector concerns the impact of CSR on financial performance. For example, other studies based on data from Thomson Reuters for 2011–2018 showed that higher CSR performance did not guarantee better financial results, as demonstrated by both the market and accounting results [40]. Another study found that the three individual dimensions of environmental responsibility (product innovation, resource reduction, and emission reduction) were positively related to the financial performance of companies. Still, the impact of the third dimension (emission reduction) was not significant [41]. A positive effect of CSR on corporate FP was also found in a study that examined the data of 210 energy firms worldwide. These results were measured as a market capitalization value [42]. A study that used a case study approach demonstrated the link between socially responsible corporate performance and profitability [43]. Although this study did not investigate the direction of causation, the results nevertheless indicated that CSR was positively associated with a better financial performance (profitability) and the relationship was statistically significant.

In conclusion, while many efforts have been made to understand the effect of CSR on FP, the accessible empirical evidence remains ambiguous. Research on the impact of CSR activities on FP can be divided into those who support positive correlation and those whoclaim the opposite. While some studies showed that the additional revenues generated by companies from CSR exceeded the expenditures incurred, the other studies argued that the costs incurred to conduct CSR activities exceeded the profits [44].

Although the existing research considers CSR in the context of the energy industry from many different perspectives, and despite its general importance in the energy sector, most studies concern the analysis of the impact of CSR on FP. The inverse link of FP–CSR in the energy sector is underrepresented in this context; we found only one item treating CSR as a dependent variable. The analysis of panel data for 14 companies from the energy sector for the years 1991 and 2009 carried out by Pätäri et al. [25] aimed to examine whether investments in CSR affected corporate financial performance and the reverse relationship. CSR was measured here using two separate constructs: strengths and concerns of CSR used in the ratings provided by MSCI ESG Research. The results did not support bidirectional causality between CSR and FP. According to these results, changes in two FP indicators, ROA and return on invested capital (ROIC), did not cause Granger causality in the total number of CSR strengths or concerns. According to Nelling and Webb [5], Corporate Social Performance (CSP) seemed to derive from the unobservable characteristics of companies rather than their financial performance.

Despite a great interest in CSR, and especially its relationship with financial results, the results of previous studies are inconclusive. It is difficult to determine whether CSR influences FP, or whether companies that achieved financial success are more proactive in sustainable development. This means that the field is full of ambiguities. Moreover, previous researchers focused mainly on the various dimensions of CSR rather than its adoption, and, to date, little research has been conducted in the energy sector. The impact of the FP on CSR adoption in the energy sector has not been investigated so far. Therefore, our goal is to reduce this research gap.

We would like to address the research gap in the literature by examining the relationships between energy sector companies' financial performances and CSR adoption. It is worth emphasizing that the degree of linkage between CSR and FP may vary depending on the measurement of specific financial ratios. Thus, this study aims to investigate whether and to what extent various financial indicators affect the implementation of CSR strategy in companies in the energy sector. This study proposes that six different indicators can measure financial performance: ROA, ROE, EBIT, Enterprise Value to EBITDA, EBITDA per Share, and Beta coefficient. These indicators are considered the potential factors that may impact CSR adoption. The study further examines whether each of the five FP indicators positively influences the implementation of the CSR strategy in energy companies. Therefore, our research hypotheses are as follows:

Hypothesis 1 (H1). *ROA has a positive impact on CSR adoption among energy sector companies.*

Hypothesis 2 (H2). *ROE has a positive impact on CSR adoption among energy sector companies.*

Hypothesis 3 (H3). *EBIT has a positive impact on CSR adoption among energy sector companies.*

Hypothesis 4 (H4). *The Enterprise Value to EBITDA ratio has a positive impact on CSR adoption among energy sector companies.*

Hypothesis 5 (H5). *EBITDA per Share has a positive impact on CSR adoption among energy sector companies.*

Hypothesis 6 (H6). *The Beta coefficient has a positive impact on CSR adoption among energy sector companies.*

This survey provides energy sector company managers with a clear insight into which kinds of financial performance are conducive to implementing a CSR strategy. The rest of the article is structured as follows. The second section describes the data collection and methodology, the third section presents the empirical results, and the final section discusses the findings and conclusions.

2. Materials and Methods

The methodology of the study involves a three-stage approach: (i) identification of the variables that may impact CSR adoption, (ii) investigation of the descriptive statistics and correlation between identified independent variables (iii) estimation of a logit model to examine the impact of identified independent variables on CSR adoption. The research focused on energy sector companies.

2.1. Data and Sample

This study uses non-probability sampling with a purposive sampling method which determines research samples using defined criteria. The criteria used by researchers for sampling are as follows: public companies from the energy sector listed in the Thomson Reuters Eikon (TR EIKON) database in 2020 who have a number of full-time employees higher than 150 with complete financial and non-financial data related to the research variables. We made this choice based on the assumptions that Corporate Social Responsibility (CSR) was associated mainly with big companies; they are often better resourced and are more able to invest in CSR. They attract more media attention, and they are particularly concerned with protecting and enhancing their reputations with the broader public and key stakeholders. The coverage of the TR EIKON database extends worldwide, reaching 99% of the market capitalization. The database contains more than 72,000 firm-level data of publicly traded companies from 150 countries. TR EIKON provides company fundamentals also published for the energy sector, including companies that produce, supply, and distribute energy. The analysis covered 219 companies from 32 countries (Table 1).

Table 1. The number of analyzed firms by country.

Country	*n*	%
United States of America	90	41.10
Canada	38	17.35
China	16	7.31
Bermuda	10	4.57
United Kingdom	8	3.65
Norway	7	3.20
Australia	6	2.74
Thailand	6	2.74
Indonesia	5	2.28
Brazil	4	1.83
France	3	1.37
Netherlands	3	1.37
Greece	3	1.37

Table 1. *Cont.*

Country	n	%
Italy	2	0.91
Papua New Guinea	1	0.46
Russia	1	0.46
Hungary	1	0.46
Turkey	1	0.46
Austria	1	0.46
Spain	1	0.46
Switzerland	1	0.46
Israel	1	0.46
Poland	1	0.46
Sweden	1	0.46
South Africa	1	0.46
Luxembourg	1	0.46
Belgium	1	0.46
Finland	1	0.46
Jersey	1	0.46
Japan	1	0.46
Portugal	1	0.46
Monaco	1	0.46
Total	219	100

As can be seen from Table 1, most of the firms covered by the analysis have their headquarters in the USA (41.10%), Canada (17.35%), and China (7.31%). The headquarters of almost 13% of the analyzed companies are in Europe. The majority of countries (18) are represented only by one company.

2.2. Key Variables

The dependent variable in the model is CSR adoption. CSR is a complex construct to measure due to its multidimensionality and invisibility. Various empirical and theoretical studies measure CSR in many distinct methods. In our research, CSR adoption is measured by a dummy variable, which takes the value of 1 if the company has implemented CSR and 0 if it has not. Measuring CSR by using a dichotomous variable was already measured in several previous studies [45–49].

FP indicators used in the previous studies can be roughly split into market and accounting measures. Market measures (e.g., Tobin Q) are calculated from the investors' points of view on a particular date. Accounting measures (e.g., EPS, ROA, ROE) are calculated on time results [50]. Unlike market measures, accounting measures can reflect managerial performance and the internal decision-making process [4]. Moreover, identifying the relationship between CSR and FP using accounting measures rather than market measures is more appropriate for detection [51]. Therefore, based on the previous research, we consider six independent variables (ROA, ROE, EBIT, Enterprise Value to EBITDA, EBITDA per Share, and Beta coefficient) as the possible factors that may influence the adoption of CSR strategy.

Return on Assets (ROA) is the first variable related to the company's FP. Profitability ratios show the company's ability to generate profit. One of the main profitability indicators is the ROA used by investors for investment decisions depending on potential returns [52]. Some studies show that ROA has a positive effect on goodwill. In this way, higher returns and rates of return force investors to invest, thereby increasing share prices and goodwill [53]. ROA is a measure of financial performance, commonly used in analyzing the effect of CSR on company finance [48,54–57].

Return on Equity (ROE) is an FP measure calculated by dividing the net income by equity. ROE demonstrates a company's ability to turn capital investments into profits. In other words, it measures the returns made on each monetary unit of equity. It is one of the all-time favorite and most widely used general measure of corporate financial performance [58], also confirmed by Monteiro [59]. ROE is popular with investors because it combines the income statement (net profit/loss) with the balance sheet (equity). The fact that ROE results from a structured financial ratio analysis, known as the Du Pont analysis, also contributes to its popularity with analysts, financial managers and shareholders [60]. ROE is already used in research on this topic [47,50].

Earnings Before Interest and Taxes (EBIT) is the company's net income before taxes and interest costs. EBIT is used to analyze the effectiveness of the company's core business without capital structure costs and tax expenses affecting profit. The approach of using EBIT in terms of CSR adoption is already presented in the literature [55,61,62].

EBITDA stands for earnings before interest, taxes, depreciation, and amortization, and is used to evaluate a company's operating performance. It can be seen as a representative of the cash flow of the business of the entire company. EBITDA is already analyzed in the literature as a financial performance measure regarding CSR adoption [63,64].

In terms of the enterprise value to earnings before interest, taxes, depreciation, and amortization (EV EBITDA), the enterprise value (EV) includes both debt and equity, and EBITDA is the profit available to investors. Since a change in capital structure has no systematic effect on company performance, this ratio is less susceptible to manipulation by changes in capital structure. Only when such a change lowers the cost of capital will it lead to a higher multiple [65].

The Beta coefficient (BETA) measures an investment's volatility and risk compared to the overall market. Beta is a statistical tool, which gives an idea of how a fund will move in relation to the market. In other words, it is a statistical measure that shows how sensitive a fund is to market moves. EV EBITDA and the Beta coefficient have not been previously used in the literature to analyze the FP–CSR relationship.

2.3. Research Model

In the second stage of our research, we estimated the correlation between the independent variables. Following that, we specified and estimated a logit model in order to examine the impact of identified independent variables on CSR adoption. Several authors already used the logit model in similar research studies [36,66,67].

The model is specified as follows:

$$\begin{aligned} logit\ (\text{Probability of CSR adoption}) = \beta 0 + \beta 1(\text{ROA}) + \beta 2(\text{ROE}) + \beta 3(\text{EBIT}) \\ + \beta 4(\text{EBITDA per share}) + \beta 5(\text{EV EBITDA}) + \beta 6(\text{BETA}) + e \end{aligned} \quad (1)$$

Data were analyzed based on descriptive statistics, such as means, standard deviations, medians, and interquartile ranges. Differences were calculated with the Mann–Whitney test. A p-value of less than 0.05 was considered statistically significant. We used STATISTICA, (TIBCO Software INC., Statsoft Polska, version 13.3, Palo Alto, CA, USA).

3. Results

3.1. Descriptive Statistics and Correlation Matrix

Table 2 provides a condensed view of various descriptive statistics for all independent variables. The average values of ROA and ROE were −8.317% and 27.39%, respectively. The Beta coefficient was 1.447. The average earnings before interest were EUR 467,626,496.818. Descriptive statistics are presented in Table 3 separately for companies that have implemented CSR and for other companies.

Table 2. Descriptive statistics of the sample firms.

Variables	Mean (SD)	Median (Q1; Q3)
ROA	−8.317 (16.223)	−3.266 (13.571; 1.408)
ROE	−27.39 (363.88)	2.940 (−5.714; 8.724)
Beta	1.447 (0.586)	1.342 (1.057; 1.814)
EBIT	467,626,496.818 (1,934,173,648.88)	46,310,761.980 (−40,537,996.858; 371,448,973.763)
EBITDA Per Share	2.956 (6.628)	1.270 (0.388; 3.105)
EV EBITDA	14.601 (25.344)	9.585 (6.572; 14.14)

SD—Standard Deviation; Q1—first quartile; Q3—third quartile.

Table 3. Comparison of variables of companies that confirmed and did not confirm CSR adoption.

Variables	CSR Adoption (n = 159)	CSR Non-Adoption (n = 60)	p-Value
	Mean (SD) Median (Q1; Q3)		
ROA	−6.500 (13.503) −2.720 [−11.16; 2.089]	−13.131 (21.254) −5.124 (−20.661; 0.866)	0.075
ROE	−35.942 (426.212) 3.770 [−4.661; 8.706]	−4.748 (39.204) 0.303 (−7.068; 8.855)	NS
Beta	1.436 (0.567) 1.336 [1.097; 1.791]	1.475 (0.640) 1.383 (0.990; 1.905)	NS
EBIT	636,045,952.00 (2,243,399,749.418) 125,499,113.600 [−58,544,200; 536,477,760]	21,314,940.955 (256,879,626.43) 3.754.670,965 (−14,541,069.2; 52,592,469.19)	0.06
EBITDA per Share	2.764 (5.106) 1.466 [0.409; 3.159]	3.464 (9.597) 0.898 (0.260; 2.815)	NS
EV EBITDA	11.163 (7.554) 9.179 [6.291; 13.604]	23.713 (45.871) 9.974 (7.473; 19.348)	NS

SD—Standard Deviation; Q1—first quartile; Q3—third quartile.

Based on the analysis of data, 72.6% of the companies adopted CSR. We found statistically significant differences between companies that confirmed and did not confirm CSR adoption in terms of ROA and EBIT. The average ROA in companies that confirmed CSR adoption was two times higher than among companies that did not confirm CSR adoption. The average values of ROA were −6.15% and −13.131% for companies that adopted and did not adopt CSR, respectively. In terms of the value of EBIT, we observed that it was almost 30 times higher among companies that adopted CSR compared to the other companies covered by the analysis. We did not observe a statistically significant difference between companies that adopted and did not adopt CSR in terms of ROE, Beta coefficient, EBITDA per Share, and EV EBITDA.

Table 4 presents correlations between the analyzed variables. We found a positive correlation between ROA and ROE, ROA and EBIT, as well as the Beta coefficient and EBIT. A positive correlation indicated that the variables increased or decreased together. The pairwise correlation coefficients were less than 0.4, indicating multicollinearity, and were not observed in the model.

Table 4. Spearman correlations coefficient between independent variables.

Variables	ROA	ROE	BETA	EBIT	EBITDA per Share	EV to EBITDA
ROA	1	0.0653 *	0.1387 *	0.0293 *	0.0058	0.0004
ROE	0.0653 *	1	0.0045	0.0001	0.0004	0.0001
BETA	0.1387 *	0.0045	1	0.0231 *	0.0024	0.0021
EBIT	0.0293 *	0.0001	0.0231 *	1	0.0001	0.0031
EBITDA Per Share	0.0058	0.0004	0.0024	0.0001	1	0.0106
EV EBITDA	0.0004	0.0001	0.0031	0.0031	0.0106	1

* Correlation is significant at the 0.05 level (2-tailed).

3.2. Binary Logit Model

To analyze the indicators affecting CSR adoption, we used the binary logit regression model. The dependent variable (CSR adoption) was a binary variable reaching the value of 1 (the company adopted CSR) or 0 (the company did not adopt CSR). The logit model, based on cumulative logistic probability functions, was computationally easier to use and could predict the probability of CSR adoption in the company. The results are presented in Table 5.

Table 5. The multivariate logistic regression model.

Variables	β-Coefficient	OR	IC	p-Value
ROA	0.023	1.023	(1.002–1.044)	<0.001
ROE	−0.002	0.998	(0.990–1.005)	0.556
BETA	0.099	1.104	(0.617–1.976)	0.738
EBIT	0.000	1.000	(1.000–1.000)	0.121
EBITDA Per Share	−0.027	0.974	(0.931–1.018)	0.238
EV EBITDA	−0.034	0.967	(0.937–0.997)	<0.001

The results confirm a statistically significant positive impact of ROA on CSR adoption. From Table 5, we found that, with a 95% confidence level, only ROA and EV EBITDA ratios had a significant effect on the CSR adoption with a *p*-value of less than the significance level alpha = 0.05, although these effects were minor. In the logistic model, if the odds ratio was greater than 1, the higher the value of the variable, and the higher the odds were of implementing CSR. Only ROA had a positive effect on CSR adoption. That is, increasing the ROA ratio level increased the probability of adopting CSR. In terms of the Enterprise Value to EBITDA ratio, a statistically significant negative effect was observed. We did not confirm the influence on CSR adoption in terms of the other analyzed variables. Thus, only Hypothesis 1 was supported.

The area under the ROC curve was found to be 0.685 (Figure 1). Since the area under the curve was more than 0.5, and the closer the curve followed the left-hand border and then the top border of the ROC space, the more acceptable the model.

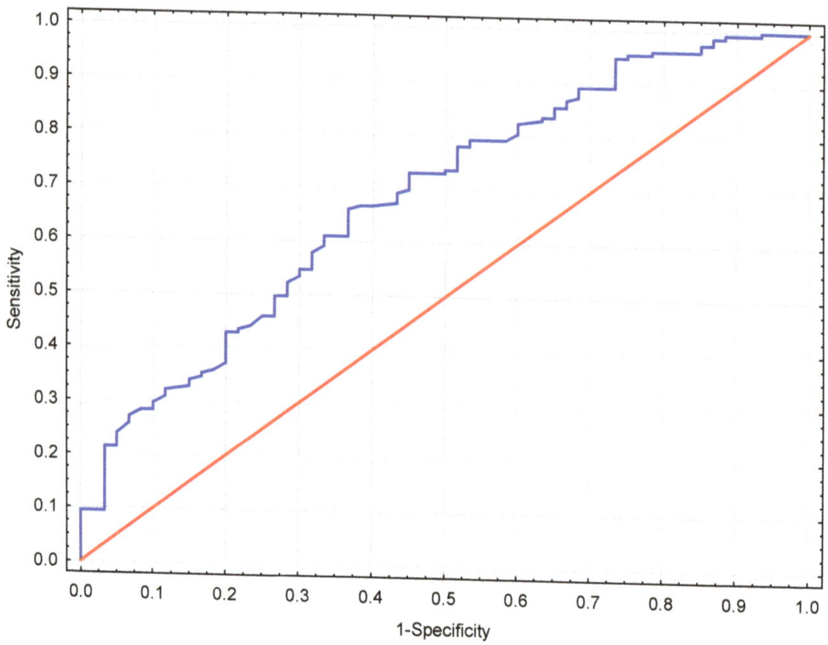

Figure 1. ROC curves for model results with AUC of 0.685.

4. Discussion

One aspect of understanding the relationship between CSR and financial performance was understanding the direction of causation. This meant to understand what factor acts as a predecessor and the consequence of such a relationship. Most studies treated financial performance as a dependent variable. In the context of this study, we checked whether better financial performance led to CSR adoption. Our study, however, went beyond the boundaries of linking general financial performance to CSR and proceeded to assess the impact of specific financial indicators on CSR adoption.

This study applies the binary logit regression model to examine the impact of selected financial indicators on CSR adoption in companies from the energy sector. The ratios include ROA, ROE, EBIT, Enterprise Value to EBITDA, EBITDA per Share, and the Beta coefficient. The analysis results show that the only indicator that increases the probability of CSR adoption is ROA. The increase in the Enterprise Value to EBITDA has a negative impact. We do not find any relationship between CSR adoption and ROE, EBIT, EBITDA per Share, and the Beta coefficient. Compared to other metrics, companies with high returns on assets show the highest likelihood of CSR adoption. Such companies are more willing to invest in social initiatives than others. A way to encourage companies with a high ROA to adopt CSR as their primary means of improving their public image and long-term performance may be through society's use of moral persuasion. This means to motivate companies to implement quality-of-life practices in the community in which they conduct business to contribute to that community's educational, social, and economic development [68].

In this context, our research confirms the Slack Resources Theory, which explains the positive impact of FP on CSR. This theory indicates that better FP results in slack resources for companies mean that they can invest them in social ventures, thus emphasizing that better FP would cause better social performance [69]. Other studies in the literature confirm that better financial results translate into CSR adoption and better CSR activities. For instance, research conducted based on data from large American corporations showed that this also positively influenced the following year's financial results [33]. Another study explored the relationship between CSR disclosure and financial performance, and vice versa, using different approaches, i.e., statistical through multiple regression modeling techniques. The study results showed that the financial results based on the company's profitability had a cause-and-effect relationship with the disclosure of CSR, and vice versa [70]. Recent research also showed a significant causal link between FP and CSR adoption, as spending on social activities depended on financial outcomes. Profitability motivated an investment in social activities and inspired investor confidence [71]. Recent studies also showed that FP affected the company's CSR in the short term and the long term [72].

The variable that was most often used to reflect financial performance in the FP–CSR relationship research was ROA; this could be found in at least 22 other studies [73]. Our analysis showed that ROA indicators had a positive and statistically significant impact on CSR adoption, thus supporting Hypothesis 1 (H1). In other words, ROA indicators were the only financial factor determining the ability of companies to engage in CSR activities. Similar results were reported by other researchers [74–76]. Dewi's [77] research showed not only a direct positive impact of ROA on CSR but also ROE on CSR. The positive correlation between the financial ratios of ROA and ROE and the CSR showed that companies with social and financial performance tended to have wide-ranging social disclosure [29]. In our research, the impact of ROE on CSR adoption turned out to be statistically insignificant, so the Hypothesis 2 was not confirmed. The estimated probabilities from the model showed that profitability had little effect on CSR adoption (low odds ratio—OR). This important finding contradicted some existing research that showed that profitability had a significant direct impact on investment in CSR.

Other hypotheses were also not confirmed in this study. The relations between EV EBITDA and CSR adoption among energy sector companies were negative. We believe that the negative impact of the Enterprise Value on EBITDA was because high-value companies tended to develop and remain competitive. Hence, they devotde most of their resources to maintaining that value rather than engaging them in social endeavors. Other relationships between financial indicators and CSR adoption were not statistically significant.

Additionally, previous research showed that CSR adoption appeared to be positively related to profitability ratios. However, the links between CSR and profitability were studied using simple statistical methods and linear regression [43]. A similar analysis, which also used a regression approach, was carried out on data from 30 publicly listed Nigerian companies [78]. Another study explored the additional effect of leverage on CSR disclosures using data from 41 listed firms [79]. As in other studies, it was found that profitability and company size positively affected CSR adoption. Importantly, it was also found that highly leveraged firms were less likely to engage in CSR. The regression model was also used to study the CSP-CFP relationship in the context of emerging markets [80]

The previous research which showed that CSR adoption appeared to be positively related to profitability ratios did not consider that many different financial indicators. The profitability measure used in another regression model used the financial data of 40 listed companies; in addition to the Return on ROA assets, was the Return on equity (ROE) [56]. Accounting indicators such as the Return on assets (ROA), return on capital (ROE) and return on sales (ROS) were also used as indicators of the financial results. However, these studies did not show a significant relationship between the examined variables. Other methods used in the analysis of FP–CSR included Dynamic Circulation Viewpoint and Multivariate Analysis of Variance (MANOVA). These studies showed a positive and mutual relationship between the variables [81].

According to Reverte [82], neither profitability nor other financial indicators explain the differences in CSR disclosure practices between companies. Thus, it is worth considering the moderating influence of other variables on these relationships. Control variables should be included in the study when there is reason to believe that they may play a role in analyzing the relationships between CSR and FP. The most influential variables explaining the differentiation of companies in CSR assessments are those related to public or social visibility. The effect of visibility has a significant, positive relationship between visibility and CSR assessment, which is confirmed in other studies [83]. Other authors [84] argue that to investigate CSR's impact on the company's financial performance, the moderating role of corporate governance should be examined. In turn, other researchers [85] investigate the role of CEO power (measured by the relative pay of the director) and find that CEO power positively moderates the relationship between CSR and financial performance. Influential CEOs have considerable freedom in determining expenditure on social and environmental activities of enterprises. So, for example, they can suspend social and environmental activities to demonstrate a better financial performance, or they can, in other cases, increase spending to gain a personal reputation for being socially responsible. There are also studies providing empirical evidence of the relationship between board attributes and CSR engagement, as well as CSR engagement and financial performance in the global energy sector. These results indicate that board diligence and CSR committees are strong drivers of CSR performance [40].

5. Conclusions and Implications

Our study makes some contributions to the literature regarding CSR. First of all, this article is one of the few attempts to use specific financial indicators to examine the decisions of companies from the energy sector regarding the implementation of CSR. While other researchers attempted to investigate the impact of overall financial performance on CSR, they were not able to discern different forms of response to financial performance feedback, because they did not analyze the effect of individual financial metrics separately. Secondly, while the energy sector has a significant environmental and social impact, no empirical cross-sectional study specific to the industry has been carried out so far from an international database of large companies. As far as we know, no previous studies on this topic have focused on a sample of international firms. Third, although the analyzed studies examined the links between companies' financial results and CSR, they are insufficiently investigated in the energy sector. Analyzing CSR adoption as a dependent variable is used to a limited extent and is not the subject of any research in the energy sector. By using this dummy variable, the results for CSR could be easily compared with other studies. Due to the variety of CSR measures, the possibility of comparing the results concerning CSR is limited. Therefore, this study adds new evidence to the existing literature by providing an empirical analysis of the relationship between FP and CSR adoption in the energy sector. The impact of the FP on CSR adoption in the energy sector has not been investigated so far. Therefore, our goal is to reduce this research gap.

Our findings also have some management implications. The results of our study may be helpful in the further understanding of the motivation of companies' decisions in the field of CSR implementation. CSR activities are becoming more and more important for the sustainable business of companies, ensuring the legitimacy and facilitating exchange relations with their stakeholders [1,25]. We show that companies are more likely to engage in CSR activities when they achieve better financial results, as measured by the ROA ratio. The relationship between other financial measures and CSR adoption is not so clear.

Our study also offers a methodological improvement through the use of the binary logit regression model, which allows the determining of the likelihood that a company will engage in CSR, taking into account its financial characteristics. This approach avoids measurement errors encountered in studies aiming to determine whether there is a positive relationship between CSR investments and the financial results [74].

CSR has become an important research area for researchers looking into its relationship with other variables, such as financial performance. This study shows that, compared to other financial ratios, the size of the ROA has the most significant impact on shaping the company's CSR policy. This evidence is based on the financial characteristics of energy sector companies with a number of full-time employees higher than 150. Other financial ratios examined include ROE, EBIT, Enterprise Value to EBITDA, EBITDA per

Share, and Beta coefficient. However, most of the studied variables turned out to be statistically insignificant.

The approach in this study differs from previous studies, as CSR is measured as a binary variable, assuming values of 1 for companies that have implemented CSR and 0 for other cases. CSR is then linked to the size of the financial ratios under study using the binary logit regression model in this categorical form. The analysis results show that CSR is a positive function only for ROA and a negative growth function of the Enterprise Value to EBITDA. Clarifying the relationship between FP and CSR adoption is critical to promoting the implementation of CSR in all business companies and communities in every country worldwide [50].

6. Limitation and Future Research

This study has several limitations, which should be considered when evaluating the results. The study sample is based on firms listed in the Thomson Reuters EIKON database whose shares are traded in stock markets. Consequently, the results may not be generalizable to other firms not listed in a stock market. As far as we know, no previous studies on this topic focus on a sample of international firms. Our research contrasts with past research conducted in individual countries. Since the studied sector comprises large capital-intensive businesses, frequently operating as natural monopolies on national markets, choosing a single jurisdiction for analysis would not yield sufficient empirical material for quantitative analysis.

On the other hand, it would be interesting to undertake future research, including a complete sample to consistently support our hypotheses. A comparative analysis between countries in different cultural and geographical areas could generate an interesting line of future research. We also believe that interesting conclusions could be drawn by introducing various control variables into the analysis. Therefore, in future research, it would be worth also focusing on the relationship between financial performance and CSR, considering the control variables. The possible channels through which FP could affect CSR, or vice versa, could be the location of the firms, corporate governance, corporate visibility, the role of board diligence and the CSR committee, the gender of the director, corporate governance mechanisms, and the location of the firms.

In our study, we used only accounting-based indicators. We chose the six accounting variables presented in this paper because they permitted us, on the one hand, and were more appropriate for detection rather than market measures, on the other hand. According to some researchers, market results may be an interesting factor in assessing and understanding CSR implementation, which could be checked in subsequent studies.

To analyze the indicators affecting CSR adoption, we used the binary logit regression model. The dependent variable (CSR adoption) was a binary variable reaching the value 1 (company adopted CSR) or 0 (the company did not adopt CSR). The logit model, based on the cumulative logistic probability functions, was computationally easier to use and could predict the probability of CSR adoption in the company. Another widely recognized alternative approach to the analysis of CSR adoption was the Environmental, Social, and Governance (ESG) score measure, used with a linear regression model. The study of the degree of applicability of CSR implementation was also a possible extension of future work.

Author Contributions: Conceptualization, M.C. and M.K.-A.; methodology, M.C. and M.K.-A.; validation, M.C. and M.K.-A.; formal analysis, M.C. and M.K.-A.; investigation, M.C. and M.K.-A.; resources, M.C. and M.K.-A.; data curation, M.C.; writing—original draft preparation, M.C. and M.K.-A.; writing—review and editing, M.C. and M.K.-A.; visualization, M.C. and M.K.-A.; supervision, M.C. and M.K.-A.; project administration, M.C. and M.K.-A.; funding acquisition, M.K.-A. All authors have read and agreed to the published version of the manuscript.

Funding: This research received no external funding.

Institutional Review Board Statement: Not applicable.

Informed Consent Statement: Not applicable.

Data Availability Statement: The data are available at the Thomson Reuters Eikon database and require a Thomson Reuters Eikon account login.

Conflicts of Interest: The authors declare no conflict of interest.

References

1. Commission of the European Communities. *Promoting a European Framework for Corporate Social Responsibility: Green Paper*; Office for Official Publications of the European Communities: Luxembourg, 2001.
2. Chamberlain, N.W. *The limits of Corporate Responsibility*; Institute of National Affairs: Port Moresby, Papua New Guinea, 1979.
3. Carroll, A.B. Corporate Social Responsibility. *Bus. Soc.* **1999**, *38*, 268–295. [CrossRef]
4. McWilliams, A.; Siegel, D. Corporate Social Responsibility: A Theory of the Firm Perspective. *Acad. Manag. Rev.* **2001**, *26*, 117. [CrossRef]
5. Nelling, E.; Webb, E. Corporate social responsibility and financial performance: The "virtuous circle" revisited. *Rev. Quant. Financ. Acc.* **2008**, *32*, 197–209. [CrossRef]
6. Barnett, M.L. Stakeholder influence capacity and the variability of financial returns to corporate social responsibility. *Acad. Manag. Rev.* **2007**, *32*, 794–816. [CrossRef]
7. Glavopoulos, E.; Bersimis, S.; Georgakellos, D.; Sfakianakis, M. Investigating the factors affecting companies' attitudes towards CSR and CER during the fiscal crisis in Greece. *J. Environ. Plan. Manag.* **2013**, *57*, 1612–1641. [CrossRef]
8. Roszko-Wójtowicz, E.; Grzelak, M.M.; Laskowska, I. The impact of research and development activity on the TFP level in manufacturing in Poland. *Equilibrium* **2019**, *14*, 711–737. [CrossRef]
9. Piersiala, L. The usage pattern of development method to assess the functioning of special economic zones: The case of Poland. *Equilibrium* **2019**, *14*, 167–181. [CrossRef]
10. Rollnik-Sadowska, E.; Dąbrowska, E. Cluster analysis of effectiveness of labour market policy in the European Union. *Oeconomia Copernic.* **2018**, *9*, 143–158. [CrossRef]
11. Kijek, A.; Matras-Bolibok, A. Technological convergence across European regions. *Equilibrium* **2020**, *15*, 295–313. [CrossRef]
12. Kijek, T.; Matras-Bolibok, A. The relationship between TFP and innovation performance: Evidence from EU regions. *Equilibrium* **2019**, *14*, 695–709. [CrossRef]
13. Wood, D.J. Corporate Social Performance Revisited. *Acad. Manag. Rev.* **1991**, *16*, 691–718. [CrossRef]
14. Scholtens, B. A note on the interaction between corporate social responsibility and financial performance. *Ecol. Econ.* **2008**, *68*, 46–55. [CrossRef]
15. Margolis, J.D.; Walsh, J.P. *People and Profits?: The Search for a Link between a Company's Social and Financial Performance*; Psychology Press: Hove East Sussex, UK, 2001; ISBN 1135642265.
16. Donaldson, T.; Preston, L.E. The Stakeholder Theory of the Corporation: Concepts, Evidence, and Implications. *Acad. Manag. Rev.* **1995**, *20*, 65. [CrossRef]
17. Hillman, A.J.; Keim, G.D. Shareholder value, stakeholder management, and social issues: What's the bottom line? *Strateg. Manag. J. Strat. Mgmt. J.* **2001**, *22*, 125–139. [CrossRef]
18. Allouche, J.; Laroche, P.; Allouche, J.; Laroche, P. A Meta-analytical investigation of the relationship between corporate social and financial performance. *Rev. Gest. Ressour. Hum.* **2005**, *57*, 18.
19. Wang, Q.; Dou, J.; Jia, S. A Meta-Analytic Review of Corporate Social Responsibility and Corporate Financial Performance. *Bus. Soc.* **2016**, *55*, 1083–1121. [CrossRef]
20. Waddock, S.A.; Graves, S.B.; Carroll, W.E. The corporate social performance-financial performance link. *Strateg. Manag. J.* **1997**, *18*, 303–319. [CrossRef]
21. Ferreira, E.J.; Sinha, A.; Varble, D. Long-run performance following quality management certification. *Rev. Quant. Financ. Acc.* **2007**, *30*, 93–109. [CrossRef]
22. Cheng, C.A.; Collins, D.; Huang, H.H. Shareholder rights, financial disclosure and the cost of equity capital. *Rev. Quant. Financ. Acc.* **2006**, *27*, 175–204. [CrossRef]
23. Parmar, B.L.; Freeman, R.E.; Harrison, J.S.; Wicks, A.C.; Purnell, L.; De Colle, S. Stakeholder Theory: The State of the Art. *Acad. Manag. Ann.* **2010**, *4*, 403–445. [CrossRef]
24. Albinger, H.S.; Freeman, S.J. Corporate Social Performance and Attractiveness as an Employer to Different Job Seeking Populations. *J. Bus. Ethics* **2000**, *28*, 243–253. [CrossRef]
25. Pätäri, S.; Arminen, H.; Tuppura, A.; Jantunen, A. Competitive and responsible? The relationship between corporate social and financial performance in the energy sector. *Renew. Sustain. Energy Rev.* **2014**, *37*, 142–154. [CrossRef]
26. Kurucz, E.C.; Colbert, B.A.; Wheeler, D. The Business Case for Corporate Social Responsibility. *Oxford Handb. Corp. Soc. Responsib.* **2009**, 83–112. [CrossRef]
27. Berens, G.; Van Riel, C.B.M.; Van Rekom, J. The CSR-Quality Trade-Off: When can Corporate Social Responsibility and Corporate Ability Compensate Each Other? *J. Bus. Ethic* **2007**, *74*, 233–252. [CrossRef]
28. Margolis, J.D.; Walsh, J.P. Misery Loves Companies: Rethinking Social Initiatives by Business. *Adm. Sci. Q.* **2003**, *48*, 268. [CrossRef]
29. López, M.V.; Garcia, A.; Rodriguez, L. Sustainable Development and Corporate Performance: A Study Based on the Dow Jones Sustainability Index. *J. Bus. Ethics* **2007**, *75*, 285–300. [CrossRef]

30. Shen, C.-H.; Chang, Y. Ambition Versus Conscience, Does Corporate Social Responsibility Pay off? The Application of Matching Methods. *J. Bus. Ethics* **2008**, *88*, 133–153. [CrossRef]
31. Cai, Y.; Jo, H.; Pan, C. Doing Well While Doing Bad? CSR in Controversial Industry Sectors. *J. Bus. Ethics* **2011**, *108*, 467–480. [CrossRef]
32. Groza, M.D.; Pronschinske, M.R.; Walker, M. Perceived Organizational Motives and Consumer Responses to Proactive and Reactive CSR. *J. Bus. Ethics* **2011**, *102*, 639–652. [CrossRef]
33. Preston, L.E.; O'Bannon, D.P. The Corporate Social-Financial Performance Relationship. *Bus. Soc.* **1997**, *36*, 419–429. [CrossRef]
34. Bowen, F.E. Environmental visibility: A trigger of green organizational response? *Bus. Strat. Environ.* **2000**, *9*, 92–107. [CrossRef]
35. Agudelo, M.A.L.; Johannsdottir, L.; Davidsdottir, B. Drivers that motivate energy companies to be responsible. A systematic literature review of Corporate Social Responsibility in the energy sector. *J. Clean. Prod.* **2019**, *247*, 119094. [CrossRef]
36. Georgopoulou, E.; Sarafidis, Y.; Mirasgedis, S.; Zaimi, S.; Lalas, D. A multiple criteria decision-aid approach in defining national priorities for greenhouse gases emissions reduction in the energy sector. *Eur. J. Oper. Res.* **2003**, *146*, 199–215. [CrossRef]
37. Omer, A.M. Green energies and the environment. *Renew. Sustain. Energy Rev.* **2008**, *12*, 1789–1821. [CrossRef]
38. Frynas, J.G. The false developmental promise of Corporate Social Responsibility: Evidence from multinational oil companies. *Int. Aff.* **2005**, *81*, 581–598. [CrossRef]
39. Streimikiene, D.; Simanaviciene, Z.; Kovaliov, R. Corporate social responsibility for implementation of sustainable energy development in Baltic States. *Renew. Sustain. Energy Rev.* **2009**, *13*, 813–824. [CrossRef]
40. Shahbaz, M.; Karaman, A.S.; Kilic, M.; Uyar, A. Board attributes, CSR engagement, and corporate performance: What is the nexus in the energy sector? *Energy Policy* **2020**, *143*, 111582. [CrossRef]
41. Lee, S.P. Environmental responsibility, CEO power and financial performance in the energy sector. *Rev. Manag. Sci.* **2021**, 1–20. [CrossRef]
42. Pätäri, S.; Jantunen, A.; Kyläheiko, K.; Sandström, J. Does Sustainable Development Foster Value Creation? Empirical Evidence from the Global Energy Industry. *Corp. Soc. Responsib. Environ. Manag.* **2011**, *19*, 317–326. [CrossRef]
43. Ekatah, I.; Samy, M.; Bampton, R.; Halabi, A. The Relationship Between Corporate Social Responsibility and Profitability: The Case of Royal Dutch Shell Plc. *Corp. Reput. Rev.* **2011**, *14*, 249–261. [CrossRef]
44. Cho, S.J.; Chung, C.Y.; Young, J. Study on the Relationship between CSR and Financial Performance. *Sustainability* **2019**, *11*, 343. [CrossRef]
45. Cardebat, J.-M.; Sirven, N. Responsabilité sociale et rendements boursiers: Une relation négative? *Manag. Avenir* **2009**, *29*, 363–378. [CrossRef]
46. Chetty, S.; Naidoo, R.; Seetharam, Y. The Impact of Corporate Social Responsibility on Firms' Financial Performance in South Africa. *Contemp. Econ.* **2015**, *9*, 193–214. [CrossRef]
47. El Yaagoubi, J. Impact of CSR on financial performance of Casablanca Stock Exchange companies: A longitudinal study-ProQuest. *Int. J. Innov. Appl. Stud.* **2020**, *29*, 1142–1152.
48. Sun, W.; Zhao, C.; Cho, C.H. Institutional transitions and the role of financial performance in CSR reporting. *Corp. Soc. Responsib. Environ. Manag.* **2018**, *26*, 367–376. [CrossRef]
49. Dhaliwal, D.S.; Li, O.Z.; Tsang, A.; Yang, Y.G. Voluntary Nonfinancial Disclosure and the Cost of Equity Capital: The Initiation of Corporate Social Responsibility Reporting. *Acc. Rev.* **2011**, *86*, 59–100. [CrossRef]
50. Lin, C.-S.; Chang, R.-Y.; Dang, V.T. An Integrated Model to Explain How Corporate Social Responsibility Affects Corporate Financial Performance. *Sustainability* **2015**, *7*, 8292–8311. [CrossRef]
51. Moore, G. Corporate Social and Financial Performance: An Investigation in the U.K. Supermarket Industry. *J. Bus. Ethics* **2001**, *34*, 299–315. [CrossRef]
52. Obradovich, J.; Gill, A. The Impact of Corporate Governance and Financial Leverage on The Impact of Corporate Governance and Financial Leverage on the Value of American Firms. *Int. Res. J. Financ. Econ.* **2013**, *91*, 1–14.
53. Biger, N.; Mathur, N.A. The effects of capital structure on profitability: Evidence from United States. *Int. J. Manag.* **2011**, *28*, 3.
54. Angelia, D.; Suryaningsih, R. The Effect of Environmental Performance and Corporate Social Responsibility Disclosure Towards Financial Performance (Case Study to Manufacture, Infrastructure, And Service Companies that Listed at Indonesia Stock Exchange). *Procedia-Soc. Behav. Sci.* **2015**, *211*, 348–355. [CrossRef]
55. Qamar, R.; Pet, M. Relationship between Corporate Social Responsibility (CSR) and Corporate Financial Performance (CFP): Literature review approach. *Elixir Fin. Mgmt.* **2012**, *46*, 8404–8409.
56. Uadiale, O.; Fagbemi, T. Corporate Social Responsibility and Financial Performance in Developing Economies: The Nigerian Experience. *J. Econ. Sustain. Dev.* **2012**, *3*, 44–55.
57. Wu, M.-W.; Shen, C.-H. Corporate social responsibility in the banking industry: Motives and financial performance. *J. Bank. Financ.* **2013**, *37*, 3529–3547. [CrossRef]
58. Rappaport, A. *Creating Shareholder Value: A Guide for Managers and Investors*; The Free Press: New York, NY, USA, 1998; ISBN 0-684-84456-7.
59. Monteiro, A. A quick guide to financial ratios: Education. *Pers. Financ.* **2006**, *307*, 8–10.
60. Ahsan, A.M.; Mainul Ahsan, A.F.M. Can Return on equity be used to predict portfolio performance? Can Roe be used to predict portfolio performance? *Manag. Financ. Mark.* **2012**, *7*, 132–148.

undertaken in the implementation of the sustainable development goals. In the document "Next steps for a sustainable European future European action for sustainability" [10], the European Commission (EC) has outlined goals to be achieved by 2030. Among the 17 goals, there are two that relate directly to the energy sector. The first one concerns ensuring access to energy sources that are, above all, affordable, reliable, and sustainable. The second of these goals concerns acting to mitigate climate change and its impacts. In another document, "Clean energy for all Europeans" [11], the EC has set a target of 50% of electricity in total EU energy consumption in 2050. Furthermore, 80% of the electricity is to be obtained from renewable sources or nuclear energy. This means that electricity will gain importance, and ultimately, it will be the key source of energy in the EU. The policy of moving towards low-carbon economies also means that the main burden of electricity production will be on renewable sources.

Eurostat's statistics clearly show that while electricity consumption has not increased significantly in recent years, the changes that have taken place in electricity production are significant. Firstly, the share of renewable sources in electricity production has increased significantly, from 16.3% in 2005 to 34.6% in 2019. Secondly, the diversification of renewable sources used has increased across the EU countries. Currently, wind energy has the largest share among renewable energy sources. In line with "EU Strategy to harness the potential of offshore renewable energy for a climate neutral future" [12], further intensive development of wind energy technology is planned in particular in marine areas. Solar energy is also gaining importance. From the perspective of the consumer, internal business processes, the development and financial aspects, solar and wind energy are considered the most competitive renewable sources in electricity production [13].

Therefore, the question arise what changes have occurred in the use of renewable sources in the production of electricity in the EU countries. In particular, the period of interest is 2005–2019. The beginning of this period was selected for two reasons. First, 2005 was the first full year in the EU after its largest enlargement. Second, when analyzing the statistics, since 2005, the greatest progress in the use of renewable sources in the EU can be observed. The end of the study period is related to the availability of data at the time of the analysis. Due to the fact that the composition of the EU underwent changes, in the presented analysis, it is assumed that the research sample includes countries from the composition from 2020 (EU-27) and the UK as the EU member until 2019. It should also be noted that the aggregate statistics for the EU area provided in the presented analysis concern the composition of the EU-27. These statistics are also used as a benchmark for national statistics. However, in more detailed analyses, considering individual countries, the UK is also added.

The aim of the conducted analysis is to assess changes in the use of renewable sources for electricity production in the EU countries in 2005–2019. This goal is carried out in two steps. In the first step, changes in the concentration of renewable sources are assessed in the production of electricity from renewable sources, using the Gini coefficient. In the second step, we apply k-means algorithm for clustering of EU countries (EU-27 + UK). The conducted study allows us to verify the following hypotheses:

1. Activities related to energy policy reduce the concentration of the use of renewable sources for electricity production.
2. There is a large diversity between EU countries in the use of renewable sources for the production of electricity, while the development of individual energy sources in specific countries is to a large extent supported by government bodies.

The literature presents numerous studies on the use of renewable sources, which are presented in international cross-sections. However, they mostly refer to several issues. The first is the analysis of the use of RES (or types of RES) in energy consumption or production (without dividing this energy into its types). The second is the analysis of share of total RES in different types of energy consumption or production. The third is the analysis of only one type of RES. In contrast, there is a lack of studies that present an analysis of the use of different types of renewable sources in the production of renewable electricity and examine their concentration. The approach presented in this paper is fulfilling the research gap. It should also be noted that we aimed to show changes in the composition of the energy portfolio composed only of RES used for electricity generation in the presented study. Therefore, this analysis refers to "renewable electricity".

2. Renewable Energy Sources in Literature and EU Directives

Renewable energy sources (RESs) are in line with the concept of the "Sustainable Development Strategy of the European Union" adopted in June 2001. The very notion of sustainable development was defined in the Brundtland Report (WCED, 1987) as development meeting the needs of the present generations without limiting the same possibilities for the future ones [14]. Such development is then to be applied at both the social and environmental levels. The verification of the 2001 strategy that took place in 2006 allowed to pursue a long-term concept of sustainable development. Article 3 of the Treaty on the Functioning of the European Union (TFEU) urges the EU to disseminate the principles of sustainable development and to guard climate change and a low-carbon economy, inter alia, by lobbying multilevel actions improving the quality of the environment [15]. It is being implemented on multiple levels, both in the long and short terms [16]. In the case of energy policy, this takes place on the basis of various directives or strategies of the European Energy Union.

Moreover, the TFEU, and in particular, its Article 194, can be indicated as a point of reference for formulating strategies related to the EU energy policy in general. The first point of Article 194 presents the objectives of the EU energy policy by calling all EU countries, among others, to promote energy efficiency, to save energy, to develop new and renewable forms of energy and to ensure security of energy supply in the EU. The first milestone indicated in the process of formalizing the EU's energy strategy is said to be the White Paper on renewable energy, adopted in 1997 [17]. The White Paper included goals that the production of electricity from renewable sources was to increase to 23.5% by 2010 (from 14.3% at that time). The next step was, the already mentioned, issuance of Directive 2001/77/EC of the European Parliament and of the Council of 27 September 2001 on supporting the production of electricity produced from renewable sources on the internal market [18]. This Directive indicated country-specific targets to be met by the year 2010 with regard to the use of RES in the production of electricity. For the entire EU, this indicator was 22%, and for individual countries it ranged from 5.7% for Luxembourg to 78.1% for Austria. The biggest increase was expected for Denmark—by 20.3%, from 8.7% in 1997 to 29% in 2010. On 10 January 2007, the European Commission issued a communication entitled "Renewable energy road map—Renewable energies in the 21st century: building a more sustainable future" [19]. In that document, the Commission indicated that the greatest progress with regard to the use of renewable sources has been made in the production of electricity. In addition, there were suggestions that in 2020, electricity production from renewable sources could increase up to 34% (from 15% at the time of the release). "Directive 2009/28/EC of the European Parliament and of the Council on the promotion of the use of energy from renewable sources" was issued in 2009, amending and subsequently repealing Directives 2001/77/EC and 2003/30/EC [20]. The Directive 2009/28/EC set a target that by 2020 the share of total energy consumption from renewable sources should amount to 20%. This goal was mentioned in the Europe 2020 strategy [21] as one of the five main priorities determining the development of the EU. Furthermore, the directive 2009/28/EC also contained guidelines related to the electricity production, in detail it stressed that energy produced from renewable sources should be prioritized and use of support schemes for electricity production from RES. In 2018, the Renewable Energy Directive was amended again [22] and showed that support schemes for renewable electricity have proved to be useful tools. In addition, a target was set to increase the share of renewable energy consumption in the total energy consumption up to 32% by 2030. In March 2019, the EC published the "Clean energy for all Europeans package" [11] presenting numerous strategic proposals in the field of energy. According to this document, electricity will gain strategic importance in the EU. Additionally, estimates are given that electricity will be responsible for more than half of the EU's energy needs by 2050, with RES and nuclear energy expected to account for 80% of the electricity generated in the EU. A discussion of the factors affecting electricity is given in Reference [23].

Due to the fact that one of the priorities of the EU's energy policy is to increase energy efficiency and increase the share of renewable energy, there is a strong synergy of renewable energy concept with some of the goals of sustainable development [24]. Research on various renewable energy sources in terms of sustainable development indicators (e.g., price of generated electricity, availability of renewable sources, gas emissions, land requirements, efficiency of energy conversion, water consumption, and social impacts) rank individual energy sources. Assuming that all factors are of equal importance, wind, hydro, solar and geothermal energy are in the first place. Wind energy has the lowest greenhouse gas emissions but requires a large land area and high investments [25].

The EU put great emphasis on various types of activities and promotion of renewable energy [26–28]. However, due to its high cost, generating energy from renewable sources on a large scale is not possible without support [29]. These support strategies differ from one Member State to another [30]. There are many ways to promote ecological solutions, as well as their various combinations. Some of the EU countries decide to promote one support system, others promote hybrid solutions [31]. However, as it turns out, one of the most beneficial actions is financial support [32,33]. Next to it, there are also tax incentives, feed-in tariffs and tenders [34]. Important aspects related to the promotion of renewable energy are social pressure, environmental impact, and the level of development of the country. The concern for the environment and the pressure of subjective norms have also an indirect influence [35].

The increased focus on sustainability in the energy field is a response to dwindling natural resources and high CO_2 emissions. Required technological changes supporting sustainable development strategies include replacing fossil fuels with renewable energy, saving energy, or improving the efficiency of its production [36]. The EU is taking a number of actions to reduce CO_2 emissions such as supporting the development of the renewable energy sector and supporting research on innovations within this sector. Restrictive measures are also taken, for examples introduction of the "European Union Emission Trading" [37,38]. RES are considered a solution to an environmental degradation [39], depletion of non-renewable resources, destruction of the ozone layer or increasing energy consumption [40]. Another reason for using renewable energy and striving to increase its share in energy consumption by the EU is the awareness of high dependence on energy imports [41] and the shortage of energy reserves [42]. The higher variety of energy sources ensures higher energy security that should take into account the security of energy supply and demand for it, as well as the existing energy shortages and its surplus [43]. The EU countries apply equal strategies in this field. In the case of coal-based countries, the coal is converted into a growing share of gas and a slow increase in the share of renewable energy [44]. Renewable energy can also be a factor in supporting economic growth. Environmentally friendly companies and institutions also receive a positive perception [45]. In the long run, there is a two-way relationship between economic growth and renewable energy consumption [46,47]. Renewable energy production also expands spatially to neighboring countries. This is due to the spread of knowledge and the similar potential of renewable energy [48].

It is also worth noting, in the context of considering electricity generation, that this aspect (getting electricity) is considered when evaluating regions in terms of attractiveness to investors [49]. In turn, given the strong focus on creating socially responsible businesses, sourcing electricity from renewable sources gains an additional dimension.

According to Eurostat, RES include hydro (RA100), wind (RA300), Solar (including RA410—Solar thermal and RA420—Solar photovoltaic), biofuels (R5110-5150_W6000RI—primary solid biofuels, R5220P—pure biodiesels, and R5290—other liquid biofuels), biogases (R5300), renewable municipal waste (W6210), other (RA200—geothermal and RA500—

tide, wave, ocean). In the literature, individual sources are mentioned in different level of detail. For example, Reference [50] mentions sun, wind, waves, tides, or biomass fuels. Meanwhile, the main sources are solar, wind, and biomass energy [51].

RES are used directly to heat or light homes, as well as to produce fuel and electricity [52]. The importance of the transport sector is increasing, the increase in renewable energy consumption reduces CO_2 emissions in this sector by around 12% [53]. This occurs through, inter alia, use of biofuels [54] that are combined with other technologies [55].

The EU promotes the direct use of renewable energy for both heating and cooling [56]. Research on transforming the heating sector into solutions using renewable energy is focused, for example, on smart grid or smart energy systems [57]. In urban environments, district heating and cooling systems (5GDHC) are proposed, among others concepts [58]. However, in the case of cooling, the RES Directive does not contain a definition of renewable cooling, and therefore it may be difficult to directly include cooling from renewable sources [59]. Nevertheless, the future lies in various types of integrated energy systems that will ensure high energy efficiency. In the case of cooling, these include, for example, district cooling systems (DSC) used in the construction sector for drying and cooling rooms. The most suitable RESs for such systems are biomass, solar, geothermal, surface water, solar, and waste heat energy [60]. The adaptive energy supply systems under development try to also consider the changeable availability of renewable energy. Finally, thanks to new technological solutions, renewable energy has a chance to be cheaper [61].

The development of new renewable energy technologies could be reflected in a decrease in electricity production costs [62]. This, in turn, could translate into lower electricity prices. Lower electricity prices can be equated with achieving the EU's affordable and clean energy goal (Sustainable Development Goal (SDG) 7). The prices of energy carriers have a wider dimension. They affect the general price level in the economy and thus have an impact on economic growth (see, e.g., References [63,64]).

The production of electricity by using renewable energy is confirmed by green certificates. They are used by energy companies, which are obliged to include renewable energy in the overall energy balance. They are therefore traded, which can help to meet the renewable energy target [65]. Green certificates also support producers of renewable energy. Apart from green certificates, feed-in tariffs are another form of support [66].

In connection with the new proposals for the EU on increasing the share of renewable energy in general energy consumption, considerations of 100% of renewable energy share arise in the literature. Multistage analysis, considering the impact in terms of energy, environment and economy indicates that such system is achievable by combining heating, electricity, cooling, and transport sectors [67,68]. Such considerations can also be found in relation to specific countries. A Danish study showed that the pursuit towards 100% of the share of renewable energy from local sources is possible. A decision on the participation of biomass and wind energy is said to be crucial [69]. According to [70] a total transition to renewable energy and closure of nuclear energy is also possible in Germany by 2050. An overview of other literature on different energy systems in terms of 100% renewable energy can be found in Reference [71]. The literature also points out the need for electricity storage in the case of transition to renewable energy and including it in intelligent energy systems [72]. The renewable energy defects include the lack of continuity of its production, which is often associated with the climate. There are different solutions in the field of optimization methods [73].

Due to the subsequent objectives placed by the EU, renewable energy will play more and more importance. It is anticipated that the share of renewable energy in gross final energy consumption in the EU in 2050 will amount to 55–75% [74]. Furthermore, the achievement of climate neutrality is associated with an increase in renewable energy by 2050 by over 80% [75]. Therefore, the national and local level implementation of these goals, as well as the similarity among EU countries in terms of achieving the targets of sustainable development, concentration of renewable sources or their shares are crucial. Recent research on a relationship between RES and sustainable EU development carried out on the basis of a hierarchical method of cluster (Ward's method) gave division of countries into five clusters. Countries that best deal with the use of renewable energy and make progress in sustainable development are Denmark, Finland, Latvia, and Sweden. Meanwhile, at the other end, there are Belgium, Cyprus, Lithuania, Luxembourg, and Malta [76].

Pacesila, Burcea and Colesca [77] examined the similarity in terms of the share of renewable energy in total consumption for EU countries, however the share of renewable energy considered was treated jointly for all sources. The research, carried out by using the k-means method for data from 2013, resulted in three clusters: the first cluster included countries with energy dependence of up to 30% (Denmark, Sweden, Estonia, Romania, and Czech Republic), the second one consisted of countries with energy dependence between 30% and 70% (Latvia, Austria, Slovenia, Finland, Germany, France, Bulgaria, Croatia, Greece, Slovakia, Hungary, Poland, Netherlands, and United Kingdom), and the last cluster (Lithuania, Portugal, Spain, Italy, Belgium, Cyprus, Ireland, Malta, and Luxembourg) with energy dependence higher than 70%. Additionally, a ranking was created in terms of the characteristics of renewable energy. The results showed that RES can help reduce energy dependency.

On the other hand, Reference [78] determined the concentration of consumption of RES in 28 EU countries in 2016. The concentration factor was 0.59. The high value of the coefficient was due to the fact that there were several countries that have high consumption of renewable energy, while all the rest have low consumption. Countries with high consumption of clean energy were Germany, Italy, and France.

The share of renewable energy in total energy consumption in 2004–2016 and the concentration of total energy consumption and renewable energy were examined in Reference [79]. The Gini coefficient was calculated for 2004 and 2016 and a high concentration of renewable energy consumption was found in several countries. For the energy in total, the concentration factor was 0.62 in 2004 and 2016. In the case of renewable energy, 0.58 (2004) and 0.59 (2016). The highest total energy consumption was in Germany, France, the UK, Italy, and Spain. The structure of renewable energy consumption was very similar; however, Sweden took the place of the UK. The concentration factor was also determined in Reference [80], but it only referred to the level of primary production, export, import, and total energy supply in the EU.

Due to the fact that the existing research focuses only on the share of renewable energy in total energy production, there is a need for extended research in terms of the use of renewable energy in electricity production. This paper will fill the research gap in this area.

3. Data and Methodology

The main analysis focused on the share of each renewable energy source in gross electricity production from renewable sources, which can be written as follows:

$$X_{ijt} = \frac{GEP\_RES_{ijt}}{GEP\_RES_{jt}} \cdot 100\% \quad (1)$$

where GEP_RES_{ijt} is the amount of electricity production from the i-th renewable source or biofuel in the j-th country in the period t, (GWh); and GEP_RES_{jt} is the total amount of electricity production from renewable sources and biofuels in the j-th country in the period t (GWh), where $GEP\_RES_{jt} = \sum_{i=1}^{n} GEP\_RES_{ijt}$.

We use publicly available Eurostat's data in the presented study [81]. In the analysis the EU countries (2020 composition) and United Kingdom are included. United Kingdom is counted since this state is the EU member until the end of 2019 and the analysis cover the period 2005–2019.

The Gini coefficient (G) is applied to the concentration analysis in Reference [80]:

$$G_{jt} = \frac{\sum_{i=1}^{n}(2i - n - 1)GEP\_RES_{ijt}}{n^2\overline{GEP\_RES}_{jt}} \quad (2)$$

where GEP_RES_{ijt} is the amount of electricity production (GWh) from i-th renewable source in j-th country in period t, and $n = 7$; and $\overline{GEP\_RES}_{jt} = \frac{1}{n}\sum_{i=1}^{n} GEP\_RES_{ijt}$.

The considered sources of renewable energy are as follows: i = 1 (hydro), 2 (wind), 3 (solar), 4 (biofuels), 5 (biogases), 6 (renewable municipal waste), and 7 (other: geothermal and tide, wave, and ocean).

The k-means is the research tool applied for data clustering. This algorithm is introduced by Reference [82], (see also the description of the algorithm presented in References [83,84]). A procedure scheme for the application of the k-means is presented by Reference [85], among others. The calculations are prepared by using STATISTICA 13 software (TIBCO Software Inc. Palo Alto, CA, USA). In the first stage, all variables are standardized. Then Euclidean distance is used as a distance measure. The clustering is conducted for a different number of clusters, $k = 2, \ldots, 12$. The number of clusters is selected by using the silhouette index (SI index [86]; see also References [87,88]. The highest value of SI index indicates the best division. In turn Reference [89] or [88] reports that acceptable divisions are characterized by the values of the SI index at least 0.5 (then the structure of the clustering is considered reasonable).

In the data clustering, seven variables constructed according the Formula (1) are considered. The list of variables is as follows:

X_{1jt}—the share of electricity production in hydro power plants in total electricity production from renewables and biofuels (GEP_RES_{jt}) in j-th country in period t;

X_{2jt}—the share of electricity production in wind power plants in total electricity production from renewables and biofuels (GEP_RES_{jt}) in j-th country in period t;

X_{3jt}—the share of electricity production from solar power (solar thermal and solar photovoltaic) in total electricity production from renewables and biofuels (GEP_RES_{jt}) in j-th country in period t;

X_{4jt}—the share of electricity production from biofuels (primary solid biofuels, pure biodiesels, and other liquid biofuels) in total electricity production from renewables and biofuels (GEP_RES_{jt}) in j-th country in period t;

X_{5jt}—the share of electricity production from biogases in total electricity production from renewables and biofuels (GEP_RES_{jt}) in j-th country in period t;

X_{6jt}—the share of electricity production from renewable municipal waste in total electricity production from renewables and biofuels (GEP_RES_{jt}) in j-th country in period t;

X_{7jt}—the share of electricity production from other sources in total electricity production from renewables and biofuels (GEP_RES_{jt}) in j-th country in period t (other sources are geothermal and tide wave, ocean).

4. Use of Renewable Energy Sources and Biofuels in the Electricity Production in the European Union

This section presents selected issues related to the use of renewable sources for electricity production in the European Union. In the first part, we present the share of electricity production from renewables and biofuels in the total electricity production. This part of the analysis covers the European Union (EU27, for the period 1990–2019) and the individual EU countries, including the UK (for the years 2005 and 2019). In the second part, we describe the types of renewable sources (RESs, according to the Eurostat's classification) used for the electricity production in the EU and characterize the changes that have occurred in the shares of the five most popular RES in renewable energy production from RES (*GEP_RES*). In the third part, we report an analysis of changes in the level of concentration (measured by the Gini coefficient) of individual RES in *GEP_RES* production. We conduct this part of the analysis for the EU area (EU27) for the period 1990–2019 and for the individual EU countries, including the UK, for the years 2005 and 2019).

4.1. Renewable Energy in the Electricity Production in the EU

The main determinant of electricity production is the demand for electricity created by consumers. According to Eurostat data [90], in 2019, final energy consumption in the EU accounted for 10,879,807.319 GWh and was 5.1% lower than in 2005. The share of electricity in total final energy consumption is 22.8%, which is higher than the corresponding rate from 2005 by 1.6 percentage point (pp). In general, the electricity final consumption increased by 2.1% in the analyzed period. Increasing the share of electricity in the EU's total energy consumption is included in the Clean energy for all Europeans package [11].

Figure 1 shows the gross electricity production (GEP), GEP from renewables and biofuels (*GEP_RES*) and share of gross electricity production from RES in total GEP in the EU (EU-27). As the presented data dates back to the year 1990, it can be noted that until around year 2005, gross electricity production in the EU had been steadily growing. During that period (1990–2005) GEP increased by 28.2%. As in the period of 2005–2019, one can observe a relatively constant level of GEP in the EU. In 2019, there was even a slight decrease in GEP compared to 2005—by 0.5%. However, analyzing the changes of the production of electricity from renewable sources (*GEP_RES*) shows that, in the period 1995–2005, there was an increase in its production by 49.2%, while, in the period between 2005 and 2019, there was an intensification and increase amounted by 110.8%. The total increase in the production of *GEP_RES* in the extended period (1990–2019) accounts for 214.4%. The vast development is also visible from the share of RES in GEP production (columns in Figure 1). In 1990, this share was 14.1%; in 2005, it slightly increased to 16.3%, but in 2019, it was already 34.6%. One of the reasoning behind such increase is the fact that the new member states joining the EU in its largest expansions in 2004 undertook many actions to adopt the guidelines related to the transition to low-emission economies (e.g., reduction of CO_2 and other harmful substances emissions and the use of renewable sources for energy production to a greater extent).

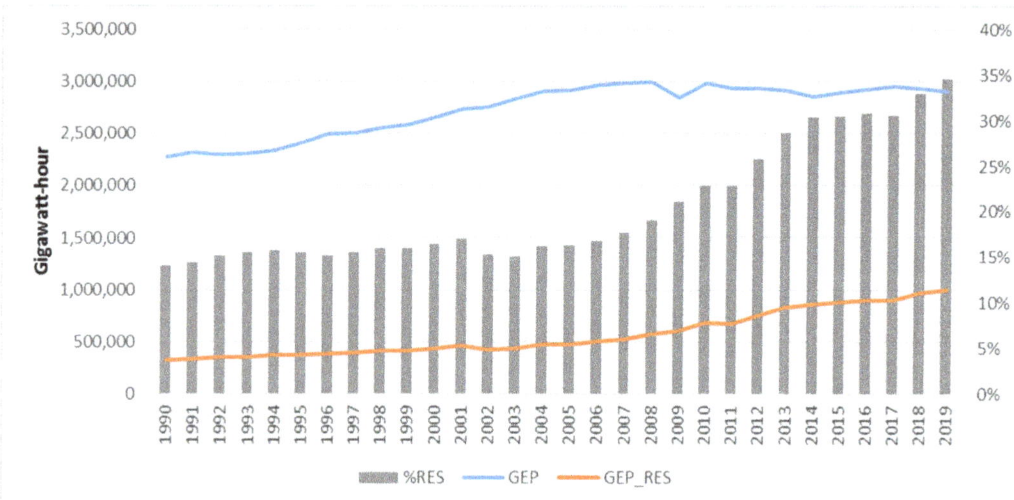

Figure 1. Gross electricity production (GEP), gross electricity production from renewables and biofuels (*GEP_RES*), and share of gross electricity production from RES in the total gross electricity production (%RES) in 1990–2019 in the European Union (EU-27). Source: Reference [81].

Table 1 shows the shares *GEP_RES* in the total GEP in individual EU countries (including UK) in 2005 and 2019. In 2005, these shares ranges from 0% in Malta to 69.6% in Latvia. In 17 out of 28 analyzed countries, in 2005, the share of RES in electricity production was below the EU level (16.3%). The RES shares in GEP ranges from 10% in Cyprus to 85.9% in Luxembourg in 2019. In three more countries, the share of RES in GEP is greater than 70%; Lithuania (81.9%), Denmark (78.2%), and Austria (77.8%). Numerous countries have recorded a significant increase in the share of RES in GEP. In 11 countries, it is higher by

20 pp, and in three by as much as 50 pp: in Denmark (by 51.1 pp), Lithuania (by 76.3 pp), and Luxembourg (by 61.8 pp). This means that the policy of increasing RES for energy production has brought visible effects, especially in the case of electricity production. As a result, there is a noticeable reduction in the differentiation between EU countries in terms of this feature.

Table 1. Share of electricity produced from renewable sources and biofuels in the total electricity production in individual EU countries in 2005 and 2019.

Country	Share of RES_GEP in GEP (%)		Change 2019–2005 (pp)	Country	Share of RES_GEP in GEP (%)		Change 2019–2005 (pp)
	2005	2019			2005	2019	
BE—Belgium	3.9	21.9	17.9	LT—Lithuania	5.7	81.9	76.3
BG—Bulgaria	10.7	18.0	7.3	LU—Luxembourg	24.1	85.9	61.8
CZ—Czech Republic	4.6	12.9	8.3	HU—Hungary	5.2	13.8	8.5
DK—Denmark	27.1	78.2	51.1	MT—Malta	0.0	10.5	10.5
DE—Germany	11.3	40.9	29.6	Netherlands	7.5	18.9	11.4
EE—Estonia	1.1	28.1	27.0	AT—Austria	64.7	77.8	13.1
IE—Ireland	8.5	38.9	30.3	PL—Poland	3.5	16.0	12.5
EL—Greece	11.7	33.2	21.5	PT—Portugal	18.6	54.2	35.7
ES—Spain	16.2	37.8	21.6	RO—Romania	34.0	42.0	8.0
FR—France	10.6	20.7	10.0	SI—Slovenia	23.6	32.6	8.9
HR—Croatia	54.4	66.2	11.8	SK—Slovakia	15.2	24.2	8.9
IT—Italy	18.3	40.1	21.9	FI—Finland	33.4	46.6	13.3
CY—Cyprus	0.0	10.0	10.0	SE—Sweden	51.3	58.7	7.4
LT—Latvia	69.6	49.6	−20.0	UK—United Kingdom	5.0	37.8	32.8
EU27	16.3	34.6	18.3				

Source: Own study based on data [81].

4.2. Types of Renewable Energy Sources Used for Electricity Production in the EU

Energy data are collected by Eurostat according to a strictly defined methodology [91]. Data are collected in areas that allow to assess, firstly, the origin of energy, secondly, the degree of dependence on energy imports, and thirdly, the types and costs of energy consumed. A key element of the EU's energy policy is increasing the use of renewable sources, in particular regarding electricity production. As previously mentioned, the Clean energy for all Europeans package [11] assumes that, by 2050, electricity will account for over 50% of the energy consumption in the EU, with a significant share of renewable energy sources. Reliable and comparable statistics are therefore essential to be able to evaluate activities and progress in this area. In Table 2, we present the types of renewable sources and biofuels used for electricity production listed by Eurostat. While in 2019 the amount of electricity produced from all sources decreased slightly compared to 2005 (by 0.5%), the production of electricity from renewable sources and biofuels increased by 100.8% and exceeded 1 M GWh. This stands for an increase in the share of RES in the total electricity production by 18.3 pp (from 16.3% in 2005 to 34.6% in 2019).

Table 2 also shows the shares of individual sources used in gross electricity production from renewables and biofuels (X_i^{EU}) in 2005 and 2019 (columns three and five) in the EU-27 area. Figure 2, additionally, presents the changes in the shares of selected sources in an extended period of 1990–2019. In 2005, hydropower constituted the largest share of the RES_GEP (71.4%). Hydropower [92] noted a great decrease compared to 1990 (by 22.9 pp), when hydropower was responsible for over 94% of electricity generated from renewable sources (see Figure 2). Continuously, this share significantly decreased by 2019—by 37 pp. (to the level of 34.3%). Even if the actual amount of the electricity produced from hydropower has slightly increased since 2005 (by 1.4%), the highly decreased trend is due to other emerging technologies enabling the use of other renewable sources. In the study period (2005–2019), energy produced from wind; kinetic energy of wind exploited for electricity generation in wind turbines [93] gained a lot of importance, and increased by 439.1%. Its share in the production of electricity from renewable sources in 2019 was 36.5% and is higher by 22.2 pp from that in 2005. In 2019, both wind and hydro were responsible for 70.9% of electricity produced from renewable sources. Thus, these two sources are currently the main RESs used for electricity production. Another source that has gained in importance in recent years is solar energy. Eurostat distinguishes two types of solar power; solar photovoltaic (sunlight converted into electricity employing solar cells which exposed to light will generate electricity [94]) and solar thermal (heat from solar radiation (sunlight) exploited for useful energy purposes [95]). The second type of energy is produced by using, for example, solar thermal–electric plants, and its technology for the production of electricity is currently under development. According to Eurostat data, in 2005 this source was not used, and in 2019 it accounted for 0.6% of electricity production. In total, in 2019, solar energy was responsible for 12.5% of electricity produced from renewable sources. This indicator was higher than in 2005 by 12.2%. Since 2007, which is the year of the technology development, there has been an increase in the share of this type of energy (see Figure 2). Among the other technologies for the production of electricity from renewable sources, biofuels (solid and liquid biofuels) and biogases are a significant source. Electricity production from solid and liquid biofuels increased by 102.9% in the period 2005–2019, and the share of GEP production from RES slightly decreased (from 8.9% in 2005 to 8.5% in 2019). Furthermore, biogas significantly increased its importance in the production of electricity. In their case, the X_i^{EU} ratio increased by 3.8 percentage points in the analyzed period, to the level of 5.5%, while the production of electricity from this source increased by 581.5%. RES of minor importance in the entire EU-27 are renewable municipal waste, which in 2019 was responsible for about 2% of electricity produced from RES and geothermal and tide, wave, and ocean. The latter two sources are used by only a few countries. Geothermal is most used in Italy; and tide, wave, and ocean are used in France. While in 2005 their share in *GEP_RES* was 2%, in 2019, it was only 0.7%. Thus, it is not a technology of strategic importance in the production of electricity, and its importance is marginalized in the scale of the entire EU.

Table 2. Gross electricity production from RES and the total gross electricity production in EU-27 in 2005 and 2019.

Energy Product-Source (Eurostat's Codes Included)	2005		2019		2019/2005
	GWh	X_i^{EU}(%)	GWh	X_i^{EU}(%)	$dGEP\_RES_i^{EU}$
(1)	(2)	(3)	(4)	(5)	(6)
RA000—Renewables and biofuels	476,989.593	100	1,005,271.556	100	110.8%
RA100—Hydro	340,546.184	71.4	345,264.887	34.3	1.4%
RA200—Geothermal	5397.673	1.1	6725.806	0.7	24.6%
RA300—Wind	68,094.587	14.3	367,115.301	36.5	439.1%
RA410—Solar thermal	0.000	0.0	5683.000	0.6	x
RA420—Solar photovoltaic	1458.688	0.3	120,034.721	11.9	8129.0%
RA500—Tide, wave, ocean	480.895	0.1	498.964	0.0	3.8%
R5110-5150_W6000RI—Primary solid biofuels	40,583.528	8.5	80,720.546	8.0	98.9%
R5220P—Pure biodiesels	0.000	0.0	29.541	0.0	x
R5290—Other liquid biofuels	1767.730	0.4	5170.842	0.5	192.51%
R5300—Biogases	8063.642	1.7	54,951.305	5.5	581.47%
W6210—Renewable municipal waste	10,596.666	2.2	19,076.643	1.9	80.02%
TOTAL—Total	2,917,663.780	x	2,904,012.166	x	−0.5%

Source: Own elaboration based on Reference [81]; $dGEP\_RES_i^{EU} = \left(\frac{GEP\_RES_{i,2019}^{EU}}{GEP\_RES_{i,2005}^{EU}} - 1\right) \cdot 100\%$ —means a change in the production of electricity from the *i*-th source in the period 2005–2019.

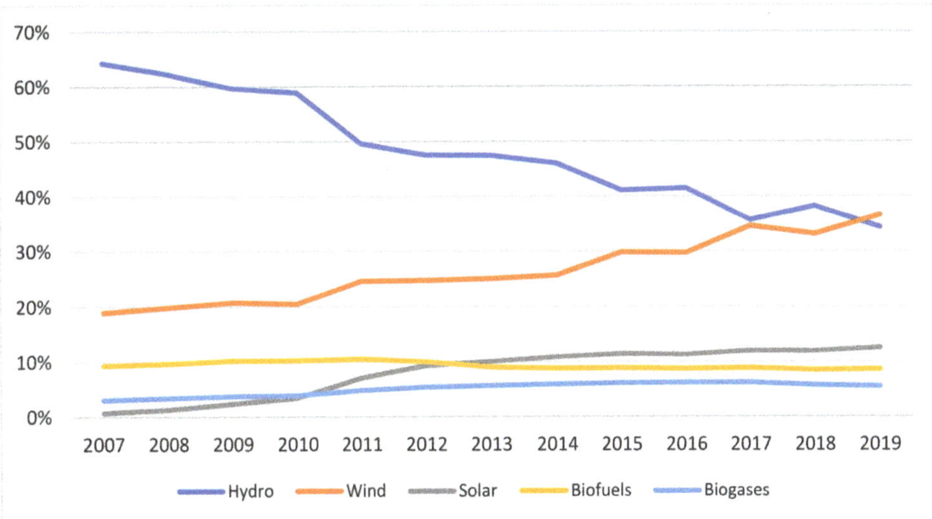

Figure 2. Shares of selected renewable sources (X_i^{EU}) in the production of electricity from renewable sources (*GEP_RES*). Source: Own calculation based on Reference [81].

4.3. Concentration of Renewable Sources in Electricity Production

In our analysis, we determine the Gini coefficient (see Formula (2)) by dividing renewable sources into seven categories (see Section 3, Data and Methodology). Figure 3 shows the evolution of this coefficient for the EU-27 area in 1990–2019. In 1990, the value of this coefficient was 0.83. This means that, in the EU, there was a high concentration of renewable sources used to produce electricity, and hydroelectric power plants were mainly used during this period. The energy produced by this method accounted for almost 95% of electricity production from renewable sources. In the following years, we observe a decrease in the value of the Gini coefficient. The pace of its changes is firstly slow, till around 2001, and then it accelerates. This is the result of measures taken to use more diverse sources of renewable energy. After 2001, we observe the use of wind energy and biofuels to a greater extent. In turn, after 2007, we can see that the importance of solar energy was increasing. In 2005, the concentration of renewable fuel sources was 0.704 and was lower than in 1990 by about 15%. In the following years, an even greater decline in the Gini coefficient occurred. In 2019, it was 0.512 and was lower than in 2005 by over 27%. This was influenced by several factors. Firstly, it refers to the largest enlargement of the EU in 2004. The EU-27 area for which we calculate the Gini coefficient includes the countries currently constituting the EU. Before accession, they were not obliged to implement measures for low-carbon economies on the scale that followed. The newly admitted member states had to comply with the introduced rules concerning the use of renewable sources for energy production. It is worth noting that, in the years 1990–2005, the average change in the concentration coefficient of the use of renewable sources for electricity production was higher in the EU-15 countries, and it was 0.11 (refer to the formula from the Methodology section), and for the new coming countries in 2004 or later, this change was only 0.02. In the period 2005–2019, the situation was different. It is in the new member states that the changes intensified (the Gini coefficient dropped by 0.16 on average, and for the EU-15 countries decreased by 0.08). Therefore, it is visible that the greatest progress in this area was recorded by the states of the EU-15 before 2005, and the newly admitted states only after joining the EU structures.

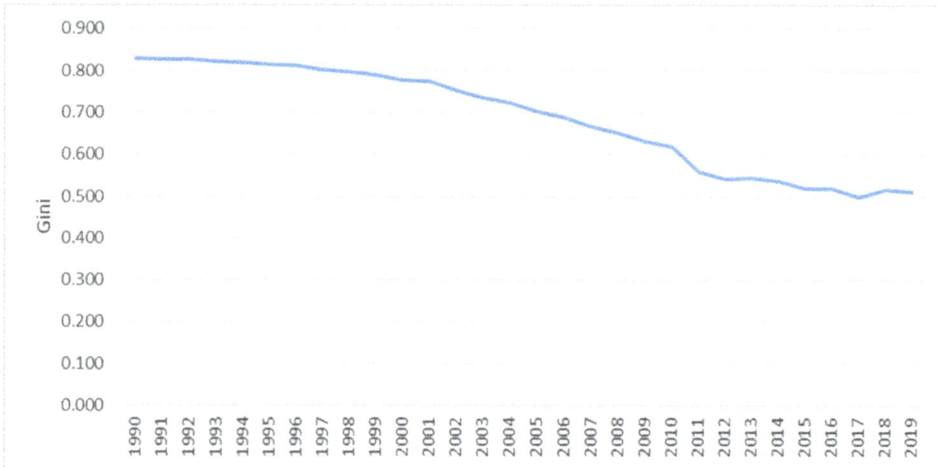

Figure 3. Changes in concentration of analyzed types of renewable sources in the production of electricity from renewable sources (*GEP_RES*) in the EU-27 in 1990–2019. Source: Own calculation based on Reference [81].

Table 3 presents the values of the Gini coefficients for 2005 and 2019 for the EU and UK. In 2005, the highest concentration of RES appeared in Bulgaria, Croatia, Lithuania, Romania, and Slovakia (in their case, the G coefficient exceeded 0.85), as well as France, Latvia, Austria, Slovenia, and Sweden (0.85 > G > 0.8). These are the countries that used mainly hydroelectric power at that time, and the share of this type of source in the generation of electricity from RES was about 90% or more. In 2005, only in eight countries, the concentration level was lower than 0.7. During this period, we notice the greatest diversity in the use of renewable sources in the case of the UK (0.51), Germany (0.562), Estonia (0.583), Belgium (0.584), the Netherlands (0.626), Portugal (0.664), Spain (0.685), and Ireland (0.697). In the UK, the distribution of renewable energy use was as follows: hydro (40%), biogases and biofuels (41%), wind (14%), and other (5%). Germany, on the other hand, made the greatest use of wind (39.6%), hydro (37.6%), biofuels (10.8%), and all others (12%). It is also worth taking a closer look at the diversity of the use of individual sources for the production of electricity. Assuming that the i-th source can be considered significant in the production of renewable electricity, we see that the limit of $X_i > 1\%$ is set, then in 2005 for two countries the limit of 1% was exceeded in the case of six sources (Germany and Italy). Furthermore, for four countries, $X_i > 1\%$ was recorded for five sources (Belgium, Luxemburg, Netherlands, and the UK). For 14 countries, there were only three or fewer sources. In 2019, the concentration level above 0.8 was recorded only in Malta (0.849), which mainly uses solar energy (97%), and Slovenia (0.807), where hydro is mainly used (89.3%). In nine countries, the concentration level was below 0.6. The lowest values of the Gini coefficients were recorded for the Czech Republic (0.409) and Italy (0.448). The Czech Republic used mostly hydro (28.3%), biogases and biofuels (43.9%), and solar (20.6). On the other hand, in Italy, hydro (40.9%), solar (20.1%), and winds (17.2%) are used the most. Considering the diversity of the use of sources, the shares of $X_i > 1\%$ for each of the seven sources, this was the case only for Italy. For seven other countries (Belgium, Germany, France, Lithuania, Luxembourg, Hungary, and the UK), $X_i > 1\%$ were recorded for six sources. For only two countries (Malta and Cyprus), the number of valid sources was three (Cyprus) or less (two—Malta). In Cyprus, mainly 46.3% winds and 42.4% solar and 11.3% biogases were used, while in Malta, the main source of renewable electricity was solar energy (97.04%) and, to a much lesser extent, biogases (2.93%). Looking at the changes in the concentration factor, it is clear that the concentration of renewable sources in electricity production increased in the analyzed period (2005–2019) in three countries. It is most noticeable in Estonia, where the value of the Gini coefficient increased by 0.104, due to an increase in the use of biofuels by 58.8%. An increase in concentration coefficient was also recorded in the case of Ireland ($\Delta G = 0.074$) and the UK ($\Delta G = 0.057$). The largest drops in the concentration level were recorded in the Czech Republic ($\Delta G = -0.373$), Bulgaria ($\Delta G = -0.342$), Hungary ($\Delta G = -0.274$), Italy ($\Delta G = -0.265$), and Lithuania ($\Delta G = -0.244$). The Czech Republic has significantly reduced the share of hydropower (to 28.3% in 2019) in favor of solar energy and biofuels and biogases. In Hungary, in 2005, the largest share was recorded for energy produced from biofuels; in 2019, the importance of this source was reduced in favor of a greater use of solar and winds.

Table 3. Concentration of types of sources in the production of electricity from renewable sources in the EU countries—the values of Gini coefficients.

	2005	2019			2005	2019	
Country	G_{2005}	G_{2019}	ΔG	Country	G_{2005}	G_{2019}	ΔG
Belgium	0.584	0.536	−0.048	Lithuania	0.852	0.608	−0.244
Bulgaria	0.857	0.515	−0.342	Luxembourg	0.790	0.594	−0.196
Czech Republic	0.782	0.409	−0.373	Hungary	0.791	0.517	−0.274
Denmark	0.716	0.711	0.005	Malta	x	0.849	x
Germany	0.562	0.552	−0.01	Netherlands	0.626	0.591	−0.035
Estonia	0.583	0.687	0.104	Austria	0.816	0.737	−0.079
Ireland	0.697	0.771	0.074	Poland	0.752	0.655	−0.097
Greece	0.795	0.614	−0.181	Portugal	0.664	0.621	−0.043
Spain	0.685	0.646	−0.039	Romania	0.857	0.722	−0.135
France	0.804	0.625	−0.179	Slovenia	0.845	0.807	−0.038
Croatia	0.856	0.708	−0.148	Slovakia	0.853	0.688	−0.165
Italy	0.713	0.448	−0.265	Finland	0.731	0.603	−0.128
Cyprus	0.747	0.672	−0.075	Sweden	0.820	0.711	−0.109
Latvia	0.846	0.700	−0.146	UK	0.510	0.567	0.057
EU27	0.704	0.512	−0.192				

Source: Own calculation based on Reference [81]. $\Delta G = G_{2019} - G_{2005}$.

In summary, we note that there has been a significant reduction in the concentration of renewable source types used for electricity production in almost all EU countries over the period analyzed. In those countries with slightly higher levels of concentration, wind-generated electricity in particular has gained in importance. In general, we are now seeing trends across the EU where two sources in particular are gaining in importance: wind and solar. Supporting these sources is part of the EU's energy policy.

5. Classification of the EU Countries by the Usage of Renewable Sources for Electricity Production

To examine similarities and differences in the use of renewables for electricity generation, we conducted the classification of the EU countries (including the UK) by applying the k-means algorithm. As in previous parts of the paper, the year 2019 was set as the reference year. The selection of the number of clusters was made based on the values of the silhouette coefficient (SI) presented in Table 4. The highest value of SI = 0.603 in the 2019 classification was obtained for 10 groups, and thus it was adopted as final. This is a satisfactory result because, with SI > 0.5, it is considered that the obtained division is characterized by a strong class structure. In the 2005 classification, the SI value for 10 groups is 0.81 and is slightly lower than the highest score for 12 groups (0.852). With an SI score > 0.7, the obtained division is considered to have a strong class structure. In addition, for the classification of data from 2011, the best division turns out to be the one into 12 groups (SI = 0.832). However, to ensure the comparability of the results, further analysis considered the division into 10 clusters, which is considered satisfactory, because the value of SI = 0.554 exceeds the limit of 0.5.

Table 4. Silhouette coefficients for the 2005, 2011, and 2019 classifications and the selected number of clusters.

Number of Clusters	Silhouette Coefficient		
	2005	2011	2019
8	0.713	0.598	0.519
9	0.790	0.480	0.590
10	0.810	0.554	0.603
11	0.830	0.649	0.545
12	0.852	0.832	0.537

Source: Own calculations.

The breakdown for 2005 (see Table 5) shows, first, numerous of one-object (one-element) clusters—as many as 7 out of 10. These are the following groups: 1 (UK), 3 (Hungary), 4 (Denmark), 5 (Netherlands), 8 (Malta), 9 (Cyprus), and 10 (Italy). Those clusters constitute countries classified as standing out from the others in terms of the use of renewable sources for the production of electricity.

Table 5. The results of the classification of EU countries according to the shares of individual renewable energy sources in the production of electricity (clusters averages, %)—data from 2005.

#	Country	Hydro	Wind	Solar	Biofuels	Biogases	Waste	Other
		−2005 X_1	−2005 X_2	−2005 X_3	−2005 X_4	−2005 X_5	−2005 X_6	−2005 X_7
1	UK	39.52	14.62	0.04	16.98	23.99	4.85	0.00
2	BG, CZ, EL, FR, HR, LV, LT, LU, AT, RO, SI SK, SE	93.34	2.44	0.14	2.53	1.00	0.49	0.06
3	HU	10.81	0.54	0.00	84.18	1.31	3.15	0.00
4	DK	0.23	67.41	0.02	19.30	2.86	10.17	0.00
5	NL	1.18	27.75	0.48	49.63	3.96	17.00	0.00
6	DE, EE, IE, ES	37.62	45.84	0.48	8.35	6.30	1.40	0.00
7	BE, PL, PT, FI	58.61	7.59	0.02	27.77	2.28	3.52	0.20
8	MT	0.00	0.00	0.00	0.00	0.00	0.00	0.00
9	CY	0.00	38.54	61.45	0.00	0.00	0.00	0.00
10	IT	77.62	4.24	0.06	3.92	2.17	2.37	9.63

Source: Own calculation in STATISTICA based on Reference [81]. Country abbreviations refer to those used by Eurostat; see Table 3.

In 2005, Malta did not use renewable sources to produce electricity. Therefore, this country is naturally classified as a separate group. However, to maintain the consistency of the samples with the samples used in the other classifications (for 2011 and 2019), it is also included in the analysis for 2005. Italy in 2005 is distinguished primarily by the fact that it used sources that for the purposes of the presented classifications are categorized as other (variable X_7). In Italy, geothermal is a popular source of energy. In 2005, this source contributed to the generation of almost 10% of renewable electricity. In the case of Cyprus, the main characteristic is that, in 2005, solar energy was mainly used to produce renewable electricity (over 61%). Moreover, in 2005, only two renewable sources were used in Cyprus—apart from solar energy, Cyprus used wind energy (over 38% share). Hungary is distinguished as a single-element group due to the fact that biofuels (over 84%) has a significant share in the production of renewable electricity. On the other hand, in the case of Denmark, the distinguishing factor is the share of wind energy (over 67%). The Netherlands and the UK are distinguished from other EU countries by the considerable variety of renewable sources they use. In the case of the Netherlands, these are biofuels (almost 50%), wind (over 27%), and waste (17%). The Netherlands is the only country where waste is classified as meaningful. In turn, the UK uses mainly hydro (almost 40%), as well as wind, biofuels, and biogases, the share of which in the production of renewable electricity is greater than 10% (for each of the sources). The UK is distinguished by the share of biogases (almost a quarter of renewable electricity generated). Great diversification of renewable sources in the case of the UK is confirmed by the low value of the Gini coefficient (see Table 3).

The most numerous cluster is cluster #2. The algorithm classified 13 countries into it (46% of the analyzed objects). This cluster is distinguished by a high share of energy produced in hydroelectric plants. The group mean for this feature is $\overline{X}_1^{2005}$ = 93.3%, and the group included countries for which these shares (X_1) are at least 80%.

Cluster number six joins four countries with large share of wind energy: Germany, Estonia, Ireland, and Spain. The group mean for this feature is $\overline{X}_2^{2005}$ = 45.844, and the individual values of this coefficient (X_2) for these countries ranged from 39% to 51%. The algorithm assigns Belgium, Poland, Portugal, and Finland to Cluster #7. These countries are characterized by a similar level of hydropower consumption (X_1 between 46% and 70%) with a simultaneous significant consumption of biofuels (group average $\overline{X}_4^{2005}$ = 27%, and the individual values of the X_4 feature are between 15% and 40%).

In the 2011 classification (Table 6), only four clusters are single-object. As in the previous classification (from 2005), Malta (#5), Italy (#10), and the Netherlands (#9) are classified as single-object clusters. In Malta, in 2011, two sources of renewable electricity were used: solar (X_3 = 50.4%) and biogases (X_5 = 49.6%). It is worth noting that, compared to other countries, Malta has the largest share of solar energy use. Italy, as in the previous classification, is distinguished due to the high level of use of other sources (X_7 = 6.66%) compared to other countries. However, this share is lower than in 2005, as, at that time, Italy began to use wind and solar panels on a larger scale. In the case of the Netherlands, there is a significant share of waste (X_6 = 16.5%), comparable to the previous classification. Other sources with a high share of renewable electricity production are wind and biofuels, but in 2011 their proportions changes in favor of greater use of the wind. Finland is also classified in a separate cluster, which is distinguished by the fact that the main sources used in the production of renewable electricity are hydro (X_1 = 51.48%) and biofuels (X_4 = 44.75%), totaling 96.23%.

In the 2005, Finland is classified together with Belgium, Poland, and Portugal. In the case of Belgium and Poland, in 2011, a much smaller share of renewable electricity production in hydroelectric plants is recorded, and in the case of Portugal, the importance of using biofuels decreased. The values of the coefficients have changed so significantly that these countries are no longer characterized as similar. This time Poland joined the group together with Estonia and Hungary (cluster #6). This cluster is distinguished by the significant use of biofuels (average $\overline{X}_4^{2011}$ = 58%). Furthermore, Belgium is classified together with the Czech Republic (#3). Characteristic for this cluster is the use of various sources. Cluster means greater than 10% are observed for the following traits: X_1 (hydro), X_2 (wind), X_3 (solar), and X_4 (biofuels). Due to the increased production of renewable electricity from wind in Ireland (the share increased from 50% to 80%), the algorithm classifies it together with Denmark (#4). The clusters' mean of this coefficient ($\overline{X}_2^{2011}$), in this case, accounted for almost 75%. Ten countries remain classified in the largest cluster (#8). Their main source of renewable electricity production is hydro, with the mean is $\overline{X}_1^{2011}$ = 84.5% and range from 71% to over 95%. Cluster #7 (Greece, Spain, Portugal, and Lithuania) is distinguished by the largest shares of two sources: hydro ($\overline{X}_1^{2011}$ =49.78%) and wind ($\overline{X}_2^{2011}$ = 38.11%). The last cluster (#2) includes Germany, UK, and Cyprus, with the main sources in the production of renewable electricity being wind ($\overline{X}_2^{2011}$ = 48.20%), biogases ($\overline{X}_5^{2011}$ = 20.64%), and hydro ($\overline{X}_1^{2011}$ = 13.55%). It is worth adding that, in the case of Cyprus, the role of the solar source has significantly decreased, from 61.45% in

2005 to 6.7% in 2011, with a simultaneous large increase in energy production from these two sources.

Table 6. Results of the classification of EU countries according to the shares of individual renewable energy sources in the production of electricity (clusters averages, %)—data from 2011.

#	Country	Hydro	Wind	Solar	Biofuels	Biogases	Waste	Other
		–2011 X_1	–2011 X_2	–2011 X_3	–2011 X_4	–2011 X_5	–2011 X_6	–2011 X_7
1	FI	51.48	1.99	0.02	44.75	0.66	1.11	0.00
2	DE, CY, UK	13.55	48.20	7.48	7.59	20.64	2.54	0.01
3	BE, CZ	24.20	14.59	19.84	28.16	8.79	4.41	0.00
4	DK, IE	6.57	74.84	0.06	12.11	3.06	3.35	0.00
5	MT	0.00	0.00	50.40	0.00	49.60	0.00	0.00
6	EE, HU, PL	10.37	25.99	0.02	58.00	4.16	1.47	0.00
7	EL, ES, LT, PT	49.78	38.11	4.71	5.13	1.53	0.52	0.21
8	BG, FR, HR, LV, LU, AT, RO, SI, SK, SE	84.89	6.67	1.68	4.09	1.71	0.89	0.08
9	NL	0.46	41.40	0.85	32.38	8.40	16.51	0.00
10	IT	56.25	11.61	12.72	6.15	4.01	2.60	6.66

Source: Own calculation in STATISTICA based on Reference [81]. For country abbreviations, refer to those used by Eurostat (see Table 3).

The compositions of clusters change again for the 2019 classification (see Table 7). Five countries are classified into single-object clusters. As in the previous classification, these are Malta (#8) and Italy (#1), as well as the Czech Republic (#3), Hungary (#5), and Estonia (#6). Italy, as in the previous cases, is distinguished primarily by a high share of other sources (X_7 = 5.16%) compared to other countries. Although it decreased compared to 2011, the amount of electricity generated with this method has increased. It is also worth noting that, in Italy, the importance of the use of solar and wind energy has increased. In 2019, in Malta, solar is the dominant source used for the production of renewable electricity, with the share of X_3 = 97.04%. This is a significant increase compared to 2011, by over 45 pp. Estonia is distinguished by a high consumption of biofuels, with a significant consumption of wind energy and a significant reduction in the share of hydropower compared to that in 2011. In Hungary, significant shares of biofuels (X_4 = 37.74%) and solar energy (X_3 = 31.94%) are recorded. On the other hand, the Czech Republic still stands out due to the significant—compared to other countries—use of biogases in the mix of renewable sources (X_5 = 22.54%) and the burden of electricity production being distributed among four sources: in addition to the mentioned biogases, hydro (X_1 = 40.92%), biofuels (X_4 = 21.38), and solar energy (X_3 = 20.13%). Thus, a low level of concentration of RES in the production of electricity.

The cluster with the highest share of hydro is Cluster #4: Croatia, Austria, Romania, and Slovenia. The clusters' mean for this variable was as high as X_1 = 74.81%. Another cluster with high hydro consumption is Cluster #2 (Bulgaria, France, Latvia, Luxembourg, Slovakia, Finland, and Sweden). At the same time, in this cluster, there is a significant consumption of biofuels (X_4 = 16.7%), which distinguished it from #4. Eight countries are classified to the largest Cluster #9, distinguished by the significant use of wind energy (X_1 = 57.35%). The other two clusters are Clusters #8 (Belgium and the Netherlands) and #10 (Germany and Cyprus), which are also characterized by significant use of wind energy (group averages for this variable being, respectively, 49.16% and 48.52%). However, significant use of other sources is also important for the breakdown. In the case of Belgium and the Netherlands, these are solar (X_3 = 22.15%) and biofuels (X_4 = 14.49%). Menawhile, in the case of Germany and Cyprus, these are solar (X_3 = 30.54%) and biogases (X_5 = 12.25%).

Table 7. Results of the classification of EU countries according to the shares of individual renewable energy sources in the production of electricity (clusters averages, %)—data from 2019.

#	Countries	Hydro	Wind	Solar	Biofuels	Biogases	Waste	Other
		−2011 X_1	−2011 X_2	−2011 X_3	−2011 X_4	−2011 X_5	−2011 X_6	−2011 X_7
1	IT	40.92	17.17	20.13	7.58	7.03	2.01	5.16
2	BG, FR, LV, LU, SK, FI, SE	55.77	15.29	6.61	16.70	4.20	1.34	0.07
3	CZ	28.30	6.24	20.61	21.38	22.54	0.93	0.00
4	HR, AT, RO, SI	74.81	14.36	4.20	4.24	1.95	0.15	0.27
5	HU	4.67	15.55	31.94	37.74	6.78	2.92	0.38
6	EE	0.89	32.07	3.43	58.81	1.81	2.99	0.00
7	BE, NL	3.06	49.16	22.15	14.49	4.29	6.86	0.00
8	MT	0.00	0.03	97.04	0.00	2.93	0.00	0.00
9	DK, IE, EL, ES LT, PL, PT, UK	17.93	57.35	8.41	11.48	2.10	1.73	0.10
10	DE, CY	5.17	48.52	30.54	2.32	12.25	1.17	0.04

Source: Own calculation in STATISTICA based on Reference [81]. Country abbreviations refer to those used by Eurostat (see Table 3).

6. Discussion

On one hand, the EU members are obliged to increase the share of renewable sources in the total energy consumption, but on the other hand, they have certain freedom in shaping the energy policy and selecting the sources according to their own possibilities. This is why the EU countries differ significantly in terms of the types of renewable sources used to produce electricity. The presented data clearly show that all EU countries are increasing the share of renewable sources for electricity production, which is in line with the guidelines contained in EU directives. Furthermore, the diversity of these sources is increasing, which is indicated by the decreasing values of the Gini coefficients (for the vast majority of countries). It is also worth noting the fact that the EU-15 countries have already started this transition process at the beginning of the 21st century, the new EU member states followed only after their accession.

In 2005, at the beginning of the analyzed period, it is noticeable that the energy produced by hydroelectric power plants is of the highest importance in the production of electricity from renewable sources. Its share in the production of renewable electricity (*GEP_RES*) accounted for over 80% and is recorded for as many as half (14) of the analyzed countries, and in the case of 21 countries, the share is greater than 50%. During the next 14 years, the importance of hydroelectric power plants in the production of electricity did not increase, although this method is said to have a high potential [96].

Although hydropower has an established position in the production of electricity and belongs to the so-called renewable sources, the amount of electricity produced by this method (in GWh) in the scale of the entire EU (EU-27) increased by only 1.4% in the period 2005–2019. In 15 countries (out of 26 analyzed), its production is even reduced (Malta and Cyprus are not included in this list, as, in 2005, electricity was not produced by this method in these countries). This is justified by the fact that this type of electricity production is not environmentally neutral [97–99], and the degree of its impact depends on the scale of the production [96,97]. There is little chance of a large-scale hydropower plant in the European Union, mainly due to the fact that most of the areas have already been taken into use [96]. Nevertheless, hydropower plays an important role in providing flexibility to the electricity system [100]. It is indicated that the technology used in the hydropower plants allows meeting sudden fluctuations in supply or demand of other renewable sources, such as solar and wind power. Therefore, the EU support hydropower innovation. It is worth noting, the EU legislation stands in the way of a freer choice of RES and restricts the development of certain technologies, including hydropower (e.g., References [99,101]). In 11 countries, the increase in production by this method is mainly not significant. Only five countries recorded an increase in production by more than 15%: Portugal (increase in production by around 100%), Slovenia (by around 35%), Spain (16.7%), Ireland (16.1%), and Lithuania (by 15.6%). It is worth mentioning that, in these countries, in 2005, hydropower accounted already for a significant share in *GEP_RES*. In 2019, these shares decreased significantly due to the fact that these countries have been developing other technologies for obtaining electricity from renewable sources to a greater extent. Only in Slovenia, in 2019, the share of hydropower in the generation of *GEP_RES* stayed significant (89.3%), and in the other countries mentioned above, it became less than 36%, while in Ireland, it was less than 10%. In Portugal, the largest increase in hydropower production is recorded, as it is a country with one of the highest possible potentials to exploit this area [99]. In addition, in 2007, the Portuguese government approved the National Program of Dams with High Hydroelectric Potential [99]. As previously mentioned, the development of hydropower may be restricted by the EU legislation (e.g., see Reference [101]), due to negative environmental effects. However, as emphasized by Reference [102], the acceptability of the side effects of RES in terms of benefits related to climate protection and socioeconomic benefits lies with the national policy pursued by states as part of the development of RES and environmental protection. Portugal, as one of the few countries, has decided to invest in this type of energy on a large scale, as the contractor (Iberdrola, Bilbao, Spain) has been awarded €650 million by the European Investment Bank [103] for the expansion of the hydroelectric power plant in Portugal. This does not confirm the thesis by Reference [96] that indicates that only small projects (generating a capacity of no more than 10 MWh) can count on support from EU bodies. In the case of Slovenia, the work of Reference [104] indicates that energy needs will be best met by a mix of nuclear, water and gas technologies. The only source of renewable energy in this list is hydropower. Therefore, it is not surprising that the energy policy in Slovenia also focuses on the development of this technology. Another study [105] considered many criteria related to electricity generation, such as environmental protection and institutional—political, economic, social, and technological. The researchers indicated that hydropower, biomass, and nuclear power are the most effective RES investments. Hence, it is not surprising that the production in hydroelectric power plants increased by over 15%. However, this potential is not fully used in Lithuania. Lithuania has one of the most restricted environmental regulations related to the introduction of this type of technology, even to a small extent [106].

Wind energy is gaining importance in the renewable energy mix used for electricity production in the EU. The data we present in this paper confirm a significant intensification of electricity production, using wind during the analyzed period. Currently, it constitutes the largest share in the production of *GEP_RES* (36.5%). The amount of electricity produced in this way increased in the period 2005–2019 by almost 440%. This is not a surprise, as the literature indicates this technology as the most competitive compared to other RES [13]. That competitiveness is examined by using a balanced scorecard based on four types of variables: the perspective of the consumer, internal business processes, the development aspect, and the financial aspect. Kapitonov and Voloshin [13] describe the advantages of this technology as "the cost of electricity, safety, minimum possible power, productivity, and performance development aspect and financial aspect". It is worth mentioning that, as in the case of hydropower, also wind energy can affect the natural environment. Wang and Wang [107] and Pecesila et al. [77] mention the following effects: noise pollution, change the landscape, and impact on local to regional weather and climate if the area of turbines is large enough, and it may affect the local populations of various species of birds. Nevertheless, this type of energy is indicated as the least harmful to the environment [108]. Therefore, it is strongly supported (in addition to solar energy) by the EU bodies as a mean of achieving the sustainable development goals, and in particular achieving the so-called climate neutrality planned for 2050 (see References [109,110]. At the same time, wind energy is mentioned as the one to support these goals to the greatest extent [111]. The European Commission notes that, thanks to the pan-European efforts to reduce greenhouse gas emissions, in 2016 (compared to 1990), it succeeded in reducing these emissions by 22% [109]. This is due to the significant increase in the share of RES in energy production, in particular wind and solar energy. It is directly linked to a significant reduction in costs related to the production of solar and on- and off-shore wind energy in the recent years (European Commission, 2018). Currently, offshore wind energy receives particular interest from the EU bodies. A strategy for the development of this type of energy was formulated, an EU Strategy to harness the potential of offshore renewable energy for a climate neutral future [12], referred to as the EU Strategy on Offshore Renewable Energy. This strategy assumes an increase in the Europe's offshore wind capacity from 12 GW (level from 2020) to at least 60 GW by 2030 and to 300 GW by 2050. Additionally, the development of several new technologies (by 2050), such as floating wind and solar, is expected.

Currently, the leaders in wind energy production are (according to data from 2019) Germany (125,894 GWh), UK (64,334 GWh), Spain (55,647 GWh) and France (34,721 GWh). As for the share of wind energy in the production of *GEP_RES*, the largest is recorded for Ireland (83%), Denmark (70%), Poland (58%), Spain (54%), UK (53%), Germany, and the Netherlands (51%). Countries with shares between 40% and 50% are Lithuania, Belgium, Portugal, Cyprus, and Greece. In the 2019 cluster classification, all these countries are classified into three clusters: #7, #9, and #10 (see Table 6). These three clusters are characterized by a high share of wind energy in *GEP_RES*, and they are differentiated by significant shares of other sources. The UK, Sweden, Denmark, and Ireland are considered the most efficient countries in terms of wind energy use [112]. It is worth noting that the countries from the abovementioned clusters have favorable conditions for the development of this type of energy because they are either large in terms of area or have the possibility of developing offshore wind farms. At the same time, the countries using offshore wind energy are considered to be the most effective [112]. Therefore, large-scale investments are being made in many countries to develop offshore wind farms. An example is the support system for the construction of offshore wind farms in Poland, which is approved by the European Commission in May 2021 [113]. According to the assumptions of the Polish energy strategy, offshore wind farms are to be the main pillar of the energy system in Poland. Government support systems for offshore wind energy can be found also in the UK ([114]), Ireland ([115,116]), and Denmark ([117]).

Another renewable source used for electricity production that has gained in importance in recent years is solar energy. In the EU-27, the production of this type of electricity increased from 1458 to 125,717 GWh, i.e., by over 8500%. Solar technology is relatively new and in 2005 it is the least used resource for the production of renewable electricity (its share is only 0.3% in *GEP_RES*). However, in 2019, it is the third most important source, and its share in *GEP_RES* is 12.5%. The increase in popularity of this source may be due to several reasons. Firstly, solar energy is the second technology, after wind energy, considered the most competitive in the group of renewable energy technologies used for energy production [13]. Secondly, the development of this technology has contributed to a significant reduction in the cost of electricity production, and therefore it will continue to be of interest to the EU bodies as a technology to be supported. As such, more investments are planned for its further development, e.g., in the form of offshore solar energy (floating solar panels) (European Commission, 2020). Thirdly, this type of energy is included in the EU strategies for reducing CO_2 emissions and ensuring energy security for the EU area.

There are different technologies for using the sun to produce energy [118]. Eurostat's data for gross electricity production include two technologies: solar photovoltaic and solar thermal, with electricity production using the latter in 2019 only in Spain (its share was 5.5% in *GEP_RES*).

The environmental impacts are discussed at the level of the photovoltaic panels' production technology [118]. The first issue being the use of allium arsenide or cadmium telluride to produce more energy-efficient photovoltaic panels. In the event of a leak, those compounds are said to not be harmful to the environment. Silicon used for the production of photovoltaic panels is said not to be harmful to the environment, however characterized by relatively lower energy converting efficiency. In the case of a technology called concentrated solar power techniques, coolant and lubricant are harmful, also in the event of a leak. The methods of neutralizing the harmfulness of these substances are included to be considered in the further development of these technologies.

The country which, in 2019, produced renewable electricity almost entirely by using solar technology was Malta (97% share in *GEP_RES*). It is also the only country where this share is greater than 50%. Therefore, in the 2019 cluster classification, it is assigned to a separate cluster. The second country in this respect is Cyprus, where 42.7% of *GEP_RES* is produced using this technology. Cyprus is classified in one cluster together with Germany, and this is due to the similarity of the use of other sources at a similar level (wind and biogases). The third country in this ranking is Hungary with 31.9% of solar energy in *GEP_RES*. For five other countries, the share is greater than 20% (but less than 30%): Greece, the Netherlands, Belgium, Czech Republic, and Italy. Belgium and the Netherlands (cluster #7) have similar levels of solar energy use, but this is not surprising, as these countries are adjacent to each other and have similar climatic conditions for using RES to produce GEP.

All in all, the increased importance of solar energy in the production of electricity is recorded in many countries. However, only in the case of Malta is it considered as the main source. In other countries, where this share is also significant (but not leading), this source can be described as complementary to the energy mix.

The paper of Reference [119] presents the possibility of developing solar technologies for energy production with respect to geographic location. Without a surprise, the countries of Southern Europe are characterized by the greatest potential. By comparing these results with those presented in this paper, it can be concluded that so far only few of the EU countries are developing their solar energy potential. These are Malta, Cyprus, Italy, and Greece, as well as Hungary. On the other hand, Spain, Portugal, and Romania are examples of countries with significant potential for the development of this technology, but only to a limited extent (favoring wind and hydropower).

It is worth noting that the use of solar energy is quite important in countries that have not been previously named as those with such a high potential in this particular technology. These are the Czech Republic, Germany, the Netherlands, and Belgium. In their electricity mix, the share of solar energy accounted for at least 18%. In the case of the Czech Republic, support systems, which Reference [120] define as generous, are of great importance for the development of photovoltaics. They see this as the cause of the massive boom for the construction of the photovoltaic power plant in the period of 2009–2012. However, it is pointed out that these systems are only slightly in line with the potential of solar resources in the Czech Republic. Furthermore, in Germany, support systems and national strategies for the development of solar energy played a significant role in its development, despite the limited domestic potential (compared to the countries of Southern Europe). In the case of Germany, the most important factor is the feed in tariffs (preferential tariffs) [121].

Biofuels, biogases, and renewable municipal waste are collectively classified under the biomass category. Currently, it is estimated that biomass contributes as much as 60% to total renewable energy production in the EU, including electricity, heat, and energy used in transport [122]. In the production of renewable electricity, this share is lower, at the level of 15.9%. Thus, in its production, these sources play a smaller role than in the case of other types of energy. Of these three sources, biofuels and biogases are used to the greatest extent. In 2019, biofuels accounted for 8.55% of *GEP_RES* production, biogases—5.47%, and renewable municipal waste only 1.9%. In the period 2005–2019, there was an increase in electricity production with these sources, by 102.9% and over 580% and 80%, respectively. These numbers show that the use of biofuels is already well established in the production of electricity in the EU, while the importance of biogas has grown significantly.

Biofuels used for electricity production according to the Eurostat category are divided into primary solid biofuels (fuelwood, wood residues, wood pellets, animal waste, and vegetal material), pure biodiesel and other liquid biofuels. The latter two are of marginal importance. Solid biofuels are used to the greatest extent in heating energy, but they have also gained popularity in the production of electricity. In this matter, biofuels are most often used in Estonia (share of 58.8% in *GEP_RES*), Finland (38.5%) and Hungary (37.7%). Estonia and Hungary in the 2019 classification are classified as single-element clusters, due to, among others, such a dominant share of biofuels in the energy mix used for the production of electricity from renewable sources. Furthermore, Estonia is the only country among the analyzed countries where biofuels are the basis for the production of *GEP_RES* (share greater than 50%). Finland, despite such a significant share of this source, was classified in the second cluster, as in addition to biofuels, hydropower, and wind power are used to a large extent. Other countries that use this source to a large extent are Poland (24.6% shares in *GEP_RES*) and the Czech Republic (21.4%). For the following nine countries, this share ranges from 10% to 20%: Bulgaria, Denmark, Latvia, Belgium, Slovakia, the Netherlands, Sweden and Lithuania, and the UK. All in all, apart from Estonia, for the other countries mentioned, biofuels complement the portfolio of renewable sources used in electricity production as one of three (or four as in the case of the Czech Republic and Bulgaria) sources. The presented results also indicate that this is the source typical for countries located mainly in the northern part of Europe, where solar energy does not have as high potential as in the case of southern countries. Bulgaria is the southernmost country on this list, followed by Hungary. In the case of Hungary, the share of biofuels has significantly decreased—from almost 85% in 2005 to almost 38% in 2019 It is due to the fact that nowadays in Hungary more use is made of other sources (solar and winds). The production of electricity from biofuels in this country in the analyzed period slightly increased—by about 12%.

Biogases are reported to be as less common source than biofuels. They are used to the bigger extent only in the Czech Republic (22.5%) as well as in Germany, Cyprus, and Latvia (share between 11% and 14%). While in the case of the Czech Republic, Germany and Cyprus, the share of biogas in the energy mix distinguishes them from other countries, in Latvia, greater shares of hydro and biofuels meant that it was assigned to cluster #2.

7. Conclusions

The purpose of the presented analysis was to assess the changes that have occurred in the use of RES in the production of electricity in the EU and UK, and the main research period was set as 2005–2019. As different countries have different levels of use of renewable energy sources (RESs) in electricity production (GEP), in our main analyses, we focused only on the electricity generated from RES (*GEP_RES*) and its amount generated from each RES. It is this approach that distinguishes the presented study from others presented in the literature that focus primarily on GEP-related analyses. In the presented study, we analyzed the shares of seven types of different sources (hydro, wind, solar, biofuels, biogases, renewable municipal waste, and others) in the production of *GEP_RES*. The main research methods are the Gini concentration coefficient and the k-means algorithm.

The analysis shows that the Gini coefficients decreased for almost all countries in the period 2005–2019. This means that the concentration of renewable sources used for electricity production has decreased significantly across the EU. As indicated in the discussion, one of the main drivers of change in this respect (increased use of RES) has been the EU energy policy targets and national energy policies to adjust national energy sectors to these targets. This inclines us to accept Hypothesis (1). This phenomenon is in line with the recommendations of the European Commission regarding the diversification of energy sources, which is to support the energy security. It is worth noting that, while in 2005, in most countries, the predominant source of renewable electricity (with a share of over 80% in *GEP_RES*) was hydroelectric power plants, in 2019, a significant increase in the importance of other sources occurred.

The level of electricity production in hydroelectric power plants did not increase significantly, but with the simultaneous significant increase in the production of renewable electricity, the share of this source decreased. Hydropower production is still of great importance and is the main renewable source for electricity production in many European countries (with a share of more than 50% in *GEP_RES*). However, since this method of obtaining energy requires specific geographical conditions and the fact that it is not completely neutral to the environment, its development focuses primarily on more efficient use of already existing facilities, in particular those large ones. Wind energy is gaining importance. In 2019 in the EU-27, the share of this type of source in the production of *GEP_RES* was already over 36% and was greater than the share of hydropower. This source is recognized as the most effective among renewable sources in the production of electricity. It is also a resource promoted by the EU bodies. In particular, the emphasis is on the development of offshore wind energy, which is included in the directive [12].

The significant shares of the abovementioned sources (hydro and wind) in the production of *GEP_RES* in the vast majority of EU countries make those two the most important sources of renewable electricity. The remaining sources are usually treated as complementary. Among those sources of renewable electricity, solar energy, in particular photovoltaic energy, is important in the EU scale. It plays a key role in most of the countries of Southern Europe that is related to the level of insolation [119]. This technology, considered the second most effective renewable source of electricity after wind energy, is also developed and promoted in countries with less favorable climatic conditions for it: the Czech Republic, Germany, the Netherlands, and Belgium. The ecological aspects add to the importance of the solar technology and influence a support from governments of many countries (e.g., Germany and the Czech Republic). In northern countries, biomass— particularly biofuels— plays an important role as a complementary resource to the renewable energy mix.

Two sources that were categorized in this paper as "other" are of importance in only two countries: energy from the geothermal source produced in Italy and tide wave ocean used in France. As the first technology increased in importance (both in Italy and to small extent in other countries), the shares of tide wave ocean technology began to lose their importance. In the analyzed period, this source was not used to a greater extent.

The performed cluster analysis shows that, firstly, there are EU countries with a very individual structure of using the renewable sources, such as Malta (the only country where solar energy is used to a higher extent) or Estonia (a leading country in the use of biofuels). Secondly, there are countries where the use of renewable sources is highly diversified. An example is the Czech Republic, where as many as four sources had shares in *GEP_RES* at the level of more than 10%. Thirdly, the vast majority of countries has been assigned to multi-element clusters, indicating that they have similar structures of consumption of renewable sources in *GEP_RES*. Importantly, the clusters constitute countries often not closely located geographically to each other. As it turns out, geographical factors are not the only determinants of the amount of energy consumed, or the sources used for its production. The key factors here are the national energy policies and strategies that formulate the national goals and publicly supported technologies. Therefore, the presented study can be used as a basis for comparisons of impacts of national policies on the promotion and use of renewable sources for electricity production.

The presented study clearly shows that all EU countries implement the assumptions of the energy policy regarding both increasing the share of renewable sources in energy production—in particular, electricity—and increasing the diversity of these sources. The results indicate a vast diversity within the EU countries in terms of the use of the renewable sources for the production of electricity. This shows individual country-oriented approaches to implement the energy transformation towards low-carbon economies. The EU's support and restrictive measures (e.g., the already mentioned EU ETS) are of considerable importance for this transformation. Currently, the main source of renewable energy in the EU is wind energy, which is increasingly used by most EU countries, and the offshore renewable energy strategy (European Commission, 2020) is guiding the development of this type of energy. Maintaining the pace of the energy transformation allows the achievement of the goals set for 2050 in the Clean Energy for all Europeans document [11].

Author Contributions: Conceptualization, A.M.-J., D.Ż.-S., U.A.-K. and M.H.-K.; methodology, A.M.-J.; validation A.M.-J. and U.A.-K.; formal analysis, A.M.-J., D.Ż.-S. and U.A.-K.; investigation, A.M.-J. and D.Ż.-S.; resources, D.Ż.-S. and A.M.-J.; data curation, A.M.-J. and U.A.-K.; writing—original draft preparation, A.M.-J. and D.Ż.-S.; writing—review and editing, D.Ż.-S., U.A.-K., A.M.-J. and M.H.-K.; visualization, D.Ż.-S.; supervision, A.M.-J., D.Ż.-S., U.A.-K. and M.H.-K.; project administration, D.Ż.-S. All authors have read and agreed to the published version of the manuscript.

Funding: This research received no external funding.

Institutional Review Board Statement: Not applicable.

Informed Consent Statement: Not applicable.

Data Availability Statement: Not applicable.

Conflicts of Interest: The authors declare no conflict of interest.

References

1. EIA—U.S. Energy Information Administration. International Energy Outlook 2019 with Projections to 2050, #IEO2019, U.S. Energy Information Administration Office of Energy Analysis. September 2019. Available online: https://www.eia.gov/outlooks/ieo/ (accessed on 7 August 2021).
2. Aklin, M.; Urpelainen, J. *Renewables: The politics of a global Energy Transition*; MIT Press: Cambridge, MA, USA, 2018.
3. Child, M.; Koskinen, O.; Linnanen, L.; Breyer, C. Sustainability guardrails for energy scenarios of the global energy transition. *Renew. Sustain. Energy Rev.* **2018**, *91*, 321–334. [CrossRef]
4. Davidson, D.J. Exnovating for a renewable energy transition. *Nat. Energy* **2019**, *4*, 254–256. [CrossRef]
5. Watari, T.; McLellan, B.C.; Giurco, D.; Dominish, E.; Yamasue, E.; Nansai, K. Total material requirement for the global energy transition to 2050: A focus on transport and electricity. *Resour. Conserv. Recycl.* **2019**, *148*, 91–103. [CrossRef]
6. Pietrzak, M.B.; Igliński, B.; Kujawski, W.; Iwański, P. Energy Transition in Poland—Assessment of the Renewable Energy Sector. *Energies* **2021**, *14*, 2046. [CrossRef]
7. Kuc-Czarnecka, M.E.; Olczyk, M.; Zinecker, M. Improvements and Spatial Dependencies in Energy Transition Measures. *Energies* **2021**, *14*, 3802. [CrossRef]
8. UNFCCC. United Nations Framework Convention on Climate Change. 1992. Available online: https://unfccc.int/resource/docs/convkp/conveng.pdf (accessed on 7 August 2021).
9. Kyoto Protocol. 1997. Available online: https://www.eea.europa.eu/policy-documents/kyoto-protocol-to-the-un (accessed on 3 June 2021).
10. European Commission. Communication from the Commission to the European Parliament, the Council, the European Economic and Social Committee and the Committee of the Regions Next Steps for a Sustainable European Future European action for Sustainability, COM/2016/0739 Final, Strasbourg, 22.11.2016. 2016. Available online: https://eur-lex.europa.eu/legal-content/EN/TXT/?uri=COM%3A2016%3A739%3AFIN (accessed on 5 August 2021).
11. European Commission. *Clean Energy for All Europeans*; European Commission: Brussels, Belgium, 2019. [CrossRef]
12. European Commission. Communication from the Commission to the European Parliament, the Council, the European Economic and Social Committee and the Committee of the Regions, An EU Strategy to Harness the Potential of Offshore Renewable Energy for a Climate Neutral Future {SWD(2020) 273 Final, Brussels, 19.11.2020 COM(2020) 741 Final. 2020. Available online: https://eur-lex.europa.eu/legal-content/EN/TXT/?uri=COM:2020:741:FIN (accessed on 30 July 2021).
13. Kapitonov, I.A.; Voloshin, V.I. International Journal of Energy Economics and Policy Strategic Directions for Increasing the Share of Renewable Energy Sources in the Structure of Energy Consumption. *Int. J. Energy Econ. Policy* **2017**, *7*, 90–98.
14. WCED (World Commission on Environment and Development). Our Common Future Report of the World Commission on Environment and Development. 1987. Available online: https://digitallibrary.un.org/record/139811 (accessed on 29 July 2021).
15. The Treaty on European Union. 2007. Available online: http://data.europa.eu/eli/treaty/teu_2012/oj (accessed on 29 July 2021).
16. Balcerzak, A.P.; Pietrzak, M.B. Application of TOPSIS Method for Analysis of Sustainable Development in European Union Countries. In Proceedings of the 10th International Days of Statistics and Economics: Conference Proceedings, Prague, Czech Republic, 8–10 September 2016.
17. Communication from the Commission Energy for the Future: Renewable Sources of Energy. White Paper for a Community Strategy and Action Plan. 1997. Available online: https://europa.eu/documents/comm/white_papers/pdf/com97_599_en.pdf (accessed on 29 July 2021).
18. Directive 2001/77/EC of the European Parlament and of the Council. September 2001. Available online: https://eur-lex.europa.eu/legal-content/EN/TXT/PDF/?uri=CELEX:32001L0077&from=en (accessed on 29 July 2021).
19. Communication from the Commission to the Council and the European Parliament—Renewable energy road map—Renewable Energies in the 21st Century: Building a More Sustainable, COM(2006)848. 2007. Available online: https://www.eumonitor.eu/9353000/1/j4nvke1fm2yd1u0_j9vvik7m1c3gyxp/vkcweehp63z8/v=s7z/f=/com(2006)848_en.pdf (accessed on 29 July 2021).

20. Directives 2009/28/EC of the European Parlament and of the council. April 2009. Available online: https://eur-lex.europa.eu/legal-content/EN/TXT/PDF/?uri=CELEX:32009L0028&from=PL (accessed on 29 July 2021).
21. Europe June 2020. Available online: https://ec.europa.eu/eu2020/pdf/COMPLETENBARROSO007-Europe2020-ENversion.pdf (accessed on 3 June 2021).
22. Directive 2018/2001 of the European Parliament and of the Council of 11 December 2018 on the Promotion of the use of Energy from Renewable Sources. December 2018. Available online: https://eur-lex.europa.eu/legal-content/EN/TXT/PDF/?uri=CELEX:32018L2001&from=PL (accessed on 29 July 2021).
23. Bojnec, Š.; Križaj, A. Electricity Markets during the Liberalization: The Case of a European Union Country. *Energies* **2021**, *14*, 4317. [CrossRef]
24. Swain, R.B.; Karimu, A. Renewable electricity and sustainable development goals in the EU. *World Dev.* **2020**, *125*, 104693. [CrossRef]
25. Evans, A.; Strezov, V.; Evans, T.J. Assessment of sustainability indicators for renewable energy technologies. *Renew. Sustain. Energy Rev.* **2009**, *13*, 1082–1088. [CrossRef]
26. Streimikiene, D.; Šivickas, G. The EU sustainable energy policy indicators framework. *Environ. Int.* **2008**, *34*, 1227–1240. [CrossRef]
27. Bohringer, C.; Loschel, A. Promoting Renewable Energy in Europe: A Hybrid Computable General Equilibrium Approach. *Energy J.* **2006**, SI2006. [CrossRef]
28. Özdemir, Ö.; Hobbs, B.F.; van Hout, M.; Koutstaal, P.R. Capacity vs energy subsidies for promoting renewable investment: Benefits and costs for the EU power market. *Energy Policy* **2020**, *137*, 111166. [CrossRef]
29. Radović, U. Support of the electricity generation from renewable energy sources in Poland: Is the additional system cost justified? *Polityka Energ. Energy Policy J.* **2005**, *8*, 469–482.
30. Reiche, D.; Bechberger, M. Policy differences in the promotion of renewable energies in the EU member states. *Energy Policy* **2004**, *32*, 843–849. [CrossRef]
31. Poullikkas, A.; Kourtis, G.; Hadjipaschalis, I. An overview of the EU Member States support schemes for the promotion of renewable energy sources. *Int. J. Energy Environ.* **2012**, *3*, 553–566.
32. Haas, R.; Panzer, C.; Resch, G.; Ragwitz, M.; Reece, G.; Held, A. A historical review of promotion strategies for electricity from renewable energy sources in EU countries. *Renew. Sustain. Energy Rev.* **2011**, *15*, 1003–1034. [CrossRef]
33. Nicolini, M.; Tavoni, M. Are renewable energy subsidies effective? Evidence from Europe. *Renew. Sustain. Energy Rev.* **2017**, *74*, 412–423. [CrossRef]
34. Kilinc-Ata, N. The evaluation of renewable energy policies across EU countries and US states: An econometric approach. *Energy Sustain. Dev.* **2016**, *31*, 83–90. [CrossRef]
35. Liobikienė, G.; Dagiliūtė, R. Do positive aspects of renewable energy contribute to the willingness to pay more for green energy? *Energy* **2021**, *231*, 120817. [CrossRef]
36. Lund, H. Renewable energy strategies for sustainable development. *Energy* **2007**, *32*, 912–919. [CrossRef]
37. EU ETS. EU ETS Handbook. 2015. Available online: https://ec.europa.eu/clima/policies/ets_en#tab-0-0 (accessed on 7 August 2021).
38. Chovancová, J.; Tej, J. Decoupling economic growth from greenhouse gas emissions: The case of the energy sector in V4 countries. *Equilibrium. Q. J. Econ. Econ. Policy* **2020**, *15*, 235–251. [CrossRef]
39. Akadiri, S.S.; Alola, A.A.; Akadiri, A.C.; Alola, U.V. Renewable energy consumption in EU-28 countries: Policy toward pollution mitigation and economic sustainability. *Energy Policy* **2019**, *132*, 803–810. [CrossRef]
40. Dincer, I. Renewable energy and sustainable development: A crucial review. *Renew. Sustain. Energy Rev.* **2000**, *4*, 157–175. [CrossRef]
41. Marques, A.C.; Fuinhas, J.A.; Manso, J.R.P. Motivations driving renewable energy in European countries: A panel data approach. *Energy Policy* **2010**, *38*, 6877–6885. [CrossRef]
42. Gökgöz, F.; Güvercin, M.T. Energy security and renewable energy efficiency in EU. *Renew. Sustain. Energy Rev.* **2018**, *96*, 226–239. [CrossRef]
43. Blum, H.; Legey, L.F.L. The challenging economics of energy security: Ensuring energy benefits in support to sustainable development. *Energy Econ.* **2012**, *34*, 1982–1989. [CrossRef]
44. Jonek-Kowalska, I. Transformation of energy balances with dominant coal consumption in European economies and Turkey in the years 1990–2017. *Oeconomia Copernic.* **2019**, *10*, 627–647. [CrossRef]
45. Filimonova, I.; Komarova, A.; Mishenin, M. Impact of the global green factor on the capitalization of oil companies in Russia. *Oeconomia Copernic.* **2020**, *11*, 309–324. [CrossRef]
46. Saad, W.; Taleb, A. The causal relationship between renewable energy consumption and economic growth: Evidence from Europe. *Clean Technol. Environ. Policy* **2018**, *20*, 127–136. [CrossRef]
47. Al-Mulali, U.; Ozturk, I.; Lean, H.H. The influence of economic growth, urbanization, trade openness, financial development, and renewable energy on pollution in Europe. *Nat. Hazards* **2015**, *79*, 621–644. [CrossRef]
48. Shahnazi, R.; Shabani, Z.D. Do renewable energy production spillovers matter in the EU? *Renew. Energy* **2020**, *150*, 786–796. [CrossRef]
49. Kisel'áková, D.; Šofranková, B.; Onuferová, E.; Čabinová, V. The evaluation of competitive position of EU-28 economies with using global multi-criteria indices. *Equilibrium. Q. J. Econ. Econ. Policy* **2019**, *14*, 441–462. [CrossRef]

50. Kothari, R.; Tyagi, V.V.; Pathak, A. Waste-to-energy: A way from renewable energy sources to sustainable development. *Renew. Sustain. Energy Rev.* **2010**, *14*, 3164–3170. [CrossRef]
51. Benedek, J.; Sebestyén, T.T.; Bartók, B. Evaluation of renewable energy sources in peripheral areas and renewable energy-based rural development. *Renew. Sustain. Energy Rev.* **2018**, *90*, 516–535. [CrossRef]
52. Johansson, T.B.; Kelly, H.; Reddy, A.K.N.; Williams, R.H. *Renewable Energy: Sources for Fuels and Electricity*; Island Press: Washington, DC, USA, 1993.
53. Amin, A.; Altinoz, B.; Dogan, E. Analyzing the determinants of carbon emissions from transportation in European countries: The role of renewable energy and urbanization. *Clean Technol. Environ. Policy* **2020**, *22*, 1725–1734. [CrossRef]
54. Jäger-Waldau, A.; Szabó, M.; Scarlat, N.; Monforti-Ferrario, F. Renewable electricity in Europe. *Renew. Sustain. Energy Rev.* **2011**, *15*, 3703–3716. [CrossRef]
55. Mathiesen, B.V.; Lund, H.; Nørgaard, P. Integrated transport and renewable energy systems. *Util. Policy* **2008**, *16*, 107–116. [CrossRef]
56. Cansino, J.M.; Pablo-Romero, M.d.; Román, R.; Yñiguez, R. Promoting renewable energy sources for heating and cooling in EU-27 countries. *Energy Policy* **2011**, *39*, 3803–3812. [CrossRef]
57. Lund, H. Renewable heating strategies and their consequences for storage and grid infrastructures comparing a smart grid to a smart energy systems approach. *Energy* **2018**, *151*, 94–102. [CrossRef]
58. Boesten, S.; Ivens, W.; Dekker, S.C.; Eijdems, H. 5th generation district heating and cooling systems as a solution for renewable urban thermal energy supply. *Adv. Geosci.* **2019**, *49*, 129–136. [CrossRef]
59. Braungardt, S.; Bürger, V.; Zieger, J.; Bosselaar, L. How to include cooling in the EU Renewable Energy Directive? Strategies and policy implications. *Energy Policy* **2019**, *129*, 260–267. [CrossRef]
60. Inayat, A.; Raza, M. District cooling system via renewable energy sources: A review. *Renew. Sustain. Energy Rev.* **2019**, *107*, 360–373. [CrossRef]
61. Scholz, Y. Renewable Energy Based Electricity Supply at Low Costs: Development of the REMix Model and Application for Europe. 2012. Available online: http://nbn-resolving.de/urn:nbn:de:bsz:93-opus-76350 (accessed on 17 July 2021).
62. The International Renewable Energy Agency (IRENA). Renewable Power: Sharply Falling Generation Costs. 2017. Available online: https://www.irena.org/-/media/Files/IRENA/Agency/Publication/2017/Nov/IRENA_Sharply_falling_costs_2017.pdf (accessed on 23 September 2021).
63. Rubene, I.; Koester, G. Recent dynamics in Energy Inflation: The Role of Base Effects and Taxes. ECB Economic Bulletin. 2021. Available online: https://www.ecb.europa.eu/pub/economic-bulletin/focus/2021/html/ecb.ebbox202103_04~{}0a0c8f0814.en.html (accessed on 23 September 2021).
64. Galchynskyi, L. Estimation of the price elasticity of petroleum products' consumption in Ukraine. *Equilibrium. Q. J. Econ. Econ. Policy* **2020**, *15*, 315–339. [CrossRef]
65. Aune, F.R.; Dalen, H.M.; Hagem, C. Implementing the EU renewable target through green certificate markets. *Energy Econ.* **2012**, *34*, 992–1000. [CrossRef]
66. Ringel, M. Fostering the use of renewable energies in the European Union: The race between feed-in tariffs and green certificates. *Renew. Energy* **2006**, *31*, 1–17. [CrossRef]
67. Connolly, D.; Lund, H.; Mathiesen, B.V. Smart Energy Europe: The technical and economic impact of one potential 100% renewable energy scenario for the European Union. *Renew. Sustain. Energy Rev.* **2016**, *60*, 1634–1653. [CrossRef]
68. Mathiesen, B.V.; Lund, H.; Connolly, D.; Wenzel, H.; Østergaard, P.A.; Möller, B.; Nielsen, S.; Ridjan, I.; Karnøe, P.; Sperling, K.; et al. Smart Energy Systems for coherent 100% renewable energy and transport solutions. *Appl. Energy* **2015**, *145*, 139–154. [CrossRef]
69. Lund, H.; Mathiesen, B.V. Energy system analysis of 100% renewable energy systems—The case of Denmark in years 2030 and 2050. *Energy* **2009**, *34*, 524–531. [CrossRef]
70. Hansen, K.; Mathiesen, B.V.; Skov, I.R. Full energy system transition towards 100% renewable energy in Germany in 2050. *Renew. Sustain. Energy Rev.* **2019**, *102*, 1–13. [CrossRef]
71. Hansen, K.; Breyer, C.; Lund, H. Status and perspectives on 100% renewable energy systems. *Energy* **2019**, *175*, 471–480. [CrossRef]
72. Lund, H.; Østergaard, P.A.; Connolly, D.; Ridjan, I.; Mathiesen, B.V.; Hvelplund, F.; Thellufsen, J.Z.; Sorknæs, P. Energy Storage and Smart Energy Systems. *Int. J. Sustain. Energy Plan. Manag.* **2016**, *11*, 3–14.
73. Baños, R.; Manzano-Agugliaro, F.; Montoya, F.G.; Gil, C.; Alcayde, A.; Gómez, J. Optimization methods applied to renewable and sustainable energy: A review. *Renew. Sustain. Energy Rev.* **2011**, *15*, 1753–1766. [CrossRef]
74. Ionescu, G.H.; Firoiu, D.; Pîrvu, R.; Enescu, M.; Rădoi, M.-I.; Cojocaru, T.M. The Potential for Innovation and Entrepreneurship in EU Countries in the Context of Sustainable Development. *Sustainability* **2020**, *12*, 7250. [CrossRef]
75. European Environmental Agency. Renewable Energy in Europe—Recent Growth and Knock-on Effects. pp. 1–65. 2016. Available online: http://www.eea.europa.eu/publications/renewable-energy-in-europe-2016/download (accessed on 29 July 2021).
76. Włodarczyk, B.; Firoiu, D.; Ionescu, G.H.; Ghiocel, F.; Szturo, M.; Markowski, L. Assessing the Sustainable Development and Renewable Energy Sources Relationship in EU Countries. *Energies* **2021**, *14*, 2323. [CrossRef]
77. Pacesila, M.; Burcea, S.G.; Colesca, S.E. Analysis of renewable energies in European Union. *Renew. Sustain. Energy Rev.* **2016**, *56*, 156–170. [CrossRef]
78. Gromada, A.; Trebska, P.; Wysokinski, M. Use of renewable energy in the European Union—Trends of change. In Proceedings of the Economic Science for Rural Development Conference, Jelgava, Latvia, 9–10 May 2019. [CrossRef]

79. Rokicki, T.; Michalski, K.; Ratajczak, M.; Szczepaniuk, H.; Golonko, M. Use of Renewable Energy Sources in European Union Countries. *Annu. Set Environ. Prot.* **2018**, *20*, 1318–1334.
80. Rokicki, T.; Perkowska, A. Diversity and Changes in the Energy Balance in EU Countries. *Energies* **2021**, *14*, 1098. [CrossRef]
81. Eurostat. Production of Electricity and Derived Heat by Type of Fuel[nrg_bal_peh]. 2021. Available online: https://appsso.eurostat.ec.europa.eu/nui/show.do?dataset=nrg_bal_peh&lang=en (accessed on 6 June 2021).
82. McQueen, J.B. Some methods of classification and analysis of multivariate observations. In Proceedings of the Fifth Berkeley Symposium on Mathematical Statistics and Probability, Los Angeles, CA, USA, 21 June–18 July 1967; pp. 181–197.
83. Han, J.; Kamber, M.; Pei, J. *Data Mining Concepts and Techniques*, 3rd ed.; Elsevier: Amsterdam, The Netherlands, 2012.
84. Yadav, J.; Sharma, M. A Review of K-mean Algorithm. *Int. J. Eng. Trends Technol.* **2013**, *4*, 2972–2976.
85. Bieszk-Stolorz, B.; Dmytrów, K. Spatial diversity of effectiveness of forms of professional activisation in Poland in years 2008–2014 by poviats. *Oeconomia Copernic.* **2019**, *10*, 113–130. [CrossRef]
86. Kaufman, L.; Rousseeuw, P.J. *Finding Groups in Data-An Introduction to Cluster Analysis*; John Wiley & Sons: New York, NY, USA, 1990.
87. Dudek, A. Silhouette Index as Clustering Evaluation Tool. In *Classification and Data Analysis. SKAD 2019. Studies in Classification, Data Analysis, and Knowledge Organization*; Jajuga, K., Batóg, J., Walesiak, M., Eds.; Springer: Berlin/Heidelberg, Germany, 2020.
88. Roszko-Wójtowicz, E.; Grzelak, M.M. Multi-dimensional analysis of regional investment attractiveness in Poland. *Equilibrium. Q. J. Econ. Econ. Policy* **2021**, *16*, 103–138. [CrossRef]
89. Walesiak, M. Rekomendacje w zakresie strategii postępowania w procesie klasyfikacji zbioru obiektów (ang. Recommendations for strategies to deal with the classification process of a set of objects). In *Przestrzenno-Czasowe Modelowanie i Prognozowanie Zjawisk Gospodarczych (Spatial-Temporal Modelling and Forecasting of Economic Phenomena)*; Zeliaś, A., Ed.; CeON Repository: Unspecified, Poland, 2006; pp. 185–203.
90. Eurostat. Simplified Energy Balances [nrg_bal_s]. Final Consumption—Energy Use. 2021. Available online: https://appsso.eurostat.ec.europa.eu/nui/show.do?dataset=nrg_bal_s (accessed on 29 July 2021).
91. Eurostat. Access to Methodology Information on Energy data. 2021. Available online: https://ec.europa.eu/eurostat/web/energy/methodology (accessed on 29 July 2021).
92. Eurostat. RAMON—Concepts and Definitions, Eurostat's Concepts and Definitions Database—Term Hydro (Code RA100). 2021. Available online: https://ec.europa.eu/eurostat/ramon/nomenclatures/index.cfm?TargetUrl=DSP_GLOSSARY_NOM_DTL_VIEW&StrNom=CODED2&StrLanguageCode=EN&IntKey=16493335&RdoSearch=&TxtSearch=&CboTheme=&IsTer=&ter_valid=0&IntCurrentPage=1 (accessed on 29 July 2021).
93. Eurostat. RAMON—Concepts and Definitions, Eurostat's Concepts and Definitions Database—Term Wind (Code RA300). 2021. Available online: https://ec.europa.eu/eurostat/ramon/nomenclatures/index.cfm?TargetUrl=DSP_GLOSSARY_NOM_DTL_VIEW&StrNom=CODED2&StrLanguageCode=EN&IntKey=16623435&RdoSearch=&TxtSearch=&CboTheme=&IsTer=&ter_valid=0&IntCurrentPage=121 (accessed on 29 July 2021).
94. Eurostat. RAMON—Concepts and definitions, Eurostat's Concepts and Definitions Database—Term Solar Photovoltaic (code RA420). 2021. Available online: https://ec.europa.eu/eurostat/ramon/nomenclatures/index.cfm?TargetUrl=DSP_GLOSSARY_NOM_DTL_VIEW&StrNom=CODED2&StrLanguageCode=EN&IntKey=22543617&RdoSearch=BEGIN&TxtSearch=solar&CboTheme=&IsTer=&IntCurrentPage=1&ter_valid=0 (accessed on 29 July 2021).
95. Eurostat. RAMON—Concepts and Definitions, Eurostat's Concepts and Definitions Database—Term Solar Thermal (Code RA410). 2021. Available online: https://ec.europa.eu/eurostat/ramon/nomenclatures/index.cfm?TargetUrl=DSP_GLOSSARY_NOM_DTL_VIEW&StrNom=CODED2&StrLanguageCode=EN&IntKey=22543667&RdoSearch=BEGIN&TxtSearch=solar&CboTheme=&IsTer=&IntCurrentPage=1&ter_valid=0 (accessed on 29 July 2021).
96. Balat, M. Hydropower Systems and Hydropower Potential in the European Union Countries. *Energy Sources Part A Recover. Util. Environ. Eff.* **2006**, *28*, 965–978. [CrossRef]
97. Sternberg, R. Hydropower's future, the environment, and global electricity systems. *Renew. Sustain. Energy Rev.* **2010**, *14*, 713–723. [CrossRef]
98. Renöfält, B.M.; Jansson, R.; Nilsson, C. Effects of hydropower generation and opportunities for environmental flow management in Swedish riverine ecosystems. *Freshw. Biol.* **2010**, *55*, 49–67. [CrossRef]
99. Botelho, A.; Ferreira, P.; Lima, F.; Pinto, L.M.C.; Sousa, S. Assessment of the environmental impacts associated with hydropower. *Renew. Sustain. Energy Rev.* **2017**, *70*, 896–904. [CrossRef]
100. European Commission. Hydropower, Why the EU Supports Hydropower Research and Innovation. 2021. Available online: https://ec.europa.eu/info/research-and-innovation/research-area/energy-research-and-innovation/hydropower_en (accessed on 5 August 2021).
101. Directive 2000/60/EC of the European Parliament and of the Council Establishing a Framework for Community Action in the Field of Water Policy. 2000. Available online: https://eur-lex.europa.eu/legal-content/en/ALL/?uri=CELEX%3A32000L0060 (accessed on 5 August 2021).
102. Fruhmann, C.; Tuerk, A.; Kulmer, V.; Gubina, A.F. Balancing environmental benefits and damages of small hydropower generation in policy-making: Assessing the implementation of a contradicting EU policy framework in Austria and Slovenia. *Int. J. Sustain. Energy* **2019**, *38*, 37–49. [CrossRef]
103. International Water Power & Dam Construction. EIB Funds New Dams in Portugal, 23 July 2018. Available online: https://www.waterpowermagazine.com/news/newseib-funds-new-dams-in-portugal-6261561 (accessed on 29 July 2021).

104. Bohanec, M.; Trdin, N.; Kontić, B. A qualitative multi-criteria modelling approach to the assessment of electric energy production technologies in Slovenia. *Cent. Eur. J. Oper. Res.* **2017**, *25*, 611–625. [CrossRef]
105. Štreimikiene, D.; Šliogeriene, J.; Turskis, Z. Multi-criteria analysis of electricity generation technologies in Lithuania. *Renew. Energy* **2016**, *85*, 148–156. [CrossRef]
106. Jatautas, J.; Kasiulis, E. The effect of legislation on hydropower development: Case study of Lithuania. *Investig. Manag. Financ. Innov.* **2016**, *13*, 300–309. [CrossRef]
107. Wang, S.; Wang, S. Impacts of wind energy on environment: A review. *Renew. Sustain. Energy Rev.* **2015**, *49*, 437–443. [CrossRef]
108. Dragomir, G.; Şerban, A.; NĂstase, G.; Brezeanu, A.I. Wind energy in Romania: A review from 2009 to 2016. *Renew. Sustain. Energy Rev.* **2016**, *64*, 129–143. [CrossRef]
109. European Commission. Communication from the Commission to the European Parliament, the European Council, the Council, the European Economic and Social Committee, the Committee of the Regions and the European Investment Bank: A Clean Planet for All. A European Strategic Long-Term Vision for a Prosperous, Modern, Competitive and Climate Neutral Economy COM/2018/773 Final, Brussels. 2018. Available online: https://eur-lex.europa.eu/legal-content/EN/TXT/?uri=CELEX:52018DC0773 (accessed on 30 July 2021).
110. European Commission. 2050 Long-Term Strategy. 2021. Available online: https://ec.europa.eu/clima/policies/strategies/2050_en#tab-0-0 (accessed on 30 July 2021).
111. European Commission. Why the EU Supports Wind Energy Research and Innovation. 2021. Available online: https://ec.europa.eu/info/research-and-innovation/research-area/energy-research-and-innovation/wind-energy_en (accessed on 30 July 2021).
112. Papież, M.; Śmiech, S.; Frodyma, K. Factors affecting the efficiency of wind power in the European Union countries. *Energy Policy* **2019**, *132*, 965–977. [CrossRef]
113. Renewable Energy Association. The European Commission Has Approved the Polish Offshore Wind Farm Support Program. 2021. Available online: http://seo.org.pl/en/ke-zatwierdzila-polski-program-wsparcia-morskich-farm-wiatrowych (accessed on 30 July 2021).
114. Higgins, P.; Foley, A. The evolution of offshore wind power in the United Kingdom. *Renew. Sustain. Energy Rev.* **2014**, *37*, 599–612. [CrossRef]
115. Kandrot, S.; Cummins, V.; Jordan, D.; Murphy, J. Economic and employment impacts of offshore wind for Ireland: A value chain analysis. *Int. J. Green Energy* **2020**, *17*, 687–696. [CrossRef]
116. O'Hanlon, Z.; Cummins, V. A comparative insight of Irish and Scottish regulatory frameworks for offshore wind energy—An expert perspective. *Mar. Policy* **2020**, *117*, 103934. [CrossRef]
117. Johansen, K. Wind Energy in Denmark: A Short History [History]. *IEEE Power Energy Mag.* **2021**, *19*, 94–102. [CrossRef]
118. Chu, Y.; Meisen, P. *Review and Comparison of Different Solar Energy Technologies*; Global Energy Network Institute (GENI): San Diego, CA, USA, 2011.
119. Šúri, M.; Huld, T.A.; Dunlop, E.D.; Ossenbrink, H.A. Potential of solar electricity generation in the European Union member states and candidate countries. *Sol. Energy* **2007**, *81*, 1295–1305. [CrossRef]
120. Hofierka, J.; Kaòuk, J.; Gallay, M. The Spatial Distribution of Photovoltaic Power Plants in Relation to Solar Resource Potential: The Case of the Czech Republic and Slovakia. 2014. Available online: http://sunbird.jrc.it/pvgis/ (accessed on 30 July 2021).
121. Timilsina, G.R.; Kurdgelashvili, L.; Narbel, P.A. Solar energy: Markets, economics and policies. *Renew. Sustain. Energy Rev.* **2012**, *16*, 449–465. [CrossRef]
122. European Commission. Why the EU Supports Bioenergy Research and Innovation. 2021. Available online: https://ec.europa.eu/info/research-and-innovation/research-area/energy-research-and-innovation/bioenergy_en (accessed on 30 July 2021).

energies

Article

Comparison of Renewable Energy Sources in 'New' EU Member States in the Context of National Energy Transformations

Agnieszka Wałachowska [1,*] and Aranka Ignasiak-Szulc [2]

1 Faculty of Management and Economics, Gdansk University of Technology, Gabriela Narutowicza 11/12 St., 80-233 Gdańsk, Poland
2 Faculty of Economic Sciences and Management, Nicolaus Copernicus University in Toruń, Gagarina 11, 87-100 Toruń, Poland; aranka@umk.pl
* Correspondence: agnwalac@pg.edu.pl

Abstract: The European Union strives to create sustainable, low-carbon economies; therefore, energy policies of all member states should move towards renewable energy sources (RES). That concerns also the so-called new EU member states. These countries, on the one hand, are characterized by significant historical similarities in terms of post-communist legacy and adopted development strategies linked with the EU membership, and on the other hand, by significant social, economic and environmental differences resulting from different transformation and development paths and conditions. The question remains how the selected countries should cope with actions in the field of national energy transformations to confront the multiple challenges linked to assuring a significant level of sustainable development. In order to be successful, it is necessary to conduct an effective and rapid changes in the energy industry, which should be preceded by an analysis of the differentiation of countries in terms of their potentials. The results of such analyses should be helpful in selecting the most appropriate strategies for transformation of the described industry. Therefore, the purpose of the article is to assess the new EU member states for RES diversification and identify similar subgroups of countries using cluster analysis, taking into account the percentage share of individual renewable energy sources in total renewable energy production. This was done for the years 2010, 2015 and 2019 which should allow us to demonstrate the differences between them as a group and also reveal changes recorded over time for a single country. Ward's method was used for the analysis. The presented approach to the analysis of energy production enabled the acquisition of new knowledge in this field and supported the assessment of the current state of RES. The results obtained can be used in countries of comparable specificity to undertake activities of similar nature in relation to internal energy production, technological development or common energy policy.

Keywords: renewable energy sources (RES); energy transformation; the new EU member states; cluster analysis; Ward's method

Citation: Wałachowska, A.; Ignasiak-Szulc, A. Comparison of Renewable Energy Sources in 'New' EU Member States in the Context of National Energy Transformations. *Energies* **2021**, *14*, 7963. https://doi.org/10.3390/en14237963

Academic Editor: Jin-Li Hu

Received: 25 October 2021
Accepted: 25 November 2021
Published: 29 November 2021

Publisher's Note: MDPI stays neutral with regard to jurisdictional claims in published maps and institutional affiliations.

Copyright: © 2021 by the authors. Licensee MDPI, Basel, Switzerland. This article is an open access article distributed under the terms and conditions of the Creative Commons Attribution (CC BY) license (https://creativecommons.org/licenses/by/4.0/).

1. Introduction

In the currently ongoing processes of globalization, closely related with the dynamics of socio-economic changes, one of the main problems faced by countries in almost every region of the world is the growing importance of energy resources [1,2].

The rate of economic growth, unprecedented in history, forces the participants of the global economy to pay special attention to the uninterrupted satisfaction of energy needs [3]. The decades-long increase in demand for energy raises a number of challenges in terms of its acquisition, transmission, processing and distribution [4,5]. This gives a rise to complex contemporary problem which is based on the issue of energy security of individual countries and which is the foundation of the policy of most countries in the world. Such a broad subject matter, in fact, covering an infinite number of issues in the field of geopolitics, international economic relations, economics or technology, is the context for the issues discussed in this study related to the use of renewable energy by the post-communist countries—the so-called new member states that joined the structures of the European Union in 2004 (Czechia/Czech Republic, Estonia, Hungary, Lithuania, Latvia, Poland, Slovakia and Slovenia) and in 2007 (Bulgaria and Romania).

The area of Central and Eastern Europe (CEE) is one of the most important regions of Eurasia in geopolitical terms. The countries of this region have made a civilization leap over the past decades, carrying out an impressive political, social and economic transformation [6,7]. To a large extent, this process was achieved through integration with the European Union. This study focuses on the new EU member states due to the fact that these countries, thanks to their accession to the EU, received great impetus for the energy transformation. Along with the progressive changes in individual countries of the discussed region, traditional energy sources based on oil, coal, and natural gas are considered the most popular and effective drivers for economic development, but at the same time they are also harmful to the environment and to human health [8,9]. In the era of climate change, environmental degradation, and also growing public awareness of environmental concerns [10], there is a need to diversify traditional energy sources that rely on fossil fuels by new ecological sources [11,12]. This condition may be met by renewable energy sources (RES), that according to glossary used by EUROSTAT are energy sources that replenish (or renew—that is why they are sometimes called renewables) themselves naturally, i.e., those generated by natural resources that are not finite (exhaustible). They include, for example, biofuels (fuels from biomass) and renewable municipal waste (i.e., combustible renewables), and non-combustible renewables as wind, solar, hydropower, and geothermal energy sources, etc. Renewable energy may have some disadvantages or limitations (high upfront costs, storage capabilities, intermittency, etc.) but it can serve as a potential way to restore balance between economic growth and environmental quality [13,14]. In the countries located in the CEE, energy security issues (as ability to act as a unified bloc) may be still relatively more important compared with the countries located in the West (especially considering their relations with Russia and having in mind so called gas crisis in Ukraine in 2006) but with growing concern of climate change, the issues related with environmental protection and sustainability affect rapidly growing world economies with increasing energy demand, including economies of the European Union member states and CEE countries. The situation in individual member states differs significantly but the EU (treated as one entity) may be seen clearly as one of the largest greenhouse gas emitters [15], and coordination of climate policy is needed also on the community level. The basis for the European energy policy was introduced by the Treaty of Lisbon in Article 194 of the Treaty on the Functioning of the European Union [16]. Pursuing a green economy, mainly understood as a low-emission economy, means the EU's climate policy is largely focused on RES on the way to climate neutrality. Climate neutrality refers to zeroing greenhouse gas emissions, i.e., reducing their emissions from industry, transport, and energy sectors as much as possible and offsetting the emissions that could not be eliminated by increasing their removal. Under the 2015 Paris Agreement (COP21), the EU promotes an energy union aimed at building energy security and solidarity as well as a fully integrated internal market, supporting research and competitiveness, accelerating energy efficiency, and climate-oriented actions for a carbon-neutral EU economy by 2050. The latter is an objective of the European Green Deal (COM(2019)640 final). This ambitious package includes not just suggestions, member states have to follow distinctive paths when it comes to meeting their obligations under the renewable energy directives, including legally binding 2020 targets. A general target for increasing the share of renewable energy sources has been set, according to which it should reach at least 32% by 2030 in line with the EU climate and energy framework (20% in 2020). The main reason for these actions is to provide EU consumers with safe, sustainable, competitive, and affordable energy. Charles Michel, the President of the European Council, said that "climate neutrality is no longer a question of choice, it is beyond doubt a necessity" [17].

To be successful in this field, it is necessary to carry out an effective and fast transformation of the energy industry process. Even though the EU as a whole is on track to meet its targets, the question is whether the generation of renewable energy at a given level is a challenge for selected new member states countries dependent on fossil fuels. It can be expected that some countries from the research area that generate energy from nuclear sources (Hungary, Czech Republic), i.e., from sources with zero emission of CO_2, will be less inclined towards the development of RES. Undoubtedly, the new member states include countries whose share of energy from renewable sources was much more than 20% already in 2019 (Estonia 32%, Latvia 41%) (Figure 1). Nonetheless, the Central and Eastern Europe countries also include those for which RES is still not sufficiently important, with the share of energy from renewable sources remaining low (Poland 12.2%, Hungary 12.6%) (Figure 1).

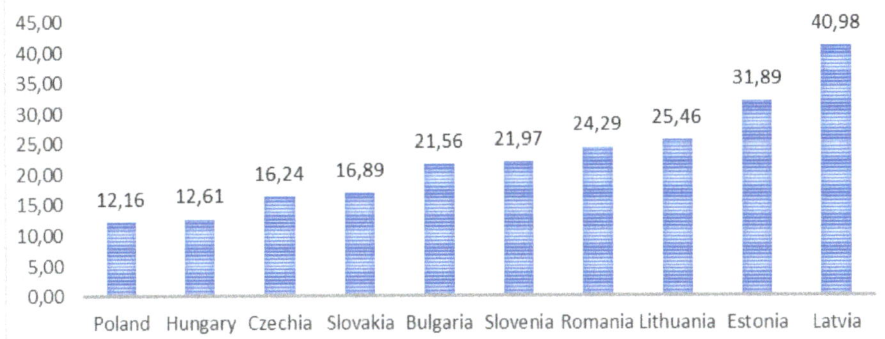

Figure 1. Share of energy from renewable sources in the new member states countries in 2019. Source: own calculations based on data from Eurostat (online data code: NRG_IND_REN_custom_716443).

The progressive European integration in the political, economic and social dimensions is still functioning in isolation from the energy transformation process, which as the foundation of security is implemented in various ways by individual member states. On the one hand, the inflow of capital from the EU helps to introduce innovations, new consumption patterns, diversification of energy sources with an emphasis on the largest possible share of energy from renewable sources. On the other hand, this process comes at a huge social cost. This element can be mitigated in the process of globalization, which affects the exchange of information between countries influencing the increase in public awareness and international integration of economies. That creates favorable conditions for more dynamic development, mostly for economic centers, both at national and regional level [18–22].

To take action for RES development in the examined countries, it is necessary to analyze their differences in terms of potential. These differences depend on the energy policies of individual countries as well as the environmental awareness of societies. The economic aspect is also important. It is undeniable that energy production from RES is considerably more expensive than from conventional sources. As a result of the high upfront costs, many new member states countries seem unable to cover them from their own resources, what limits the potential for renewable energy development in the area. The potential and availability of individual renewable energy sources are also of great importance. In order to find the most effective solutions to common problems, analyses have been carried out to identify similarities between new member states countries in terms of the structure of RES energy production. They usually include the eight major renewable energy sources: hydro, geothermal, wind, and solar as well as primary solid biofuels, biogases, renewable municipal waste, and liquid biofuels. The analysis indicates groups of the CEE countries with the highest similarly identified. So far, not a lot of analyses focusing on the new member states countries with these factors have been conducted in the literature, there are just single and very recent papers that include analyses of the EU indicating the specifics of the countries (or their groups) in the region [23]. Therefore, the approach used can be still be called new and the results of such analyses should support the implementation of the adopted strategies in the countries of the region. It is known that one of the main factors affecting the pace of changes is the amount of public spending on energy transition [24,25]; therefore, the pro-environment policy implemented and required to be continued by the EU must be adapted to the uniqueness of the new EU member states. In addition, the awareness of mutual similarities among new member states should promote cooperation and acceleration of actions towards energy transition.

Due to the exchange of information, the development of the energy system, including RES, is possible. In order to be able to take a holistic view of the energy transformation process in the discussed region, it is worth seeing the similarities and differences in RES production in surveyed countries. The aim of the paper is to analyze the level of diversification of renewable energy sources in three selected years for analysis (2010, 2015, 2019) and to identify groups of similar countries due to the structure of the percentage share of individual renewable energy sources in the total production of renewable energy. The implementation of the goal allowed the countries to be organized in terms of the dominant role of selected energy sources. The assessment of the state of RES in the new member states countries in these three years also indicated possible directions of the state's policy in the field of the ongoing energy transformation. In addition, the obtained research results can be related to groups of countries where, due to the similarities in the nature of RES, they can take actions of a similar nature at the EU level with regard to internal energy production, technological development or common energy policy. The adopted research approach is proprietary due to the classification of countries in terms of the percentage share of individual renewable energy sources in the total production of renewable energy. Most of the research is carried out on the values of energy produced from selected RES (per capita or per GDP product), which may lead to erroneous conclusions. The paper presents Figure 1, which shows that selected countries currently differ significantly in terms of the share of energy production from RES in the total energy production. This means that the analysis of the similarity of countries in the context of comparing the value of energy production from selected RES should be performed carefully.

The paper is organized as follows: an overview of the related literature is provided in Section 2; the description of methods is included in Section 3, data are described in the Section 4, and results in the Section 5. Two final sections: include discussion (Section 6) and conclusions (Section 7).

2. Literature Review

The importance of renewable energy in the energy mix is increasingly recognized. Traditionally, literature reviews of the papers devoted to the issue have been based on the problem of sustainable development (SD) and strategies or policies of achieving it in the context of international or national goals [26,27]. SD is today a well-established term both in all territorial perspectives as it had been introduced already by 1970 by the United Nations and the most common definition was proposed by the Brundtland Commission of 1987 and the interdisciplinary discussions that followed the publication of *Our Common Future* [28–30]. They allowed SD policy to be distinguished from environmental policy as it can only be achieved by integration of environmental, social, and economic concerns and objectives, and it has given a fresh impetus for advancing both theoretical and practical aspects of SD. It became a guiding institutional principle and a policy goal, implemented also in the EU.

The advantages of renewable energy for the environment are undeniable and widely discussed in the world literature as RES are usually seen as an element that will contribute to mitigating climate change and the opportunity to achieve also other sustainability targets as RES technologies can facilitate social and economic development [31]. Therefore, there are also analyses concerning issues related to climate change prepared for the EU [32] and the CEE countries [33]. In order to address the sustainability challenges the considerations connected with environmental concerns are not only concentrated on ecosystem complexity but accompanied by social and economic impacts of RES in territorial [34] or general perspective [35]. As mentioned, energy policy is often perceived also in the energy security context [36–38] also as RES has the potential to improve energy security by making the system more resilient to disruptions.

As the energy sector is often seen as a key to economic development, there are numerous studies on the link between RES and various macroeconomic variables [39,40]. They often focus on the impact of RES on economic growth. In their studies, Ohler and Fetters [41] identify a two-way relationship between aggregate renewable energy and real GDP. Simultaneously, they argue that biomass, hydroelectric, and waste electricity generation have the largest impact on real GDP in the long-run. The findings of Kula [42] also support the existence of a long-term balanced relationship between renewable electricity consumption and GDP, while Anwar, Arshed and Kousar [43] argue that the impact of renewable electricity consumption on economic growth is positive and significant. As for measuring SD it is necessary to go beyond the quantitative indicators, a set of aggregate indicators of SD is usually suggested and widely discussed in the literature [31,44,45]. They have gained a lot of criticism due to the arbitrary character of procedures used in their construction, but one of the advantages of the development of such indicators is that they allow international comparability.

An international perspective is important here as issues related to climate change call for cross-border cooperation. The cooperation in the EU connected with Green Deal described above is important but in terms of the expansion of renewable energy is also significant and results in a possible increase in the efficiency of individual countries. Nevertheless, the studies show that the majority of European countries are yet to pursue any cooperation and wish to achieve the expansion objective in terms of renewable energy primarily through expansion within their own national borders [46,47]. On the other hand, history shows that new member states' energy security meant the pursuit of the highest self-sufficiency and independence compared to other entities. For several years, these countries have been transforming their economies at a different pace, with the aim of the marketisation of economic processes. Many changes are underway in the industry area, which is still energy-intensive and dependent on primary energy sources. Although transition economies are trying to become greener through the reduction of greenhouse gas emissions, they are doing so at different rates. Some of them, such as Estonia, Latvia or Lithuania, have undergone a drastic transformation, whereas others, such as Hungary or Slovenia, are making changes at a slower pace [48]. As noted by others, the large area of forestry land in Central Europe makes forest biomass and waste from the wood processing industry (sawdust, chips) or in a processed form (pellet, briquette) suitable for energy purposes, while the involvement of agriculture in building bioenergy production capacity, in particular the development and distribution of multiannual crops and their potential in new member states countries, is crucial for bioenergy in the long term [49].

The development of RES brings about challenges in terms of how to run energy policy as countries with a diversified level of RES offer incentives to advance RES [31] and obtain goals set at international, national, regional and local level. It must be more stable and, thus, more credible and clearer to citizens. This is the EU energy policy that is characterized by a high degree of social involvement in the ownership, management, and benefits of energy projects [50,51]. In addition, econometric analysis reveals a positive correlation between subsidies and the generation of motivated energy as well as installed capacity [24,52]. Bersalli, Menanteau and El-Methni [53] showed that promotional policies have a positive and statistically significant effect on RES investment.

Papers addressing the issue of renewable energy sources in the context of Europe mainly focus on the European Union [54–56]. Few of them focus on the uniqueness of Central and Eastern Europe countries [23], for which, for historical reasons, it is a challenge to change the regulatory system in order to allow entities not controlled by their governments to produce and transmit electricity. Recently, Pakulska [57] proved that for most of the CEE countries transformation towards a climate-neutral economy is challenging task. Papers focusing on the new member states countries often address a single country in terms of renewable energy for the near future [58,59] or a specific renewable energy source of a selected country [60].

What is proved in the research presented above is that energy transition, aimed at replacing conventional energy sources with RES, what is more, is a complex process that depends on many factors, and the reasons for the uneven distribution of energy production from RES in new member states countries are diverse and include social, economic, and environmental (like issues connected with climate, hydrology and geology). The results also support the need to conduct a study that should focus not only on the absolute values of renewable energy, but also on other factors that may shape the volume and structure of such RES production in the future. Therefore, such an approach to this analysis appears to be fully justified.

The analysis of new member states countries in this study fills a gap in the existing literature and contributes to the discussion on the European Strategy 2020–2030 for the whole EU and especially for the countries of the region of Central and Eastern Europe—new member states. The approach adopted in this paper gives a broader view and considers the uniqueness of each new member states countries.

3. Methods

A wide group of methods that are used in the problems of multivariate comparative analyses are taxonomic methods, also commonly found under the term "cluster analysis" [61,62]. They deal with the rules and procedures for the classification of various types of objects. Taxonomies have many applications that span a variety of fields. For example, in economics and finance, taxonomies are used to group countries based on sets of development indicators or to recognize the level of regions development [63,64]. The most general division allows taxonomic methods to be converted to hierarchical methods (agglomeration and division) and grouping by the k-means method, where objects are assigned to k clusters and the number of clusters is determined by the researcher [65]. Among the agglomeration algorithms, Ward's method is widely used [66–68].

The research method applied in this paper is cluster analysis. Cluster analysis is a method used in multivariate comparative analyzes that breaks down a large group of objects into relatively homogeneous groups called clusters. In general, cluster analysis is used to classify n objects, while these objects are described with k statistical features. In the analysis of clusters, the similarity or dissimilarity of objects is taken into account, and on this basis groups of objects (clusters) that are mutually exclusive are distinguished. The objects assigned to each cluster are similar to each other in terms of the values of all k variables.

The article uses the method of Ward [69], because it is the most frequently used method in economic research [61,70,71]. In Ward's method, the sum of squares within groups is minimized, where in the first stage of grouping, each object forms an independent cluster. In the next steps, the standalone clusters are grouped into superior clusters based on the selected distance measure. In the last step, all statistical objects are combined into one cluster [61].

The use of cluster analysis has allowed us to separate homogeneous subsets of population objects, which are new member states countries, based on variables describing the examined countries, i.e., the value of energy produced from RES commonly used in Eurostat analyses. The main idea behind cluster analysis is to group objects (countries) in such a way that the objects included in the same group are characterized by a significant similarity and at the same time they differ from objects from other groups as much as possible. To do so, the Euclidean distance was used as a measure of distance, which is given by:

$$d(x,y) = \sqrt{\sum_{i=1}^{p}(x_i - y_i)^2}, \quad (1)$$

where $x = (x_1, \ldots, x_p)$ and $y = (y_1, \ldots, y_p)$, and in this case $p = 8$, which is the number of variables that characterise a country. The greater the distance between two countries, the more diverse they are. As a result, a cluster includes countries close to each other and far away from others that form separate clusters.

Before the determination of distance matrices, the variables were standardized using the formula:

$$z_i = \frac{x_i - \overline{x}}{s_x}, \quad (2)$$

where $\overline{x}$ and s_x refer to the mean and standard deviation of the sample.

The agglomerative hierarchical clustering algorithm was applied in the first step of analysis. The agglomeration method was Ward's method, which is used to minimize the sum of the squares of within-cluster variance. This resulted in a graphical illustration of the agglomeration pattern in the form of a diagram referred to as a dendrogram and the suggestion of the number of clusters to which the countries are to be assigned. In the second step of the analysis k-means non-hierarchical clustering was used. The optimal number of clusters was determined with the use of the Silhouette index [72–74]:

$$S(u) = \frac{1}{n}\sum_{i=1}^{n}\frac{b(i) - a(i)}{\max\{a(i), b(i)\}}, \; S(u) \in [-1, 1], \quad (3)$$

where u is the number of classes, n is the object (country) number, $a(i)$ is the mean distance of the country with index i from other countries belonging to class number r, $r = 1, \ldots, u$. $b(i)$ is the mean distance of the country with index i from other countries belonging to class number s, $s = 1, \ldots, u$. The criterion based on the Silhouette index indicates the selection of the number of classes u, for which index $S(u)$ takes the maximum value.

The results obtained on the basis of Ward's method are most often presented in the form of a dendrogram. At the top of the dendrogram, all objects form one shared cluster. Moving to lower levels, successive clusters with a smaller number of objects are distinguished, where at the lowest level all objects form separate clusters [71,75,76].

4. Data

Data on the structure of RES energy production for the new member states was taken from the Eurostat database for 2010 (Table 1), 2015 (Table 2) and 2019 (Table 3).

Table 1. Structure of energy production from renewable energy sources (RES) in the new member states in 2010.

Countries	Hydro	Geothermal	Wind	Primary Solid Biofuels	Biogases	Renewable Municipal Waste	Liquid Biofuels	Solar	Total
				Gigawatt-Hour					
Bulgaria	5692.52	0.00	681.37	19.74	15.50	0.00	0.00	14.87	6423.99
Czechia	3380.58	0.00	335.49	1492.24	634.66	35.59	0.00	615.70	6494.26
Estonia	26.88	0.00	276.99	729.82	10.19	0.00	0.00	0.00	1043.88
Latvia	3520.51	0.00	49.06	8.41	56.69	0.00	0.00	0.00	3634.66
Lithuania	1295.00	0.00	224.00	116.00	31.00	0.00	0.00	0.00	1666.00
Hungary	188.38	0.00	533.79	2034.28	117.32	144.86	0.00	0.85	3019.49
Poland	3488.14	0.00	1664.34	5905.21	398.38	0.00	0.90	0.00	11,456.97
Romania	20,242.98	0.00	306.35	109.84	0.65	0.00	0.00	0.04	20,659.86
Slovenia	4702.96	0.00	0.00	119.53	97.35	0.00	0.00	12.86	4932.69
Slovakia	5649.00	0.00	6.00	606.00	34.00	22.00	0.00	16.94	6333.94

Table 2. Structure of energy production from RES in the new member states in 2015.

Countries	Hydro	Geothermal	Wind	Primary Solid Biofuels	Biogases	Renewable Municipal Waste	Liquid Biofuels	Solar	Total
				Gigawatt-Hour					
Bulgaria	6146.54	0.00	1451.51	151.08	119.11	0.00	0.00	1383.27	9251.51
Czechia	3070.77	0.00	572.61	2091.50	2610.96	86.65	0.00	2263.85	10,696.34
Estonia	26.60	0.00	715.00	710.00	50.00	58.00	0.00	0.00	1559.60
Latvia	1860.36	0.00	147.14	377.78	391.71	0.00	0.00	0.24	2777.22
Lithuania	1024.00	0.00	810.00	318.00	86.00	42.00	0.00	73.00	2353.00
Hungary	233.71	0.00	693.32	1660.96	292.99	207.29	0.00	141.00	3229.27
Poland	2435.20	0.00	10,858.37	9026.64	906.40	0.00	3.82	56.64	23,287.06
Romania	17,006.55	0.10	7062.93	462.27	60.78	0.00	0.00	1982.00	26,574.63
Slovenia	4090.18	0.00	6.03	131.28	132.30	0.00	0.00	274.23	4634.02
Slovakia	4137.00	0.00	6.00	1099.00	541.00	22.00	0.00	506.00	6311.00

Table 3. Structure of energy production from RES in the new member states in 2019.

Countries	Hydro	Geothermal	Wind	Primary Solid Biofuels	Biogases	Renewable Municipal Waste	Liquid Biofuels	Solar	Total
				Gigawatt-Hour					
Bulgaria	3382.75	0.00	1316.99	1545.79	230.72	44.30	0.00	1442.47	7963.01
Czechia	3174.69	0.00	700.01	2398.73	2528.08	104.85	0.00	2311.57	11,217.94
Estonia	19.00	0.00	687.00	1259.79	38.84	64.07	0.00	73.50	2142.20
Latvia	2107.55	0.00	154.00	575.02	352.40	0.00	0.00	3.14	3192.11
Lithuania	947.70	0.00	1499.40	330.70	154.40	48.10	0.00	91.10	3071.40
Hungary	219.00	18.00	729.00	1769.00	318.00	137.00	0.00	1497.00	4687.00
Poland	2664.88	0.00	15,106.76	6441.15	1135.01	104.83	1.99	710.67	26,165.30
Romania	16,005.70	0.00	6772.81	450.34	53.81	0.00	0.00	1777.62	25,060.27
Slovenia	4682.54	0.00	6.15	151.46	94.36	0.00	5.22	303.04	5242.76
Slovakia	4571.00	0.00	6.00	1130.00	534.00	29.00	0.00	589.00	6859.00

Then, the percentage share of eight individual variables describing the volume of energy production from selected renewable sources in total renewable energy was calculated for each year—2010 (Table 4), 2015 (Table 5) and 2019 (Table 6). Initial calculations pointed to strong differences in the levels of percentage share of individual renewable energy sources in total renewable energy from selected renewable sources. The significant differences of the new member states in terms of RES production structure are reflected by the high coefficient of variation of individual variables. Therefore, the presented average values for the percentage share of individual renewable energy sources in the total production of renewable energy, calculated for all countries, have a low cognitive value and only the cluster analysis performed in the further part of the article will allow for a proper comparison of selected countries. In addition, the correlation of the variables describing the countries was not high, which supported the applicability of the Euclidean distance to conduct cluster analysis.

Table 4. Percentage share of individual RES in total production in 2010.

Countries	Hydro	Geothermal	Wind	Primary Solid Biofuels	Biogases	Renewable Municipal Waste	Liquid Biofuels	Solar
				%				
Bulgaria	88.61	0.00	10.61	0.31	0.24	0.00	0.00	0.23
Czechia	52.05	0.00	5.17	22.98	9.77	0.55	0.00	9.48

Table 4. Cont.

Countries	Hydro	Geothermal	Wind	Primary Solid Biofuels	Biogases	Renewable Municipal Waste	Liquid Biofuels	Solar
				%				
Estonia	2.58	0.00	26.53	69.91	0.98	0.00	0.00	0.00
Latvia	96.86	0.00	1.35	0.23	1.56	0.00	0.00	0.00
Lithuania	77.73	0.00	13.45	6.96	1.86	0.00	0.00	0.00
Hungary	6.24	0.00	17.68	67.37	3.89	4.80	0.00	0.03
Poland	30.45	0.00	14.53	51.54	3.48	0.00	0.01	0.00
Romania	97.98	0.00	1.48	0.53	0.00	0.00	0.00	0.00
Slovenia	95.34	0.00	0.00	2.42	1.97	0.00	0.00	0.26
Slovakia	89.19	0.00	0.09	9.57	0.54	0.35	0.00	0.27

Table 5. Percentage share of individual RES in total production in 2015.

Countries	Hydro	Geothermal	Wind	Primary Solid Biofuels	Biogases	Renewable Municipal Waste	Liquid Biofuels	Solar
				%				
Bulgaria	66.44	0.00	15.69	1.63	1.29	0.00	0.00	14.95
Czechia	28.71	0.00	5.35	19.55	24.41	0.81	0.00	21.16
Estonia	1.71	0.00	45.85	45.52	3.21	3.72	0.00	0.00
Latvia	66.99	0.00	5.30	13.6	14.1	0.00	0.00	0.01
Lithuania	43.52	0.00	34.42	13.51	3.65	1.78	0.00	3.10
Hungary	7.24	0.00	21.47	51.43	9.07	6.42	0.00	4.37
Poland	10.46	0.00	46.63	38.6	3.89	0.00	0.02	0.24
Romania	64.00	0.00	26.58	1.74	0.23	0.00	0.00	7.46
Slovenia	88.26	0.00	0.13	2.83	2.86	0.00	0.00	5.92
Slovakia	65.55	0	0.10	17.41	8.57	0.35	0.00	8.02

Table 6. Percentage share of individual RES in total production in 2019.

Countries	Hydro	Geothermal	Wind	Primary Solid Biofuels	Biogases	Renewable Municipal Waste	Liquid Biofuels	Solar
				%				
Bulgaria	42.48	0.00	16.54	19.41	2.90	0.56	0.00	18.11
Czechia	28.30	0.00	6.24	21.38	22.54	0.93	0.00	20.61
Estonia	0.89	0.00	32.07	58.81	1.81	2.99	0.00	3.43
Latvia	66.02	0.00	4.82	18.01	11.04	0.00	0.00	0.10
Lithuania	30.86	0.00	48.82	10.77	5.03	1.57	0.00	2.97
Hungary	4.67	0.38	15.55	37.74	6.78	2.92	0.00	31.94
Poland	10.18	0.00	57.74	24.62	4.34	0.40	0.01	2.72
Romania	63.87	0.00	27.03	1.80	0.21	0.00	0.00	7.09
Slovenia	89.31	0.00	0.12	2.89	1.80	0.00	0.10	5.78
Slovakia	66.64	0.00	0.09	16.47	7.79	0.42	0.00	8.59

5. Results

Using the algorithms discussed in the previous section, an analysis was conducted to select subgroups among the 10 selected countries (the new EU member states) based on eight variables characterizing percentage share of individual renewable energy sources in total renewable energy. The number of clusters was determined with the use of the Silhouette index. The index points to the adoption of two classes, it also seems reasonable to consider four clusters, because for $u = 4$, $S(u)$ reaches the second maximum.

Next, the countries were assigned to four clusters with the use of hierarchical clustering and Ward's method for each year—2010 (Figure 2), 2015 (Figure 3) and 2019 (Figure 4).

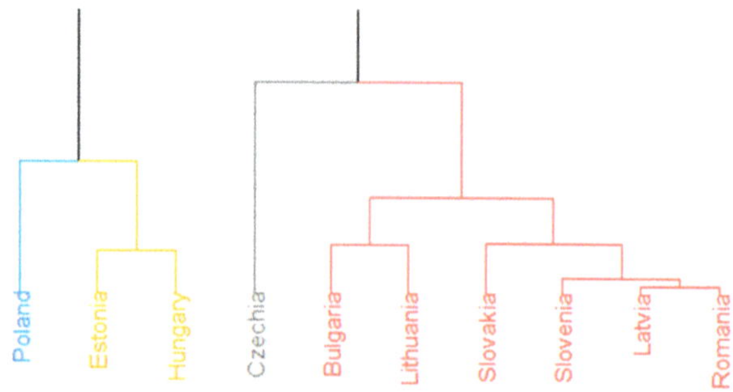

Figure 2. Results of the hierarchical grouping of similarities between the new member states in 2010.

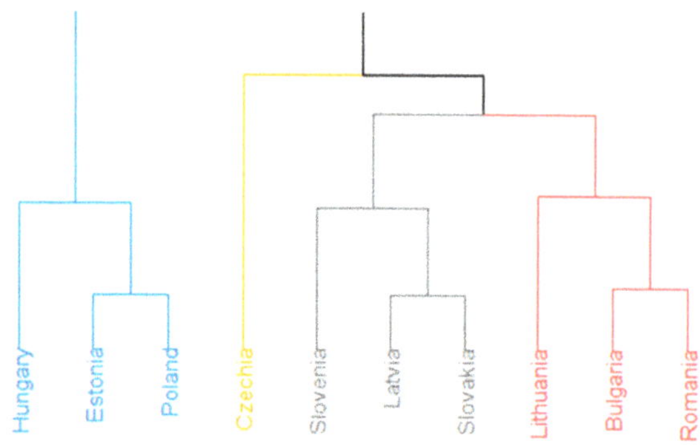

Figure 3. Results of the hierarchical grouping of similarities between the new member states in 2015.

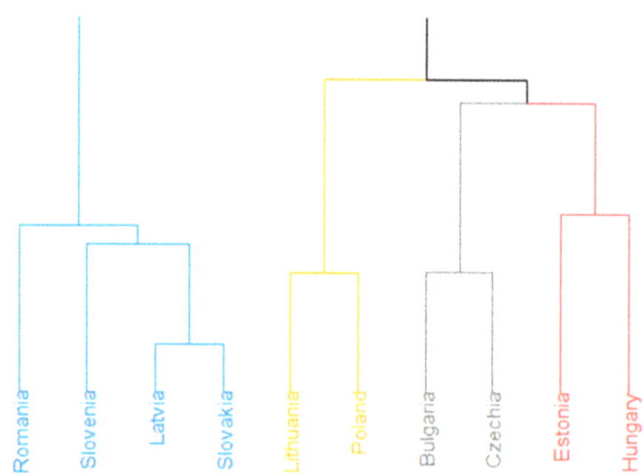

Figure 4. Results of the hierarchical grouping of similarities between the new member states in 2019.

In 2010 Poland was the only country assigned to the first group. In Poland the structure of renewable energy was determined by three sources: primary solid biofuels (51.54%), hydro energy (30.45%) and wind (14.53%). In the second group were classified Estonia and Hungary. Their renewable energy sources were based mostly on primary solid biofuels and wind. The third group consists of Czechia (with 52.02% of hydro energy and 22.98%

on primary solid biofuels). It is worth noting that in Czechia the share of solar energy was 9.48%, which was the highest value among all analyzed countries in this year. In the fourth cluster, Bulgaria, Lithuania, Slovakia, Slovenia, Latvia and Romania were classified. In these countries the hydro energy dominated over other considered energy sources.

The structure of renewable energy sources used in CEE new member states has been changing over time. This is due to the fact that these countries make new investments into renewable energy and create own new energetic policies. It is visible in the energy structure in 2015. According to the diagram presented in Figure 3, Czechia constitutes a separate group. In comparison to 2010, the renewable energy sources in Czechia are almost equally distributed among hydro energy, biogases, solar energy and primary solid biofuels. The first group, i.e., Hungary, Estonia and Poland, based their energy on primary solid biofuels and wind. Third group consists of Slovenia, Latvia and Slovakia. These countries used mainly hydro energy and primary solid biofuels and biogases. The last group: Lithuania, Bulgaria and Romania were primarily using hydro energy and wind.

In 2019 the first cluster included countries with over 60% share of renewable energy production from hydropower in the total renewable energy production (it must be added that taking into account the total electricity production from RES, hydropower is very important for the whole EU—currently it is the second largest RES; the source in this category that is more important is wind; wind and hydro power accounted for two-thirds of the total electricity generated from renewable sources in the EU). These countries include: Romania, Slovenia, Latvia and Slovakia. This group generally is very diversified taking into account energy from RES in its gross final consumption of energy as it includes Latvia, in which the share of energy production from RES in total energy production was the highest in the whole group selected for analysis, i.e., among the new member states countries in 2019. For analyses conducted for the whole EU, Latvia usually is placed among leaders (only Sweden and Finland could boast with better performance in that field) with its electricity sector is dominated by large hydropower plants on the Daugava River and cogeneration plants [77]. Slovakia is below the average in the EU-27 and Slovenia and Romania just above it, but still exceeding 20% (Figure 1). Lithuania and Poland are the second cluster with the highest use of wind energy, but at the same time their share of renewable energy production in total energy production is highly diversified—for Poland it is the lowest among the new member states countries, while Lithuania is in the top three in this respect. Treating this part of the analysis as a kind of indicator of climate awareness, it has been found that Poland still has a lot of catching up to do, because as the study shows [50] high community involvement in ownership and management of energy projects can bring many benefits in RES development. The third cluster consisted of countries with a high percentage of the production of individual renewable energy sources in the production of total renewable energy, mainly from hydro, primary solid biofuels, solar and winds. Bulgaria and Czechia were included in this group. When taking into account the share of energy production from RES in total energy production the situation of both countries is different as Czechia is below the EU-27 average and Bulgaria exceeded it and the level of 20% as well (Figure 1). The fourth cluster includes countries where the percentage share of renewable energy from primary solid biofuels in the total renewable energy production is the highest among the analyzed countries. These include Estonia and Hungary. This is consistent with what Koppel and Ots [78] note in their paper claiming that Estonia has been successful in implementing biofuel programs. In the study [79], Hungary was treated as the country with the greatest potential among new member states countries in terms of geothermal development, for which it was projected about 19% share of geothermal in gross final consumption of RES in 2020. Even though this threshold was not reached, the successful development of projects in Hungary has shown that the relative ease of access to geothermal resources in the region means that there are real opportunities to develop this RES source. At the same time, as in the case of the second cluster, these countries occupy extremely different positions when it comes to their share of energy production from RES in total energy production (Figure 1).

6. Discussion

In the energy production or consumption patterns and especially in the share of energy from RES one can find significant dissimilarities characterizing the countries of the CEE that from outside are still very often treated as a quite homogeneous region. Taking into account historical background of these countries, on the one hand, they are (or rather were) characterized by significant economic similarities (like significant development gap towards the EU, reflected in the differences in GDP per capita between most of 'old' and 'new' member states and its regions; followed by development ambitions and strategies based on the integration with the EU), and by social, environmental or economic differences, on the other hand. Undoubtedly, the connecting point for all countries selected for analysis is their membership in the EU, which results in the need for the adoption of many regulations in the area of environmental protection and organization of the energy sector, which has and will clearly translate into the energy sector.

The analysis covered 10 new EU member states, among which Poland and Romania had the highest absolute values of energy generated from RES in 2019 (Table 1). However, these values do not mean that these countries had a distinctive large share of RES in final energy compared to the remaining countries. For example, this share was only 12.2% for Poland in 2019 (Figure 1). It must be added that the data prove that the situation among the new members states concerning not only the share of RES is diversified, but the changes are recorded at a different pace (Table 7). The illustration of the situation for 2019 has been included in Figure 1.

Table 7. Percentage share of energy from renewable sources in the new member states in 2010, 2015, 2019. Source: data from Eurostat (online data code: NRG_IND_REN__custom_1595802).

Country	Year		
	2010	2015	2019
Bulgaria	13.93	18.26	21.56
Czechia	10.51	15.07	16.24
Estonia	24.60	28.53	31.89
Latvia	30.38	37.54	40.98
Lithuania	19.64	25.75	25.46
Hungary	12.74	14.50	12.61
Poland	9.30	11.89	12.16
Romania	22.83	24.79	24.29
Slovenia	21.08	22.88	21.97
Slovakia	9.10	12.88	16.89

The leader for the 2010 was Latvia with the share of energy from RES of over 30% and this country stayed at the first position also in 2019 exceeding 40%, which makes the change more than 10 p.p. Not a single country from the analyzed group of the new member stated reached such a result. There was also a country that recorded a decrease in the share of energy from RES in 2019 compared to 2010 and it was Hungary. The share calculated for Lithuania, Romania and Slovenia in 2019 was slightly worse for those countries compared with their performance in 2015. It proves that changes in the analyzed sector in some countries require a lot of time and effort as generally their pace in quite slow. Lack of continuous improvement and high variability recorded for some countries proves that there is a need for a more consistent policy of support for RES in those countries.

To reflect the actual state of renewable energy in new member states more accurately, the percentage share of individual renewable energy sources in total production from RES were analyzed in the above study. Such an approach gives a better and more objective insight into the state of this energy sector and should allow to undertake activities of similar nature in relation to internal energy production, technological development or common energy policy (national action plans). Taking into account the percentage share of individual renewable energy sources, new member states countries were divided into four clusters that were characterized by some similarity taking into account share of individual renewable energy sources in total renewable energy. In other words, countries of comparable specificity were defined, as the analysis showed those clusters included diversified countries taking into account the overall performance (understood as share of energy from RES as % of gross final energy consumption). The results proved that the group of the new member states was not homogeneous. What is more, the structure of energy sources from RES is very unstable over time and even small changes were reflected in cluster analysis. resulting in a different grouping. The most current analysis has been carried out based on data for the 2019. In this year for Lithuania and Poland almost 50% of their RES was based on wind. Romania, Slovenia, Latvia and Slovakia use mainly hydro energy and primary solid biofuels and biogases, although Romania is more widely using wind energy. Bulgaria and Czechia constitute a separate group, as they rely mainly on hydro energy, biogases, solar energy and primary solid biofuels. The last group, consisting of Estonia and Hungary use mainly primary solid biofuels, wind, and additionally Hungary was using solar energy most intensively (31.94%).

The analyses presented should support this process by considering the needs and opportunities of the identified clusters of similar countries. This creates the possibility of a more efficient use of resources than if all countries were treated according to the same criterion. The problem of the proper energy transformation of the selected countries is important because many studies show a bidirectional relationship between aggregate renewable energy and real GDP [41,80,81]. Energy is one of the key factors shaping an economy's ability to grow. Therefore, it is important to maintain stability in the operation of this sector.

It must be stated that the renewable energy policy conducted by the EU takes into account differences in potentials of its member states. Member states should follow the obligations set for them in renewable energy directives. Until the end of June 2021 it was Directive 2009/28/EC of the European Parliament and of the Council of 23 April 2009 on the promotion of the use of energy from renewable sources and amending and subsequently repealing Directives 2001/77/EC and 2003/30/E that were repealed by Directive (EU) 2018/2001 of the European Parliament and of the Council of 11 December 2018 on the promotion of the use of energy from renewable sources. Long-term goals were set for the whole community (a 20% share of its gross final energy consumption from renewable sources by 2020 and a 32% by 2030) but their 'distribution' between EU member states is based on national actions plans that express the individual pathways to development of RES in each member state. As shown, individual countries have different resources, their energy markets also differ in terms of size and characteristics. The performance in the field recorded by the countries of the CEE is therefore diversified, but it must be stressed once again that among 14 member states that had reached their national targets, half of the countries are located in the region of analysis (Bulgaria, Czechia, Estonia, Latvia, Lithuania, Romania and Slovakia). Hungary was very close to the target, and the countries that are furthest from meeting their targets belong to the group of old member states—France, Ireland and the Netherlands [82].

7. Conclusions

Energy production is moving toward renewables. Fossil fuels are gradually taking a step back giving way to more environmentally friendly resources. The European Union has chosen to be an active player in this transition, and although there are difficulties along the way, all member states are moving in the same direction, which is the creation of a new and efficient energy system. A common feature of the new member states analyzed in the article is that winter heating still involves heavy use of fossil fuels such as coal or gas, although it is known that less pollution leads to a healthier atmosphere. The benefits are positive for both restoring the local ecosystem and improving human health. In this context, RES offers great opportunities, but requires proper use and support from the government. This is because CEE countries diverge the most from the most developed

EU countries [54,83]. Therefore, it is an area that should be of particular interest within the framework of energy policy, although the implementation of these goals will require very high investment outlays. However, increasing use of renewable energy will help economies in transition achieve both economic growth and clean environment goals [84].

It is quite understandable that moving to renewable energy will not be possible in each country totally or/and at the same pace. The disadvantages of RES (especially high costs of such technologies upfront and storage compared to traditional energy and geographical limitations), as well as some other economic, social or political issues (like those connected with local labor markets). The circumstances of increasing the share of RES in the energy system are very complex and the evaluation of the energy sector is capital intensive and includes expensive installations. Their rapid replacement does not pay off until there is a return on invested capital. This may also explain why of some of the economies studied are so reluctant to the proposed changes.

The results of the research presented in this article are intended to see the structure of renewable energy production in new member states in the context of their national energy transformations. The presented approach to the analysis of energy production, taking into account the demographic and economic potential of individual countries, enabled the acquisition of new knowledge in this field and supported the assessment of the current state of RES. It also showed the diversity of these countries. Countries with similar problems need to take action of a similar nature in relation to internal energy production, technological development or community policy. Investments should be made to promote the purchase and use of RES installations, such as a system of subsidies or regulatory mechanisms. In addition, commonality across countries in the region may increase the available potential for RES development in these countries combined, improve their energy security and consolidate sustainable development. It seems that cooperation in the form of joint projects or joint support systems should be intensified.

Recently, in the statement for the UN Climate Change Conference in Glasgow (COP26) President von der Leyen underlined that "there is an encouraging message that shows that you can cut emissions and prosper" and that a number of new actions will be launched with other countries worldwide, also in order to develop renewables earlier and faster [85]. This proved that the EU incessantly aims at becoming a role model in the analyzed area. Some of the CEE countries can indeed boast of being good examples in the field of RES development.

Author Contributions: Conceptualization, A.W., A.I.-S.; methodology, A.W., A.I.-S.; formal analysis, A.W.; writing—original draft preparation, A.W., A.I.-S.; writing—review and editing, A.W., A.I.-S.; visualization, A.W., A.I.-S. All authors have read and agreed to the published version of the manuscript.

Funding: This research received no external funding.

Institutional Review Board Statement: Not applicable.

Informed Consent Statement: Not applicable.

Data Availability Statement: All data are contained within the article.

Conflicts of Interest: The authors declare no conflict of interest.

References

1. Ellabban, O.; Abu-Rub, H.; Blaabjerg, F. Renewable Energy Resources: Current Status, Future Prospects and Their Enabling Technology. *Renew. Sustain. Energy Rev.* **2014**, *39*, 748–764. [CrossRef]
2. Türkoğlu, S.P.; Kardoğan, P.S.Ö. The Role and Importance of Energy Efficiency for Sustainable Development of the Countries. In *Proceedings of the 3rd International Sustainable Buildings Symposium (ISBS 2017)*; Fırat, S., Kinuthia, J., Abu-Tair, A., Eds.; Springer International Publishing: Cham, Switzerland, 2018; pp. 53–60. Available online: https://www.researchgate.net/publication/324110795_The_Role_and_Importance_of_Energy_Efficiency_for_Sustainable_Development_of_the_Countries (accessed on 15 October 2021).

3. Sasana, H.; Ghozali, I. The Impact of Fossil and Renewable Energy Consumption on the Economic Growth in Brazil, Russia, India, China and South Africa. *Int. J. Energy Econ. Policy* **2017**, *7*, 194–200.
4. Ahuja, D.; Tatsutani, M. Sustainable energy for developing countries. *SAPIENS Surv. Perspect. Integr. Environ. Soc.* **2009**, *2*, 1–16.
5. Staffell, I.; Scamman, D.; Abad, A.V.; Balcombe, P.; Dodds, P.E.; Ekins, P.; Shah, N.; Ward, K.R. The Role of Hydrogen and Fuel Cells in the Global Energy System. *Energy Environ. Sci.* **2019**, *12*, 463–491. [CrossRef]
6. Goldstone, J.A. The Rise of the West-Or Not? A Revision to Socio-Economic History. *Sociol. Theory* **2000**, *18*, 175–194. [CrossRef]
7. Androniceanu, A.-M.; Kinnunen, J.; Georgescu, I.; Androniceanu, A. A Multidimensional Approach to Competitiveness, Innovation and Well-Being in the EU Using Canonical Correlation Analysis. *J. Compet.* **2020**, *12*, 5–21. [CrossRef]
8. Ntanos, S. Renewable Energy Consumption, Carbon Dioxide Emissions and Economic Growth: Evidence from Europe and Greece. *Proc. e-RA* **2015**, *10*, 46–56.
9. Zou, C.; Qun, Z.; Zhang, G.; Xiong, B. Energy Revolution: From a Fossil Energy Era to a New Energy Era. *Nat. Gas Ind. B* **2016**, *3*, 1–11. [CrossRef]
10. Rousseau, S.; Deschacht, N. Public Awareness of Nature and the Environment during the COVID-19 Crisis. *Environ. Resour. Econ.* **2020**, *76*, 1149–1159. [CrossRef]
11. Ramanathan, V.; Carmichael, G. Global and Regional Climate Changes Due to Black Carbon. *Nat Geosci.* **2008**, *1*, 221–227. [CrossRef]
12. Bölük, G.; Mert, M. Fossil & Renewable Energy Consumption, GHGs (Greenhouse Gases) and Economic Growth: Evidence from a Panel of EU (European Union) Countries. *Energy* **2014**, *74*, 439–446. [CrossRef]
13. Arndt, C.; Arent, D.; Hartley, F.; Merven, B.; Mondal, A.H. Faster Than You Think: Renewable Energy and Developing Countries. *Annu. Rev. Resour. Econ.* **2019**, *11*, 149–168. [CrossRef]
14. Inglesi-Lotz, R. The Impact of Renewable Energy Consumption to Economic Growth: A Panel Data Application. *Energy Econ.* **2016**, *53*, 58–63. [CrossRef]
15. Fernández González, P.; Landajo, M.; Presno, M.J. Tracking European Union CO_2 Emissions through LMDI (Logarithmic-Mean Divisia Index) Decomposition. The Activity Revaluation Approach. *Energy* **2014**, *73*, 741–750. [CrossRef]
16. Huhta, K. The Scope of State Sovereignty under Article 194(2) TFEU and the Evolution of EU Competences in the Energy Sector. *Int. Comp. Law Q.* **2021**, *70*, 991–1010. [CrossRef]
17. Available online: https://www.consilium.europa.eu/en/press/press-releases/2020/07/05/climate-transition-transformation-and-convergence-europe-s-path-towards-robust-resilience-speech-of-president-charles-michel/ (accessed on 7 November 2021).
18. Pietrzak, M.; Balcerzak, A.; Gajdos, A.; Arendt, L. Entrepreneurial Environment at Regional Level: The Case of Polish Path towards Sustainable Socio-Economic Development. *Entrep. Sustain. Issues* **2017**, *5*, 190–203. [CrossRef]
19. Chocholatá, M.; Furková, A. The Analysis of Employment Rates in the Context of Spatial Connectivity of the EU Regions. *Equilib. Q. J. Econ. Econ. Policy* **2018**, *13*, 181–213. [CrossRef]
20. Grybaitė, V.; Stankevičienė, J. An Empirical Analysis of Factors Affecting Sharing Economy Growth. *Oeconomia Copernic.* **2018**, *9*, 635–654. [CrossRef]
21. Kuc, M. Is the Regional Divergence a Price for the International Convergence? *J. Compet.* **2017**, *9*, 50–65. [CrossRef]
22. Kijek, A.; Matras-Bolibok, A. Technological Convergence across European Regions. *Equilib. Q. J. Econ. Econ. Policy* **2020**, *15*, 295–313. [CrossRef]
23. Kacperska, E.; Łukasiewicz, K.; Pietrzak, P. Use of Renewable Energy Sources in the European Union and the Visegrad Group Countries—Results of Cluster Analysis. *Energies* **2021**, *14*, 5680. [CrossRef]
24. Nicolini, M.; Tavoni, M. Are Renewable Energy Subsidies Effective? Evidence from Europe. *Renew. Sustain. Energy Rev.* **2017**, *74*, 412–423. [CrossRef]
25. Kuzemko, C.; Bradshaw, M.; Bridge, G.; Goldthau, A.; Jewell, J.; Overland, I.; Scholten, D.; Van de Graaf, T.; Westphal, K. COVID-19 and the Politics of Sustainable Energy Transitions. *Energy Res. Soc. Sci.* **2020**, *68*, 101685. [CrossRef]
26. Lund, H. Renewable energy strategies for sustainable development. *Energy* **2007**, *32*, 912–919. [CrossRef]
27. Włodarczyk, B.; Firoiu, D.; Ionescu, G.H.; Ghiocel, F.; Szturo, M.; Markowski, L. Assessing the Sustainable Development and Renewable Energy Sources Relationship in EU Countries. *Energies* **2021**, *14*, 2323. [CrossRef]
28. World Commission on Environment and Development. Report of the World Commission on Environment and Development: Our Common Future. Available online: https://sustainabledevelopment.un.org/content/documents/5987our-common-future.pdf (accessed on 18 November 2021).
29. Sneddon, C.; Howarth, R.B.; Norgaard, R.B. Sustainable development in a post-Brundtland world. *Ecol. Econ.* **2006**, *57*, 253–268. [CrossRef]
30. Choy, Y.K. 28 years into "Our Common Future": Sustainable development in the post-Brundtland world. *Sustain. Dev.* **2015**, *2*, 1197–1211.
31. Sathaye, J.; Lucon, O.; Rahman, A.; Christensen, J.; Denton, F.; Fujino, J.; Heath, G.; Mirza, M.; Rudnick, H.; Schlaepfer, A. Potsdam Institute for Climate Impact Research Renewable Energy in the Context of Sustainable Development. Physics Faculty Publications. Paper 1. 2011. Available online: https://ecommons.udayton.edu/phy_fac_pub/1 (accessed on 18 November 2021).
32. Wang, Q.; Yang, X. Investigating the sustainability of renewable energy—An empirical analysis of European Union countries using a hybrid of projection pursuit fuzzy clustering model and accelerated genetic algorithm based on real coding. *J. Clean. Prod.* **2020**, *268*, 121940. [CrossRef]

33. Anders, I.; Stagl, J.; Auer, I.; Pavlik, D. Climate Change in Central and Eastern Europe. In *Managing Protected Areas in Central and Eastern Europe under Climate Change. Advances in Global Change Research*; Rannow, S., Neubert, M., Eds.; Springer: Dordrecht, The Netherlands, 2014; Volume 58. [CrossRef]
34. Bagliani, M.; Dansero, E.; Puttilli, M. Territory and energy sustainability: The challenge of renewable energy sources. *J. Environ. Plan. Manag.* **2010**, *53*, 457–472. [CrossRef]
35. Kumar, M. *Social, Economic, and Environmental Impacts of Renewable Energy Resources, Wind Solar Hybrid Renewable Energy System, Kenneth Eloghene Okedu, Ahmed Tahour and Abdel Ghani Aissaou*; IntechOpen: London, UK, 2020; Available online: https://www.researchgate.net/publication/339064877_Social_Economic_and_Environmental_Impacts_of_Renewable_Energy_Resources (accessed on 15 October 2021). [CrossRef]
36. Sheikh, N.J.; Kocaoglu, D.F.; Lutzenhiser, L. Social and Political Impacts of Renewable Energy: Literature Review. *Technol. Forecast. Soc. Chang.* **2016**, *108*, 102–110. [CrossRef]
37. Butturi, M.A.; Lolli, F.; Sellitto, M.A.; Balugani, E.; Gamberini, R.; Rimini, B. Renewable Energy in Eco-Industrial Parks and Urban-Industrial Symbiosis: A Literature Review and a Conceptual Synthesis. *Appl. Energy* **2019**, *255*, 113825. [CrossRef]
38. Dudin, M.; Frolova, E.; Protopopova, O.; Mamedov, O.; Odintsov, S. Study of Innovative Technologies in the Energy Industry: Nontraditional and Renewable Energy Sources. *Entrep. Sustain. Issues* **2019**, *6*, 1704–1713. [CrossRef]
39. Cucchiella, F.; D'Adamo, I.; Gastaldi, M.; Koh, S.L.; Rosa, P. A Comparison of Environmental and Energetic Performance of European Countries: A Sustainability Index. *Renew. Sustain. Energy Rev.* **2017**, *78*, 401–413. [CrossRef]
40. García-Álvarez, M.T.; Moreno, B.; Soares, I. Analyzing the Environmental and Resource Pressures from European Energy Activity: A Comparative Study of EU Member States. *Energy* **2016**, *115*, 1375–1384. [CrossRef]
41. Ohler, A.; Fetters, I. The Causal Relationship between Renewable Electricity Generation and GDP Growth: A Study of Energy Sources. *Energy Econ.* **2014**, *43*, 125–139. [CrossRef]
42. Kula, F. The Long-Run Relationship between Renewable Electricity Consumption and GDP: Evidence from Panel Data. *Energy Sources Part B Econ. Plan. Policy* **2014**, *9*, 156–160. [CrossRef]
43. Anwar, A.; Arshed, N.; Kousar, N. Renewable Energy Consumption and Economic Growth in Member of OIC Countries. *Eur. Online J. Nat. Soc. Sci.* **2017**, *6*, 111–129.
44. Stiglitz, J.E.; Sen, A.; Fitoussi, J.-P. Report by the Commission on the Measurement of Economic Performance and Social Progress. 2009. Available online: https://ec.europa.eu/eurostat/documents/8131721/8131772/Stiglitz-Sen-Fitoussi-Commission-report.pdf (accessed on 18 November 2021).
45. Fleurbaey, M. Beyond GDP: The quest for the measure of social welfare. *J. Econ. Lit.* **2009**, *47*, 1029–1075. [CrossRef]
46. Unteutsch, M.; Lindenberger, D. Promotion of Electricity from Renewable Energy in Europe Post 2020. The Economic Benefits of Cooperation. In *Interdisziplinäre Aspekte der Energiewirtschaft*; von Weizsäcker, C.C., Lindenberger, D., Höffler, F., Eds.; Energie in Naturwissenschaft, Technik, Wirtschaft und Gesellschaft; Springer Fachmedien: Wiesbaden, Germany, 2016; pp. 61–78. ISBN 978-3-658-12726-8.
47. Cucchiella, F.; D'Adamo, I.; Gastaldi, M. Future Trajectories of Renewable Energy Consumption in the European Union. *Resources* **2018**, *7*, 10. [CrossRef]
48. Gurkov, I. Transition Economy. In *Wiley Encyclopedia of Management*; American Cancer Society: Chichester, UK, 2015; ISBN 978-1-118-78531-7.
49. Faaij, A.P.C. Bio-Energy in Europe: Changing Technology Choices. *Energy Policy* **2006**, *34*, 322–342. [CrossRef]
50. Walker, G.; Devine-Wright, P. Community Renewable Energy: What Should It Mean? *Energy Policy* **2008**, *36*, 497–500. [CrossRef]
51. Burke, M.J.; Stephens, J.C. Political Power and Renewable Energy Futures: A Critical Review. *Energy Res. Soc. Sci.* **2018**, *35*, 78–93. [CrossRef]
52. Shao, X.; Fang, T. Performance Analysis of Government Subsidies for Photovoltaic Industry: Based on Spatial Econometric Model. *Energy Strategy Rev.* **2021**, *34*, 100631. [CrossRef]
53. Bersalli, G.; Menanteau, P.; El-Methni, J. Renewable Energy Policy Effectiveness: A Panel Data Analysis across Europe and Latin America. *Renew. Sustain. Energy Rev.* **2020**, *133*, 110351. [CrossRef]
54. Brodny, J.; Tutak, M. Analyzing Similarities between the European Union Countries in Terms of the Structure and Volume of Energy Production from Renewable Energy Sources. *Energies* **2020**, *13*, 913. [CrossRef]
55. Armeanu, D.Ş.; Vintilă, G.; Gherghina, Ş.C. Does Renewable Energy Drive Sustainable Economic Growth? Multivariate Panel Data Evidence for EU-28 Countries. *Energies* **2017**, *10*, 381. [CrossRef]
56. Simionescu, M.; Strielkowski, W.; Tvaronavičienė, M. Renewable Energy in Final Energy Consumption and Income in the EU-28 Countries. *Energies* **2020**, *13*, 2280. [CrossRef]
57. Pakulska, T. Green Energy in Central and Eastern European (CEE) Countries: New Challenges on the Path to Sustainable Development. *Energies* **2021**, *14*, 884. [CrossRef]
58. Nikolaev, A.; Konidari, P. Development and Assessment of Renewable Energy Policy Scenarios by 2030 for Bulgaria. *Renew. Energy* **2017**, *111*, 792–802. [CrossRef]
59. Al-Mansour, F.; Sucic, B.; Pusnik, M. Challenges and Prospects of Electricity Production from Renewable Energy Sources in Slovenia. *Energy* **2014**, *77*, 73–81. [CrossRef]

60. Câmpeanu, V.; Pencea, S. Renewable Energy Sources in Romania: From a "Paradise" of Investors to a Possible Abandon or to Another Boom? The Impact of a New Paradigm in Romanian Renewable Sources Policy. *Procedia Econ. Financ.* **2014**, *8*, 129–137. [CrossRef]
61. Everitt, B.S.; Landau, S.; Leese, M.; Stahl, D. *Cluster Analysis*, 5th ed.; John Wiley & Sons, Ltd.: Chichester, UK, 2011; ISBN 978-0-470-97844-3.
62. Hartigan, J.A. *Clustering Algorithms*; Wiley: John Wiley and Sons, Inc.: New York, NY, USA, 1975; ISBN 978-0-471-35645-5.
63. Semenenko, I.; Halhash, R.; Sieriebriak, K. Sustainable Development of Regions in Ukraine: Before and after the Beginning of the Conflict. *Equilib. Q. J. Econ. Econ. Policy.* **2019**, *14*, 317–339. [CrossRef]
64. Kuc, M. The Use of Taxonomy Methods for Clustering European Union Countries Due to the Standard of Living. *Oeconomia Copernic.* **2012**, *3*, 5–24. [CrossRef]
65. MacKay, D. *Information Theory, Inference, and Learning Algorithms*; Cambridge University Press: Cambridge, UK, 2003; Volume 50, ISBN 978-0-521-64298-9. Available online: https://inst.eecs.berkeley.edu//~{}ee121/sp08/handouts/it.pdf (accessed on 18 October 2021).
66. Miłek, D. Spatial Differentiation in the Social and Economic Development Level in Poland. *Equilib. Q. J. Econ. Econ. Policy* **2018**, *13*, 487–507. [CrossRef]
67. Szymańska, A. National Fiscal Frameworks in the Post-Crisis European Union. *Equilib. Q. J. Econ. Econ. Policy* **2018**, *13*, 623–642. [CrossRef]
68. Rollnik-Sadowska, E.; Dąbrowska, E. Cluster Analysis of Effectiveness of Labour Market Policy in the European Union. *Oeconomia Copernic.* **2018**, *9*, 143–158. [CrossRef]
69. Ward, J.H. Hierarchical Grouping to Optimize an Objective Function. *J. Am. Stat. Assoc.* **1969**, *58*, 236–244. [CrossRef]
70. Małkowska, A.; Głuszak, M. Pro-investment Local Policies in the Area of Real Estate Economics—Similarities and Differences in the Strategies Used by Communes. *Oeconomia Copernic.* **2016**, *7*, 269–283. [CrossRef]
71. Reiff, M.; Surmanová, K.; Balcerzak, A.P.; Pietrzak, M. Multiple criteria analysis of European Union agriculture. *J. Int. Stud.* **2016**, *9*, 62–74. [CrossRef] [PubMed]
72. Starczewski, A.; Krzyżak, A. Performance Evaluation of the Silhouette Index. In *Proceedings of the Artificial Intelligence and Soft Computing*; Rutkowski, L., Korytkowski, M., Scherer, R., Tadeusiewicz, R., Zadeh, L.A., Zurada, J.M., Eds.; Springer International Publishing: Cham, Switzerland, 2015; pp. 49–58.
73. Dudek, A. Silhouette Index as Clustering Evaluation Tool. In *Proceedings of the Classification and Data Analysis*; Jajuga, K., Batóg, J., Walesiak, M., Eds.; Springer International Publishing: Cham, Switzerland, 2020; pp. 19–33.
74. Mamat, A.R.; Mohamed, F.S.; Mohamed, M.A.; Rawi, N.M.; Awang, M.I. Silhouette Index for Determining Optimal K-Means Clustering on Images in Different Color Models. *Int. J. Eng. Technol.* **2018**, *7*, 105–109. [CrossRef]
75. Ivaničová, Z.; Kalužák, L. Dynamika vývoja krajín Eurozóny na základe vybraných odsovateľov. *Ekon. Rozhľady* **2015**, *44*, 476–492.
76. Reiff, M.; Surmanová, K. Cluster Analysis of EE Agriculture. In Proceedings of the International Scientific Conference Quantitative Methods in Economics Multiple Criteria Decision Making XVIII, Vrátna, Slovakia, 25–27 May 2016; pp. 322–326.
77. Kasiulis, E.; Punys, P.; Kvaraciejus, A.; Dumbrauskas, A.; Jurevičius, L. Small Hydropower in the Baltic States-Current Status and Potential for Future Development. *Energies* **2020**, *13*, 6731. [CrossRef]
78. Koppel, M.; Ots, M. Use of renewable energy sources in Estonia. In *Investigation and Usage of Renewable Energy Sources. First Conference Proceedings*; Estonian Agricultural University: Tartu, Estonia, 4 November 1999; pp. 10–20. ISBN 9985-882-63-6.
79. Pająk, L.; Kępińska, B.; Kasztelewicz, A.; Bujakowski, W.; Tomaszewska, B.; Hołojuch, G. Some Factors Determining the Geothermal Energy Uses' Development in the Central and Eastern European Countries in Coming Years. *Zesz. Nauk. Inst. Gospod. Surowcami Miner. Energią PAN* **2016**, *92*, 337–357.
80. Matei, I. Is There a Link between Renewable Energy Consumption and Economic Growth? A Dynamic Panel Investigation for the OECD Countries. *Rev. Econ. Polit.* **2017**, *127*, 985–1012. [CrossRef]
81. Liu, W.-C. The Relationship between Primary Energy Consumption and Real Gross Domestic Product: Evidence from Major Asian Countries. *Sustainability* **2020**, *12*, 2568. [CrossRef]
82. Available online: https://www.eea.europa.eu/data-and-maps/indicators/renewable-gross-final-energy-consumption-5/assessment (accessed on 7 November 2021).
83. Boltho, A. Southern and Eastern Europe in the Eurozone: Convergence or Divergence? *Balt. J. Econ.* **2020**, *20*, 74–93. [CrossRef]
84. Adedoyin, F.; Abubakar, I.; Bekun, F.V.; Sarkodie, S.A. Generation of Energy and Environmental-Economic Growth Consequences: Is There Any Difference across Transition Economies? *Energy Rep.* **2020**, *6*, 1418–1427. [CrossRef]
85. Available online: https://ec.europa.eu/commission/presscorner/detail/en/STATEMENT_21_5643 (accessed on 7 November 2021).

energies

Article

The Potential of Ecological Distributed Energy Generation Systems, Situation, and Perspective for Poland

Piotr Senkus [1,*], Waldemar Glabiszewski [2], Aneta Wysokińska-Senkus [3], Szymon Cyfert [4] and Roman Batko [5]

1 Faculty of Social Sciences, Siedlce University of Natural Sciences and Humanities, 08-110 Siedlce, Poland
2 Faculty of Economic Sciences and Management, Nicolaus Copernicus University, 87-100 Torun, Poland; waldemar.glabiszewski@umk.pl
3 Faculty of Management and Command, War Studies University Warsaw, 00-910 Warsaw, Poland; a.wysokinska-senkus@akademia.mil.pl
4 Institute of Management, Poznan University of Economics and Business, 61-875 Poznan, Poland; s.cyfert@ue.poznan.pl
5 Faculty of Management, Military University of Land Forces, 51-150 Wroclaw, Poland; roman.batko@awl.edu.pl
* Correspondence: piotr.senkus@uph.edu.pl; Tel.: +48-604-645-154

Abstract: Poland needs to fulfill its climate goals and become "climate neutral" by 2050. The plan is intricate for the Polish Government because coal-powered power plants generate about 80 percent of electricity in the country. Although policymakers are making an effort to redesign the energy sector, a lot still remains to be done. The viral trend in that transformation involves installing photovoltaic (PV) panels by private, corporate, and self-government investors. For example, the "My energy" support program of the National Fund for Environmental Protection and Water Management has helped finance 220,000 micro-PV installations. The achievement is significant but constitutes only partial success. PV powerplants will not simply replace coal powerplants. That is why the research on the ecological distributed energy generation systems has to be executed. The article presents the research results on ecological distributed energy generation systems, making the transformation of the Polish energy sector possible. The study's primary objectives were to review the energy situation with particular attention paid to the technologies that could be used as the ecological distributed energy generation systems and draw the scenarios for the sector development. The authors used Desk research, the Delphi method supported with the Computer Assisted-Web Interview (CAWI) technique, and the Weighted SWOT analysis to fulfill the objectives. The findings showed that photovoltaic (PV) systems would be the fastest-growing energy sector even in the perspective of doubling the energy consumption by 2050. Private investors investing in ecological distributed energy generation systems, especially the PV systems mentioned above, and biomass or biogas systems, would significantly help policymakers, including those in Poland, fulfill the climate goals.

Keywords: alternative energy sources; photovoltaic systems; wind systems; hydropower systems; biogas; biomass systems

Citation: Senkus, P.; Glabiszewski, W.; Wysokińska-Senkus, A.; Cyfert, S.; Batko, R. The Potential of Ecological Distributed Energy Generation Systems, Situation, and Perspective for Poland. *Energies* **2021**, *14*, 7966. https://doi.org/10.3390/en14237966

Academic Editors: Bartlomiej Iglinski, Antonio Rosato and Michał Bernard Pietrzak

Received: 4 October 2021
Accepted: 23 November 2021
Published: 29 November 2021

Publisher's Note: MDPI stays neutral with regard to jurisdictional claims in published maps and institutional affiliations.

Copyright: © 2021 by the authors. Licensee MDPI, Basel, Switzerland. This article is an open access article distributed under the terms and conditions of the Creative Commons Attribution (CC BY) license (https://creativecommons.org/licenses/by/4.0/).

1. Introduction

> *"Energy cannot be created or destroyed; it can only be changed from one form to another."* (Albert Einstein)

> *"Nothing is lost, nothing is created, everything is transformed."* (Antoine-Laurent de Lavoisier [1])

Energy has become one of the most strategic resources for countries to retain their sustainable development ability. Human beings require more and more electricity and heat to perform their daily activities, which leads to increased energy consumption. If the current trends continue, global energy demand will double by 2050 [1]. Polish policymakers face that demand-driven challenge and must also fulfill the climate goals, including becoming "climate neutral" by 2050. Achieving this is even harder because coal-powered facilities have about eighty percent of electricity in Poland, and the country is under growing pressure from the EU to reduce emissions as soon as possible.

Luckily, science and practice show that a strong innovation wave is growing in the energy sector, resulting in a disruptive change in the efficiency of clean energy facilities and cutting the unit cost of energy production [2–7].

Some energy sector analytics, such as Lovins A., declare that achieving the energy transformation goal is possible. He said that "dependence on oil and coal could be eliminated by 2050 while switching to efficient use and renewable supply", implementing only existing technologies [3,7]. Other inventors and scientists such as McDonough W. and Braungart M. claim that it can be achieved even without reducing the quality of life. The authors have declared using clean and renewable energy as one of the fundamental issues of their "Cradle to cradle" concept: "Living things thrive on the energy of current solar income. Similarly, human constructs can utilize clean and renewable energy in many forms—such as solar, wind, geothermal, gravitational energy and other energy systems being developed today—thereby capitalizing on these abundant resources while supporting human and environmental health".

The article presents the research results on ecological distributed energy generation systems, making the transformation of the Polish energy sector possible.

The main objectives of the study included:

1. To review the global energy situation with particular attention on the technologies that could be used as the ecological distributed energy generation systems;
2. To review the technologies that could be used as ecological distributed energy generation systems from the perspective of Poland.

The following research questions were asked:

(a) If the renewable energy sources could provide sufficient energy to replace the conventional energy sources (from global and Polish perspectives)?
(b) What renewable energy sources are suitable as ecological distributed energy generation systems in Poland?
(c) What are the most Strengths, Weaknesses, Opportunities, and Threats for the essential technologies to replace conventional energy sources in Poland?
(d) What are the scenarios for energy production in Poland?

The novelty of this study is the use of the weighted SWOT analysis method to analyze the Polish energy sector and individual electricity production systems. When one enters the phrase "weighted SWOT" into the Google Scholar search engine, one only finds about 70 results, of which about 30% are associated with energy. "Google Scholar" returned more than one and a half million results after searching "SWOT". The application of the above analysis was the basis for achieving the research goals set by the authors. The following study applied a weighted SWOT to Poland, but it can also be used on a voivodeship, city, or commune scale.

This study may be of interest to politicians, local government officials, managers of public, private, and non-profit organizations, and environmental activists, and engineers in the field of environmental protection. Genuine care for the planet, expressed, among other things, through the optimization of energy production processes, should be part of the typical activities of the stakeholders mentioned above and, just like ordinary people, should not be dominated by any field of science.

Another novelty of this study is showing the reader the so-called "big picture"; that is, a bird's-eye view, which allows people specializing in a vase in their fields to have a different look at the issues of energy production own business.

2. Materials and Methods

The authors used the following methods to achieve the objectives and outlines in the introduction:

- Desk research—a literature review was performed on reports and academic publications obtained mainly from the digital libraries. The desk research results were used to prepare the global energy situation review, characteristics of the different energy sources that could be used in distributed power plants, and partly the current state of production and prognosis of energy in Poland. The desk research results were also used to prepare the CAWI survey used in the Delphi method to collect the information from participants.
- The Delphi method—was developed by RAND Corporation in 1950 [7]. The Delphi approach makes use of an interactive iterative technique to look for professionals in assess or behavior predictive research. The classical Delphi approach consists of 4 key features: anonymity, iteration, managed feedback, and statistical summary [8–11]. The CAWI method became used to guide the Delphi approach. The CAWI technique was used to support the Delphi method.
- The Computer-Assisted-Web Interview (CAWI) technique—authors used the method to collect experts' opinions on energy sources in ecological distributed energy generation systems, especially their strengths, weaknesses, opportunities, and threats. A mostly five-point scale was used. The data were collected in two rounds. In the first round of the survey, mainly open-ended questions were included. Then in the second round, the survey included five-point scale questions to evaluate the most frequent factors from the first round. Finally, the authors analyzed the collected data to prepare the weighted SWOT
- The Weighted SWOT analysis—SWOT analysis is a widely used technique to map out the present Strengths (S), Weaknesses (W), Opportunities (O), and Threats (T). The SWOT Analysis has two main steps: listing the key internal strengths and weaknesses and listing the key external opportunities and threats. Because SWOT analysis is often subjective, the authors prepared a weighted SWOT score matrix, where the metrics are defined as below:
    - Weight (W)—the relative importance of a given factor in the range from 0.00 to 1.00 (0.00—insignificant factor; 1.00—significant factor). The total weight −1.00;
    - Rating (R)—on a scale from 1 to 5. The higher rating, the more like to attain;
    - Weighted score—that was computed by multiplying each factor's weight by its rating;
    - Total Weighted Score (TWS) = (W) × (R);
    - Internal Factors Ratio (IFR) = (S TWS)/(W TWS).
    - External Factors Ratio (EFR) = (O TWS)/(T TWS) [12–15].

  The template of the Weighted SWOT analysis is presented in Table 1.

Table 1. The example of the Weighted SWOT analysis.

		Weight	Rating	Score
Strengths		1.00		
1	Strength 1	0.40	5	2.00
2	Strength 2	0.30	4	1.20
	Total weighted score			3.20
Weaknesses		1.00		
1	Weakness 1	0.30	5	1.50
2	Weakness 2	0.30	5	1.50
	Total weighted score			3.00
Internal Factors Ratio (IFR)				1.06
Opportunities		1.00		
1	Opportunity 1	0.60	5	3.00
2	Opportunity 2	0.40	5	2.00
	Total weighted score			5.00
Threats		1.00		
1	Threat 1	0.50	5	2.50
2	Threat 2	0.30	5	1.50
	Total weighted score			4.00
External Factors Ratio (EFR)				1.25

The authors have described the most popular energy sources in the following aspects:

- General characteristics;
- Current usage;
- Possibility to use as a distributed powerplant;
- Usage potential;
- Levelized cost of energy (LCOE).

The LCOE is calculated and discussed in this section from an economic point of view (excluding transfers such as grants, taxes, or other financial incentives (e.g., the LCOE calculated from a financial or investor's point of view include these transfers)). The LCOE is calculated as follows: (3)

$$LCOE = ACC + VOMC + FC + FOMC \quad (1)$$

- "ACC, VOMC, FOMC, and FC—annualized capital costs, variable O & M costs, fixed O & M costs, and fuel costs.

All these costs are expressed in terms of energy ($/MWh). ACC and FOMC are calculated as presented below" (4):

$$ACC = (OC * CRF * 1000)/(CAF * 24 * 365) \quad (2)$$

$$FOMC = (FXC * 1000)/(CAF * 24 * 365) \quad (3)$$

- OC means a lump-sum investment of a facility ($/kW), and FXC is the annual fixed costs $/kW. CRF is the capacity recovery factor that converts the costs expressed in terms of capacity to the related costs in terms of energy. CAF is the capacity availability factor.

The fuel cost does not apply to renewable technologies, except biomass. It is determined based on the heat content of a fuel (HC), fuel prices (FP), and the heat rate of the system (HR). Fuel prices are in USD per metric ton of coal. Heat content—the amount of heat, in kilocalories (Kcal) or megajoules (MJ), divided by one physical unit of the fuel (MJ/kg). "Heat rate is the inverse of the thermal efficiency of a power plant; it refers to the amount of heat needed to produce one unit of electric power (MJ/kWh). Thus, the fuel cost (FC) was calculated as:

$$FC = (FP/HC) * HR \tag{4}$$

Finally, the CRF is derived by using the discount rate (r) and the economic life (n) of a plant as follows" [15,16]:

$$CRF = \{r * (1 + r)n\}/[\{(1 + r)n\} - 1] \tag{5}$$

3. Global Energy Situation Review

Human beings have always used energy to fulfill needs and to transform the world around them. When considering energy consumption from 1800 to 2019, its growth is exponential and could be described with a function. e = 2.71828 is Euler's number. The "x" occurs as an exponent (1). The R^2 indicator—0.9671 is very high in that case and shows that the curve equation almost perfectly fits the analyzed data (2).

$$y = 0.1449e^{0.1329x} \tag{6}$$

$$R^2 = 0.9671 \tag{7}$$

When considering the energy sources' mix, traditional biomass gave 5653 TWh in 1800, and 11,111 TWh in 2020. That means over 100% growth, but it was the slowest growing energy source (Figure 1).

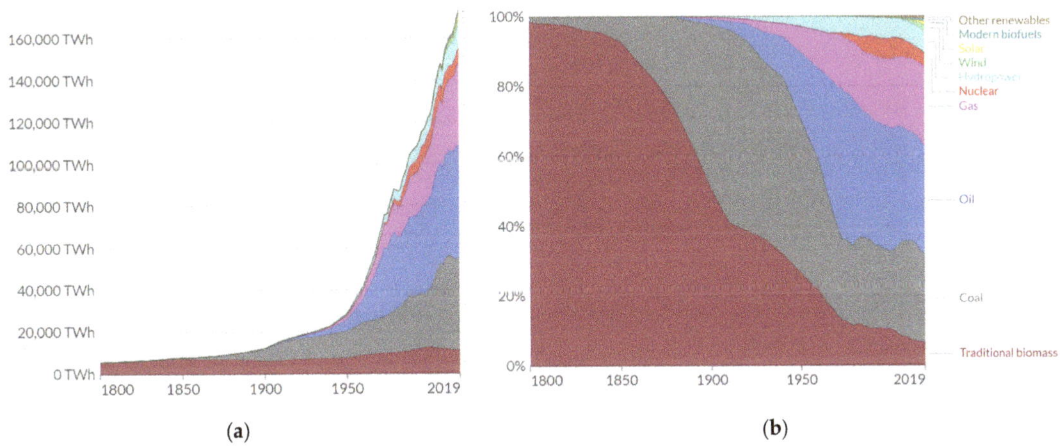

Figure 1. Global primary energy consumption by source: (**a**) In Twh; (**b**) relative (%) [17].

Coal was the second traditional primary energy source. It delivered about 97 TWh in 1800, and 44,109 TWh in 2019. In the 19th century, coal was the primary energy source to enable the "1st Industrial Revolution" and was 55.23% of the global energy mix in 1910. Since then, no primary energy source achieved such a significant share in the world energy mix. The percentage of coal in the global energy mix has been declining and now is about 25.7%.

Oil was the third primary energy source that has appeared in the 19th century and has become the essential energy source. The history of oil began in occupied Poland. The Polish pharmacist, engineer, businessman, inventor, and philanthropist Ignacy Łukasiewicz was the pioneer who designed the world's first modern oil refinery in 1856. "His achievements included discovering the methods of distilling kerosene from seep oil, the invention of the modern kerosene lamp, the introduction of the first modern street lamp in Europe, and the construction of the world's first modern oil well" [18,19]. Oil has provided 53,620 TWh in 2019, and was the first most crucial energy source, but its share was about 31% at that time. It obtained its highest share—43.3% in the global energy mix in 1973.

Natural gas was the fourth most crucial energy source in 2019. The history of gas also began in the 19th century. The popularity of natural gas is still growing. In 2019, gas provided 39,292 TWh, which constituted 22.5% of the global energy mix.

Hydropower was the fifth most crucial energy source in 2019. It has provided 10,455 TWh, making 6.05% of the global primary energy mix—the most significant share of the renewables.

Nuclear fusion was the sixth most important primary energy source in 2019. It provided 6923 TWh and about 3.99%.

Recently, the most popular renewables: wind, solar, and modern biofuels, and other sources provided 4.64%, merely 1.64% more is considered a statistical error in the analysis. It has to be mentioned that all of the four primary energy sources used in 2019 were fossil fuels that have provided around 147,876 TWh, making 85.31% of the global energy mix.

This is the perfect time to ask about the possibility of the total reconstruction of the global energy mix by 2050. However, please wait a little bit for the answer and scenarios for the future related to Poland.

Other statistics are even more interesting, such as the energy consumption per capita. Qatar was the first place according to primary energy consumption in 2020—200.00 MWh per capita. The rest included: Iceland—181.38; Singapore—171.23; Trinidad and Tobago—143.26; United Arab Emirates—138.42; Kuwait—108.99; Canada—106.38; Norway—91.97; Saudi Arabia—90.17; and Oman—85.21. The average energy consumption was 21.19, and the median value—28.71 MWh per capita. That means that the average use of primary power sources is about 9.4× lower than the consumption in Qatar. Poland with 31.61 was between median and average. Almost all countries in that ranking acquire their energy primarily from fossil fuels they possess and excavate. That means that the countries' governments could not be particularly interested in energy transformation. According to the UN, there is one exception—Iceland, which could be the model energy transformation for the world because over 85% of the energy consumed there was acquired from renewable energy sources [20,21].

The factor that encourages the investment in renewable energy sources is the combination of the cost of energy generation and the market price of the electricity. According to the authors, this relationship is essential in encouraging potential investors to invest in distributed power plants. Figure 2 presents the average wholesale baseload electricity prices for the first quarter of 2021 [21].

Compared to the previous year, the average electricity prices in the EU have grown 57% compared to the preceding 12 months. The highest growth has been observed in (a) Norway (about 190%), (b) Sweden (about 160%), (c) Denmark (about 130%), and (d) Finland (about 100%).

Although that situation means higher prices and enabling inflation, it will still boost investment in the eco-distributed power plants.

Will the growing trend continue? It depends. EU citizens might remember when in 2009, the EU called for withdrawing traditional halogen light bulbs [22]. The only developed lightning technology at that time was compact fluorescent lamps (CFL). Unfortunately, that technology was much more expensive than traditional halogen lightbulbs. Light emitting diode (LED) technology was marginal at that time. Nowadays, consumers prefer LED instead of CFL [22–24]. The same case could be with energy generation, and that is why relying only on statistical data referring to changes in the technology, including energy technology, is useless. Statistical information has nothing to do with "destructive technologies" that appear on the market and are changing it dramatically. The exact mechanism could be seen in other sectors, too [25,26]. This is why the authors treat the statistical analysis as background for further additional analysis like expert panels with the Delphi method and CAWI surveys.

Figure 2. Average wholesale electricity prices, for the first quarter of 2021.

For those who think that the status quo in which fossil fuels burned traditionally will still be the leading source of energy, the words of Don Huberts should be recalled. He was "convinced that fuel cells, which generate clean energy from hydrogen, would soon replace power stations and cars that mostly burn coal, oil, or natural gas" [26]. The Royal Dutch/Shell manager said: "The stone age did not end because the world ran out of stones, and the oil age will not end because we run out of oil".

4. Characteristics of the Different Energy Sources That Could Be Used in Ecological Distributed Energy Generation Systems

The following energy generation systems are characterized in that section: • Photovoltaic systems; • Concentrating Solar Power (CSP) system; • Wind systems; • Natural Gas; • Geothermal systems; • Hydropower systems; • Coal systems; • Nuclear systems; and • Biomass and biogas systems.

Photovoltaic systems are designed to provide usable solar energy through photovoltaic cells. The system consists of various modules, including • solar panels that absorb sunlight and convert it into electricity; • solar inverter to convert direct current to alternating current; • plumbing; • cables; and • other equipment. PV systems range from an integrated roof or building systems with several kilowatts to large power plants of several hundred megawatts. Nowadays, most photovoltaic systems are on-grid, while off-grid systems make up a small market [27–37]. In the publication year, PV systems are mostly made from crystalline silicon (c-Si), and the most popular on the market are:

- Monocrystalline silicon (mono-Si) solar cells feature a single-crystal composition;
- Polycrystalline silicon cells that are made from cast square ingots—large blocks of molten silicon carefully cooled and solidified.

It has been estimated that the silicon-based PV systems efficiency is limited to about 30%, which outputs about 330 (Pmax) W/m^2. According to respondents who took part in the survey, PV systems have become "disruptive technologies" and could be most suitable for distributed power plants [38,39]. According to the respondents or literature sources, the popularity of PV systems will grow because the systems are becoming cheaper and cheaper over time [40,41].

The photovoltaic systems are very elastic in terms of their application, for example, (a) rooftop and building integrated systems; (b) photovoltaic thermal hybrid solar collector; (c) power stations; (d) agrivoltaics; (e) rural electrification; (f) spacecraft applications; (g) indoor photovoltaics (IPV); and (h) spacecraft applications [41]. In the future, this technology will be even cheaper and more accessible due to research on Perovskite solar cells (PSC). "PSC is a type of solar cell that includes a perovskite-structured compound, most commonly a hybrid organic–inorganic lead or tin halide-based material, as the light-harvesting active layer. The researchers have estimated that solar systems based on PSC could gain efficiency about 66%" [42–49]. That means the potential output is about 1200 (Pmax) W/m^2. It is even more important that perovskite PV systems, according to Saule Technologies, can be placed on almost every surface that makes them perfect for building distributed powerplants [44,45]. The advantages of PV systems include: (a) high reliability; (b) strong persistence; (c) low maintenance costs; (d) zero fuel consumption; and (e) strong independence. The most important disadvantages of photovoltaic systems are (a) high start-up cost, (b) available solar radiation instability, and (c) have energy storage requirements. According to the IRENA photovoltaic system, even based on silicone, in 2019, it had the lowest LCOE—618.00 USD/kW (Table 2). That makes it one of the most economic-efficient energy systems. The cost per unit in 2019 was about 0.25 USD/kWh and competed with hydropower systems and biofuels. The experts predict that with upcoming perovskite and mix silicon-perovskite in no more than five years, photovoltaic systems will be the most economically effective from the others discussed in the paper.

Concentrating Solar Power (CSP) systems—sunlight is focused by mirrors or lenses on the receiver generating a large amount of energy. When sunlight is converted into heat (solar heat), electricity is generated, connecting a heat engine (usually a steam turbine) with an electricity generator, or carrying out a thermochemical reaction. "The CSP had a total installed capacity of 5500 MW worldwide in 2018, compared to 354 MW in 2005" [34–36]; in 2020, 133 powerplants in 22 countries [45]. According to the experts' system, concentrating solar power (CSP) is also a "disruptive technology", but technical aspects are not accessible for individuals or small businesses. This is why the potential of using solar power (CSP) as distributed powerplants is more limited than photovoltaic systems. It is not very popular technology.

Wind systems—onshore wind is the type that blows straight from the sea. On the other hand, an offshore wind is composed of waves that blow from the ground. In 2019, wind power provided 1430 TWh of electricity, equivalent to 5.3% of global electricity production, with a worldwide installed wind capacity of over 651 GW, an increase of 10% compared to 2018. According to the experts, the onshore and offshore wind systems are also "disruptive technologies." Offshore wind systems, due to technical aspects, are not so accessible for individuals or small businesses. Onshore wind systems are very scalable and more accessible for smaller investors. The experts predict that due to ongoing research on wind turbines, they will be more effective. Figure 3 presents different wind turbines that could have been in various industrial, commercial, or residential installations. The LCOE of the onshore wind turbine systems is second-lowest from analyzed energy generation systems—Table 2.

Figure 3. Different types of small wind turbines.

Natural Gas systems—Natural gas belongs to fossils but burns more cleanly than others. The systems that transform energy are mainly gas turbines and steam turbines. Natural gas burns. Natural gas could support future renewable energy sources. Natural gas is used in nearly every country and is traded on every continent. The LCOE of the natural gas made it the most attractive energy—Table 2. Technologies like micro-combined heat and power (CHP), which produce power and heat from natural gas, could be the disruptive technology that keeps natural gas as one of the most important energy sources, and makes the technology very important in ecological-distributed powerplants [50,51].

Geothermal systems—is the thermal energy from Earth's crust. In 2019, 13,900 MW (MW) of geothermal energy were retrieved worldwide. Since 2010, 28 gigawatts of direct geothermal heating systems have been installed. Geothermal power generators have been used in 26 countries and heating facilities in 70 countries [52–55]. Although the LCOE of the geothermal systems was above the average for renewables (Table 2), they are promising alternatives for fossil fuels depending on the individual counties' physical conditions. According to the experts, small-scale geothermal energy systems could build ecological distributed powerplants in countries that are rich in geothermal sources. This point of view has also be seen in recent publications [56,57].

Hydropower systems—Hydropower systems installed capacity was 1308 GW in 2020, making over 71% of all RES in 2020. Hydropower generated [2] and was expected to increase by about 3.1% each year for the next 25 years. As can be seen by LCOE analysis, the cost of hydroelectricity was low. When natural conditions are available, hydropower systems are the most compatible of RES. Traditionally, hydropower systems are related to rivers or water reservoirs [58–60]. According to the experts, small-scale hydropower energy systems could even be used as power accumulation for PV or wind turbines to store energy when more energy from renewables is produced than used and retrieved when it is needed.

Table 2. The cost of different RES (USD/kW, 2019 price).

No.	Source	Value	Solar PV Systems	CSP	Wind Systems		Natural Gas Systems	
					Onshore	Offshore	Gas-CC	Gas-GT
1	EIA		1331.00	7191.00	1319.00	5446.00	1017.00	710.00
2	IEA	Min	1005.00	3831.00	1287.00	3973.00	673.00	536.00
		Avg	1877.50	6283.00	2252.00	5169.00	1028.00	768.50
		Max	2750.00	8735.00	3217.00	6365.00	1383.00	1001.00
3	IRENA	Min	618.00	3704.00	1039.00	2677.00	n.a.	n.a
		Avg	1706.00	5415.50	1760.50	4114.00	n.a.	n.a
		Max	2794.00	7127.00	2482.00	5551.00	n.a.	n.a
4	Lazard	Min	900.00	6000.00	1100.00	2350.00	700.00	700.00
		Avg	1000.00	7550.00	1300.00	2950.00	1000.00	825.00
		Max	1100.00	9100.00	1500.00	3550.00	1300.00	950.00
5	NREL	Min	1142.00	6574.00	1678.00	3145.00	944.00	937.00
		Avg	1142.00	6574.00	1678.00	4231.50	944.00	937.00
		Max	1142.00	6574.00	1678.00	5318.00	944.00	937.00

No.	Data Source	Value	Geothermal Systems	Hydropower Systems	Coal Systems	Nuclear Systems	Biomass Systems
1	EIA	Min	2680.00	2752.00	3661.00	6317.00	2831.00
2	IEA	Min	1602.00	1282.00	1072.00	2805.00	630.00
		Avg	4355.00	n.u.	2181.00	4736.50	4964.00
		Max	7108.00	n.u.	3290.00	6668.00	9298.00
3	IRENA	Min	2020.00	680.00	n.a	n.a	422.00
		Avg	4650.00	2409.00	n.a	n.a	4582.00
		Max	7280.00	4138.00	n.a	n.a	8742.00
4	Lazard	Min	3950.00	n.a.	3000.00	6900.00	n.a.
		Avg	5275.00	n.a.	4600.00	3456.10	n.a.
		Max	6600.00	n.a.	6200.00	12.20	n.a.
5	NREL	Min	4557.00	3974.00	3867.00	6460.00	3988.00
		Avg	n.u.	5696.00	4046.00	6460.00	4085.00
		Max	n.u.	7418.00	4225.00	6460.00	4182.00

n.a.—not available because the study does not include the particular technology, n.u.—not used as the values are outliers. Data from the sources were transformed to 2019 prices using US GDP deflators.

Coal systems—traditional coal transformation is conducted by burning coal to generate heat or electricity. Coal-fired power plants are responsible for one-third of the world's electricity, "but cause hundreds of thousands of early deaths each year, mainly from air pollution" [60,61]. According to the experts and current literature, there are technologies to burn coal cleanly [62,63]. There are also technologies to transform coal into gas underground [64,65]. Coal in the future may be treated as a clean energy source because of the proper transformation process that coal will undergo. The possibility of using coal as ecological distributed powerplants strongly would depend on the technology applied to transform coal into energy.

Nuclear systems—atomic energy can now be produced by nuclear fusion and decay. At present, the great majority of electricity is generated through nuclear fission of uranium and plutonium. In 2019, civilian atomic power generated 2586 TWh, or in other words, nearly 10% of worldwide energy output. That made nuclear systems the second-largest low-carbon power source. At the beginning of 2021, there were 442 civilian facilities, with 392 GW.

Moreover, there were 53 facilities under construction and 98 planned. There is a popular hypothesis regarding cold fusion [66–68]. That it is only a matter of time that clean fusion generators would be in nearly every home [68]. Temporary in many countries, there is a fear of implementing this type of technology because of spectacular nuclear catastrophes such as the Chernobyl Nuclear Power Plant—1986, and the Fukushima Daini Nuclear Power Plant—2011.

Biomass and biogas systems—The facilities use green or animal wastes to produce heat or electricity. Although biomass technically can be used directly as a fuel, some people use biomass and biofuel interchangeably [69,70]. According to the experts, biomass systems are excellent solutions for the ecological distributed powerplants, especially in rural areas.

5. The Current State of Production and Prognosis of Energy in Poland

5.1. Polish Energy Sector—Overview and Limitations

Every analysis should start with the demand side, followed by the supply analysis. It is good to put the study in context to understand the upcoming trends better. This idea highlighted the following analysis. The data from Poland were presented in the context of Germany and the EU or the world average. The previous research performed by the authors on different economic branches ensured that analyzing trends in Germany was helpful to predict changes in Poland.

The primary energy consumption in MWh per capita changed from 1965 to 2019. While the consumption of primary energy per capita in the world was growing over the analyzed period, the energy consumption per capita in Germany and Europe (average) was maximum in the early 1990 of the XX century and then started to fall. In 2019, the consumption in Germany was 78.3% of its maximum and 85.1% in Europe. The consumption in Poland dropped rapidly from 1990 to 2001 due to the fall of communism (Figure 4). From 2002, energy consumption had started to grow again. The period from that year should be taken into consideration to predict the consumption for the future. In 2019, the consumption in Poland was 71.7% of the consumption of Germany. The authors have estimated that the consumption in Poland and Germany would be equal in the statistical scenario in about 2040.

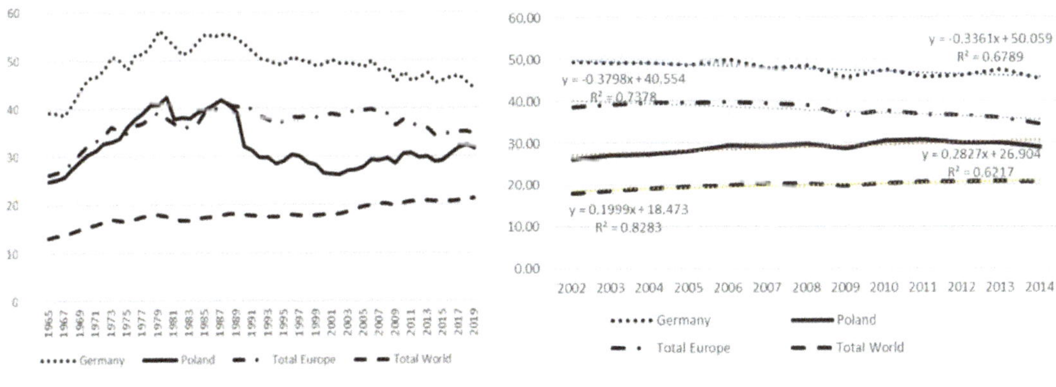

Figure 4. Energy consumption per capita in 2019 in Poland (in MWh), based on [17].

The change in primary consumption in Poland can be described by the equation below. The R-Squared (R^2) indicator over 0.5 is acceptable.

$$y = 0.2827x + 26.904 \tag{8}$$

$$R^2 = 0.6217 \tag{9}$$

The costs of fuel and energy in Poland constitute a significant share of household expenditure. According to Jurdziak and others [71–73], this share may be over 20%. That means that an increase in energy prices contributes to the impoverishment of society.

When analyzing the energy balance of Poland (Figure 5) from 2014 (except 2015), it turns out that electricity production in Poland did not cover the demand. Unfortunately, the experts predict that without a restructuring program of the Polish energy sector, the gap between demand and supply will increase, which is dangerous for the country's energy security.

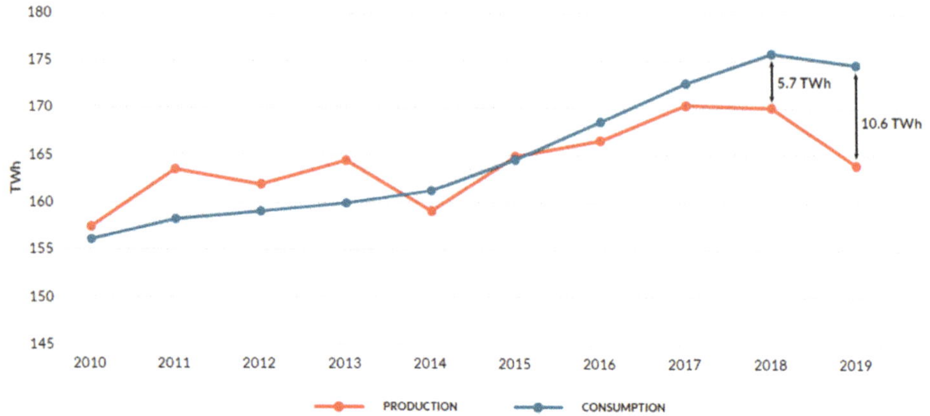

Figure 5. Balance of domestic electricity production and consumption [73].

The supply of the Polish energy system in 2019 was one of the largest in Europe [71,72] was covered mainly by coal-fired power plants—about 70% of the installed capacity [71,72]. Unfortunately, many outdated power plants and combined heat and power plants operate in Poland.

Agencja Rynku Energii S.A. (ARE) has reported in 2019 that hard coal systems with 23.9 GW had 50.4% of Poland's total installed capacity, which output 78.9 TWh of energy production and was 48.1% of total energy production in Poland. The second biggest share in the installed capacity was lignite—9.3 GW; 19.6% made 41.7 TWh; 25.5% of energy production. The other were natural gas systems—2.7 GW (5.7%) installed capacity and 14.5 TWh (8.8%) production; other industrial systems 0.6 GW (1.2%) installed capacity and 3.0 TWh (1.8%) production; pumped-storage systems 1.4 GW (3.0%) installed capacity and 0.7 TWh (0.4%) production. The total renewable energy systems had 9.5 GW (20.1%) installed capacity and 25.2 TWh (15.4%) production (hydropower systems 1.0 GW (2.0%) installed capacity and 2.0 TWh (1.2%) in production; onshore wind systems 5.9 GW (12.5%) installed capacity and 15.1 TWh (9.2%) production; biogas systems 0.2 GW (0.5%) installed capacity and 1.2 TWh (0.7%) production; biomass systems 0.9 GW (1.9%) installed capacity and 6.3 TWh (3.9%) production (including co-firing); and photovoltaics systems 1.5 GW (3.2%) installed capacity and 0.7 TWh (0.4%) production [73–76].

The structure of the energy facilities in Poland correlated to the structure of primary energy consumption by source (Figure 6).

Based on the opinion of the experts and the previous studies, the weighted SWOT analysis of the Polish energy sector has been made (Table 3). It has to be said in terms of the world energy transformation that the Polish energy sector does not fit those trends. A lot has to be done to build a future Polish energy sector that fulfills the EU policy regulations and demand requirements.

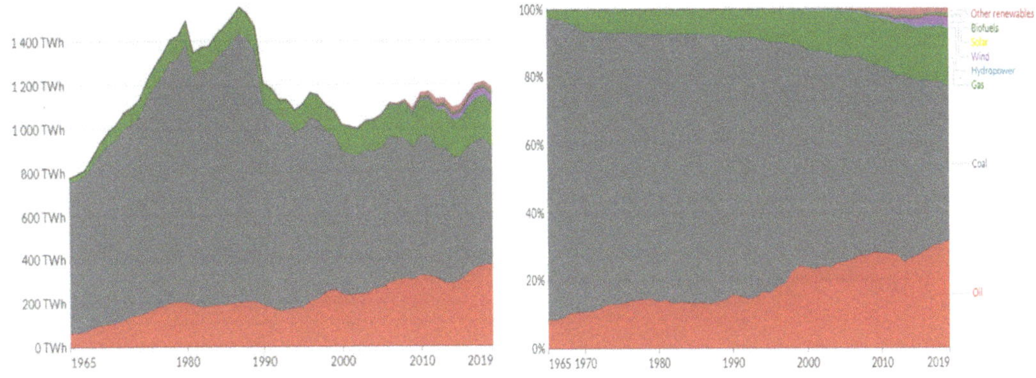

(**a**) (**b**)

Figure 6. Poland primary energy consumption by source: (**a**) in TWh; (**b**) relative (%) [1].

Table 3. Weighted SWOT matrix for the Polish energy sector.

		Weight	Rating	Score
	Strengths	1.00		
1	Little diversification of the Polish energy sector	0.40	5	2.00
2	Polish energy sector is one of the largest in Europe	0.30	4	1.20
3	Sufficient number of potential clients	0.20	5	1.00
4	Research centers and universities are researching energy issues	0.10	4	0.40
	Total weighted score			4.60
	Weaknesses	1.00		
1	Equipment to produce and to distribute energy is old	0.30	5	1.50
2	Polish power grids require general modernization	0.30	5	1.50
3	The Polish energy system is not ready to absorb energy from distributed sources on a larger scale	0.20	5	1.00
4	The sector is not currently self-sufficient	0.10	4	0.40
5	Power plants in Poland are not well distributed to cover demand needs	0.10	4	0.40
	Total weighted score			4.80
	Internal Factors Ratio (IFR)			0.96
	Opportunities	1.00		
1	Citizens of Poland are eagle to invest in the ecological distributed powerplants	0.60	5	3.00
2	When making buying decisions related to home appliances, consumers consider energy efficiency as one of the main factors	0.40	5	2.00

Table 3. *Cont.*

		Weight	Rating	Score
	Total weighted score			5.00
	Threats	1.00		
1	Unstable domestic law	0.50	5	2.50
2	The most crucial power source is coal; the ban on coal would be a vital issue for Polish energy security	0.30	5	1.50
3	Small micro-sources at the level of one circuit, without appropriate technical solutions, may harm energy parameters	0.20	5	1.00
	Total weighted score			5.00
	External Factors Ratio (EFR)			1.00

The conclusion from the weighted SWOT matrix for the Polish energy sector (Table 3) was not very optimistic. The internal factors ratio (IFR) was below 1, meaning the weaknesses were more critical than strengths. Opportunities and threats were balanced, which means that development perspectives of the Polish energy sector are not favorable. The ratio between internal factors ratio (IFR) and external factors ratio EFR) was 0.96:1.00. The strategy of the Polish policymakers should be focused on building or better allowing private investors to build distributed energy generation systems, especially photovoltaic systems, wind systems, hydropower systems, biogas, and biomass systems [27,77–79]. The Polish policymakers' strategic activity should also modernize and keep good quality energy distribution grids.

In Poland, the sources of distributed energy generation systems are classified as follows according to the "installed capacity:

- Microgeneration systems—from 1 W to 5 kW;
- Small generation systems—from 5 kW to 5 MW;
- Medium generation systems—from 5 MW to 50 MW;
- Large generation—from 50 to 150 MW" [70].

5.2. Photovoltaic Systems

Private investors, especially prosumers, have noticed the advantage having of PV systems in Poland. Installed capacity in photovoltaics at the end of 2020 was almost 4 GW, which means a 200% increase year on year. According to experts, PV is one of the most flexible systems to establish many ecological distributed powerplants. Its growth was exponential and could be described with the Equation (10). The R^2 indicator—0.9672 was very high in that case and showed that the curve equation had described the analyzed data very well (11) (Figure 7).

$$y = 8E - 5e^{1.0858x} \tag{10}$$

$$R^2 = 0.9672 \tag{11}$$

The experts make a remark that, unfortunately, the situation on the PV systems market is highly dependent on politicians' decisions. Currently, there is information that the rules for collecting energy from prosumers will change dramatically; investing in PV systems will be unprofitable for the prosumers.

According to the Energy Regulatory Office, the number of photovoltaic installations at the end of 2020 was around 460,000. The average capacity of the installed PV system was 5.78 kWp. Photovoltaic installations account for 99.89% of all RES micro-installations. Since 2018, the share of monocrystal PV systems has hit 98% in the year-to-year sales.

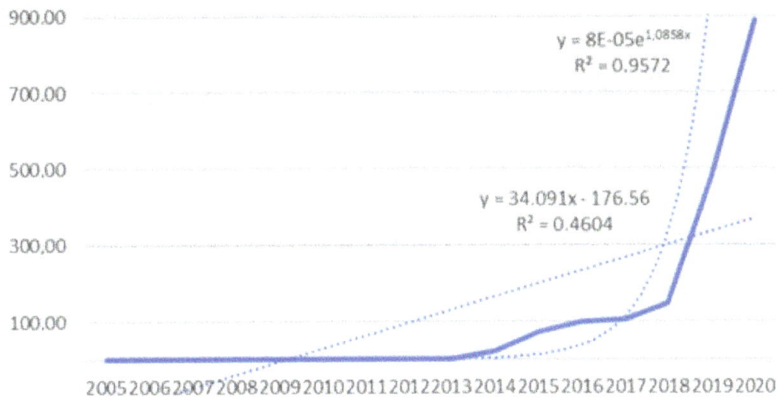

Figure 7. Photovoltaic systems—installed capacity (MW) and trends.

The PV systems have tremendous potential for growth. It is estimated that there are 5,522,000 residential buildings in Poland that are suitable for PV systems. The size of the average roof of a residential building is about 100 m^2. That gives about 552,200,000 m^2 of estimated area for PV systems to be installed. The total area could be from 182.23 GWp capacity installed from silicone PV systems to 662.24 GWp from hybrid or perovskite PV systems. That potential could be even more considerable because PV systems can also be installed on facades.

It is estimated that the PV system could fulfill from 4.0 to 10.0% of electricity demand in Poland.

The efficiency of PV systems could be upscaled 2–3 times by combining them with heat pumps, depending on the COP of the heat pumps.

The conclusion from the weighted SWOT matrix for the photovoltaic systems in Poland (Table 4) was also not very optimistic as for the weighted SWOT matrix for the Polish energy sector presented in Table 3. The internal factors ratio (IFR) was above one, meaning that the strengths were more critical than the weaknesses. The opportunities to threats ratio was 0.86, which means that the photovoltaic systems' development perspectives are not favorable because of extreme dependence on political decisions. The balance between the IFR and EFR was 1.10:0.86. The Polish policymakers' only strategy should be to not disturb the private sector and to provide energy distribution grids to handle the new installations. As shown in Figure 7, Polish private investors are very eager to build PV systems and understand the need to convert to a "zero-emissions" policy, but they have to see an economic reason to do it.

This vital activity should also be able to provide affordable and effective solutions to store energy, which requires further research and funding.

5.3. Wind Systems

Out of 6.35 GW of the installed capacity, onshore wind power supplies 12.6 percent of electricity in Poland. New onshore and offshore wind energy development will lead to at least double these figures in the coming years.

The Energy Regulatory Office reported 1239 wind farms in operation in the country in 2020 (including 1111 with a capacity below 10 MW (89.7%) and 128 with a capacity greater than or equal to 10 MW).

The electricity produced from wind sources in the Polish power system has also been systematically growing. In 2020, wind systems delivered 14,174 GWh of energy (compared to 13,903 GWh in 2019).

Wind energy accounted for ca. 8.2 percent of the energy consumed in the country in 2019.

Igliński believes that considering all the limiting criteria, the area excluded from the possibility of locating wind energy is 311.657 km^2, i.e., 99.92% of the land area of Poland, i.e., 311.904 km^2. That means that only 247 km^2 is available for the construction of wind farms, i.e., 0.02% of the country's territory. Therefore, since introducing the distance act, wind energy systems in Poland have practically stopped being developed [75]. The analysis of Figure 8 clearly shows that the slowdown in the development of wind systems is abnormal in Poland and confirms the claims of Igliński's thesis claims.

Table 4. Weighted SWOT analysis for photovoltaic systems in Poland.

		Weight	Rating	Score
	Strengths	1.00		
1	Roi from 8–17%	0.30	5	1.50
2	Very low operating costs for the installation	0.10	5	0.50
3	Its use does not emit harmful gases (including CO2)	0.10	5	0.50
4	Very low maintenance and operating costs for the installation	0.10	5	0.50
5	Can be used in off-grid locations	0.10	5	0.50
6	Scalable efficiency in combination with other installations, e.g., heat pump	0.10	4	0.40
7	Relatively low failure rate of the solar installation	0.10	4	0.40
8	Ability to combine with energy storage systems	0.10	3	0.30
	Total weighted score			4.60
	Weaknesses	1.00		
1	More energy is generated in spring–summer than autumn–winter	0.50	5	2.50
2	Energy from PV systems is produced only when the sun shines	0.20	4	0.80
3	Big initial costs of photovoltaic installations	0.30	3	0.90
	Total weighted score			4.20
	Internal Factors Ratio (IFR)			1.10
	Opportunities	1.00		
1	High social acceptance	0.30	5	1.50
2	Big potential of growth	0.30	4	1.20
3	Perovskite based solar cells are produced in Poland	0.20	5	1.00
4	Market is adopting new technologies very fast	0.20	3	0.60
	Total weighted score			4.30
	Threats	1.00		
1	Unstable domestic law	0.50	5	2.50
2	Small micro-sources at the level of one circuit, without appropriate technical solutions, may have a negative impact on energy parameters	0.30	5	1.50
3	Bad condition of national grid	0.20	5	1.00
	Total weighted score			5.00
	External Factors Ratio (EFR)			0.86

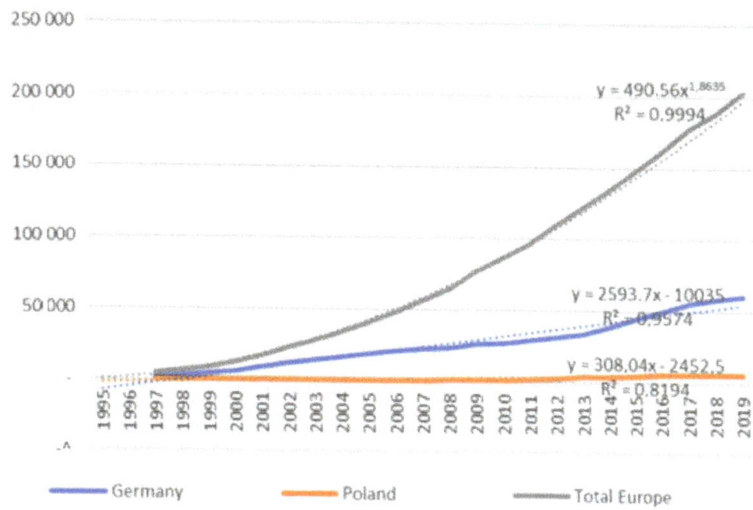

Figure 8. Wind systems—installed capacity (MW) and trends.

Table 5 presents the the Weighted SWOT Analysis for Wind Systems in Poland.

Table 5. Weighted SWOT analysis for wind systems in Poland.

		Weight	Rating	Score
	Strengths	1.00		
1	Well known technology	0.40	5	2.00
2	Wind is a source of clean energy	0.20	5	1.00
3	Wind farm require less space than solar farms or hydroelectric generation	0.20	3	0.60
4	Availability of resource in Poland	0.20	4	0.80
	Total weighted score			4.40
	Weaknesses	1.00		
1	The speed of wind is dependent on the metrological conditions	0.30	5	1.50
2	High initial investment costs in wind systems	0.30	4	1.20
3	Wind turbines could be harmful for migrating birds	0.20	3	0.60
4	The wind turbine blades are made of had to recycle material	0.10	4	0.40
5	Noise pollution is one of the setbacks of the wind technology	0.10	3	0.30
	Total weighted score			4.00
	Internal Factors Ratio (IFR)			1.10
	Opportunities	1.00		
1	Big potential of growth	0.50	5	2.50
2	Many producers of the systems are located in Poland	0.30	4	1.20
3	Research and development in the sector boost the efficiency of the wind systems	0.20	4	0.80
	Total weighted score			4.50
	Threats	1.00		
1	Unstable domestic law, unfavorable for the wind systems	0.40	5	2.00
2	Bad condition of national grid	0.40	4	1.60
3	Low social acceptance	0.20	3	0.60
	Total weighted score			4.20
	External Factors Ratio (EFR)			1.07

The conclusion from the weighted SWOT matrix for the wind systems in Poland (Table 5) was also quite optimistic. The IFR was above one, meaning that the strengths were more critical than weaknesses. The opportunities to threats ratio was 1.07, which means that the wind systems' development perspectives are pretty favorable, but one must remember extreme dependence on political decisions. The balance between the IFR and EFR was 1.10:1.07. This means moderately positive prospects for the development of these systems. The only strategy of the Polish policymakers should be, like in the case of the PV systems, not to disturb the private sector and to provide energy distribution grids that can handle the new installations. The Polish policymakers did precisely the opposite, as was mentioned by Igliński and others [75–81].

5.4. Hydropower Systems

In Poland, compared to other countries, the use of the water energy potential is incomparably lower, which is primarily due to the climatic conditions, average rainfall, and topography. Traditional hydroelectric power plants could be built in places with large natural slopes or where water could be artificially dammed. Hydropower systems require high initial investment costs (mainly due to construction costs). It was estimated that the share of hydropower in primary energy was about 1.5%. Poland's hydropower resources alone are estimated at 13.7 TWh per year. If it were possible to fully use Poland's hydropower systems potential, it would be possible to achieve even ca. 11 GW of capacity in big, and ca. 1.2 GW in small and medium, hydropower plants. In 2020, there were only 18 hydroelectric plants in the country with more than 5 MW.

The conclusion from the weighted SWOT matrix for the hydropower systems in Poland (Table 6) was also quite optimistic. The IFR was above 1–1.45, meaning that the strengths were more critical than weaknesses. The opportunities to threats ratio was 0.98, which means that the hydropower systems' development perspectives are not favorable because of extreme dependence on political decisions. The balance between the IFR and EFR was 1.45:0.98. This means a favorable situation because strengths and opportunities are more critical than weaknesses and threats. The things that Polish policymakers should do are to manage the potential resistance of society and to promote the building of micro- to medium hydropower power stations primarily in the places where watermills operated over 100 years ago.

5.5. Biogas and Biomass Systems

Poland has considerable resources of biomass and biogas available in every region. It has been estimated that biomass and biogas could fulfill about 15% of the demand for electricity in Poland. Additionally, the facilities could supply 28% of the heat demand. Figure 9 shows the actual number of biogas and biomass systems in Poland in 2020. Their distribution was correlated to the level of agricultural technology.

The conclusion from the weighted SWOT matrix for the analysis for biogas and biomass Systems (Table 7) was also, such as in the case of wind and hydropower systems, quite optimistic. The IFR was above 1–1.04, meaning the strengths were more critical than weaknesses. The opportunities and threats ratio was 1.27, which means that development perspectives of the biogas and biomass systems are the most favorable of the analyzed cases. The balance between the IFR) and EFR was 1.04:1.27. That means an extraordinary situation because strengths and opportunities are more critical than weaknesses and threats. The things that Polish policymakers should do is not to disturb and allow farmers to use the waste they produce as the source of clean energy.

Table 6. Weighted SWOT analysis for hydropower systems in Poland.

		Weight	Rating	Score
	Strengths	1.0		
1	Well known technology	0.20	5	1.00
2	Availability of resource	0.20	5	1.00
3	Easy and precise control of energy produced	0.15	4	0.60
4	Much lower operating costs than in conventional power plants	0.15	4	0.60
5	Means the access to a stable energy source (unlike, e.g., solar and wind energy electricity production is independent of weather and time)	0.10	5	0.50
6	Possibility of using energy locally or sending it to grid	0.10	4	0.40
7	Its use does not emit harmful gases (including CO2)	0.10	4	0.40
	Total weighted score		4.50	
	Weaknesses	5.00		
1	Construction of hydropower plants constitutes a serious interference with the natural environment	0.20	4	0.80
2	Inability to work during prolonged drought	0.20	5	1.00
3	The construction costs of this type of power plant are 2–3 times higher than the expenditure for the construction of conventional power plants	0.20	3	0.60
4	Reduction in reservoir capacity through sediment accumulation	0.10	3	0.30
5	Noise and pollution during construction	0.10	4	0.40
	Total weighted score		3.10	
	Internal Factors Ratio (IFR)			1.45
	Opportunities	1.00		
1	Water reservoirs could be used for tourist and recreational purposes	0.50	5	2.50
2	Pumped storage plants could be widely used energy storage	0.30	5	1.50
3	New directions of owe development	0.20	5	1.00
	Total weighted score		5.00	
	Threats	1.00		
1	Unstable domestic law	0.50	5	2.50
2	May cause secondary effects in the form of barrage cracking and water disasters	0.30	4	1.20
3	Moderate social acceptance	0.20	4	0.80
4	Resistance of ecological groups (large hydroelectric plants)	0.10	3	0.30
5	Damming of water in reservoirs may lead to flooding of settlements and agricultural lands, which may result in the necessity of displacement of the population	0.10	3	0.30
	Total weighted score		5.10	
	External Factors Ratio (EFR)			0.98

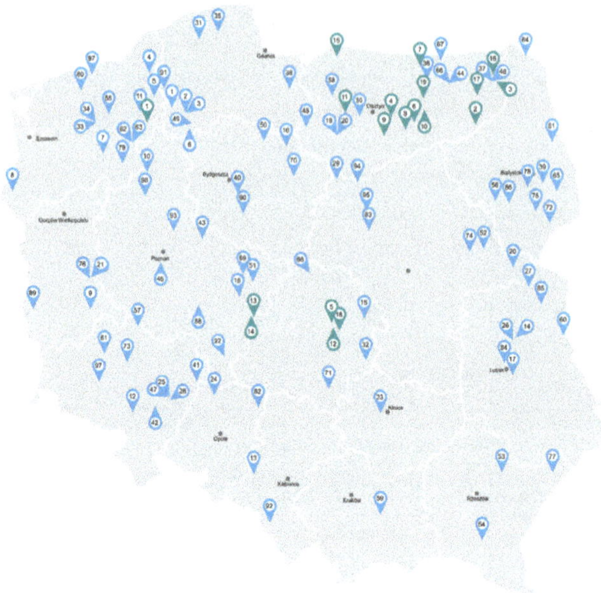

Figure 9. Location of biogas and biomass systems in Poland.

Table 7. SWOT analysis for biogas and biomass systems in Poland.

		Weight	**Rating**	**Score**
	Strengths	1.00		
1	Well known technology	050	5	2.50
2	Availability of resource	0.30	5	1.50
3	Possibility of using energy locally or sending it to grid	0.20	4	0.80
	Total weighted score			4.80
	Weaknesses	1.00		
1	Heat unused directly from the biogas is waste	0.60	5	3.00
2	Big initial costs of biogas and biomass systems	0.40	4	1.60
	Total weighted score			4.60
	Internal Factors Ratio (IFR)			1.04
	Opportunities	1.00		
1	High social acceptance	0.30	5	1.50
2	Combining high energy-intensive greenhouse cultivation could make the technology more efficient	0.20	5	1.00
2	Big potential of growth	0.30	3	0.90
3	Fast development of the technology	0.20	4	0.80
	Total weighted score			4.20
	Threats	1.00		
1	Volatility in the prices of resources from agriculture	0.50	3	1.50
2	Unstable domestic law unfavorable for the biomass and biogas systems	0.30	4	1.20
3	No guarantee of stable feedstock supplies in agricultural biogas and biomass plants	0.20	3	0.60
	Total weighted score			3.30
	External Factors Ratio (EFR)			1.27

In Poland, in 2020, were about 110 biogas plants. That was eight biogas plants per 1 million inhabitants, while in the Czech Republic, for example, this number was 54. The massive slowdown in biomass and biogas development was the effect of administrative and legal barriers.

The thesis about bad legislation was verified positively by experts and the historical data—Figure 10. Figure 10 presents two scenarios about developing the installed capacity of biogas and biomass systems in Poland. The higher line represents the prediction based on the period from 2005 to 2016, and the lower line is based on the period from 2005 to 2018. The actual number of the biogas and biomass systems followed, unfortunately, the worst scenario.

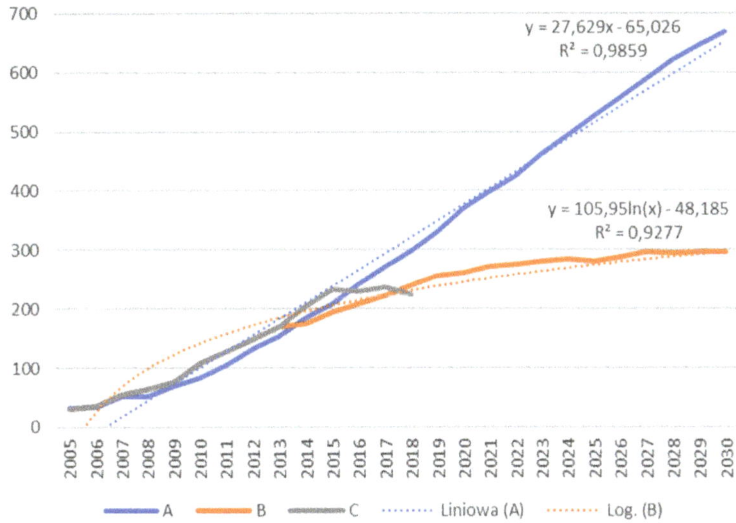

Figure 10. Biogas and biomass systems in Poland (MW) [28,82].

5.6. Other Energy Systems

Some other energy systems that were less popular or not considered as ecological-distributed energy generation systems have to be mentioned—especially geothermal and small nuclear systems.

- Geothermal systems—according to the Polish Geological Institute, National Research Institute Poland, it has great potential. According to estimates, the geothermal resources of Poland, available in 80 percent of the country, would be enough to meet the country's heat needs by 30 percent [83]. It has to be noticed that usability of geothermal resources in Poland is limited by their temperature: (1) level—low-level geotherm—down to 100 m +10 °C temperature could be obtained, which would provide an excellent possibility to set up "water to water" heat pumps, which are more effective during winter than "air to water systems"; (2) level—average temperature geotherm—down to 3 km—the temperature of about +100 °C could be obtained—which is a suitable temperature for a geothermal heating plant; (3) level—high-temperature geotherm—down to 6 km—the temperature of about +200 °C could be obtained—which is a suitable temperature for a geothermal electricity plant. The considerable advantage of geothermal systems is that the systems could be used to heat or cool the facilities.
- Nuclear systems—"As nuclear power generation has become established since the 1950s, the average size of reactor units has grown from 60 MWe to more than 1600 MWe, with corresponding economies of scale in operation" [84,85]. One can see the growing popularity of small and medium reactors (SMRs), more commonly recognized as "small modular reactors". SMRs are typically under 15 MWe. There are designed mainly for remote communities. The units can be set up independently or like LEGO

bricks, combined in a larger complex. There are also plans to construct tiny self-contained units for distant locations. SMRs are viewed as far more manageable than large ones, especially in terms of costs. Another reason for interest in SMRs is that they may be located in closed coal-fired power plants. "The World Nuclear Association outlines the characteristics of an SMR, which include (a) modest power and compact architecture; (b) modularity of construction; and (c) the possibility for sub-grade siting of the reactor unit, which provides better protection from natural or man-made risks". Once, the IBM engineers had believed that the world needed five computers a year. They were wrong. Who knows, maybe the SMR for energy generation will be the same for which the Apple II was for the information revolution and will be used instead of classic high pollutive coal power plants in Poland.

5.7. Scenarios for Renewable Energy Sources Development in Poland

Based on the information above and discussion with experts, the following main scenarios for Poland could be drawn:

- Catastrophe—due to Polish legislation, the current state of the energy sector would be preserved. The government wants to protect the national grid, so for it to change the way to enable and use small on-grid power plants to build new energy systems would be economically unjustified. Some investors decide to set up off-grid, but that would be a margin in the energy consumption scale. The electricity prices mainly due to emission prices go up. The energy to fulfill growing demand would be imported. High energy prices would make the Polish economy uncompetitive;
- Full RES—that could be treated as a mix of scenarios that the RES technologies would fulfill the demand for electricity and heat. This scenario can be believed because it is estimated that in a single hour, the amount of power from the sun that reaches the Earth is more than the entire world consumes in a year. The RES energy systems, like wind or photovoltaic, do not provide the energy constantly. Currently, there are not efficient enough energy storage systems developed to store enough energy produced to provide reliable power. Only biomass, biogas, and geothermal systems could provide energy regularly. Still, substantial technological development has to be achieved to make them more efficient and ready to fulfill the growing energy demand. For the scenario to succeed, severe investments into the national grid have to be built. The lesson learned from the Grand Ethiopian Renaissance Dam project shows that financing of the dam could be in a public–private partnership.
- Clean coal and RES—a lot of research has been done on coal gasification [86,87]. Hydrogen, for example, could be the final product of the process. The gasification process could be done underground and in a remote installation. The process would be done in distributed locations. The facilities could be placed near the demand site because the final product would be electricity or heat to provide society or industry. To succeed in this scenario, substantial investments into the national grid have to be made.
- Nuclear MIX—as time goes and as the Polish energy companies take no action, the only way to fulfill international agreements would be to install SMRs instead of disabling coal plants. To succeed in this scenario, substantial investments into the national grid would need to be made.

6. Discussion

Most Polish researchers and politicians agree that the country faces significant challenges related to the energy sector. The challenge is even more challenging because the Polish economy relies on coal as the primary energy source. Nearly everybody agrees that "something has to be done". Most of these people also agree that future technology should help to build a low-carbon economy. The papers related to the subject mainly highlight the possibilities and advantages in using RES as a substitute for coal. Of course, physicists argue that it is possible; but still, researchers do not mention the disadvantages and threads that should be considered in constructing a new energy mix for Poland. As the problem affects the entire economy, industry and economists should take part in the discussion and estimate the economic efficiency of the proposed solutions from designing to the utilization of the wastes. This should be an idea for further research. For example, the authors strongly recommend using PV systems as ecological distributed energy generation systems because of their flexibility, use, and growing efficiency. Still, it has to be mentioned that millions of solar panels installed in the last two decades will soon be ready to be disabled and recycled. Recycling is also a considerable problem of wind turbine blades, batteries used as energy storage, and nuclear waste. Making the total cost calculation could change the efficiency of the proposed solutions. This is why further research has to be carried out.

7. Conclusions

The main objectives of the paper were completed:

(1) Based on literature studies and LOCE analysis, photovoltaic systems would be game-changers to allow the globe to switch to the "zero-emission" economy. The PV systems would be supported with wind systems and biogas systems located mainly in rural areas. In some areas nuclear systems would also play an important role;
(2) The review of the technologies that could be used as the ecological distributed energy generation systems from Poland's perspective showed that the PV systems supported with wind systems and biogas systems located predominantly in the rural areas would help achieve the Agenda 2030 goals.

Additionally, the research questions were answered:

(a) Renewable energy sources could provide sufficient energy to replace the conventional energy sources (from global and Polish perspectives). As mentioned before, the only energy that comes from the sun would fulfill all current and future consumption. The only problem is developing more efficient systems that would transform solar energy and develop energy storage systems that would minimize the weaknesses, especially of the PV and wind systems. From the Polish perspective, the ideal energy mix, in the authors' opinion, would be PV systems, wind, biogas, and hydropower systems supported with energy storage systems (hydropower systems could also be used as energy storage). The remaining energy could be obtained from natural gas, which also comes from the environmentally friendly coal gasification process.
(b) PV systems, as well as wind, biogas, and hydropower systems are suitable as ecological distributed energy generation systems in Poland.
(c) In general, the main strength of the technologies to replace conventional energy sources in Poland is that all systems require low maintenance costs. Unfortunately, these systems require high initial costs, and this is the most important weakness. The most important thread of the development of alternative energy systems in Poland is government policy. However, theoretically, a society interested in financing ecological distributed energy generation systems could change the policy or policymakers.
(d) Unfortunately, the "Catastrophe scenario" mentioned above is coming true—due to Polish legislation, and the current state of the energy sector will be preserved.

Author Contributions: P.S.—Conceptualization, formal analysis, writing—original draft; W.G.—conceptualization, formal analysis, writing—review and editing, funding acquisition; A.W.-S.—conceptualization, formal analysis, writing—original draft; S.C.—conceptualization, formal analysis, writing—initial draft; R.B.—conceptualization, formal analysis, writing—original draft. All authors have read and agreed to the published version of the manuscript.

Funding: Funded by Faculty of Economic Sciences and Management, The Nicolaus Copernicus University in Toruń (NCU).

Institutional Review Board Statement: Not applicable.

Informed Consent Statement: Not applicable.

Data Availability Statement: The primary statistical data https://data.worldbank.org/; https://stat.gov.pl/; https://data.europa.eu/en.

Conflicts of Interest: The authors declare no conflict of interest.

References

1. De Lavoisier, A.L. *Traité Élémentaire de Chimie*; Forgotten Books: Maxtor, France, 2019.
2. Ghiani, E.; Serpi, A.; Pilloni, V.; Sias, G.; Simone, M.; Marcialis, G.; Armano, G.; Pegoraro, P.A. Multidisciplinary Approach for the Development of Smart Distribution Networks. *Energies* **2018**, *11*, 2530. [CrossRef]
3. Lovins, A.B.; Ürge-Vorsatz, D.; Mundaca, L.; Kammen, D.M.; Glassman, J.W. Recalibrating climate prospects. *Environ. Res. Lett.* **2019**, *14*, 120201. [CrossRef]
4. Binz, C.; Tang, T.; Huenteler, J. Spatial lifecycles of cleantech industries—The global development history of solar photovoltaics. *Energy Policy* **2017**, *101*, 386–402. [CrossRef]
5. Poulek, V.; Khudysh, A.; Libra, M. Innovative low concentration PV systems with bifacial solar panels. *Solar Energy* **2015**, *120*, 113–116. [CrossRef]
6. Boamah, F.; Rothfuß, E. From technical innovations towards social practices and socio-technical transition? Re-thinking the transition to decentralised solar PV electrification in Africa. *Energy Res. Soc. Sci.* **2018**, *42*, 1–10. [CrossRef]
7. Ceran, B.; Szczerbowski, R. Analiza techniczno-ekonomiczna instalacji fotowoltaicznej. In *Zeszyty Naukowe Instytutu Gospodarki Surowcami Mineralnymi i Energią PAN*; PAN: Warsaw, Poland, 2017.
8. Dalkey, N.; Helmer, O. An experimental application of the Delphi method to the use of experts. *Manag. Sci.* **1963**, *9*, 458–467. [CrossRef]
9. Skulmoski, G.J.; Hartman, F.T.; Krahn, J. The Delphi method for graduate research. *J. Inf. Technol. Educ. Res.* **2007**, *6*, 1–21. [CrossRef]
10. Rowe, G.; Wright, G. The Delphi technique as a forecasting tool: Issues and analysis. *Int. J. Forecast.* **1999**, *15*, 353–375. [CrossRef]
11. Dalkey, N.C. *The Delphi Method: An Experimental Study of Group Opinion*; Rand Corp.: Santa Monica, CA, USA, 1969.
12. Kurttila, M.; Pesonen, M.; Kangas, J.; Kajanus, M. Utilizing the analytic hierarchy process (AHP) in SWOT analysis—A hybrid method and its application to a forest-certification case. *For. Policy Econ.* **2000**, *1*, 41–52. [CrossRef]
13. Hill, T.; Westbrook, R. SWOT analysis: It's time for a product recall. *Long Range Plan.* **1997**, *30*, 46–52. [CrossRef]
14. Alptekin, N. Integration of SWOT analysis and TOPSIS method in strategic decision making process. *Macrotheme Rev.* **2013**, *2*, 1–8.
15. Engel, A. *Practical Creativity, and Innovation in Systems Engineering*; John Wiley & Sons: Hoboken, NJ, USA, 2018.
16. Timilsina, G.R. Are renewable energy technologies cost competitive for electricity generation? *Renew. Energy* **2021**, *180*, 658–672. [CrossRef]
17. Lovins, A. A 40 Year Plan for Energy. Available online: https://www.ted.com/talks/amory_lovins_a_40_year_plan_for_energy/transcript (accessed on 31 July 2021).
18. Global Change Data Lab. Our World in Data. Available online: https://ourworldindata.org/ (accessed on 31 July 2021).
19. Ignacy, Ł. Available online: https://EncyklopediaPWN (accessed on 31 July 2021).
20. UN. Iceland's Sustainable Energy Story: A Model for the World? UN Chronicle. 2020. Available online: https://www.un.org/en/chronicle/article/icelands-sustainable-energy-story-model-world (accessed on 31 July 2021).
21. Ministry of Industries and Innovation, Government of Iceland, Energy. Available online: https://www.government.is/topics/business-and-industry/energy/ (accessed on 31 July 2021).
22. LED. *Lighting Market Size, Share & Trends Analysis Report by End-Use (Residential, Commercial), by Product (Lamps, Luminaires), by Application (Indoor, Outdoor), by Region, and Segment Forecasts, 2021–2028*; Fortune Business Insights: Maharashtra, India, 2021.
23. Fischer, C. EU Ecodesign Regulation. In *More as Only a Light Bulb Ban*; EU-Oekodesign-Richtlinie. Mehr als ein Gluehbirnenverbot; Umweltmagazin, VDI Fachmedien GmbH & Co.: Düsseldorf, Germany, 2011.
24. Fukasawa, T. Consumer Preference for Durability and Energy Efficiency: Welfare Analysis of Light Bulb Market. Available online: https://ssrn.com/abstract=3839998 (accessed on 9 September 2021).
25. LED Light Bulb to Last More than 20 Years. Available online: https://www.bbc.com/news/technology-17992927 (accessed on 31 July 2021).
26. Senkus, P. *Zarządzanie i Dowodzenie z Wykorzystaniem Orientacji Procesowej*; Wydawnictwo Difin: Warszawa, Poland, 2013.
27. Economist. *Fuel Cells Meet Big Business*; Economist 07/24/99; The Economist Group: London, UK, 1999; pp. 59–60.
28. Govinda, R. *Demystifying the Costs of Electricity Generation Technologies*; Policy Research Working Paper, No. 9303; World Bank: Washington, DC, USA, 2020.
29. Ray, D. *Lazard's Levelized Cost of Energy Analysis—Version 13.0*; Lazard: New York, NY, USA, 2019; p. 20.
30. Smil, V. *Energy Transitions: Global and National Perspectives*; ABC-CLIO: Santa Barbara, CA, USA, 2016.
31. Ritchie, H.; Roser, M. Energy, OurWorldInData.org. Available online: https://ourworldindata.org/energy (accessed on 31 July 2021).
32. EU. *Quarterly Report on European Electricity Markets with Focus on the Impact of High Carbon Prices in the Electricity Sector Market Observatory for Energy DG Energy*; European Commission: Brussels, Belgium, 2021; Volume 14.
33. McDonough, W.; Braungart, M. *Cradle to Cradle: Remaking the Way We Make Things*; North Point Press: New York, NY, USA, 2010.
34. Boerema, N.; Morrison, G.; Taylor, R.; Rosengarten, G. High temperature solar thermal central-receiver billboard design. *Sol. Energy* **2013**, *97*, 356–368. [CrossRef]

35. Law, E.W.; Prasad, A.A.; Kay, M.; Taylor, R.A. Direct normal irradiance forecasting and its application to concentrated solar thermal output forecasting—A review. *Sol. Energy* **2014**, *108*, 287–307. [CrossRef]
36. Law, E.W.; Kay, M.; Taylor, R.A. Calculating the financial value of a concentrated solar thermal plant operated using direct normal irradiance forecasts. *Sol. Energy* **2016**, *125*, 267–281. [CrossRef]
37. Fraunhofer Institute for Solar Energy Systems, ISE. *Photovoltaics Report (PDF)*; Fraunhofer ISE: Freiburg, Germany, 2014.
38. Sabillon, C.; Franco, J.F.; Rider, M.J.; Romero, R. Joint optimal operation of photovoltaic units and electric vehicles in residential networks with storage systems: A dynamic scheduling method. *Int. J. Electr. Power Energy Syst.* **2018**, *103*, 136–145. [CrossRef]
39. Chung, J.; Shin, S.S.; Hwang, K.; Kim, G.; Kim, K.W.; Kim, W.; Ma, B.S.; Kim, Y.K.; Kim, T.S.; Seo, J. Record-efficiency flexible perovskite solar cell and module enabled by a porous-planar structure as an electron transport layer. *Energy Environ. Sci.* **2020**, *13*, 4854–4861. [CrossRef]
40. Feldman, D.; Ramasamy, V.; Fu, R.; Ramdas, A.; Desai, J.; Margolis, R. *US Solar Photovoltaic System and Energy Storage Cost Benchmark: Q1 2020*; No. NREL/TP-6A20-77324; National Renewable Energy Lab. (NREL): Golden, CO, USA, 2021.
41. Shah, S.A.A.; Valasai, G.D.; Memon, A.A.; Laghari, A.N.; Jalbani, N.B.; Strait, J.L. Techno-economic analysis of solar pv electricity supply to rural areas of Balochistan, Pakistan. *Energies* **2018**, *11*, 1777. [CrossRef]
42. Xu, L.; Wang, Y.; Solangi, Y.A.; Zameer, H.; Shah, S.A.A. Off-grid solar PV power generation system in Sindh, Pakistan: A techno-economic feasibility analysis. *Processes* **2019**, *7*, 308. [CrossRef]
43. Chen, P.Y.; Qi, J.; Klug, M.T.; Dang, X.; Hammond, P.T.; Belcher, A.M. Environmentally responsible fabrication of efficient perovskite solar cells from recycled car batteries. *Energy Environ. Sci.* **2014**, *7*, 3659–3665. [CrossRef]
44. Malinkiewicz, O. Building on trust and vision. *Nat. Mater.* **2021**, *20*, 904. [CrossRef]
45. Wojciechowski, K.; Forgács, D.; Rivera, T. Industrial opportunities and challenges for perovskite photovoltaic technology. *Sol. RRL* **2019**, *3*, 1900144. [CrossRef]
46. The National Renewable Energy Laboratory. Concentrating Solar Power Projects. Available online: https://solarpaces.nrel.gov/ (accessed on 31 July 2021).
47. Irena, I. *Future of Wind: Deployment, Investment, Technology, Grid Integration and Socio-Economic Aspects*; IRENA Innovation and Technology Center: Bonn, Germany, 2019.
48. Doerffer, P.; Doerffer, K.; Ochrymiuk, T.; Telega, J. Variable Size Twin-Rotor Wind Turbine. *Energies* **2019**, *12*, 2543. [CrossRef]
49. The National Renewable Energy Laboratory. Land Use by System Technology. Available online: https://www.nrel.gov/analysis/tech-size.html (accessed on 31 July 2021).
50. BP, Great Britain. Statistical Review of World Energy. Available online: https://www.bp.com/en/global/corporate/energy-economics/statistical-review-of-world-energy.html (accessed on 31 July 2021).
51. Maghanki, M.M.; Ghobadian, B.; Najafi, G.; Galogah, R.J. Micro combined heat and power (MCHP) technologies and applications. *Renew. Sustain. Energy Rev.* **2013**, *28*, 510–524. [CrossRef]
52. Dye, S.T. Geoneutrinos and the radioactive power of the Earth. *Rev. Geophys.* **2012**, *50*. [CrossRef]
53. REN21. *Renewables 2020: Global Status Report*; Chapter 01; Global Overview: Paris, France, 2020.
54. Geothermal Energy Association. Geothermal Energy: International Market Update May. 2010. Available online: https://geothermal.org/sites/default/files/2021-02/GEA_International_Market_Report_2010.pdf (accessed on 31 July 2021).
55. El Bassam, N.; Maegaard, P.; Schlichting, M. *Distributed Renewable Energies for Off-Grid Communities: Strategies and Technologies toward Achieving Sustainability in Energy Generation and Supply*; Elsevier: London, UK, 2013.
56. Morrone, P.; Algieri, A. Integrated Geothermal Energy Systems for Small-Scale Geothermal atlas of Europe. Combined Heat and Power Production: Energy and Economic Investigation. *Appl. Sci.* **2020**, *10*, 6639. [CrossRef]
57. Hurtig, E.; Cermak, V.; Haenel, R.; Zui, V. *Geothermal Atlas of Europe*; IAEA: Vienna, Austria, 1992.
58. REN21 Renewable Now. Available online: https://www.ren21.net/ (accessed on 31 July 2021).
59. Abdelshafy, A.M.; Jurasz, J.; Hassan, H.; Mohamed, A.M. Optimized energy management strategy for grid connected double storage (pumped storage-battery) system powered by renewable energy resources. *Energy* **2020**, *192*, 116615. [CrossRef]
60. Witt, A.; Chalise, D.R.; Hadjerioua, B.; Manwaring, M.; Bishop, N. *Development and Implications of a Predictive Cost Methodology for Modular Pumped Storage Hydropower (m-PSH) Projects in the United States*; ORNL/TM-2016/590; Oak Ridge National Laboratory: Oak Ridge, TN, USA, 2016.
61. Gençsü, I.; Whitley, S.; Roberts, L.; Beaton, C.; Chen, H.; Doukas, A.; Geddes, A.; Gerasimchuk, I.; Sanchez, L.; Suharsono, A. *G20 Coal Subsidies (PDF)*; ODI: London, UK, 2019.
62. Chang, S.; Zhuo, J.; Meng, S.; Qin, S.; Yao, Q. Clean coal technologies in China: Current status and future perspectives. *Engineering* **2016**, *2*, 447–459. [CrossRef]
63. Guan, G. Clean coal technologies in Japan: A review. *Chin. J. Chem. Eng.* **2017**, *25*, 689–697. [CrossRef]
64. Burton, E.A.; Upadhye, R.; Friedmann, S.J. *Best Practices in Underground Coal Gasification*; (No. LLNL-TR-225331); Lawrence Livermore National Lab. (LLNL): Livermore, CA, USA, 2017.
65. Perkins, G. Underground coal gasification–Part II: Fundamental phenomena and modeling. *Prog. Energy Combust. Sci.* **2018**, *67*, 234–274. [CrossRef]
66. Berlinguette, C.P.; Chiang, Y.M.; Munday, J.N.; Schenkel, T.; Fork, D.K.; Koningstein, R.; Trevithick, M.D. Revisiting the cold case of cold fusion. *Nature* **2019**, *570*, 45–51. [CrossRef]
67. Pitkänen, M. Cold Fusion Again. *Prespacetime J.* **2016**, *379-396*, 379–396.

68. Petrescu, F.I.; Calautit, J.K. About nano fusion and dynamic fusion. *Am. J. Appl. Sci.* **2016**, *13*, 261–266. [CrossRef]
69. El Naschie, M.S. From fusion algebra to cold fusion or from pure reason to pragmatism. *Open J. Philos.* **2015**, *5*, 319. [CrossRef]
70. Biofuels Explained Ethanol and Biomass-Based Diesel. Available online: https://www.eia.gov/energyexplained/index.php?page=biofuel_home (accessed on 31 July 2021).
71. Paska, J. *Wytwarzanie Rozproszone Energii Elektrycznej i Ciepła*; Oficyna Wydawnicza Politechniki Warszawskiej: Warsaw, Poland, 2010.
72. Szczerbowski, R.; Kornobis, D. The proposal of an energy mix in the context of changes in Poland's energy policy. *Polityka Energetyczna* **2019**, *22*, 5–18. [CrossRef]
73. Jurdziak, L. Czy grozi nam ubóstwo? Analiza potencjalnych skutków unijnej polityki walki z globalnym ociepleniem dla gospodarstw domowych w Polsce. *Polityka Energetyczna* **2012**, *15*, 23–50.
74. Energii, F. Energy Transition in Poland Edition. 2020. Available online: http//www.forum-energii.eu (accessed on 31 July 2021).
75. Pietrzak, M.B.; Igliński, B.; Kujawski, W.; Iwański, P. Energy transition in Poland—Assessment of the renewable energy sector. *Energies* **2021**, *14*, 2046. [CrossRef]
76. Igliński, B. Badanie Sektora Energii Odnawialnej w Polsce–Potencjał Techniczny, Badania Ankietowe, Analiza SWOT, Analiza PEST. Available online: https://www.researchgate.net/publication/338740614_Badanie_sektora_energii_odnawialnej_w_Polsce_-potencjal_techniczny_badania_ankietowe_analiza_SWOT_analiza_PEST (accessed on 30 July 2021).
77. Szczerbowski, R. Wyzwania Polskiego Sektora Wytwórczego do 2030 Roku. Available online: http://yadda.icm.edu.pl/yadda/element/bwmeta1.element.baztech-942bfb08-5d91-496a-a9ac-cccc305832ae (accessed on 31 July 2021).
78. Szczerbowski, R. The forecast of Polish power production sector development by 2050–coal scenario. *Polityka Energetyczna–Energy Policy J.* **2019**, *19*, 5–18. [CrossRef]
79. Szczerbowski, R. Strategia Zrównoważonego Rozwoju a Sektor Wytwarzania Energii w Polsce. Energetyka. Available online: https://cire.pl/filemanager/Zalczniki%20stron%20tekstowych/1f238adcf76fdc992b70b76917205c41e9b00e97ba4d041b6008c06f3ef940e2.pdf (accessed on 11 July 2021).
80. Ceran, B.; Szczerbowski, R. Energy cost analysis by hybrid power generation system. In *IOP Conference Series: Earth and Environmental Science*; IOP Publishing: Poznan, Poland, 2019; Volume 214, p. 012001.
81. Dyląg, A.; Kassenberg, A.; Szymalski, W. *Energetyka Obywatelska w Polsce–Analiza Stanu i Rekomendacje do Rozwoju*; Instytut na Rzecz Ekorozwoju: Warszawa, Poland, 2019.
82. Woszczyński, M.; Stankiewicz, K. Analiza efektywności energetycznej i założenia techniczne rozbudowy instalacji fotowoltaicznych na dachach obiektów przemysłowych. In *Innowacyjne I Przyjazne Dla Środowiska Techniki I Technologie Przeróbki Surowców Mineralnych*; KOMEKO: Gliwice, Poland, 2021; p. 67.
83. Dach, J. *Biogaz w Polsce—Raport 2020*; Biomass Media Group Sp. z o.o.S: Warsaw, Poland, 2020.
84. Small Nuclear Power Reactors. Available online: https://www.world-nuclear.org/information-library/nuclear-fuel-cycle/nuclear-power-reactors/small-nuclear-power-reactors.aspx (accessed on 31 July 2021).
85. Kapusta, K.; Stańczyk, K.; Wiatowski, M.; Chećko, J. Environmental aspects of a field-scale underground coal gasification trial in a shallow coal seam at the Experimental Mine Barbara in Poland. *Fuel* **2013**, *113*, 196–208. [CrossRef]
86. Zhang, Y.; Jia, W.; Wang, R.; Guo, Y.; Guo, F.; Wu, J.; Dai, B. Investigation of the Characteristics of Catalysis Synergy during Co-Combustion for Coal Gasification Fine Slag with Bituminous Coal and Bamboo Residue. *Catalysts* **2021**, *11*, 1152. [CrossRef]
87. Schilling, H.D.; Bonn, B.; Krauss, U. *Coal Gasification*; Graham & Trotman: London, UK, 1976.

energies

Article

The ESCO Formula as Support for Public and Commercial Energy Projects in Poland

Joanna Kurowska-Pysz [1] and Grzegorz Kunikowski [2,*]

1 Faculty of Applied Sciences, WSB University, 41-300 Dąbrowa Górnicza, Poland; jkurowska@wsb.edu.pl
2 Faculty of Management, Warsaw University of Technology, 00-661 Warszawa, Poland
* Correspondence: grzegorz.kunikowski@pw.edu.pl

Abstract: The formula of engaging an energy service company (ESCO) in Poland is not a new form of accounting for investments in improving energy efficiency. The results of our survey confirm that many entities still lack sufficient knowledge about this subject. The research problem this paper is concerned with is the conditions of applying the ESCO formula (a model of investment financing with the participation of a specialised company) to support local government units and enterprises in energy-industry project development. For the purpose of this study, the research questions were formulated to analyse of the following issues: the reasons for interest in the ESCO formula and the sources of knowledge about this solution; activities and other factors that can increase or reduce interest in the ESCO formula; services in terms of ESCO formula implementation; the attractiveness of alternative instruments for financing energy industry projects, the benefits of using the ESCO formula and the influence of current and future target groups on ESCO formula development in Poland. This paper, therefore aims to recognise the conditions under which the ESCO formula can be applied by local government units and enterprises implementing energy industry projects in Poland. The research problem was solved using a triangulation of research methods: empirical qualitative research (desk research analysis, individual in-depth interviews, computer-assisted web interview (CAWI) survey, and focus group interviews) and one of the foresight methods (an expert panel). The research revealed that the lack of knowledge amongst local government units and enterprises with regard to the ESCO formula, although not unique to Poland, is insufficient to explain the low level of interest in this solution. One of the key conclusions is the need to educate local government units and enterprises on energy efficiency. This is vital to arouse their interest in the more complex ESCO implementation solutions that they have not yet investigated. Furthermore, by following and analysing the project implementation process in the ESCO formula, we can conclude that the risk generated is primarily on the part of the energy service company itself. For this reason, it is doubtful that energy service companies will invest the equity necessary to develop this challenging market. Based on the research conclusions, we indicate some recommendations that the government and related public institutions should consider in order to boost this market and support ESCO companies.

Keywords: ESCO; EPC; energy efficiency; private sector; public sector

Citation: Kurowska-Pysz, J.; Kunikowski, G. The ESCO Formula as Support for Public and Commercial Energy Projects in Poland. *Energies* **2021**, *14*, 8098. https://doi.org/10.3390/en14238098

Academic Editor: Vincenzo Bianco

Received: 25 October 2021
Accepted: 1 December 2021
Published: 3 December 2021

Publisher's Note: MDPI stays neutral with regard to jurisdictional claims in published maps and institutional affiliations.

Copyright: © 2021 by the authors. Licensee MDPI, Basel, Switzerland. This article is an open access article distributed under the terms and conditions of the Creative Commons Attribution (CC BY) license (https://creativecommons.org/licenses/by/4.0/).

1. Introduction

Energy efficiency (EE) plays an essential role in accelerating clean energy transitions and achieving global climate and sustainability goals [1,2]. Therefore, it has an important place in the public policy agenda of most developed countries as they make great efforts to reduce energy consumption through climate and energy policies, programmes and targets [3,4].

In terms of EE, we have a related energy service company (ESCO) formula which is a solution that considers both energy technologies and the sharing of technological, financial and legal risks [5]. In literature, the acronym ESCO refers both to specialised companies [6] known as ESCOs and to a business model [5]. Reference to ESCO is also made in literature to describe an EE project delivery model. In addition to the above-mentioned options there are alternatives such as public–private partnership (PPP) and energy performance contracts (EPC) which provide other procurement models [7]. For the purpose of this research paper, we use a broader definition of the ESCO formula that encompasses all of the meanings above (hereafter ESCO or ESCO formula) and we refer to a model of investment financing mechanism using EPC contracts with the participation of a specialised company.

The value of the global ESCO market was estimated at USD 28.6 billion in 2017 [8] while the European ESCO market was in the region of USD 2.7 billion in 2015 [3,9]. According to ESCO market development research [10], the Polish market experienced a slow increase in the period 2015–2018. In 2020, there were 20 companies registered as energy service providers [11].

For local government units, the spending of funds under the operational programs of individual voivodeships is a priority.

A detailed description of the priority axes clearly indicates that in the thermo-modernisation of public utility and residential buildings, the reduction of low emissions and energy-saving outdoor lighting (streets, parks, squares), there is a preference for investments carried out using the ESCO formula as can been seen in the Mazowieckie Voivodeship [12]. The same is true for other voivodeships.

Therefore, the implementation of investments using the ESCO formula means the applicant is given preference. This preference confirms that the voivodeship authorities in Poland took the ESCO formula into account when preparing the rules for spending EU funds. Currently, the use of the ESCO formula facilitates the financing of the investment.

The main barrier to the ESCO market in Poland is that the legal framework is still ambiguous, and further review is seen as necessary by market actors. In addition, a lack of trust is distressing the market due to the scarcity of appropriate financing and lack of good examples to counterbalance the ample information about bad examples. Transaction costs are high as a result, and energy prices are relatively low [10].

Several studies have analysed the drivers and barriers of the ESCO market and the use of ESCO in general [9,13,14] and in the public domain [5,15,16].

The risk of projects implemented using the ESCO formula has also been described in literature [14,17–19]. Barrier assessments and recommendations for Poland are general [20–22] and formulated as part of international comparative analyses. In our opinion, due to the growing importance of local energy projects, these barriers require in-depth study.

This research paper aims to recognise the conditions necessary to apply the ESCO formula as a solution to support local government units and enterprises in the field of energy industry project development and implementation in Poland. The authors also defined three sub-goals related to:

- the identification of factors that raise the interest of Polish local government units and enterprises in implementing energy industry projects in the ESCO formula;
- the identification of solutions under which energy projects financed using the ESCO formula can be implemented in Poland;
- the preparation of recommendations to the Polish Ministry of Energy, regarding ESCO formula improvement and development in Poland.

To solve the research problem, the research questions were formulated, and the following issues were analysed: the reasons for interest in the ESCO formula and the sources of knowledge about its solution; activities and other factors that can increase or decrease interest in the ESCO formula; services in terms of the ESCO formula implementation; the attractiveness of different instruments for financing energy industry projects and the benefits resulting from using this formula, as well as the influence of current and future target groups on ESCO formula development in Poland. Growing interest in public and commercial energy projects using EU funds in 2021–2027 is also a driving force. Investment is typically financed by subsidies and either private funds or a commercial loan and private funds. This approach increases long-term debt, thus limiting interest from public sector investors. An equally important challenge is the technical knowledge about energy efficiency and the risk involved in maintaining energy parameters. Using the ESCO formula, expert knowledge is not required from the investor. Similarly, the risk is transferred to the ESCO, which guarantees and accounts for energy effects through EPC contracts. Poland needs to make use of the ESCO formula because traditional and more widespread investment financing models, those based on subsidies for example, can no longer be maintained (European funds are shrinking). Moreover, investments that are considered likely to produce returns should slowly move away from public funds. The third reason is that some local government units and enterprises do not have funds necessary to contribute to EU projects or projects based on redeemable loans. Therefore, the support of ESCOs is an alternative model for implementing such projects.

Basing on a triangulation of research methods [23], the research problem was solved using empirical qualitative research (desk research analysis, in-depth individual interviews, computer assisted web interview (CAWI) survey, focus group interviews), as well as one of the foresight methods (an expert panel).

The study was undertaken with the agreement of the Polish Ministry of Energy [24], which commissioned and funded the research.

The remainder of the paper is organised as follows: in Section 2, we present the literature review. In the Section 3, research methods and the design of the empirical research are outlined. In Section 4, we present the results. Section 5 provides a discussion and presents the limitations of our study. Finally, the paper ends with Section 6, where we present our conclusions and recommendations.

The article's novelty is to examine the assessments of the ESCO formula in Poland from the perspective of stakeholders representing the public and commercial sectors. The publications describing the current problems of Poland are international comparative analyses. Publications on ESCO in Poland are from the beginning of the 21st century and there have been very few since 2010. This article presents the results of current empirical research in which various points of view of enterprises and local government units were examined on the basis of a triangulation of research methods.

2. Literature Review

This article explains ESCO as a model investment financing mechanism (referred to forthwith as the "ESCO formula" or "ESCO"), in which one of the parties is a specialised company providing energy services based on the energy performance contracts (EPC). The two most common types of EPC are referred to as a (1) shared savings or (2) guaranteed savings model [8].

In essence, the ESCO formula is a comprehensive approach to financing investment projects, including financing using EU funds [7]. Other innovative methods of financing energy projects in the context of the challenges of the energy transformation are also used, such as energy cooperatives, analysed by Yildiz et.al [25], or crowdfunding models as the philanthropic–crowdfunding partnership model proposed by Ari and Koc [26].

Due to the complexity of ESCO projects in the public sector, the simplest and most popular projects are lighting modernisation [27] and energy efficiency in the municipal and housing sectors [6,28,29]. Advanced solutions concern the generation of energy for own needs—in CHP installations [19] for example. Advanced energy projects the industrial sector have also been realised using the ESCO formula.

The definition of EPCs depends on the country [30]. Common to all EPCs, however, is the fact that they are drawn for a specific period and detail the delivery of specific energy effects, monitored and linked to financial settlements.

In the Energy Efficiency Directive 2012/27/EU [31], EPC is defined as a contractual arrangement between the beneficiary and the provider of an energy efficiency improvement measure, verified and monitored during the whole term of the contract, where investments (work, supply or service) in that measure are paid for at a contractually agreed level of energy efficiency improvement or other agreed energy performance criterion, such as financial savings.

2.1. Drivers of Interest in the ESCO Formula

To stimulate the development of the energy services market using the ESCO formula, governments use incentives and drivers. Trianni et al. [13] classify drivers for development as follows: regulatory, economic, informative and vocational, distinguishing external and internal drivers for each category. The importance of public financing is emphasised which, apart from commercial financing, constitutes essential support, especially for small and medium-sized enterprises (SMEs) [13]. Bertoldi and Boza-Kiss [9] used the following categorisation in their international research: (1) legal and political; (2) procedural factors and tools; (3) financing; (4) information and awareness; and (5) structural and market-related changes. Those areas of the research that concerned the situation in Poland, white certificates are identified as the legal driver.

To stimulate private sector investments and promote energy efficiency through public–private partnership arrangements, the public sector uses private companies, that is to say, an energy service company (ESCO) [6].

2.2. Risks and Barriers in the Application of ESCO Formula

Barriers to implementing the ESCO formula have been well researched and described by scholars using qualitative research methods [9,13,15,32,33].

ESCO solutions raise numerous concerns and risks [34] and encounter various barriers. Lee et al. used the following risk categorisation: economic; financial; project design; installation; technology; operational, and measurement and verification [18]. Garbuzova-Schlifter and Madlener [17] distinguished the following risk factors: project preparation and execution phases; contractual; technical and operational; financial; clients; human and behavioural; political and regulatory, and market.

The barriers are: legal; unfamiliarity with ESCO contracting; public debt; ownership issues; ownership of energy savings; competition from subsidies; reluctance to enter into PPP (public-private partnership); in-house implementation of projects; reluctance among the company's engineers, and decision-making processes [22]. Other researchers classify the barriers to information and awareness: legislative and accounting; behavioural; market and external; financial, and technical and administrative [3]. According to [1], the main barriers in developing countries are institutional and financial.

Bertoldi and Boza-Kiss indicate the problem of the cost of projects implemented using the ESCO formula to the public debt by local governments. This approach follows the EUROSTAT methodology and places restrictions in EU legislation on the requirements of the budgetary frameworks of the Member States. Although the text of the legislation is ambiguous, almost all countries have interpreted the EUROSTAT methodology as a barrier to ESCO projects, including Slovakia, the Czech Republic, and Poland. Others, such as Spain did not take a stand. The second problem was the limited liquidity and creditworthiness of public administrations, especially following the financial crisis [9].

Another barrier indicated in literature is the limited knowledge of the possibilities of financing energy efficiency projects in the form of ESCOs amongst potential beneficiaries [2]. Such barriers have been ling since identified [35]. On the other hand, current research shows more detailed knowledge needs have been formulated-the publishing of model EPC contracts [27], the dissemination of good examples, the modification of tender procedures [14,18] and public–private partnership [3].

Stagnation and decline must be taken into account, particularly in developing markets. For example, in 2010, the ESCO market was perceived as a fast-growing market in Sweden [35]. However, nine years later in 2019, it was noted in an international study that it was the only market in Europe to record a steady decline [10].

It is worth paying attention to the beneficiaries' expectations that the ESCO formula will produce visible savings immediately while also avoiding risks [36]. Even though this example concerns a business model of heat sales in Finland, this expectation is universal and demonstrates the need to create business models which promote risk-taking by companies that provide ESCO services.

2.3. ESCO in Poland

The primary legal Act for the activities of ESCOs is the Energy Efficiency Act [37]. Due to the financing of crucial provincial funding programmes that directly indicate the formula of ESCO as an acceptable form of co-financing of regional operational programmes [38,39] and regional plans for a low-carbon economy, such as in the Polish city of Kielce [40]. The majority of services provided by ESCO companies in Poland involve advice/audit services (the most popular and most frequently provided), energy efficiency in buildings, district heating and cooling, lighting, CHP, energy generation and distribution [10].

The Polish ESCO market is described in the literature, but it is not a popular topic. Most of the information available comes from international review reports. In 2017, it was identified as a market in the initial stage of development in 2013–2016, and it was in a state of slow growth [3].

In 2014, actions promoting the development of this market were recommended including the continuation of the creation of dedicated funds to support the use of the ESCO formula. Central administration support was also recommended in the field of formal solutions for ESCO business, a knowledge development platform and the dissemination of sample documents, in particular EPC contracts, and criteria for selecting ESCOs in tenders [41]. The ex ante analysis also formulated recommendations for developing ESCOs and using EPC contracts as financial instruments to be used in the Operational Program Infrastructure and Environment 2014–2020 [20]. According to the EU Joint Research Centre [10], respondents to a survey conducted in Poland indicated the following barriers: insufficient promotion of energy services, mistrust, and the incompatibility of ESCOs with other financial systems.

The amendment of the Energy Efficiency Act [37] should be considered significant, where provisions were introduced to directly facilitate the ESCO formula in local governments.

From the point of view of the dynamics of the development of the ESCO services market in Poland, the slow growth should be noted. In 2005, there were eight such companies [3]; by 2020 this number had increased, but only to twenty [11].

The drivers of the energy services market include dynamic development of energy technologies (including smart grids), the relationship of the price of construction services and materials to the prices of energy carriers, the growing energy awareness of end-users, and the involvement of non-energy companies such as telecommunication operators, in the energy services market. It is expected that energy companies will continue to diversify their portfolio of services seemingly not related to their core operations [10].

In the literature analysis, we can say that in general the ESCO topic is well described in the international literature. According to the authors, the conclusions are universal and relevant for other developed and developing countries.

As mentioned above, the topic of the ESCO market in Poland is not covered in scientific studies but is present in industry reports. There are, however, visible actions aimed at popularising the recommended mechanism in the official funding rules for energy efficiency projects. The experiences in implementing energy efficiency investments in public utility institutions are questionable in terms of the results achieved and the resources spent. The results of the audit of energy efficiency investments made in public-use organisations by the Polish Supreme Audit Office [42] proved that for the audited facilities, the average basic investment return period was 65 years. This example proves insufficient knowledge or imperfect procedures in which indicators are not used to assess the energy effect when planning investments.

Therefore, the authors identify a gap in research concerning the conduct and publication of scientific results of ESCO market research in Poland regarding the state of market development, current barriers, and identification of the awareness of the mechanism among public and commercial recipients of ESCO services.

3. Research Approach and Methodology

3.1. Research Assumptions and Design

As shown above, the realisation of energy industry projects in local government units and enterprises is often a challenge and even a problem, resulting in low motivation to undertake such projects. The reasons for this situation certainly include the difficulty of selecting appropriate sources of financing for those projects, often combined with poor preparation and qualification for managing them and the need to accept deferred profit or to obtain dominant environmental and social profits with only a small margin of economic gain from such projects.

In principle, in addition to direct financial resources, the use of both EU funds and repayable and non-repayable national public funds for energy industry projects is recommended. This is, however, connected with numerous limitations described in literature, such as bureaucratic procedures, a long waiting period for project evaluation, restrictive conditions of project accounting, the risk of having to return the grant, a lack of funds for investor contribution, limited eligibility of costs and activities that can be financed from the grant. Moreover, when it comes to externally financed repayable debt (e.g., a loan or credit), the main problem is often the inability to demonstrate adequate collateral for the debt or the need to include it in the financial statements as an additional encumbrance on the entity's liquidity [43–45].

Taking the identified limitations concerning the implementation of the described forms of financing of the energy industry projects by local government units and enterprises into account, in this article, the authors formulated a research problem concerning the conditions of applying the ESCO formula as a solution supporting those entities in terms of energy industry project development.

To solve the research problem, the authors defined two research questions:

1. What factors influence the interest of local government units and enterprises in implementing energy industry projects in the ESCO formula in Poland?

2. What solutions can support the implementation of the ESCO formula in Poland?

To answer the first research questions, the authors analysed:

- the reasons for the interest of local governments units and enterprises in the ESCO formula;
    - the sources of knowledge about its solution;
    - the activities and other factors that can increase or decrease interest in the ESCO formula.

To answer the second research questions the authors analysed:

- the services offered for the local governments' units and enterprises in terms of the ESCO formula implementation;
    - the attractiveness of different instruments financing energy industry projects;
    - the benefits for the local government units and enterprises resulting from using the ESCO formula;
    - the influence of current and future target groups on the ESCO formula development in Poland.

Based on the above research questions, the authors defined the paper's goal concerning recognising the conditions for applying the ESCO formula by local government units and enterprises implementing energy industry projects in Poland.

Additionally, the authors formulated three sub-goals:

1. The identification of factors influencing the interest of Polish local government units and enterprises in implementing energy industry projects in the ESCO formula.
2. The identification of solutions under which energy projects financed using the ESCO formula can be implemented in Poland.
3. Defining the Polish Ministry of Energy recommendations, useful for the ESCO formula improvement and development in Poland.

3.2. The Research Procedure

To solve the research problem and achieve the research goals, the authors based on a triangulation of the research methods [23]. Triangulation uses several research techniques to analyse a given issue to increase the credibility and accuracy of the results, and as a result, give a thorough understanding of the phenomenon under study. They conducted the empirical qualitative research and applied the foresight method [1]. Such diverse perspectives justify approaches similar to those adopted by authors of other studies regarding ESCO [3,18,27,46,47].

The authors also realised the appropriate desk research analysis based on the literature. They found data concerning the documentation of the implementation of European programmes financing projects in the field of energy and analytical reports on the implementation of the ESCO formula in Poland and abroad.

In terms of qualitative research, a combination of the following research techniques was applied:

- Direct/online in-depth individual interviews (IDI) realised according to the scenario, including open questions. The in-depth individual interview is pervasively applied in qualitative research because it is versatile across a range of study topics, adaptable in challenging field conditions, and excellent for not just providing information but also for generating understanding as well [48]. There are some examples of the application of in-depth individual interviews in investigating social barriers to the adoption of smart homes [49]; stakeholder analysis of energy performance contract models [50].
- CAWI surveys (computer-assisted web interview) based on standardised questionnaires (the close questions, single choice or multiple choice). CAWI surveys produce savings on direct costs (logistics and interviewers) concerning phone or Internet interviews, while at the same time providing a high (or higher) level of quality in terms of sampling procedures, data collection, field monitoring, and data processing. There is an example of applying an extensive CAWI survey in the research concerning the mobility data across the EU 28 members [51].
- Focus group interview (FGI) conducted as a structured and moderated discussion in which six people participated, realised according to the scenario defining the conversation goals and guidelines for the moderator. Focus group interviews are ubiquitous in social-behavioural sciences, public policy, and political research [48], the example of application is given above [50].

In terms of the foresight method, the authors applied a panel of experts. The expert panel is one of the key foresight methods, carried out in the form of a panel discussion with the participation of a wide group of specialists in a given field. During such a meeting, various positions are confronted, knowledge is accumulated, and, for example, recommendations and proposed solutions are developed [52]. An expert panel was applied in research focused on zero-emission vehicle technologies recognition [53]. The authors organised the panel to verify the validity of the detailed findings resulting from other research techniques but also to confirm the final conclusion and recommendations.

The qualitative research was conducted between 2017 and 2019. It covered a sample of 6 experts and 68 entities implementing or intending to implement projects based on the ESCO formula, including:

- 42 local government units;
- 21 enterprises;
- 5 institutions that provide financial support for energy projects.

The selection of the research sample was non-random. We selected the representatives of the respective target groups (local government units and companies) in two ways:

- the authors conducted desk research and social media analysis to recognize what entities deal with ESCO; those entities were invited to participate in the research as respondents;

- the authors used the mailing list prepared by the Ministry of Energy; those entities were invited to participate in the research as respondents.

In total, we found about 75 representatives of the target groups, but not all we available to participate in the research (in-depth individual interviews, CAWI surveys and focus group interviews).

The distribution of entities covered by the study on the map of Poland is presented in Figure 1 below.

Figure 1. The distribution of entities covered by the study on the map of Poland (local government units-in red, enterprises-in blue).

The research procedure was carried out in four stages as presented in Table 1.

Based on the scenarios organized according to the research questions, the authors conducted both in-depth interviews and focus group interviews, recording all of them. The authors then prepared transcriptions and analysed the scope of the text based on the following procedure:

- comparing answers to the same questions;
- grouping statements into thematic threads;
- evaluating individual statements within the same thematic threads;
- identifying key words, similarities and differences in statements and relating them to the entire context of the research;
- formulation of partial conclusions;
- formulation of final conclusions.

Table 1. The research procedure. Source: own elaboration.

Stage	Target Group and Sample	Outcome
1. In-depth individual interviews	- 3 institutions that provide financial support for energy project - 3 local government units - 3 enterprises	Identification of general conditions for the functioning of the ESCO formula in Poland from three different perspectives.
2. CAWI survey	- 37 local government units - 16 enterprises	Expanding knowledge about the ESCO formula in accordance with the research questions
3. Focus group interview	- 2 institutions that provide financial support for energy industry projects - 2 local government units - 2 enterprises	Analysing the survey results and defining key findings, conclusions and recommendations
4. Expert panel	- 3 representatives from academia and 3 business representatives with knowledge of how the ESCO refundable financing formula works	Additional independent assessment of the conclusions and recommendations formulated

It should be highlighted that the above research approach was chosen in agreement with the Polish Ministry of Energy [24] that commissioned and funded the research. The research was undertaken for the purpose of a general diagnosis of the factors influencing the interest of Polish entities towards implementing energy projects using the ESCO formula, as well as the identification of business models related to implementing energy projects using the ESCO formula in Poland. These are the reasons that the authors based their research on qualitative research instead of quantitative research. Qualitative research was more efficient thanks to its methodological approach, and provided sufficient results much faster than quantitative research. Additionally, using one of the foresight methods enriched the study and provided objective results, verified not only from the perspective of the target groups but also taking the feasibility of the recommended solutions into account, which were assessed by experts.

4. Results

4.1. Factors Influencing the Interest of Local Government Units and Enterprises in the Implementation of Energy Projects Using the ESCO Formula in Poland

In-depth interviews revealed that the ESCO formula is potentially interesting, especially for those entities which do not have many other possibilities to develop energy industry projects due to reasons such as the limited amount of European funds available, the demand for innovative financial solutions and the need to be supported in energy industry projects by professional consultancies. Local government units and enterprises are interested in ESCO services either as an alternative to grants and loans or in linking such sources of financing for energy projects, but so far, this mechanism has not been sufficiently researched by the respondents.

An analysis of the reasons for interest in the ESCO formula presented in Table 2 concludes that the target groups of the ESCO formula are eager to learn more about this mechanism when they have access to the appropriate information (44.44%), when there are favourable economic conditions for projects based on the ESCO formula (13.89%) and when they receive offers from ESCO companies and other entities (11.11%).

Table 2. Reasons for interest in the ESCO formula. Source: CAWI survey.

No.	Response Option	Responses (in %)
1	Available, approachable information about ESCO	44.44
2	Possible savings on energy costs	8.34
3	Results of previous successful projects	2.78
4	Need to implement important projects	5.55
5	Project evaluation conditions when applying for funding	8.34
6	Investment tasks already underway	2.78
7	Received offers for such activities	11.11
8	Favourable economic conditions for projects based on the ESCO formula	13.88
9	Other	2.78
10	Not interested	0.00

The above findings were confirmed by the representatives of target groups that participated in the focus group interview. They stated that interest in the ESCO formula could not be linked with the innovative attitude towards new solutions supporting EE, but it results rather from access to reliable and objective but convincing information on this subject (e.g., campaigns conducted by public institutions) or from observing the positive effects of implementing the ESCO formula in other energy industry projects.

Concerning the sources of knowledge about the ESCO formula, the key conclusion coming from in-depth interviews was that the entities which can be potentially interested in implementing this mechanism are afraid whether the promises of ESCO companies and the conditions of implementation offered will be achievable in reality. The key reasons are not only the potentially improperly provided ESCO company services of but also the changing operating conditions of local governments units and enterprises (for example, changes in legal-administrative matters or to public aid rules). Considering the above-mentioned concerns, as well as a very dynamic situation in the energy market, many entities highlight that they cannot rely on the most available sources of information concerning the ESCO formula, as they may be out of date. One can say that the knowledge of the ESCO formula is quite superficial as evidenced by the respondents' identification of the Internet as a significant source of knowledge on the subject. Obviously, the Internet can be a source of information concerning the ESCO formula in the initial stage; however, the energy industry projects based on those mechanisms are very difficult and complex. They require interdisciplinary knowledge and experience, which cannot be provided based mainly on the Internet.

Table 3 shows that the largest number of respondents stated they were interested in the ESCO formula because of participation in the projects co-financed by European funds (36.01%) or participation in commercial projects. It should be stated that participation in the projects does not mean that the respondents have a positive approach to the ESCO formula. Their opinion is based on the individual experience arising from particular projects. Another very popular source of interest is the Internet (22.23%) as well as conferences and events (13.88%). These sources of information are not equal in terms of information quality. Conferences and events have a very limited range, although they can deliver professional knowledge. The Internet provides more general information, whereas experiences related to participation in the project are much more concrete, although they can be either positive or negative.

The focus group interview confirmed that the Internet should be taken into account as a primary source of information when planning information campaigns about ESCO companies targeted at potential investors, including local governments units (e.g., owners of public buildings) and enterprises. The focus group interview participants showed that a highly specialised topic such as the ESCO formula could not be effectively explained based on general information and data because its effectiveness must be related to the specific scope of investment in each case. If an entity, when calculating the effectiveness of the ESCO formula, relies on information from the Internet rather than professional advice, mistakes or misinterpretation of the results are possible. The focus group interview participants, therefore, strongly advocated encouraging local governments units and enterprises to seek expert advice from ESCO companies. One also pointed out that the low interest in the ESCO formula results mainly from the lack of recognition of this instrument among potential users and a lack of knowledge about it.

Table 3. Sources of knowledge about ESCO formula. Source: CAWI survey.

No.	Response Option	Responses (in %)
1	Participation in the EU projects	36.01
2	Participation in the commercial projects	8.34
3	Internet	22.23
4	Training	2.78
5	Information publications	5.55
6	Scientific articles	2.78
7	Participation in conferences and events	13.88
8	Do not remember	8.34
	No opinion	0.00

Regarding the activities to increase the interest in the ESCO formula, the ESCO companies interviewed strongly emphasised the need to involve government institutions in popularising this mechanism in order to increase its credibility and objectively and reliably present it to potential investors, while information about the ESCO formula by companies providing such services is perceived as marketing activity rather than as reliable information.

The CAWI survey found (Table 4) that two important actions to increase interest in the ESCO formula are to improve access to knowledge about it (30.95%) and to include it in grant projects (26.19%). Both should be subordinated to the key action of improving access to expertise and objective information about this financing model. As many respondents (21.42%) pointed out, much-needed action should be focused on the promotion and recognition of the ESCO formula. This also results from the fact that the financial engineering of an investment project based on a subsidy in connection with the ESCO formula may, in individual situations, raise doubts concerning the due amount of the subsidy and the eligibility of all costs or other conditions of using the subsidy, while the investor is unable to clarify these doubts. This may be the case particularly for companies preferring non-refundable financing, with much higher requirements for obtaining the grant compared to refundable financing. In these situations, a lack of expertise in the principles of the ESCO formula may discourage its use.

Table 4. Activities to increase interest in ESCO projects. Source: CAWI survey.

No.	Response Option	Reponses (in %)
1	Improving the competitiveness of the ESCO formula against other forms of financing	7.16
2	Improving access to knowledge about the ESCO formula	30.95
3	Integrating ESCO into grant projects	26.19
4	Regulatory changes	14.28
5	Increased promotion and recognition of the ESCO formula	21.42
6	No opinion	0.00

The interviewees pointed out the important issue of an objective presentation of the advantages and disadvantages of using the ESCO formula in energy industry projects because promoting this solution without pointing out the significant limitations and risks will not help increase credibility among local governments and enterprises in the long run. Reliable information regarding the ESCO formula should serve well to popularise it in various sectors of the energy market.

During the focus group interviews, it was clearly stated that the activities related to increasing the interest in ESCO projects are strongly linked with the process of mitigating the barriers which restrain such investments in the energy industry. There are many more obstacles than incentives for local government units, as well as the enterprises which would like to implement the ESCO formula in their projects. That is why it is necessary to indicate and analyse the key barriers of ESCO formula implementation in public and commercial projects related to the energy industry.

In the question regarding the factors reducing the interest of local governments and enterprises in investing in the ESCO formula, a rather diverse set of answers was obtained (Table 5). A significant percentage of respondents said that savings possible under the ESCO formula were not enough to tempt them into using the model in the first place (45.23%). Other factors cited by a relatively large percentage of respondents (lack of regulation–21.42% or insufficient legal regulations–26.19% as well as the issue of debt increase–23.80%) once again confirm that those potentially interested in the ESCO formula are not familiar with the details of its principles of operation.

Table 5. Factors reducing interest in ESCO investments. Source: CAWI survey (multiple choice questions).

No.	Response Option	Reponses (in %)
1	Complicated decision-making and ownership processes	11.90
2	Competition from grants	19.04
3	Competition from banks	4.76
4	Lack of adequate consulting	14.28
5	Underestimation of savings relate to the ESCO formulation	45.23
6	Increase in local government debt or corporate debt	23.80
7	Lack of familiarity with the ESCO formula	21.42
8	Insufficient legal regulations	26.19
9	Other	4.76
10	No opinion	0.00

Knowledge regarding the barriers that local governments and enterprises identify in relation to the ESCO formula was expanded through the focus group interview and in-depth interviews. The feedback collected from respondents indicates that:

- the best method of encouraging local governments units and enterprises to use the ESCO formula may be the presentation of examples of such investments in different variants and the resulting benefits—the focus should be on pilot investments using public funds in places where market solutions are not viable due to numerous risks;
- insufficient or unclear legal regulations regarding the ESCO formula are mainly related to the fear of receiving state aid, which discourages the use of this mechanism;
- inconsistent interpretations as to how ESCO-related liabilities should be recognised in the financial statements of local governments units and enterprises meaning that for many entities already carrying high debt rates, this solution cannot be taken into account;
- complicated decision-making and ownership processes regarding the ESCO formula relating primarily to the problems of determining the ownership of the components of the ESCO installation, the adoption of uniform rules for the accounting and depreciation of fixed assets and the recognition of expenses incurred and savings generated in the financial statements, further multiply the legal doubts regarding the proper inclusion of such operations in the accounting ledgers.

4.2. Solutions Supporting the Implementation of the ESCO Formula in Poland

The study found that despite many barriers limiting interest in the ESCO formula, local governments units and enterprises are generally eager to implement particular types of ESCO services in combination with various financial supports. The in-depth interviews revealed that a general problem of projects co-financed with grants or loans is the lack of comprehensive consulting services available throughout the project life-cycle. Such services are necessary in order to choose the optimal path of energy industry project implementation and to take all required standards and requirements into account. Meanwhile, local government units and enterprises often lack the expertise to independently design solutions best suited to their particular conditions and needs. They focus on those activities that are, for instance, eligible for co-financing as part of a grant, omitting those that cannot be considered eligible.

Financial support, without additional activities of a formal and informative nature, may lead to inefficient use of the funds, as exemplified by energy efficiency investments in public buildings implemented under the green investment scheme audited by the Polish Supreme Audit Office [42]. Therefore, it is crucial to correctly identify the solutions which can be possible and most beneficial to implement energy projects using the ESCO formula in Poland.

The results of the in-depth interviews were confirmed by the CAWI survey (Table 6). The majority of respondents want expert consulting (61.90%) and energy audits (52.38%), that are necessary to determine the scope and method of implementation of an energy industry project. The high demand for technical services in the project (45.23%) once again confirms the lack of competence of local government units and enterprises to handle this type of investment on their own. This is an important factor to consider when taking steps to popularise the ESCO formula. As one interviewee stated during an in-depth interview, "the financing entity does not really need an energy audit; it is in the interest of the investor".

Table 6. Services offered in the ESCO formula. Source: CAWI survey (multiple choice questions).

No.	Response Option	Reponses (in %)
1	Energy retrofits of buildings—securing financing	23.80
2	Heat source replacement—securing financing	30.95
3	Expert consulting	61.90
4	Conducting an energy audit	52.38
5	Technical support for the project at the planning and implementation stage	45.23
6	Handling the subsidy program for inhabitants (energy retrofit, heat source replacement, RES installation)	23.80
7	Other	4.76
8	No opinion	0.00

The focus group interview participants noted that utilising the wide range of services of the ESCO formula increases the chance that an energy industry project will achieve the expected savings. In their view, the framework scope of ESCO services in an energy project should include:

- conducting an energy audit and multi-criteria economic and technical analysis of the profitability of the implementation of the project in various business models,
- technical and economic consulting on choosing the optimal variant of project implementation,
- ensuring financial engineering for the project,
- participation in the design of technical solutions and appropriate selection of equipment,
- participation in investment project management and accounting settlement,
- participation in energy management after project completion, and
- maintenance and operation of equipment acquired as part of the project.

According to in-depth interviews results, the solutions related to investing in energy industry projects that involve the use of the ESCO formula should consider two equally important elements:

- the benefits expected by investors: local government units and enterprises implementing such projects, and
- sustainable public strategies of the local government units and sustainable business strategies of enterprises and ESCO companies.

Additionally, the participants of the in-depth interviews noted that ultimately, a move away from direct grant funding to a mixed mechanism, i.e., grants with forgiveness, should be recommended.

The local government units and enterprises surveyed evaluated the attractiveness of various financing instruments for energy projects in different ways (Table 7). Obviously, the most popular are European funds (33.33%), but it should be stated clearly that it is probable that this source of capital will be limited gradually in the next European financial perspectives of the cohesion policy. Traditionally, loans, either long-term and returnable (21.42%) or forgivable (7.14%) are also popular. It is very puzzling that as many as one-fifth of the respondents have no opinion at all.

Table 7. Interest in types of ESCO services. Source: CAWI survey (multiple choice questions).

No.	Response Option	Responses (in %)
1	EU grant	33.33
2	National grant	4.76
3	Long-term loan	21.42
4	Forgivable loan	7.14
5	Public-private partnership	2.38
6	Other	9.52
7	No opinion	21.45

The CAWI survey confirmed that respondents are unable to identify the solutions that work best in conjunction with the ESCO formula. Opinions on this topic are very divided. Below is the distribution of responses to the question as to what types of projects the respondents would want to associate with the ESCO formula:

- projects with repayable funding (7.14% of respondents),
- projects with non-repayable funding (7.14% of respondents),
- projects with public-private partnerships (4.76% of respondents).

A total of 80.96% of the respondents had no opinion on this issue. It confirms again that the key challenge to increasing the number of energy industry projects based on the ESCO formula is to educate target groups in terms of its usefulness. Otherwise, they are not aware of what projects can be the implemented using the ESCO formula and what financial models can be applied with such a formula.

The assessment of ESCO formula value for the potential investors should consider identifying the potential benefits expected by those who use or intend to use ESCO services. The respondents in the CAWI survey identified the following benefits:

- savings in energy consumption (80.95% of respondents);
- modernisation of existing infrastructure (52.38% of respondents);
- energy efficiency improvement consulting (28.57% of respondents);
- environmental benefits (7.14% of respondents).

Energy savings that translate into financial savings are a key benefit of the ESCO formula for the vast majority of respondents (80.95%). At the same time, such factors as the scope of investment and its potential connection with public aid (e.g., energy retrofitting, the replacement of lighting systems, the purchase of electric busses) and the type of entity implementing the investment (local government, enterprise) are factors which objectively condition the different scope of benefits possible to obtain from the energy project. It is lamentable that environmental benefits are important for only 7.14% of respondents, which indicates that sustainable public and business strategies are not core policies for the potential beneficiaries of the ESCO formula.

ESCO companies shape their offerings in relation to the expectations of different target groups in the energy market while being guided by criteria such as profitability or potential market size for a given service. The results of the in-depth interviews indicate that the current dominant investor groups in the ESCO services market are:

- local governments at all levels, and
- large-scale industrial plants where energy demand is the highest. Slightly less popular among energy service companies are:
- other public entities such as universities, hospitals, etc., and
- housing associations and cooperatives.

Due to the dispersed structure of this market, it is currently difficult to quantify its size. The CAWI survey identified the likeliest prospective target groups for ESCO services as:

- the industrial sector (23.08% of respondents),
- local governments (15.38% of respondents),
- office buildings and shopping centres (15.38% of respondents), and
- small and medium-sized enterprises (7.69% of respondents).

The in-depth interviews resulted in the formulation of key characteristics that encourage investment projects to use the ESCO formula:

1. high profitability (benefits are distributed between the investor and the ESCO)
2. the implementation of the project in an industry where a high EE of investment can be obtained
3. adequate liquidity and securing of funds by the ESCO that are attractive for investors without their own contribution
4. the possibility for the investor to verify the level of competence of the ESCO and the quality of its offer
5. sufficient knowledge of EE issues by the investor

The focus group identified the leading types of energy industry projects that are feasible under the ESCO formula while providing the potential for satisfactory savings. At the same time, they indicated the potential current and future target groups of the ESCO formula in Poland (sectors and branches). One agreed that these are projects related to:

- improving the energy efficiency of buildings and their surroundings (e.g., lighting),
- modernisation of heat sources and heating networks,
- projects implemented to reduce the energy intensity of processes or industrial infrastructure modifications.

Lighting replacement and industrial projects were found to be the most financially viable. On the other hand, investments in the energy retrofitting of public utility and residential buildings, including the so-called deep energy retrofits, are characterised by a longer payback period, which is not always acceptable for the investor. The ESCO formula is particularly suitable for projects that combine energy retrofits with other investment activities, such as energy source replacement. As the interviews show, the role of ESCO is also important in the context of construction. This is true not only in industry, where financial efficiency is higher but also in local government construction projects, where procedures for energy efficiency improvement, e.g., standards for energy audits, have already developed. Energy audits are much more expensive in industry as they have to be more extensive and are often carried out on the basis of measurements, as they are the basis for the formulation of contracts and future settlements between the energy service company and the investor. The technological complexity, but also the need to monitor the effects, substantially justify the involvement of qualified entities such as ESCO companies.

5. Discussion

Although ESCO in Poland is not a new form of accounting for investments aimed at improving energy efficiency, many entities still do not have sufficient knowledge about it, which was clearly confirmed by the results of the survey. The process of accounting for investments in the ESCO mechanism is perceived as time-consuming and complicated, and above all, has been researched to a much lower degree than, for example, the procedures of using EU funding. These factors and other conditions highlighted in the study demonstrate that the ESCO market, although in existence for nearly 30 years, is still in its early stages of development [3]. The result of our research was in line with the expected results [9]. Often this lack of knowledge means that local governments or enterprises expect deep energy retrofits to be absolutely necessary for ESCO projects and find it difficult to understand that even small, local investment projects combined with optimisation of energy management processes can significantly improve energy efficiency and thus lead to lower monthly energy costs. Meanwhile, research confirms that the benefits of energy savings and reduced energy costs are the key criteria for investors to become interested in the ESCO mechanism. The statement agrees with Trianinin et al. [13], which shows that an economic driver is crucial for companies regardless of their size.

Potential investors using the ESCO mechanism, i.e., local governments and enterprises, have an only superficial knowledge of the working principles of this mechanism. Because research has shown that in many cases, interest in this topic only arises in connection with incoming offers from ESCO companies, the common belief in this topic is that it involves incurring expenses for various types of consulting services and thus having to accept the risk of investment failure. Few entities are aware that in the ESCO mechanism, increasing energy efficiency does not have to involve capital expenditure.

Polish local government units and enterprises have become accustomed to using grant or loan formulas, so the solution of placing the burden of financing an energy project on an ESCO company is still quite innovative and even arouses incredulity. Other innovative financing models, such as energy cooperatives or crowdfunding, however common in other countries [25,26], are not known in the Polish public sector. They are the domain of activities, such as technology start-ups, also in the energy sector. The public sector remains conservative. The fact, proved by the research, that knowledge on the subject is mainly drawn from the Internet and not from professional advisors is not conducive to a proper understanding of the ESCO financing mechanism. For many investors, the method of financing energy investments by an ESCO coupled with its commitment to achieving a certain level of savings from energy utility expenses raises many questions. Innovative elements appearing in this model (e.g., an escrow account, used for settlements with the ESCO) or ambiguities (e.g., the necessity of appropriate recognition in financial statements of costs and revenues from the ESCO mechanism) also negatively influence interest in this form of financing energy investments.

Reliable and comprehensive information and promotion activities conducted by public institutions should, therefore, be considered crucial. They should serve to present the ESCO mechanism in an objective manner, considering its advantages and limitations. Above all, however, they should serve to educate investors in energy projects about energy efficiency itself and the possibilities of improving it using the ESCO mechanism. When such knowledge is missing, the simplest solutions are resorted to (e.g., grants or loans), which are also burdened with disadvantages or at least result in various types of limitations.

According to the research, the task of popularising the ESCO mechanism should fall to both the entities responsible for the national energy policy, as well as ESCO companies themselves. Barriers to the use of ESCO in Poland were identified and diagnosed several years ago, and specific recommendations were formulated. The report by the World Bank and Ministry of Development [20] also pointed out the need to develop model templates and documents, introduce changes in the budgeting rules for the procurement system to enable EPC contracts, establish nodal energy agencies and recruitment of project facilitators, launch incentive mechanisms and financing schemes, develop of targets for energy efficiency indicators for public offices and prepare measurement and verification reports, etc. This can contribute firstly to the process being promoted and popularised at all times, secondly to the gains in energy efficiency being sustained, and thirdly to measures being put in place to realise more significant potential energy savings [20]. Our research supplemented the results with the needs resulting from the specificity of EU funds and changes in the conditions in the 2021–2027 period.

It is difficult to pinpoint why, despite recognising this problem, not enough has been done to improve the promotion of ESCO companies in the energy market. ESCO companies competing for customers have found it challenging to demonstrate the attractiveness and competitiveness of this formula against, for example, grants. In the case of grants, many public and local government institutions had high budgets for information and promotional activities to attract applicants and beneficiaries and ultimately demonstrate the achievement of indicators in EU programs. As a result, respective resources were used to promote grant programs, but ESCO was not mentioned in energy projects.

In light of the research, however, it appears that the most work remains to be done in the legislative area. The legal barriers are identified by Bertoldi and Boza-Kiss [9] and Trianni, Cagno and Farné [13], and we agree with the authors cited in this regard. This concerns primarily unambiguous jurisprudence and implementation of uniform formal and legal solutions facilitating the use of ESCO both by public sector entities and commercial entities, by both housing associations, and cooperatives. Insufficient or unclear legal regulations regarding the ESCO mechanism are primarily related to the fear of receiving state aid, which is consistent with Rogić Lugarić, Dodig and Bogovac [7]. Exceeding public aid limits may mean the necessity of returning the resources with statutory interest. There is also insufficient interpretation on how to record liabilities from using the ESCO mechanism in local governments and enterprises (especially the impact of ESCO on local government debt rates. Great interpretive difficulties also pertain to accounting for savings from the ESCO mechanism. In the latter regard, it is worth citing the examples of Germany and the United States, where, in order not to exceed debt ceilings, contractual EPC payments qualify as an operating expense. In some countries, EPC contracts are included in calculating the debt ceiling of public sector entities. In contrast, in Germany and the US, for example, EPC contracts are permitted provided the following criteria are met:

- the ESCO must incur commercial debt and account for it on its balance sheet, and
- the project has a guaranteed positive cash flow to the government agency every year of the contract, backed by rigorous performance guarantees, including performance and payment bonds, certificates of insurance or surety bonds.

As long as the above conditions are met, payments to the EPC contractor are counted as operating expenses (building maintenance) and not as debt repayment [30].

The above proves that for the formula to grow on the Polish market, it is crucial to clearly define whether repayments to ESCO should be reported as debt or included in operating expenses [20]. Such solutions have been introduced in Poland in the amended Act on energy efficiency, where a provision has been added: obligations resulting from an energy efficiency improvement contract do not affect the level of public debt and the deficit of the public finance sector, where the energy services provider bears most of the construction risk and the risk of achieving a guaranteed level of average annual energy savings, taking into account the impact on these risks of factors, such as guarantees and financing by the energy services provider as well as asset allocation at the end of the contract. However, while clarifying much, this provision still does not entirely convince potential investors from the public sector to use ESCO.

Although many barriers indicated by the studies (e.g., competition from grants, the too-small scale of savings from ESCO) cannot be fully eliminated, this formula can be successfully improved, e.g., by combining it in certain variants of financial engineering with grants and forgivable loans or by transferring more innovative and effective technological solutions to the Polish market, allowing for the generation of greater savings from the use of the ESCO mechanism.

It is also worth emphasising the important role played by companies providing such services in popularising the ESCO mechanism. The research clearly showed that the elements of ESCO service that are essential in proper project planning are expert advice and professional performance of energy audits. Unfortunately, practice shows that in many projects, this element is omitted or implemented incorrectly, as a result of which even correct implementation of the investment does not guarantee obtaining the expected improvement in energy efficiency. ESCO companies should therefore do much more to advise and even educate their customers on the importance of the planning process for the final results of an energy project.

ESCO companies recognise the potential of this market and are able to identify target groups with whom they would like to carry out energy projects now, and in the future, but the principles of the functioning of this market depend on many conditions, including many which are completely independent from this market. One such factor is the international and domestic energy policy, which has a large impact on the viability of the ESCO business model. The current form of documents, such as the European Green Deal [54], NextGenerationEU (European Instrument for Reconstruction and Increasing Resilience) [55] and EU Cohesion Policy 2021–2027 (European Cohesion Policy 2021–2027) [56] indicate a very large proportion of energy projects in financing the development processes of the European Union in the coming years. This opens a number of new possibilities for the use of the ESCO mechanism and is an important argument for its further improvement. In this aspect, the current level of interest in particular types of ESCO services which emerges from the research may increase significantly in the coming years, which opens new prospects for this market.

We should also not forget the growing awareness of institutional and individual customers about the impact of industrial production and other human activities on climate change. It is to be expected that consumers will increasingly attach more importance to the way in which a product or service is produced or provided, and more specifically, will be interested in ensuring that this process takes place in the most sustainable manner possible, with the least possible damage to humans and the environment. Therefore, many entities will seek to intensify energy efficiency measures for their operations in a way that does not increase costs and maintains the existing competitiveness of their products and services.

When referring to the business models that can be applied to the ESCO mechanism, it is hard to ignore the fact that they generate risks primarily for the energy service company itself. Such a company may rely on its own or returnable capital, e.g., on repayable financing in the form of a preferential loan fund granted to it by state institutions. The company, in turn, organises an open call for proposals for ESCO projects within a specific technical and technological framework. In order for the whole project to be successful, it is necessary to introduce an appropriate formal and legal framework and for public institutions to run educational campaigns aimed at encouraging both entities (ESCO and investors) to implement the ESCO mechanism. An important element facilitating the functioning of this mechanism may be, for example, an escrow account, which serves as collateral for the repayment of the liabilities of the investor towards the ESCO. It should be emphasised that it is the energy service company that bears the entire risk of not achieving energy savings and return of funds from the loan fund to state institutions, which must be factored in its business activity and included in the contract.

The way in which the parameters of each project are approached depends primarily on its scale, size and potential environmental and financial effects. Therefore, a case-by-case approach is recommended for selecting the parameters of a specific ESCO contract. Research-based criteria for evaluating target groups should also be part of the contract. The key criterion is the profitability of the contract with the entity, which should be correlated with the level of risk the ESCO is willing to bear. Other important elements of the contract include the contribution of the investor, the interest rate on the preferential loan, the payback period, the preparation fee, and how the cost savings will be distributed in correlation with the planned payback period.

The main benefit of using the ESCO mechanism is the reduction of the need for the investor to make a contribution and the quick benefits from the savings on energy costs, which appear immediately after the completion of the investment and start of operation which is in line with the European Energy Service Initiative [34].

The application of the ESCO mechanism usually binds the owner of the facility to the ESCO for many years in terms of defining the effects (e.g., by means of an EPC contract), and during the operation–monitoring the effects, which are the basis for accounting settlements. This approach reduces mismanagement on the one hand but at the same time requires patience on both sides when it comes to the rate of return on investment. Taking into consideration the results of the audit of energy efficiency investments made in public-use organisations by the Polish Supreme Audit Office [42], the results of our research confirm the need for and gaps in specialist knowledge and ineffective investment implementation mechanisms. Major shortcomings were also identified at the initial stage of defining design assumptions and later at the stage of monitoring energy effects during operation. The popularisation of investments within the ESCO mechanism should be accompanied by procedures and formal documentation at each stage of investment (planning, tender, implementation, monitoring and settlement) so that it is not possible to repeat the mistakes indicated in the cited audit report.

6. Conclusions

The research problem of the study was to identify the determinants of the use of the ESCO mechanism by local government units and enterprises implementing energy projects. The research conducted by the authors, which was also addressed to the Polish Ministry of Energy, provided a number of interesting observations and conclusions which enabled achieving the research objectives defined in the paper.

The first objective of the study was to determine the factors affecting the interest of Polish local government units and enterprises in implementing energy projects in the ESCO mechanism. The research revealed that the lack of knowledge of local government units and enterprises regarding the ESCO mechanism, although not unique to Poland, is not a sufficient explanation for the low interest in this solution. The problem is much more complex, and its sources should be sought first in the poor preparation of Polish local governments and enterprises for the energy transition, and second in the legislative and interpretational ambiguities that have grown around the ESCO mechanism. Theoretically, investors implementing energy projects should have full knowledge of the conditions and consequences of implementing these projects in different business models (e.g., grant, forgivable loan, credit, ESCO), but in practice, they often choose the simplest, most accessible or highly advertised solutions. For this reason, interest in grants, for example, far exceeds interest in the ESCO mechanism. In the case of EU funds, for example, the rules for project implementation and criteria for the eligibility of costs, as well as conditions for project accounting settlement and maintenance of its sustainability, are clearly and comprehensively defined, and access to advice on EU grants is widespread. The rules of accounting for investments in the ESCO mechanism are much more complex and individualised, depending on multiple factors. When a local government or an enterprise is poorly informed regarding the technological solutions in the area of improving investment efficiency, the simplest solutions, generating the least risk, are usually selected. Thus, the first key conclusion to be drawn from this article is the need to educate local governments and enterprises in the field of energy efficiency, which is key to interesting them in the more complex ESCO implementation models. Such educational tasks should be primarily the responsibility of public institutions shaping the national energy policy, but also of the ESCOs themselves. The research confirmed that energy service companies pay particular attention to educating investors and offer them expert consulting during the planning and implementation stages of the project, but this is not always understood by local governments and enterprises. With regard to the results of the research, another important conclusion can be drawn—that the information addressed to investors to increase their knowledge regarding the ESCO mechanism should be formulated in terms of benefits and should be based on various examples of ESCO mechanism applications in investment projects by local governments and enterprises. While the Internet should remain the first source of information on ESCO for investors, the availability of expert consulting should be improved, and investors should be encouraged to commission professional energy audits as a first step in the planning process for energy projects. Meanwhile, investors often want to save on consulting and audits, not being aware that by following this path they lose the opportunity for a reliable analysis of different options for project implementation.

Expert consulting by ESCOs is also essential to provide investors with reliable and credible information on the terms and conditions of ESCO projects. As the research has shown, there are many stereotypical associations about the ESCO mechanism (e.g., the need to combine ESCO with a deep energy retrofit) or innovative elements that are not yet trusted by investors (e.g., the method of settlement with the energy service company through an escrow account). The ESCO mechanism connected with a grant also often raises questions from the perspective of state aid regulations, which are quite unclear. It seems that in spite of existing research and previously formulated recommendations, which are also referred to by the authors of this paper, solutions have still not been implemented to reduce barriers to the use of the ESCO mechanism, e.g., the development of model document specimens for conducting and accounting for EPC contracts, the establishment of nodal energy agencies, as well as the introduction of ESCO incentive mechanisms and financing schemes.

This last element, concerning the provision of funding sources for umbrella programs for the energy service companies, is considered by the authors as especially crucial for the further development of this market. The second research objective formulated in this study, i.e., the identification of business models in which it is possible to implement energy projects in the ESCO model in Poland, was highly related to determining:

- What mechanisms do ESCOs and investors want to implement in such contracts?
- Which target groups and types of projects are strategic in terms of marketing for the ESCOs?

By following the project implementation process in the ESCO mechanism, we can conclude that it generates risk primarily for the energy service company itself. For this reason, it is doubtful that energy service companies will invest the equity necessary to develop this challenging market. An ESCO generally relies on repayable capital from a preferential loan fund operated by state institutions. Since research has shown that the growth of the ESCO market is derived from the availability of public funds for the ESCOs servicing the contracts, a third important conclusion concerns the need for the state to provide adequate capital to such companies. The task of the energy service company is to effectively carry out the call for projects to be implemented in the ESCO mechanism and to service them in terms of finances, organisation and technology. It is in the interest of the state to create a competitive ESCO market that creates broad access to operators for the investors. However, it should be clearly emphasised that given the current perception of the novelty of the ESCO mechanism in the Polish energy market, the effective development of this market is possible only with the intensification of educational activities by state institutions and the professionalisation of the activities of ESCO companies. The amendment of the Act on energy efficiency and the activity of the Ministry of Climate to promote the ESCO mechanism should be evaluated positively.

According to the research, the new perspective for EU funds 2021–2027 will continue to allow local governments and enterprises to access grants for energy efficiency improvements, but in many cases, the conditions for obtaining grants may not meet all investor expectations. This results in an even greater necessity to promote hybrid financial engineering for projects that can combine, for example, the ESCO mechanism with grants and forgivable loans. However, this requires investors to be provided with expert advice on the planning and implementation of such projects while also giving them access to modern technological solutions offering a longer lifespan and greater savings in energy consumption.

A final element worth emphasising is the growing public awareness of the impact of industrial production and other human activities on climate change. Improving energy efficiency in the activities of public institutions or enterprises is an important element of the social communication of these organisations with their stakeholders. Energy projects that generate returns for investors while helping to mitigate the effects of climate change and prevent further change are increasingly socially desirable, and this is yet another reason to promote the ESCO mechanism.

These conclusions of the research conducted on behalf of the Polish Ministry of Energy have been reviewed by a panel of experts. The panel included representatives from academia and business with knowledge of how the ESCO refundable financing mechanism works. The problem of public debt and EPC contracts in the public sector identified in the research has also been confirmed, as evidenced by the cited amendment to the Act on energy efficiency, which clarifies the situations whereby EPC contracts will not increase the level of public debt. Nevertheless, the research presented in this article has some limitations due to its methodology. These limitations relate to the national nature of the research (which is why the title emphasises that the article concerns Poland). In addition, it is necessary to take the dynamic nature of the international and domestic energy market into account (green deal, COVID, development of prosumer installations), which results in equally dynamic changes in the preferences of investors and the conditions for ESCOs. It should also be emphasised that the research was commissioned by the then Ministry of Energy, and the research problems addressed were in response to the needs of the said ministry. Nevertheless, the identified research gap-the limited number of national academic publications addressing the topic of ESCO projects, remains timely and inspiring.

As further directions of research regarding the ESCO mechanism, the authors point to the assessment of the importance of the ESCO mechanism in the transformation of the energy sector, in particular, research on the perception of the mechanism from the perspective of stakeholders and the further development of business models related to the ESCO mechanism–for example, in the context of changes in the conditions for applying for EU funds in the 2021–2027 period.

Author Contributions: Conceptualisation, J.K.-P. and G.K.; methodology, J.K.-P. and G.K.; validation, J.K.-P. and G.K.; formal analysis, J.K.-P. and G.K.; investigation, J.K.-P. and G.K.; resources, J.K.-P. and G.K.; writing—original draft preparation; writing—review and editing J.K.-P. and G.K.; visualisation, J.K.-P. and G.K.; supervision, J.K.-P. and G.K. All authors have read and agreed to the published version of the manuscript.

Funding: The authors based on the research conducted by them as a part of the study commissioned and funded by the Polish Ministry of Energy. The study was co-financed by the Cohesion Fund. The APC costs of the article was financed by Warsaw University of Technology, the cost of language proofreading by the WSB University.

Institutional Review Board Statement: Not applicable.

Informed Consent Statement: Not applicable.

Conflicts of Interest: The authors declare no conflict of interest.

References

1. IEA. *Energy Efficiency 2020*; International Energy Agency: Paris, France, 2020.
2. Román-Collado, R.; Economidou, M. The Role of Energy Efficiency in Assessing the Progress towards the EU Energy Efficiency Targets of 2020: Evidence from the European Productive Sectors. *Energy Policy* **2021**, *156*, 112441. [CrossRef]
3. Boza-Kiss, B.; Bertoldi, P.; Economidou, M. *Energy Service Companies in the EU: Status Review and Recommendations for Further Market Development with a Focus on Energy Performance Contracting*; Publications Office of the European Union: Luxembourg, 2017; ISBN 9789279714757.
4. Patterson, M.G. What Is Energy Efficiency?: Concepts, Indicators and Methodological Issues. *Energy Policy* **1996**, *24*, 377–390. [CrossRef]
5. Bolton, R.; Hannon, M. Governing Sustainability Transitions through Business Model Innovation: Towards a Systems Understanding. *Res. Policy* **2016**, *45*, 1731–1742. [CrossRef]
6. Carbonara, N.; Pellegrino, R. Public-Private Partnerships for Energy Efficiency Projects: A Win-Win Model to Choose the Energy Performance Contracting Structure. *J. Clean. Prod.* **2018**, *170*, 1064–1075. [CrossRef]
7. Rogić Lugarić, T.; Dodig, D.; Bogovac, J. Effectiveness of Blending Alternative Procurement Models and EU Funding Mechanisms Based on Energy Efficiency Case Study Simulation. *Energies* **2019**, *12*, 1612. [CrossRef]
8. IEA. *Energy Service Companies (ESCOs)*; International Energy Agency: Paris, France, 2018.
9. Bertoldi, P.; Boza-Kiss, B. Analysis of Barriers and Drivers for the Development of the ESCO Markets in Europe. *Energy Policy* **2017**, *107*, 345–355. [CrossRef]

10. EC JRC, *Energy Service Market in the EU: Status Review and Recommendations 2019*; Publications Office of the European Union: Luxembourg, 2019; ISBN 978-927-613-093-2.
11. Ministry of Climate and Environment ESCO-Type Energy Services on the Polish Market—Open Data. Available online: https://dane.gov.pl/pl/dataset/1727,usugi-energetyczne-typu-esco-na-polskim-rynku (accessed on 21 July 2021). (In Polish)
12. ZWM. *Detailed Description of the Priority Axes of the Regional Operational Program of the Mazowieckie Voivodeship for 2014–2020*; Board of the Mazowieckie Voivodeship: Warsaw, Poland, 2017. (In Polish)
13. Trianni, A.; Cagno, E.; Farné, S. Barriers, Drivers and Decision-Making Process for Industrial Energy Efficiency: A Broad Study among Manufacturing Small and Medium-Sized Enterprises. *Appl. Energy* **2016**, *162*, 1537–1551. [CrossRef]
14. Okay, N.; Akman, U. Analysis of ESCO Activities Using Country Indicators. *Renew. Sustain. Energy Rev.* **2010**, *14*, 2760–2771. [CrossRef]
15. Hannon, M.J.; Bolton, R. UK Local Authority Engagement with the Energy Service Company (ESCo) Model: Key Characteristics, Benefits, Limitations and Considerations. *Energy Policy* **2015**, *78*, 198–212. [CrossRef]
16. Lu, Z.; Shao, S. Impacts of Government Subsidies on Pricing and Performance Level Choice in Energy Performance Contracting: A Two-Step Optimal Decision Model. *Appl. Energy* **2016**, *184*, 1176–1183. [CrossRef]
17. Garbuzova-Schlifter, M.; Madlener, R. AHP-Based Risk Analysis of Energy Performance Contracting Projects in Russia. *Energy Policy* **2016**, *97*, 559–581. [CrossRef]
18. Lee, P.; Lam, P.T.I.; Lee, W.L. Risks in Energy Performance Contracting (EPC) Projects. *Energy Build.* **2015**, *92*, 116–127. [CrossRef]
19. Pantaleo, A.; Candelise, C.; Bauen, A.; Shah, N. ESCO Business Models for Biomass Heating and CHP: Profitability of ESCO Operations in Italy and Key Factors Assessment. *Renew. Sustain. Energy Rev.* **2014**, *30*, 237–253. [CrossRef]
20. Vandycke, N. *Poland—Ex-Ante Evaluation of Financial Instruments; For the Proposed "Infrastructure and Environment" Operational Program for 2014-2020 in Selected Sectors: Energy and Solid Waste*; The World Bank and the Ministry of Regional Development: 2013. Available online: https://www.gov.pl/attachement/16c5a9af-519e-4855-835f-e2d163bd9597 (accessed on 1 December 2021).
21. Lisowski, R.; Woźniak, M.; Jastrzębski, P.; Karafolas, S.; Matejun, M. Determinants of Investments in Energy Sector in Poland. *Energies* **2021**, *14*, 4526. [CrossRef]
22. Rajek, Ł.; Liszka, S.; Rajek, Ł. *Overview of the EPC Potential and Market. National Report for Poland WP3 Combines Project Central Europe Programme/4CE499P3*; Polish Foundation for Energy Efficiency: Katowice, Poland, 2013.
23. Bogdan, R.; Biklen, S.K. *Qualitative Research for Education an Introduction to Theory and Methods*; Pearson: Boston, MA, USA, 2009; ISBN 9780205512256.
24. Kreatus Ltd. *Opportunities for the Implementation of the ESCO Mechanism under Priority Axis I of the Infrastructure and Environment Operational Programme 2014–2020*; Kreatus Ltd: Bielsko-Biała, Poland, 2018. (In Polish)
25. Yildiz, Ö.; Rommel, J.; Debor, S.; Holstenkamp, L.; Mey, F.; Müller, J.R.; Radtke, J.; Rognli, J. Renewable Energy Cooperatives as Gatekeepers or Facilitators? Recent Developments in Germany and a Multidisciplinary Research Agenda. *Energy Res. Soc. Sci.* **2015**, *6*, 59–73. [CrossRef]
26. Ari, I.; Koc, M. Philanthropic-Crowdfunding-Partnership: A Proof-of-Concept Study for Sustainable Financing in Low-Carbon Energy Transitions. *Energy* **2021**, *222*, 119925. [CrossRef]
27. Polzin, F.; von Flotow, P.; Nolden, C. What Encourages Local Authorities to Engage with Energy Performance Contracting for Retrofitting? Evidence from German Municipalities. *Energy Policy* **2016**, *94*, 317–330. [CrossRef]
28. Brown, D. Business Models for Residential Retrofit in the UK: A Critical Assessment of Five Key Archetypes. *Energy Effic.* **2018**, *11*, 1497–1517. [CrossRef]
29. Streimikiene, D.; Balezentis, T. Willingness to Pay for Renovation of Multi-Flat Buildings and to Share the Costs of Renovation. *Energies* **2020**, *13*, 2721. [CrossRef]
30. Langlois, P.; Hansen, S.J. *World ESCO Outlook*; Fairmont Press: Lilburn, GA, USA, 2012; ISBN 9781466558144.
31. 2012/27/EU Directive 2012/27/EU of the European Parliament and of the Council of 25 October 2012 on Energy Efficiency, Amending Directives 2009/125/EC and 2010/30/EU and Repealing Directives 2004/8/EC and 2006/32/EC Text with EEA Relevance 2012. *Off. J. Eur. Union L.315/1* **2012**, *4*, 204–257.
32. Akman, U.; Okay, E.; Okay, N. Current Snapshot of the Turkish ESCO Market. *Energy Policy* **2013**, *60*, 106–115. [CrossRef]
33. Hannon, M.J.; Foxon, T.J.; Gale, W.F. The Co-Evolutionary Relationship between Energy Service Companies and the UK Energy System: Implications for a Low-Carbon Transition. *Energy Policy* **2013**, *61*, 1031–1045. [CrossRef]
34. Graz Energy Agency; Berlin Energy Agency. *Facilitators Guideline for Energy Performance Contracting. European Energy Service Initiative—EESI2020*. 2014. Available online: https://guaranteeforum.eu/file/download/4286735d-af19-4397-89af-45aadb555e5d/1593786338eesi2020_epc_facilitators_guideline.pdf (accessed on 1 December 2021).
35. Soroye, K.L.; Nilsson, L.J. Building a Business to Close the Efficiency Gap: The Swedish ESCO Experience. *Energy Effic.* **2010**, *3*, 237–256. [CrossRef]
36. Suhonen, N.; Okkonen, L. The Energy Services Company (ESCo) as Business Model for Heat Entrepreneurship-A Case Study of North Karelia, Finland. *Energy Policy* **2013**, *61*, 783–787. [CrossRef]
37. Journal of Laws 2021, Item 468, Consolidated Text The Act of May 20, 2016 on Energy Efficiency 2021. Available online: https://isap.sejm.gov.pl/isap.nsf/download.xsp/WDU20160000831/U/D20160831Lj.pdf (accessed on 1 December 2021).
38. Available online: https://mazovia.pl/resource/29298/1753_1011.pdf (accessed on 1 December 2021).

39. Wiśniewska, M. *Adaptation to Climate Change, Cross-Sectional Assessment of 16 Regional Operational Programs and Detailed Descriptions of Priority Axes*; Związek Stowarzyszeń Polska Zielona Sieć: Warsaw, Poland, 2016. (In Polish)
40. Consus Carbon Engineering, 2016. Low-Emission Economy Plan for the City of Kielce (In Polish). [online] Kielce: Consus Carbon Engineering. Available online: http://www.um.kielce.pl/gfx/kielce2/userfiles/files/gospodarka-niskoemisyjna/plan_gospodarki_niskoemisyjnej_dla_miasta_kielce_2.pdf (accessed on 1 December 2021).
41. Garnier, O. *D2.6 EU Summary Report Documenting the Collected Information on Recommendations for EPC Markets*; EEVS Insight Ltd: London, UK, 2014.
42. SAU. *Energy-Saving Investments in Public Utility Buildings—Information on the Audit Results*; Supreme Audit Office (Najwyższa izba Kontroli): Warsaw, Poland, 2014. (In Polish)
43. Abdulkadyrov, A.S.; Zhigulina, E.P.; Kovnerev, M.A. A Model for Achieving Comprehensive Efficiency of the Energy Projects Implementation in the Sustainable Development Framework. In *Industry Competitiveness: Digitalization, Management, and Integration*; Bogoviz, A.V., Ragulina, Y.V., Eds.; Lecture Notes in Networks and Systems; Springer International Publishing: Cham, Switzerland, 2020; Volume 115, pp. 589–595. ISBN 9783030407483.
44. Johansen, J.P.; Isaeva, I. Developing and (Not) Implementing Radical Energy Efficiency Innovations: A Case Study of R&D Projects in the Norwegian Manufacturing Industry. *J. Clean. Prod.* **2021**, *322*, 129077. [CrossRef]
45. Thumann, A.; Woodroof, E.A. *How to Finance Energy Management Projects: Solving the "Lack of Capital Problem"*, 1st ed.; Fairmont Press: Lilburn, GA, USA, 2013; ISBN 9781466571532.
46. Turturean, C. Classifications of Foresight Methods. *Anu. Inst. De Cercet. Econ. Gheorghe Zane -Iasi* **2011**, *20*, 113–123.
47. Larsen, P.H.; Goldman, C.A.; Satchwell, A. Evolution of the U.S. Energy Service Company Industry: Market Size and Project Performance from 1990–2008. *Energy Policy* **2012**, *50*, 802–820. [CrossRef]
48. Guest, G.; Namey, E.E.; Mitchell, M.L. *Collecting Qualitative Data: A Field Manual for Applied Research*; SAGE Publications, Ltd: London, UK, 2013; ISBN 9781412986847.
49. Balta-Ozkan, N.; Davidson, R.; Bicket, M.; Whitmarsh, L. Social Barriers to the Adoption of Smart Homes. *Energy Policy* **2013**, *63*, 363–374. [CrossRef]
50. Liu, P.; Zhou, Y.; Zhou, D.K.; Xue, L. Energy Performance Contract Models for the Diffusion of Green-Manufacturing Technologies in China: A Stakeholder Analysis from SMEs' Perspective. *Energy Policy* **2017**, *106*, 59–67. [CrossRef]
51. Fiorello, D.; Martino, A.; Zani, L.; Christidis, P.; Navajas-Cawood, E. Mobility Data across the EU 28 Member States: Results from an Extensive CAWI Survey. *Transp. Res. Procedia* **2016**, *14*, 1104–1113. [CrossRef]
52. Nazarko, J.R.; Ejdys, J.R. *Metodologia I Procedury Badawcze W Projekcie Foresight Technologiczny "NT FOR Podlaskie 2020": Regionalna Strategia Rozwoju Nanotechnologii*; Oficyna Wydawnicza Politechniki Białostockiej: Białystok, Poland, 2011.
53. Kalhammer, F.R.; Kopf, B.M.; Swan, D.H.; Roan, V.P.; Walsh, M.P. *Status and Prospects for Zero Emissions Vehicle Technology*, Report of the ARB Independent Expert Panel 2007; State of California Air Resources Board: Sacramento, CA, USA, 2007.
54. EC The European Green Deal Investment Plan and JTM Explained. Available online: https://ec.europa.eu/commission/presscorner/detail/en/qanda_20_24 (accessed on 27 July 2021).
55. EC Recovery Plan for Europe. Available online: https://ec.europa.eu/info/strategy/recovery-plan-europe_en (accessed on 27 July 2021).
56. EC EU Cohesion Policy 2021–2027. Available online: https://ec.europa.eu/regional_policy/pl/2021_2027/ (accessed on 27 July 2021).

energies

MDPI

Article

Evaluation of the Level of Electricity Generation from Renewable Energy Sources in European Union Countries

Robert Huterski [1], Agnieszka Huterska [1,*], Ewa Zdunek-Rosa [1] and Grażyna Voss [2]

1 Faculty of Economic Sciences and Management, Nicolaus Copernicus University in Toruń, J. Gagarina 13a, 87-100 Toruń, Poland; robhuski@umk.pl (R.H.); ezdunek@umk.pl (E.Z.-R.)
2 Faculty of Management, Bydgoszcz University of Technology, ul. Fordońska 430, 85-790 Bydgoszcz, Poland; grazyna.voss@pbs.edu.pl
* Correspondence: huterska@umk.pl

Abstract: Changes in recent years have resulted in an increase in the ways in which renewable energy is used and shared in total electricity generation. Each type of renewable energy is characterised by its uniqueness of the physical specificity and, therefore, differences in technological solutions. In this study, one of the methods of multidimensional comparative analysis (WAP)—Hellwig's taxonomic measure of development—was used to assess the level of development of electricity production from renewable sources. Twenty-eight countries were surveyed, including 27 countries of the current European Union and the United Kingdom. Panel models were used to describe the relationship between the share of electricity production from RES in total electricity production and GDP per capita, public spending by countries on energy as a percentage of GDP as well as electricity production from water, wind, solar, and biogas per capita. The presented synthetic measures confirmed the more favourable situation of the rich northern EU countries in the production of electricity from renewable sources (solar, wind, hydro, and bio), at the same time highlighting problems with the greening of electricity production in a large group of the new EU member states. The panel study confirmed the importance of differences in economic potential and wealth between EU countries for the development of the use of RES for electricity production.

Keywords: renewable energy sources (RES); power generation; taxonomic measure of development (TMR); panel model

Citation: Huterski, R.; Huterska, A.; Zdunek-Rosa, E.; Voss, G. Evaluation of the Level of Electricity Generation from Renewable Energy Sources in European Union Countries. *Energies* **2021**, *14*, 8150. https://doi.org/10.3390/en14238150

Academic Editor: Tomonobu Senjyu

Received: 22 October 2021
Accepted: 2 December 2021
Published: 5 December 2021

Publisher's Note: MDPI stays neutral with regard to jurisdictional claims in published maps and institutional affiliations.

Copyright: © 2021 by the authors. Licensee MDPI, Basel, Switzerland. This article is an open access article distributed under the terms and conditions of the Creative Commons Attribution (CC BY) license (https://creativecommons.org/licenses/by/4.0/).

1. Introduction

The supply of energy from renewable sources (i.e., hydro, solar, photovoltaic, wind, geothermal, biomass, and others) is a fundamental element of any country's energy strategy and is driven by the concern for the local and global environment, as well as for energy security and sustainable development. Changes in recent years have resulted in an increase in the ways in which renewable energy is used and shared in total electricity generation. This change has been facilitated by environmental investments, market trends, appropriate policies (support for technology development), changes in legal regulations as well as business opportunities.

Renewable energy helps secure national resources, mitigate pollution and climate change, and provides cost-effective services [1].

Each type of renewable energy is characterised by its uniqueness of the physical specificity and, therefore, differences in technological solutions.

When describing renewable energy sources (RES), the authors focus primarily on electricity: ways of obtaining it and meeting the country's energy needs.

Hydropower storage is a well-established and commercially acceptable technology for industrial-scale electricity storage and has been used since the 1890s. Hydropower is not only a renewable and sustainable source of energy, but its flexibility and storage capacity also enable it to improve grid stability and support the deployment of other intermittent renewable energy sources such as wind and solar [2].

Wind energy is one of the cheapest renewable energy sources. The cost of producing it using wind in regions with good wind resources is comparable with producing electricity from fossil fuels. In most cases, the cost is lower or almost the same [3].

Photovoltaics, in turn, is a relatively young technology within the renewables group. Therefore, it is rather difficult to assess its environmental impacts and related costs due to uncertainties in assessing the causal impact of photovoltaic technologies on the environment and human health at each stage of the technology's life cycle [4].

Biomass, or more precisely its availability for energy purposes, is a result of the adopted forestry and agricultural model and the rate of introduction of more efficient energy crop plantation [5]. In Poland, biofuels are primarily used to produce thermal energy.

The aim of this paper is to present the use of renewable sources in electricity production in Poland and compare it with other European Union countries in selected years.

In order to achieve the aim, one of the methods of multidimensional comparative analysis—the taxonomic development measure (which allows for the ranking of countries according to the level of RES use for electricity production) was applied along with other panel models.

The paper sets out the following research hypothesis:

H: The level of RES use for electricity production in European Union countries depends on the level of development of these countries, measured by GDP per capita.

2. Literature Review

The use of renewable energy sources has been determined by legal regulations. They relate to legal conditions on the global and European Union (taking the form of Directives) levels and to individual solutions introduced by the Polish regulator. Let us mention some of them:

1. At the UN Summit in Poznań in December 2008, the 'Climate Package', which was developed by the European Commission, was adopted. Its approval obliges Poland to develop renewable energy and increase energy efficiency in accordance with the 3 × 20 Programme [6].
2. On 24 November 2010, Directive 2010/75/EU of the European Parliament and of the Council on industrial emissions (IED) was adopted, under which permissible standards for dust, sulphur oxide, and nitrogen oxide emissions must be lowered. It came into force in 2016 [7].
3. At the beginning of March 2011, the European Commission presented a document entitled 'A roadmap for moving to a competitive low-carbon economy', which includes a long-term plan for reducing carbon dioxide emissions after 2020 [8]. According to this plan, greenhouse gas emissions should be cut by 80% by 2050. The EC, by its decision of 27 April 2011, allocated for Poland approximately 477 million emission allowances for the period 2013–2020.
4. On 6 December 2012, at the UN COP 18 summit in Doha, Qatar, with participants from 194 countries, the first Kyoto Protocol was extended to 2020. At this conference, a nuclear power expert—Alan McDonald—presented the IAEA agency's report 'Climate Change and Nuclear Power 2012', which classified nuclear power as a clean source [9].
5. The document entitled 'A roadmap for moving to a competitive low-carbon economy' contains provisions according to which the reform may be carried out in the event of low prices of pollution allowances, occurring since the beginning of 2013, when the price dropped to 4 euros per tonne. On 14 March 2013, the European Parliament voted on and approved this reform, which in practice means the withdrawal of 900 million allowances and a loss of around one billion euros in budget revenues for Poland.
6. On 28 March 2014 the Government approved the draft of the Renewable Energy Sources Act [10], which sets out how their development will be supported in the country.

Renewable energy sources have been governed by legal regulations for many years. At the European Union level, these include:

- Directive 2003/87/EC of the European Parliament and of the Council of 13 October 2003, establishing a scheme for greenhouse gas emission allowance trading within the Community and amending Council Directive 96/61/EC (Text with EEA relevance) (OJ L 275, 25 December 2003) [11].
- Directive 2008/98/EC of the European Parliament and of the Council of 19 November 2008 on waste and repealing certain Directives (Text with EEA relevance) (OJ L 22 November 2008) [12].
- Directive 2010/75/EU of the European Parliament and of the Council of 24 November 2010 on industrial emissions (integrated pollution prevention and control) [7].
- Directive (EU) 2018/2001 of the European Parliament and of the Council of 11 December 2018 on the promotion of the use of energy from renewable sources [13].

Furthermore, it is worth noting the proposed changes in the regulations.

The current Renewable Energy Directive was published in the EU Official Journal only at the end of 2018. The European Commission has now signalled that changes are needed to it to bring it in line with the European Union's new, more ambitious CO_2 reduction targets. The provisions that appeared in it were supposed to lead to a 40% reduction in CO_2 emissions by 2030 compared with 1990 emissions. However, now, according to the EU, emissions reductions should be increased by 55%. This means adapting EU legislation. Moreover, EU officials stress the need to introduce additional criteria for the use of forest biomass.

The ongoing consultation will contribute to a regulatory change, which will undoubtedly be linked to EU countries' actions, aimed at further developing and increasing the use of renewable energy sources. Renewable electricity is a pillar of the transition to sustainable development, implemented through climate and energy policy strategies, and the European Green Deal provides a potential investment blueprint for this new phase of development [14]. Halkos and Gkampoura [15] analysed fossil fuel and renewable electricity generation and examined their links to CO_2 emissions and economic growth in 119 countries around the world with different income levels. The publications by Szopik-Depczyńska et al. [16] and Szopik-Depczyńska et al. [17] are the examples of articles dealing with broader aspects of the implementation of the sustainable development goals, taking into account the role of RES. The problems of energy transformation, including the development of the use of renewable sources for electricity production, were studied by Pao and Fu [18] on the example of Brazil, by Lin, Liou, and Chou [19], who compared Taiwan and Japan, and by Pietrzak, Igliński, Kujawski, and Iwański [20] in relation to Poland.

Recently, there have been a number of publications relating to various aspects of renewable energy generation and consumption in European Union countries, as well as the question of energy self-sufficiency in these countries.

A wide spectrum of motivations to switch to environment friendly green energy was described by Grosse [21] and Pietrzak, Igliński, Kujawski, and Iwański [20] in relation to Poland, by Chovancová and Tej [22] from the point of view of four countries of the Visegrad Group, by Zielenkiewicz [23] and the Institute for European Environmental Policy [24] in the context of all the EU countries. The use of RES in the EU countries from the perspective of fuel and energy consumption data was presented by Piekut [25]. The share of renewables in power generation in EU countries was analysed by Yu et al. [26], in the context of both investment, production, and consumption aspects of power generation. The authors highlight the significant progress in the penetration of renewables in Western Europe against the background of Eastern Europe making progress, but much less within the EU. At the same time, they point to the significant, as yet untapped potential for renewable development across the EU [26]. Markandya, Arto, Eguino, and Román [27] examined the impact of low-carbon technologies on the employment in the EU countries Kacperska, Łukasiewicz, and Pietrzak [28] used the clustering method to assess the use of renewable energy sources in EU countries with a focus on the Visegrad Group, i.e., the Czech Republic, Hungary, Poland, and Slovakia.

Simoes et al. [29] produced six climate projections to calculate indicators of future wind, solar, and hydro power generation capacity, as well as temperature impacts on electricity demand for heating and cooling for each EU Member State. The projected indicators for climate-dependent renewable energy sources showed relative stability in the ability to meet renewable energy production and emission reduction targets across the EU, but at the same time, significant variation at the level of individual member states, especially for wind and solar energy.

Mehedintu et al. [30] investigated the evolution and forecasting of renewable energy consumption in EU countries with a special focus on Romania, also estimating the development of renewable energy use in the energy sector. Incentives and barriers to renewable energy consumption in twelve net energy importing EU countries were presented by Marra and Colantonio [31] using panel vector autoregression, focusing on socio-technological rather than economic aspects. Chakraborty and Mazzanti [14] in a panel analysis of OECD countries showed the existence of a moderately significant positive long-run relationship between renewable electricity consumption and economic growth. According to them, economic growth per capita is a causal factor for total electricity consumption as well as for fossil electricity consumption. The links between energy transition and economic growth were addressed in their articles by Marinaș, Dinu, Socol, and Socol [32], who examined Central and Eastern European countries, by Belke, Dreger, and Dobnik [33], who examined OECD countries, and Overland [34] in the broader context of globalisation processes.

Remeikiene et al. [35] evaluated the progress in the development of the use of renewable energy sources in European Union countries for construction purposes by grouping countries according to a set of characteristics. They showed that the more developed EU countries use renewable energy sources to a greater extent than the others. Aklin [36] pointed out the danger of EU households associating the impact of renewables aggressively promoted by policymakers with the rising cost of purchasing electricity, resulting in declining public support for renewables.

The relationship between generation and consumption of electricity leads to a surplus of electricity for export or a shortage of electricity that has to be imported. The relationship between dependence on energy imports and carbon emissions in EU countries was investigated by Percebois and Pommeret [37], who showed that the best combination of low-carbon electricity production and independence from electricity imports is achieved by countries with a significant share of hydro and nuclear generation.

Much space in current publications is devoted to the proper functioning of power grids, which is necessary for the storage of electricity and the management of its surpluses and shortages. The issue of the cost of managing the stock of renewable energy produced is often raised.

Schreiner and Madlener [38] made a macroeconomic assessment of planned investments in electricity grid infrastructure in Germany. By conducting a static input–output analysis, they showed how the multiplier effects of grid investments affect macroeconomic outcomes—in terms of output, value added, employment, and fiscal revenues. They highlighted the importance of the planned significant increase in investments in electricity grid infrastructure as an important element of a sustainable energy transition in Germany. These investments are intended to ensure the flexibility of the electricity system, which is increasingly based on renewable energy sources.

Hiesl, Ajanovic, and Haas [39] highlighted the problems of long- and short-term storage of solar and hydropower and energy from other renewable sources in the countries of the European Union. According to the authors, managing surplus energy from a wide range of renewable sources requires a flexible and case-by-case approach to choosing the form of energy storage. This, however, implies potentially higher storage costs, limiting the profitability of production and encouraging the search for a mix of renewable energy sources that is optimal mainly from an economic rather than an environmental point of view.

The problem of managing surplus energy from renewable sources, in this case solar and wind, in the UK was studied by Cardenas et al. [40], concluding that providing adequate energy storage capacity from these two sources would require an investment equivalent to 7% of the country's GDP. Aqachmar et al. [41] performed an in-depth analysis of solar power generation technology options taking into account the environmental and financial performance of each option, creating global rankings of solar power generation. An important conclusion of their analysis is that many countries are not using their solar power production potential due to technological limitations, especially if the solar technology applied is accompanied by too high water consumption [41].

The prospects and cost effectiveness of combining biomass liquefaction with photovoltaics for energy storage and electricity production were presented by Perkins [42]. He concluded that the levelized cost of electricity (LCOE) of such production can be competitive with solar production combined with lithium-ion batteries in certain situations. This is another example of technical solutions to increase the economic viability of renewable power generation [42]. Based on the numerical data on solar modules, De Negri, Pezzutto, Gantioler, Moser, and Sparber [43] indicate a significant relationship between technological progress, the development of electricity generation devices, and their ever lower price, which translates into the universality of use and the amount of energy obtained.

3. Material and Methods

The article presents variables (indicators) that describe the level of use of renewable energy sources (RES) for electricity generation. Twenty-eight countries were surveyed, including 27 countries of the current European Union and the United Kingdom, which was still formally a member of the EU until the end of 2020. The following years were selected for the study: 2004, 2009, 2014, and 2019. The years 2004, 2009, 2014, and 2019 were selected to show the changes in the studied quantities at equal time intervals of five years, counting backwards to 2019, from which the most recent complete data are derived. At the same time, 2004 is the year of enlargement of the European Union by 10 new member states. According to the authors, data from a greater number of years would limit the transparency of the presentation of the problem and not increase the accuracy of the WAP method used. The data from all the years 2004–2019 were used in the panel model. The data come from the Eurostat database [44].

In this study, one of the methods of multidimensional comparative analysis (WAP)—Hellwig's taxonomic measure of development—was used to assess the level of development of electricity production from renewable sources. It is one of the methods of linear ordering, which allows for the ranking of objects in order from the best to the worst according to the level of a complex phenomenon. Multidimensional comparative analyses are a willingly used research method, as evidenced by the works of Cheba and Szopik-Depczyńska [45], Rollnik-Sadowska and Dąbrowska [46], and Ginevičius [47].

The concept of a complex phenomenon is closely related to the concept of a diagnostic variable and a synthetic (aggregate) variable. Diagnostic variables are the variables describing the examined complex phenomenon, whereas a synthetic variable is 'a variable which, based on a set of normalized diagnostic variables, determines quantitatively the level (degree of development) of the considered phenomenon in the studied objects' [48]. The synthetic measure is unitless.

The selection of diagnostic variables is based on substantive and formal criteria [48,49]. The basic substantive criterion is the importance of a given variable in the description of a complex phenomenon under study (e.g., according to expert opinion). Formal criteria include a high degree of variability and weak correlation of diagnostic features. The variables that qualify to the set of diagnostic variables, apart from having a significant impact on the studied complex phenomenon, should also be characterized by an appropriate degree of variation and should be weakly correlated among themselves (then, they do not duplicate the information transmitted by other variables).

In the next step, the diagnostic variables are identified, i.e., the nature of the impact of particular variables on a complex phenomenon is determined. In practice, it means a division of the set of diagnostic variables into two subsets: variables—stimulants (S) and variables—destimulants (D).

Next, the diagnostic variables are normalised. Normalisation aims to bring the values of individual variables to comparability (by being rid of denominators and standardising the ranges of values taken by diagnostic characteristics) [50]. In this article, the standardisation method was used to normalise the values of individual diagnostic variables. Variables normalized by this method are characterized by an arithmetic mean equal to zero and a standard deviation equal to unity.

In the next step, the so-called development pattern is determined, i.e., an 'ideal' object having the most favourable values of diagnostic variables (i.e., in the case of stimulants—the highest values, while in the case of destimulants—the lowest values).

Then Euclidean distances d_i of particular objects from the so-called development pattern were calculated according to the following formula:

$$d_i = \sqrt{\sum_{j=1}^{k} \left(z_{ij} - z_{0j}\right)^2}, \quad (i = 1,2,\ldots,n), (j = 1,2,\ldots,k), \tag{1}$$

where:

k—number of diagnostic variables;
n—number of objects (here: countries).

Subsequently, a synthetic measure was constructed, describing the level of use of renewable sources for electricity generation in each of the countries included in the study. The paper uses the following formula aggregating the normalized diagnostic variables:

$$z_i = 1 - \frac{d_i}{d_0}, \quad (i = 1,2,\ldots,n), \tag{2}$$

where:

$$d_0 = maxd_i.$$

The above formula is counted among model aggregating functions [48,49]. Formula (2) does not take into account weights, i.e., it assumes equal importance of all diagnostic features that describe the examined complex phenomenon. The synthetic measure takes values in the range [0; 1]. The level of development of renewable sources for electricity generation is the higher, the closer the synthetic measure is to one.

4. Results

The following variables were used to construct a taxonomic measure of development to describe the level of RES use in selected countries:

X_1—share of electricity generation from renewable sources in total electricity generation (in %);
X_2—electricity generation from water energy (in GWh per capita);
X_3—electricity generation from wind energy (in GWh per capita);
X_4—electricity generation from solar energy (in GWh per capita);
X_5—electricity generation from biomass (in GWh per capita).

These variables were selected based on substantive criteria and constitute a set of potential diagnostic variables.

In the next step, formal (statistical) criteria were checked, i.e., potential diagnostic variables were analysed in terms of the degree of differentiation and correlation between particular variables. In order to assess the degree of variables' diversity, the coefficient of variation (V_j) was calculated [13]. In this article, a limit value of coefficient of variation was assumed at the level of 10%. In the case when the condition is met:

$$0 \leq V_j \leq 10\% \quad (j = 1,2,\ldots,p),$$

where:

p—number of potential diagnostic variables.

The variable Xj is eliminated from the set of diagnostic variables. In this article, all proposed diagnostic variables were found to be sufficiently diverse.

Next, Pearson's linear correlation coefficients were calculated between each pair of potential diagnostic variables.

Finally, all potential diagnostic variables were used to construct a taxonomic measure of development. The nature of these variables (i.e., the impact on the level of RES use for electricity generation) was then determined. All variables adopted for the study turned out to be stimulants, which means that the higher the values of these variables, the higher the level of development of RES for electricity generation in each country.

In the next step, the values of diagnostic variables were normalised using the standardisation method, and the so-called development pattern was determined, i.e., the object (here: country) with the best values of individual diagnostic variables. Then, the Euclidean distances of individual countries from the development pattern were calculated. The smaller the distance from the development pattern, the higher the level of renewable energy use in a given country.

The last stage consisted of calculating a synthetic measure (the so-called development measure), on the basis of which it is possible to rank the 28 countries surveyed in terms of the level of a composite phenomenon (i.e., the level of RES use for electricity generation). The values of the development indicator are standardised in the interval [0; 1]. The closer the value of this measure is to unity, the higher the level of the complex phenomenon (here: the level of development of renewable energy). Figures 1–4 show the values of the development ratio for the studied countries in selected years.

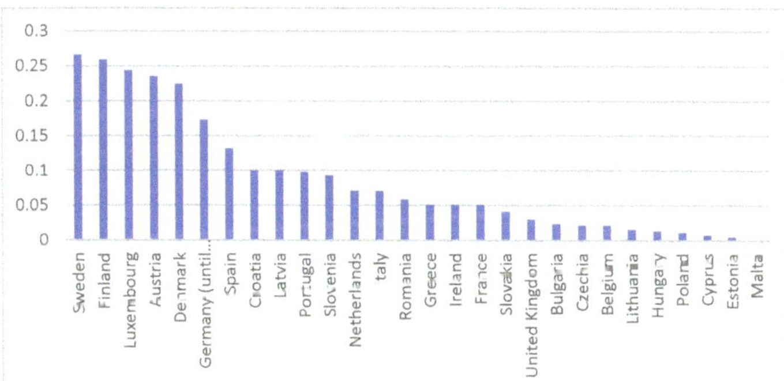

Figure 1. Development taxonomic ratio values (TMR) for the examined countries in 2004. Source: author's own elaboration.

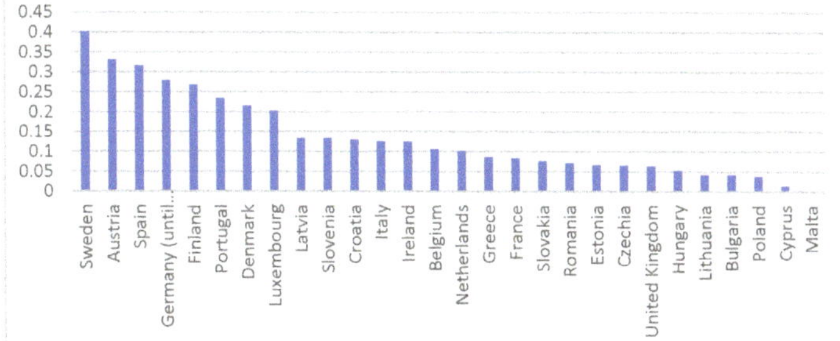

Figure 2. Development taxonomic ratio values (RMT) for the examined countries in 2009. Source: author's own elaboration.

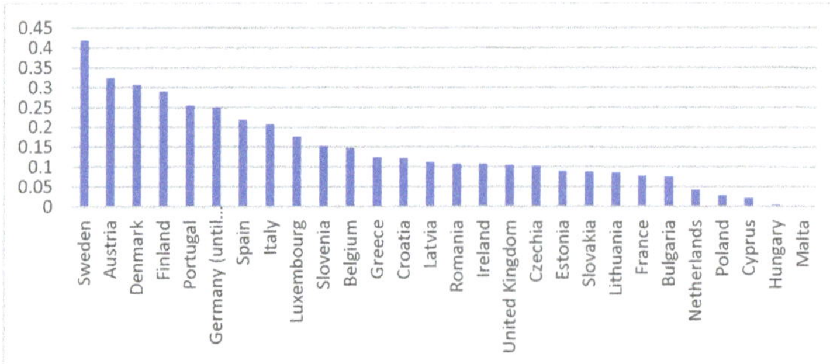

Figure 3. Development taxonomic ratio values (RMT) for the examined countries in 2014. Source: author's own elaboration.

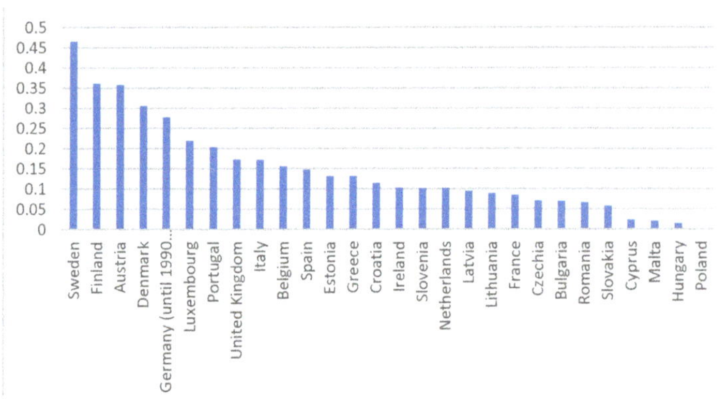

Figure 4. Development taxonomic ratio values (RMT) for the examined countries in 2019. Source: author's own elaboration.

Table 1 presents the ranking of the surveyed countries by level of RES use for electricity production for four selected years (in five-year intervals). Table 1 and Figures 1–4 intentionally present the same data in two graphical forms, as they have varying usability for the reader.

Table 1. The ranking of countries by level of RES use for electricity production for four selected years (alphabetical order).

Country	Ranking Position in the Year:			
	2004	2009	2014	2019
Austria	4	2	2	3
Belgium	22	14	11	10
Bulgaria	20	25	23	22
Croatia	8	11	13	14
Cyprus	26	27	26	25
Czechia	21	21	18	21
Denmark	5	7	3	4
Estonia	27	20	19	12
Finland	2	5	4	2

Table 1. *Cont.*

Country	Ranking Position in the Year:			
	2004	2009	2014	2019
France	17	17	22	20
Germany	6	4	6	5
Greece	15	16	12	13
Hungary	24	23	27	27
Ireland	16	13	16	15
Italy	13	12	8	9
Latvia	9	9	14	18
Lithuania	23	24	21	19
Luxembourg	3	8	9	6
Malta	28	28	28	26
Netherlands	12	15	24	17
Poland	25	26	25	28
Portugal	10	6	5	7
Romania	14	19	15	23
Slovakia	18	18	20	24
Slovenia	11	10	10	16
Spain	7	3	7	11
Sweden	1	1	1	1
United Kingdom	19	22	17	8

Source: author's own elaboration.

In each of the years surveyed, Sweden topped the ranking in terms of the level of RES use for electricity generation. Malta was ranked last in 2004, 2009, and 2014, but Poland was ranked last in the last year examined.

A synthetic measure that takes into account both sources of electricity production directly dependent on climate (hydro, solar, and wind) and biocomponents (Table 2 column A) has identified an interesting group of seven countries at the top of the ranking for 2019. Table 2 should be viewed in conjunction with Table 3. When such a measure is reduced by energy production from biocomponents, the same countries appear in the top seven of the ranking, albeit in a different order (Table 2 column B). This shows a good diversification of renewable sources for electricity generation in those countries that (apart from Finland) are in the middle of the ranking in terms of the share of biocomponents in electricity generation from renewable sources. However, in the case of Estonia, the inclusion of biocomponents as a renewable source of electricity generation (the share of 60.6%) has resulted in the country moving up in the 2019 ranking from 25th to 12th place.

The group of countries that both measures present as the best performers in the development of renewable electricity sources includes Sweden, Finland, Austria, Denmark, Germany, Luxembourg, and Portugal. It should be noted that, apart from Portugal, the other six countries in this group are rich countries in northern Europe, and all of them belong to the group of the so-called old EU member states (adopted before 2004). This group includes both net electricity exporting countries (Sweden, Germany) and significant net importers (Finland and especially Luxembourg—Table 2 column C).

Table 2. Taxonomic measures of development of renewable electricity generation in EU countries and net electricity exports—country rankings for 2019.

Taxonomic Measures of Development *						C. Net Electricity Exports ** (as % of Electricity Production)		
A. Ranking (with Bio)			B. Ranking (without Bio)					
1. O	Sweden	0.4645	1. O	Sweden	0.4792	1. O	Sweden	15.5
2. O	Finland	0.3611	2. O	Austria	0.4641	2. N	Czechia	15.1
3. O	Austria	0.3580	3. O	Germany	0.3293	3. N	Bulgaria	13.1
4. O	Denmark	0.3058	4. O	Denmark	0.3025	4. O	France	10.1
5. O	Germany	0.2784	5. O	Luxembourg	0.2778	5. O	Germany	5.4
6. O	Luxembourg	0.2186	6. O	Portugal	0.2726	6. O	Belgium	2.0
7. O	Portugal	0.2030	7. O	Finland	0.2482	7. N	Slovenia	2.0
8. O	United Kingdom	0.1726	8. O	Spain	0.2345	8. N	Cyprus	0.0
9. O	Italy	0.1717	9. O	Italy	0.2269	9. O	Netherlands	−0.7
10. O	Belgium	0.1565	10. O	Greece	0.2257	10. O	Ireland	−2.1
11. O	Spain	0.1474	11. O	Belgium	0.1861	11. O	Spain	−2.5
12. N	Estonia	0.1317	12. O	United Kingdom	0.1836	12. N	Romania	−2.5
13. O	Greece	0.1312	13. O	Ireland	0.1645	13. O	Austria	−4.2
14. N	Croatia	0.1143	14. N	Slovenia	0.1610	14. N	Slovakia	−6.0
15. O	Ireland	0.1021	15. N	Croatia	0.1583	15. O	Portugal	−6.4
16. N	Slovenia	0.1013	16. O	France	0.1412	16. N	Poland	−6.5
17. O	Netherlands	0.1011	17. O	Netherlands	0.1393	17. O	United Kingdom	−6.5
18. N	Latvia	0.0948	18. N	Lithuania	0.1305	18. O	Italy	−13.0
19. N	Lithuania	0.0889	19. N	Romania	0.1290	19. N	Latvia	−17.4
20. O	France	0.0845	20. N	Bulgaria	0.0852	20. O	Denmark	−19.7
21. N	Czechia	0.0709	21. N	Latvia	0.0791	21. O	Greece	−20.4
22. N	Bulgaria	0.0688	22. N	Malta	0.0655	22. N	Estonia	−28.3
23. N	Romania	0.0657	23. N	Slovakia	0.0599	23. O	Finland	−29.3
24. N	Slovakia	0.0575	24. N	Cyprus	0.0573	24. N	Malta	−30.6
25. N	Cyprus	0.0225	25. N	Estonia	0.0552	25. N	Hungary	−37.0
26. N	Malta	0.0205	26. N	Czechia	0.0505	26. N	Croatia	−48.1
27. N	Hungary	0.0143	27. N	Hungary	0.0168	27. N	Lithuania	−249.3
28. N	Poland	0.0000	28. N	Poland	0.0000	28. O	Luxembourg	−308.1

Source: own elaboration based on Eurostat data. O = old EU countries (pre-2004). N = new EU countries (2004 and later). * Taxonomic measures of development of electricity generation from renewable sources: A.—solar, wind, hydro, and bio; B.—solar, wind, hydro only. ** C.—net electricity exports—as the difference between the percentage ratio of electricity exports to total generation (GWh) and the percentage ratio of electricity imports to total generation (GWh) regardless of source.

At the same time, in the case of both versions of the measure, the penultimate and last place goes to Hungary and Poland, respectively.

When interpreting the results, it has to be remembered that the sizes of electricity production (in GWh) from different renewable sources included in the synthetic measures were calculated per capita, and nominal production volumes are not used here. As a result, the construction of the measure does not directly reflect the size of the population and the size of the economic potential of the countries studied.

Interestingly, five countries out of the seven identified by the synthetic measure are also among the top seven EU countries for 2019 in terms of the percentage of renewable electricity production in total electricity production (Luxembourg, Denmark, Austria, Sweden, and Portugal—Table 3 column A) and in terms of the percentage ratio of renewable electricity production to final consumption (Austria, Sweden, Denmark, Portugal, and Germany—Table 3 column B).

It is characteristic that the last places in the ranking according to synthetic measures (Table 2 column A and column B) are occupied by a compact group of the so-called new Member States (admitted to the EU in 2004 and later). The countries, whose capacity utilisation of electricity generation from renewable energy sources is considered the weakest by the synthetic measures, besides the already mentioned Hungary and Poland, are also the Czech Republic, Cyprus, Malta, Romania, Bulgaria, and Lithuania.

Table 3. Share of renewable sources in electricity production and consumption and total energy consumption (in %) in European Union countries—country rankings for 2019.

Share of Renewable Sources (in %)								
A. Electricity Production *			**B. Electricity Consumption ****			**C. Energy Consumption *****		
1. O	Luxembourg	85.9	1. O	Austria	87.5	1. O	Sweden	56.4
2. N	Lithuania	81.9	2. O	Sweden	77.7	2. O	Finland	43.1
3. O	Denmark	78.2	3. O	Denmark	71.2	3. N	Latvia	41.0
4. O	Austria	77.8	4. O	Portugal	59.1	4. O	Denmark	37.2
5. N	Croatia	66.2	5. N	Croatia	51.0	5. O	Austria	33.6
6. O	Sweden	58.7	6. N	Romania	50.5	6. N	Estonia	31.9
7. O	Portugal	54.2	7. O	Germany	48.6	7. O	Portugal	30.6
8. N	Latvia	49.6	8. N	Latvia	48.0	8. N	Croatia	28.5
9. O	Finland	46.6	9. O	Spain	42.6	9. N	Lithuania	25.5
10. N	Romania	42.0	10. O	Ireland	42.1	10. N	Romania	24.3
11. O	Germany	40.9	11. O	United Kingdom	40.5	11. N	Slovenia	22.0
12. O	Italy	40.1	12. O	Italy	39.0	12. N	Bulgaria	21.6
13. O	Ireland	38.9	13. O	Finland	38.4	13. O	Greece	19.7
14. O	United Kingdom	37.8	14. N	Slovenia	38.1	14. O	Spain	18.4
15. O	Spain	37.8	15. O	Greece	31.2	15. O	Italy	18.2
16. O	Greece	33.2	16. N	Estonia	27.5	16. O	Germany	17.4
17. N	Slovenia	32.6	17. N	Lithuania	26.9	17. O	France	17.2
18. N	Estonia	28.1	18. O	France	26.7	18. N	Slovakia	16.9
19. N	Slovakia	24.2	19. N	Slovakia	26.4	19. N	Czechia	16.2
20. O	Belgium	21.9	20. O	Luxembourg	25.6	20. N	Cyprus	13.8
21. O	France	20.7	21. N	Bulgaria	25.2	21. N	Hungary	12.6
22. O	Netherlands	18.9	22. O	Belgium	24.4	22. O	United Kingdom	12.3
23. N	Bulgaria	18.0	23. O	Netherlands	19.7	23. N	Poland	12.2
24. N	Poland	16.0	24. N	Czechia	18.7	24. O	Ireland	12.0
25. N	Hungary	13.8	25. N	Poland	17.2	25. O	Belgium	9.9
26. N	Czechia	12.9	26. N	Hungary	11.3	26. O	Netherlands	8.8
27. N	Malta	10.5	27. N	Cyprus	10.9	27. N	Malta	8.5
28. N	Cyprus	10.0	28. N	Malta	8.7	28. O	Luxembourg	7.0

Source: own elaboration based on Eurostat data. O = old EU countries (pre-2004). N = new EU countries (2004 and onwards). * A.—Percentage share of renewable sources in electricity generation (GWh). ** B.—Percentage share of renewable sources in electricity consumption (GWh). *** C.—Percentage share of renewable sources in total energy consumption (GWh).

This dichotomy of 'old versus new EU Member States' is more strongly accentuated by the synthetic measures than is apparent from the percentage share of renewable electricity generation in total electricity generation (Table 3 column A) or in final consumption (Table 3 column B) in the countries under study.

The thesis on the influence of the wealth of the surveyed countries on their current development of the use of renewable energy sources for electricity production will be verified later in this article by means of a panel model.

Of note is the low ranking of France (the 16th and 20th position for 2019) and the Netherlands (the 17th and 17th position for 2019) according to both synthetic measures (Table 2 column A and column B).

An important issue is the growing number of EU countries dependent on electricity imports. In 2004, it was 13 countries, in 2009, already 16 countries, in 2014, the number increased to 17 countries, and in 2019, it was as high as 20 countries. This means that seven countries have lost their energy independence in this way. The calculation uses the difference between the percentage ratio of electricity exports to total electricity production and the percentage ratio of electricity imports to total electricity generation in individual countries (Table 2 column C).

Interestingly, the Czech Republic and Bulgaria, which are ranked low in synthetic measures, reflecting the possibility of using renewable energy sources to produce electricity, are also countries that have a positive balance of electricity exports and imports, i.e., produce more electricity than they need domestically. This indicates the generation of surplus electricity for export through intensive production of electricity from non-renewable sources.

In the following section of this article, a panel model defined by formula (3) is used to describe the relationship between the share of RES power generation in total power generation and per capita power generation from water, wind, solar, and biomass energy in all EU countries. Data were collected for all 28 EU countries and cover the 2004–2019 period. The source of data was the Eurostat database.

The theoretical model can be written as follows:

$$Y_{jt} = \alpha_0 + \alpha_1 X_{1jt} + \alpha_2 X_{2jt} + v_{jt} \tag{3}$$

$$v_{jt} = e_t + u_j + \varepsilon_{jt}, \tag{4}$$

Table 4 presents a description of the individual variables.

Table 4. The description of variables.

Variables	Variables Description
Y_{1jt}	Share of electricity production from RES in total electricity production (in %)
Y_{2jt}	Electricity production from water per capita (in GWh)
Y_{3jt}	Electricity production from wind per capita (in GWh)
Y_{4jt}	Electricity production from solar per capita (in GWh)
Y_{5jt}	Electricity production from biogas per capita (in GWh)
	Explanatory Variables
X_{1jt}	GDP per capita
X_{2jt}	Public spending by countries on energy as a percentage of GDP
v_{jt}	The random error in the object *j*, in the time period *t*, which consists of the following components: e_t—impulses affecting all observations in the time period *t*; u_j—impulses affecting all the observations in the object *j*; ε_{jt}—impulses affecting only observations in the object *j*, in the time period *t*.

Source: elaborated by the authors.

The level of GDP per capita was used as a potential factor influencing the level of electricity production from renewable sources, assuming that richer countries with a higher level of development care more about the environment than poorer ones and, thus, invest in renewable energy sources. There is no measure of a country's wealth that is not questioned. The Stiglitz commission's report to then-French President Nicolas Sarkozy is an example of an extensive discussion on the subject. Alternative measures such as ISEW or HDI use GDP or its derivatives (GNI) in their construction. However, perhaps had there been more awareness of the limitations of standard metrics, such as GDP, there would have been less euphoria over economic performance in the years prior to the crisis; metrics that incorporated assessments of sustainability (e.g., increasing indebtedness) would have provided a more cautious view of economic performance. However, many countries lack a timely and complete set of wealth accounts—the 'balance sheets' of the economy—that could give a comprehensive picture of assets, debts, and liabilities of the main actors in the economy [51].

Panel data models (1) were estimated using the GRETL software (GNU Regression Econometrics Time-Series Library). In turn, for the estimation of panel data models, the following were used:

- Classic least squares method (KMNK) estimator;
- Fixed effect (FE) estimator;
- Random effect (RE) estimator.

The KMNK estimator is used when all objects covered by the study are homogeneous, and the differences between the empirical and theoretical values of the explained variable Y result only from the random component [52].

The FE and RE estimators are used in the case of sample heterogeneity. Individual effects are the source of sample non-homogeneity. The FE estimator assumes that the individual effects are non-random and can be estimated. In the case of the RE estimator, it is assumed that the individual effects are random and that they are part of the random component. In this case, it is not possible to estimate the value of individual effects; it is only possible to estimate their dispersion [52].

When selecting the type of panel model (simple model, i.e., without individual-dual effects or models with unidirectional individual effects, i.e., FEM—fixed effect model or REM—random effect model) the following tests are used: Wald test, Breusch–Pagan test, and Hausman test. The aforementioned tests allow to assess the correctness of the estimated model. These tests are discussed in many studies in the field of econometrics [53–55].The choice of the estimation method was based on the decision procedure presented in the econometrics literature [52–57]. First, a simple panel model (without individual effects) was estimated using the classical least squares method, and diagnostic tests of the model were conducted.

Table 5 presents the results of Wald, Breusch–Pagan, and Hausman tests, based on which a decision is made on the choice of an appropriate model. These tests allow for the verification of the assumptions about the correctness of panel model estimation.

Table 5. The result of the Wald, Breusch–Pagan, and Hausman tests.

Test	Hypotheses	Model	Test Statistics	p Value	Decision *
The Wald's test	**H1:** *the homogeneous model constant terms, independent of the item and time (OLS estimator)* **H2:** *the heterogeneous terms for individual items, but constant over time (FE estimator)*	Y_1	$F_1 = 85.3686$	$p \approx 0.00$	Rejection of **H1**
		Y_2	$F_2 = 777.714$	$p \approx 0.00$	Rejection of **H1**
		Y_3	$F_3 = 52.1408$	$p \approx 0.00$	Rejection of **H1**
		Y_4	$F_4 = 14.4418$	$p \approx 0.00$	Rejection of **H1**
		Y_5	$F_5 = 258.451$	$p \approx 0.00$	Rejection of **H1**
The Breusch–Pagan's test	**H3:** *the variance of the random component of individual effects insignificantly differs from zero (OLS estimator)* **H4:** *the variance of the random component of individual effects significantly differs from zero (RE estimator)*	Y_1	$LM_1 = 2062.31$	$p \approx 0.00$	Rejection of **H3**
		Y_2	$LM_2 = 3174.36$	$p \approx 0.00$	Rejection of **H3**
		Y_3	$LM_3 = 1514.2$	$p \approx 0.00$	Rejection of **H3**
		Y_4	$LM_4 = 437.889$	$p \approx 0.00$	Rejection of **H3**
		Y_5	$LM_5 = 2832.37$	$p \approx 0.00$	Rejection of **H3**
The Hausman's test	**H5:** *both FE and RE estimators are unbiased (RE estimator is more effective)* **H6:** *FE estimator is unbiased but RE estimator is biased (FE estimator)*	Y_1	$\chi^2_1 = 24.5178$	$p \approx 0.000$	Rejection of **H3**
		Y_2	$\chi^2_2 = 2.83383$	$p \approx 0.24246$	No grounds for rejection of **H5**
		Y_3	$\chi^2_3 = 35.5537$	$p \approx 0.000$	Rejection of **H5**
		Y_4	$\chi^2_4 = 39.4448$	$p \approx 0.000$	Rejection of **H5**
		Y_5	$\chi^2_5 = 2.96674$	$p \approx 0.22687$	No grounds for rejection of **H5**

* The adopted level of significance is 0.05 (i.e., $\alpha = 0.05$). Source: author's own calculation.

Analysing the results of the Wald test, it can be stated that a fixed effects model (FEM) is the correct model in all cases for describing the relation of the share of RES production in total electricity production and of production from water, wind, solar, and biomass and GDP per capita and government expenditure on energy per capita.

The results of the Breusch–Pagan test in each case indicate the random effects model (REM) as the better model. Finally, the results of the Hausman test allow for the conclusion, with the risk of error at the level of 5% ($\alpha = 0.05$), that in the case of power generation from water energy and biogas energy per capita, the model with random individual effects is appropriate for describing the examined relationship, while in the remaining cases, the model with fixed individual effects (FEM). However, further analysis confirmed the occurrence of heteroskedasticity of the random component in the models Y_1, Y_3, and Y_4. To address this shortcoming, the weighted least squares (WLS) method was used to estimate the parameters of share of electricity production from RES in a total electricity production model, electricity production from wind per capita, and electricity production from solar per capita. In other cases—electricity production from water per capita and electricity production from biogas per capita—the random effects models were estimated.

Table 6 presents the estimation results of the above models.

Table 6. Results of model estimation.

Dependent Variable Y_1, Share of Electricity Production from RES in Total Electricity Production (%) (WLS)					
Independent variables	Coefficient	St. Error	t-ratio	*p*-value	Significance [a]
Constant	10.6331	1.18383	8.982	<0.0001	***
X_{1jt}	0.000486	0.000042	11.68	<0.0001	***
X_{2jt}	10.7911	0.177197	6.090	<0.0001	***
Observations				448	
Standard error residuals				0.977433	
R^2				0.265472	
$F(2, 445) = 80.41575$				*p*-value for test F < 0.00001	
Dependent Variable Y_2, Electricity Production from Water per Capita (in GWh) (Model REM)					
Independent variables	Coefficient	St. Error	t-ratio	*p*-value	Significance [a]
Constant	0.00103611	0.000302482	3.425	0.0006	***
X_{1jt}	−0.00000000001099	0.00000000243479	−0.004512	0.9964	
X_{2jt}	0.0000931543	0.0000502918	1.852	0.0640	*
Observations				448	
Standard error residuals				0.001577	
Dependent Variable Y_3, Electricity Production from Wind per Capita (in GWh) (WLS)					
Independent variables	Coefficient	St. Error	t-ratio	*p*-value	Significance [a]
Constant	0.00001506	0.0000166	0.9074	0.3647	
X_{1jt}	0.00000001029	0.00000000079	12.95	<0.0001	***
X_{2jt}	−0.0000147311	0.0000212765	−0.6924	0.4891	
Observations				448	
Standard error residuals				0.942170	
R^2				0.279467	
$F(2, 445) = 86.29923$				*p*-value for test F <0.00001	
Dependent Variable Y_4, Electricity Production from Solar per Capita (in GWh) (WLS)					
Independent variables	Coefficient	St. Error	t-ratio	*p*-value	Significance [a]
Constant	−0.0000115383	0.00000506263	−2.279	0.0231	**
X_{1jt}	0.000000001475	0.000000000176	8.373	<0.0001	***
X_{2jt}	0.000103074	0.00000962641	10.71	<0.0001	***
Observations				448	
Standard error residuals				0.947601	
R^2				0.272744	
$F(2, 445) = 83.44460$				*p*-value for test F <0.00001	
Dependent Variable Y_5, Electricity Production from Biogas per Capita (in GWh) (Model REM)					
Independent variables	Coefficient	St. Error	t-ratio	*p*-value	Significance [a]
Constant	−0.0000707669	0.0000780464	−0.9067	0.3646	
X_{1jt}	0.000000011876	0.00000000104695	11.34	<0.0001	***
X_{2jt}	0.000169734	0.0000220063	7.713	<0.0001	***
Observations				448	
Standard error residuals				0.000399	

Source: author's own calculation. [a] *** The statistically significant variable at the level of 1%; ** at the level of 5%; * at the level of 10%.

In the case of the model describing the share of RES energy generation in total energy generation, the model describing solar power production per capita and the model describing biogas power production per capita, both potential explanatory variables proved to be statistically significant. In the case of the model describing electricity production from wind, only GDP per capita was statistically significant. In the case of the model describing electricity production from water, only public spending by countries on energy as a percentage of GDP was statistically significant at the 0.1 significance level. This proves large public expenditure on the development of solar electricity production in EU countries, which is significantly greater than for other RES sources.

At the same time, it confirms the significance of economic potential and wealth of countries for better use of RES for electricity generation, which was initially indicated by the analysis of country rankings according to the taxonomic measure of development.

The results obtained allow us to state that in the case of the model describing the share of electricity generation from RES in total electricity generation, two explanatory variables, i.e., GDP per capita and public expenditure on energy as percent of GDP, positively influence the explained variable (i.e., the share of renewable energy generation in total energy generation). The evaluation of the parameter with the independent variable X_{1jt} (0.000486) should be interpreted as follows: if GDP per capita increases by one percentage point, the share of electricity generation from RES in total electricity generation will increase on this account by about 0.000486 on average, assuming constancy of the other variables. The interpretation of the evaluation of the parameter with the explanatory variable X_{2jt} (10.7911) should be as follows: if public expenditure on energy per capita increases by 1 percent point, then the share of renewable energy production in total energy production will increase for this reason by approximately 10.7911 on average, under the assumption of constancy of the other variables.

5. Conclusions

The use of renewable energy sources is an important element in the functioning of the global economy due to its impact on environmental pollution. Minimising climate change and reducing CO_2 emissions is of interest to society, energy producers, government authorities, and many non-governmental organisations. Despite the great diversity in the involvement of selected countries (included in the study) in the production of electricity from water, sun, wind, and biomass energy, it should be noted that an important element is the development of social awareness and support for the actions of the governments in individual countries to make greater use of these energy carriers, which will allow for the slowdown of climate change and care for ecology.

The challenge for individual countries is to support them in their joint activities related to the use of new technologies to protect the environment. Studies have shown that there is still a lot to be done in the field of implementation, support, and use of RES.

The analyses carried out by the authors confirmed the hypothesis that the level of RES use for electricity production in the European Union countries depends on the level of development of these countries, measured by GDP per capita.

The presented synthetic measures confirmed the more favourable situation of the rich northern EU countries in the production of electricity from renewable sources (solar, wind, hydro, and bio), at the same time highlighting problems with the greening of electricity production in a large group of the new EU member states.

The panel analysis also showed a positive impact of GDP per capita on the use of RES for electricity production. Only the production of electricity from water (hydro) is not affected by the level of GDP per capita. It is the only renewable source of energy more strongly influenced by natural conditions (geological, hydrological) than economic ones.

The study also shows an increase in the number of countries in the studied period 2004–2019 that have a negative balance of electricity exports and imports, i.e., a lack of self-sufficiency in the country's electricity supply. It follows from the data analysed that there is no direct relationship between energy self-sufficiency and various measures of the use of renewable sources for electricity generation.

Limitations of the Study

The analyses carried out do not take into account the volumes of electricity generation from different sources in the countries studied, and thus, weights reflecting the variation in energy production potential between the countries studied as a derivative of differences in economic potential and wealth were not used. The latter two issues are taken into account in the panel study part of the article; however, the data on electricity production are calculated on a per capita basis. Further research is needed to find a model that would allow for extending the analysis to include aspects of differences in the volume of electricity generation potential between EU countries, which would provide a better background for considering the structure of sources used for electricity generation.

Author Contributions: Conceptualization, R.H., A.H., E.Z.-R., G.V.; methodology, R.H., A.H., E.Z.-R., G.V.; software, A.H., E.Z.-R.; validation, A.H., E.Z.-R.; formal analysis, A.H., E.Z.-R.; investigation, R.H., A.H., E.Z.-R., G.V.; resources, R.H., G.V.; writing—original draft preparation, R.H., A.H., E.Z.-R., G.V.; writing—review and editing, R.H., A.H., E.Z.-R., G.V.; visualization, R.H., A.H., E.Z.-R., G.V.; supervision, R.H., A.H., E.Z.-R., G.V.; project administration, R.H., A.H., E.Z.-R., G.V.; funding acquisition, R.H., A.H., E.Z.-R., G.V. All authors have read and agreed to the published version of the manuscript.

Funding: This research received no external funding.

Institutional Review Board Statement: Not applicable.

Informed Consent Statement: Not applicable.

Conflicts of Interest: The authors declare no conflict of interest.

References

1. Twidell, J.; Weir, T. *Renewable Energy Resources*, 3rd ed.; Taylor&Francis Ltd.: London, UK, 2015.
2. Rehman, S.; Al-Hadhrami, L.M.; Alam, M. Pumped hydro energy storage system: A technological review. *Renew. Sustain. Energy Rev.* **2015**, *44*, 586–598. [CrossRef]
3. Pramod, J. *Wind Energy Engineering*; McGraw-Hill Education: New York, NY, USA, 2016.
4. Trela, M.; Dubiel, A. Efektywność Ekonomiczna Elektrowni Fotowoltaicznych Oraz Uniknięte Koszty Zewnętrzne w Wyniku ich Funkcjonowania. 2014. Available online: https://tinyurl.com/3exstvb4 (accessed on 3 October 2021).
5. Gary, B.R.; Cagle, D.S. *Biomass Energy*; Oak Ridge National Lab. (ORNL): Oak Ridge, TN, USA, 2010. Available online: https://www.osti.gov/biblio/1050890, (accessed on 3 October 2021).
6. European Commision. Climate Package. 2008. Available online: https://ec.europa.eu/clima/eu-action/climate-strategies-targets/2020-climate-energy-package_en (accessed on 5 September 2021).
7. Directive 2010/75/EU of the European Parliament and of the Council of 24 November 2010 on Industrial Emissions (Integrated Pollution Prevention and Control). 2010. Available online: https://eur-lex.europa.eu/LexUriServ/LexUriServ.do?uri=OJ:L:2010:334:0017:0119:en:PDF (accessed on 5 September 2021).
8. European Commision. A Roadmap for Moving to a Competitive Low-Carbon Economy in 2050. 2011. Available online: https://www.eea.europa.eu/policy-documents/com-2011-112-a-roadmap (accessed on 3 October 2021).
9. IAEA. Climate Change and Nuclear Power. 2012. Available online: https://www.iaea.org/sites/default/files/climate_change.pdf (accessed on 5 September 2021).
10. The Draft of Renewable Energy Sources Act. 2014. Available online: https://wysokienapiecie.pl/18095-analysis-draft-amendment-renewable-energy-sources-act/ (accessed on 5 September 2021).
11. Directive 2003/87/EC of the European Parliament and of the Council of 13 October 2003 Establishing a Scheme for Greenhouse Gas Emission Allowance Trading within the Community and Amending Council Directive 96/61/EC. 2003. Available online: https://eur-lex.europa.eu/legal-content/EN/ALL/?uri=celex:32003L0087 (accessed on 5 September 2021).
12. Directive 2008/98/EC of the European Parliament and of the Council of 19 November 2008 on Waste and Repealing Certain Directives (Text with EEA Relevance) (OJ L 22 November 2008). 2008. Available online: https://eur-lex.europa.eu/LexUriServ/LexUriServ.do?uri=OJ:L:2008:312:0003:0030:en:PDF (accessed on 5 September 2021).
13. Directive (EU) 2018/2001 of the European Parliament and of the Council of 11 December 2018 on the Promotion of the Use of Energy from Renewable Sources. 2018. Available online: https://eur-lex.europa.eu/legal-content/EN/TXT/PDF/?uri=CELEX:32018L2001&from=EN (accessed on 5 September 2021).
14. Chakraborty, S.K.C.; Mazzanti, M. Renewable electricity and economic growth relationship in the long run: Panel data econometric evidence from the OECD. *Struct. Chang. Econ. Dyn.* **2021**, *59*, 330–341. [CrossRef]
15. Halkos, G.; Gkampoura, E.-C. Examining the Linkages among Carbon Dioxide Emissions, Electricity Production and Economic Growth in Different Income Levels. *Energies* **2021**, *14*, 1682. [CrossRef]

16. Szopik-Depczynska, K.; Kędzierska-Szczepaniak, A.; Szczepaniak, K.; Cheba, K.; Gajda, W.; Ioppolo, G. Innovation in sustainable development: An investigation of the EU context using 2030 agenda indicators. *Land Use Policy* **2018**, *79*, 251–262. [CrossRef]
17. Szopik-Depczynska, K.; Cheba, K.; Bąk, I.; Stajniak, M.; Simboli, A.; Ioppolo, G. The study of relationship in a hierarchical structure of EU sustainable development indicators. *Ecol. Indic.* **2018**, *90*, 120–131. [CrossRef]
18. Pao, H.-T.; Fu, H.-C. Renewable energy, non-renewable energy and economic growth in Brazil. *Renew. Sustain. Energy Rev.* **2013**, *25*, 381–392. [CrossRef]
19. Lin, M.-X.; Liou, H.M.; Chou, K.T. National Energy Transition Framework toward SDG7 with Legal Reforms and Policy Bundles: The Case of Taiwan and Its Comparison with Japan. *Energies* **2020**, *13*, 1387. [CrossRef]
20. Pietrzak, M.; Igliński, B.; Kujawski, W.; Iwański, P. Energy Transition in Poland—Assessment of the Renewable Energy Sector. *Energies* **2021**, *14*, 2046. [CrossRef]
21. Grosse, T.G. Low Carbon Economy Policy in Poland: An Example of the Impact of Europeanisation. *Equilibrium* **2011**, *6*, 9–39. [CrossRef]
22. Chovancová, J.; Tej, J. Decoupling economic growth from greenhouse gas emissions: The case of the energy sector in V4 countries. *Equilib. Q. J. Econ. Econ. Policy* **2020**, *15*, 235–251. [CrossRef]
23. Zielenkiewicz, M. Institutional Environment in the Context of Development of Sustainable Society in the European Union Countries. *Equilibrium* **2014**, *9*, 21–38. [CrossRef]
24. Sustainable Development Solutions Network and the Institute for European Environmental Policy. 2019 Europe Sustainable Development Report, Sustainable Development Solutions Network and Institute for European Environmental Policy. 19 November 2019. Available online: https://www.sdgindex.org/reports/ (accessed on 5 September 2021).
25. Piekut, M. The Consumption of Renewable Energy Sources (RES) by the European Union Households between 2004 and 2019. *Energies* **2021**, *14*, 5560. [CrossRef]
26. Yu, B.; Fang, D.; Yu, H.; Zhao, C. Temporal-spatial determinants of renewable energy penetration in electricity production: Evidence from EU countries. *Renew. Energy* **2021**, *180*, 438–451. [CrossRef]
27. Markandya, A.; Arto, I.; González-Eguino, M.; Román, M.V. Towards a green energy economy? Tracking the employment effects of low-carbon technologies in the European Union. *Appl. Energy* **2016**, *179*, 1342–1350. [CrossRef]
28. Kacperska, E.; Łukasiewicz, K.; Pietrzak, P. Use of Renewable Energy Sources in the European Union and the Visegrad Group Countries—Results of Cluster Analysis. *Energies* **2021**, *14*, 5680. [CrossRef]
29. Simoes, S.G.; Amorim, F.; Siggini, G.; Sessa, V.; Saint-Drenan, Y.-M.; Carvalho, S.; Mraihi, H.; Assoumou, E. Climate proofing the renewable electricity deployment in Europe—Introducing climate variability in large energy systems models. *Energy Strat. Rev.* **2021**, *35*, 100657. [CrossRef]
30. Mehedintu, A.; Soava, G.; Sterpu, M.; Grecu, E. Evolution and Forecasting of the Renewable Energy Consumption in the Frame of Sustainable Development: EU vs. Romania. *Sustainability* **2021**, *13*, 10327. [CrossRef]
31. Marra, A.; Colantonio, E. The path to renewable energy consumption in the European Union through drivers and barriers: A panel vector autoregressive approach. *Socio-Economic Plan. Sci.* **2020**, *76*, 100958. [CrossRef]
32. Marinaș, M.-C.; Dinu, M.; Socol, A.-G.; Socol, C. Renewable energy consumption and economic growth. Causality relationship in Central and Eastern European countries. *PLoS ONE* **2018**, *13*, e0202951. [CrossRef]
33. Belke, A.; Dobnik, F.; Dreger, C. Energy consumption and economic growth: New insights into the cointegration relationship. *Energy Econ.* **2011**, *33*, 782–789. [CrossRef]
34. Overland, I. Energy: The missing link in globalization. *Energy Res. Soc. Sci.* **2016**, *14*, 122–130. [CrossRef]
35. Remeikiene, R.; Gasparenien, L.; Fedajev, A.; Szarucki, M.; Ðeki´c, M.; Razumiene, J. Evaluation of Sustainable Energy Development Progress in EU Member States in the Context of Building Renovation. *Energies* **2021**, *14*, 4209. [CrossRef]
36. Aklin, M. Do high electricity bills undermine public support for renewables? Evidence from the European Union. *Energy Policy* **2021**, *156*, 112400. [CrossRef]
37. Percebois, J.; Pommeret, S. Efficiency and dependence in the European electricity transition. *Energy Policy* **2021**, *154*, 112300. [CrossRef]
38. Schreiner, L.; Madlener, R. A pathway to green growth? Macroeconomic impacts of power grid infrastructure investments in Germany. *Energy Policy* **2021**, *156*, 112289. [CrossRef]
39. Hiesl, A.; Ajanovic, A.; Haas, R. On current and future economics of electricity storage. *Greenh. Gases Sci. Technol.* **2020**, *10*, 1176–1192. [CrossRef]
40. Cárdenas, B.; Swinfen-Styles, L.; Rouse, J.; Hoskin, A.; Xu, W.; Garvey, S. Energy storage capacity vs. renewable penetration: A study for the UK. *Renew. Energy* **2021**, *171*, 849–867. [CrossRef]
41. Aqachmar, Z.; Ben Sassi, H.; Lahrech, K.; Barhdadi, A. Solar technologies for electricity production: An updated review. *Int. J. Hydrog. Energy* **2021**, *46*, 30790–30817. [CrossRef]
42. Perkins, G. Perspectives and economics of combining biomass liquefaction with solar PV for energy storage and electricity production. *Energy Sources Part B Econ. Plan. Policy* **2020**, *16*, 118–134. [CrossRef]
43. De Negri, J.F.; Pezzutto, S.; Gantioler, S.; Moser, D.; Sparber, W. A Comprehensive Analysis of Public and Private Funding for Photovoltaics Research and Development in the European Union, Norway, and Turkey. *Energies* **2020**, *13*, 2743. [CrossRef]
44. Eurostat. Database. Available online: https://ec.europa.eu/eurostat/web/main/data/database (accessed on 3 October 2021).

45. Cheba, K.; Szopik-Depczyńska, K. Multidimensional comparative analysis of the competitive capacity of the European Union countries and geographical regions. *Oeconomia Copernic.* **2017**, *8*, 487–504. [CrossRef]
46. Rollnik-Sadowska, E.; Dąbrowska, E. Cluster analysis of effectiveness of labour market policy in the European Union. *Oeconomia Copernic.* **2018**, *9*, 143–158. [CrossRef]
47. Ginevičius, R. Multi-criteria assessment of socioeconomic systems' conditions based on hierarchically structured indicator systems. *Econ. Sociol.* **2020**, *13*, 256–266. [CrossRef] [PubMed]
48. Kukuła, K. *Metoda Unitaryzacji Zerowanej*; Wydawnictwo Naukowe PWN: Warszawa, Poland, 2000.
49. *Statystyczne Metody Analizy Danych*; Ostasiewicz, W. (Ed.) Wydawnictwo Akademii Ekonomicznej im. Oskara Langego we Wrocławiu: Wrocław, Poland, 1999.
50. Walesiak, M. *Uogólniona Miara Odległości GDM w Statystycznej Analizie Wielowymiarowej z Wykorzystaniem Programu R*; Wydawnictwo Uniwersytetu Ekonomicznego we Wrocławiu: Wrocław Poland, 2016.
51. Stiglitz, J.E.; Sen, A.; Fitoussi, J.-P. Report by the Commission on the Measurement of Economic Performance and Social Progress. Executive Summary. 2009. Available online: https://ec.europa.eu/eurostat/documents/8131721/8131772/Stiglitz-Sen-Fitoussi-Commission-report.pdf (accessed on 10 November 2021).
52. Osińska, M.; Kośko, M. Modele danych panelowych. In *Ekonometria Współczesna*; Osińska, M., Ed.; Towarzystwo Naukowe Organizacji i Kierownictwa: Toruń, Poland, 2007; pp. 416–418.
53. Kufel, T. *Ekonometria. Rozwiązywanie Problemów z Wykorzystaniem Programu Gretl*; Wydawnictwo Naukowe PWN: Warszawa, Poland, 2007.
54. Maddala, G. *Introduction to Econometrics*, 2nd ed.; Macmillan Publishing Company: New York, NY, USA, 1992.
55. Baltagi, B.H. *Econometric Analysis of Panel Data*; John Wiley & Sons Ltd.: Chichester, UK, 2001.
56. Dańska-Borsiak, B. *Dynamiczne Modele Panelowe w Badaniach Empirycznych*; Wydawnictwo Uniwersytetu Łódzkiego: Łódź, Poland, 2011.
57. Huterski, R.; Huterska, A.A.; Łapińska, J.; Zdunek-Rosa, E. The problem of savings exclusion and gross savings in the new European Union member states. *Entrep. Sustain. Issues* **2020**, *7*, 2470–2480. [CrossRef]

energies

Article

Renewable Energy in the Pomerania Voivodeship—Institutional, Economic, Environmental and Physical Aspects in Light of EU Energy Transformation

Bartłomiej Iglinski [1], Karol Flisikowski [2], Michał Bernard Pietrzak [2,*], Urszula Kiełkowska [1], Mateusz Skrzatek [1], Anas Zyadin [3] and Karthikeyan Natarajan [3]

1 Faculty of Chemistry, Nicolaus Copernicus University in Toruń, Gagarina 11, 87-100 Toruń, Poland; iglinski@chem.umk.pl (B.I.); ulak@chem.umk.pl (U.K.); teusz@doktorant.umk.pl (M.S.)

2 Faculty of Management and Economics, Gdańsk University of Technology, Gabriela Narutowicza 11/12, 80-233 Gdańsk, Poland; kfi@zie.pg.gda.pl

3 School of Forest Sciences, University of Eastern Finland, P.O. Box 111, 801014 Joensuu, Finland; anas.zyadin@uef.fi (A.Z.); kaartheekeyan@rediff.com (K.N.)

* Correspondence: michal.pietrzak@zie.pg.gda.pl

Citation: Iglinski, B.; Flisikowski, K.; Pietrzak, M.B.; Kiełkowska, U.; Skrzatek, M.; Zyadin, A.; Natarajan, K. Renewable Energy in the Pomerania Voivodeship—Institutional, Economic, Environmental and Physical Aspects in Light of EU Energy Transformation. *Energies* **2021**, *14*, 8221. https://doi.org/10.3390/en14248221

Academic Editors: Adrian Ilinca and Dalia Štreimikienė

Received: 21 October 2021
Accepted: 3 December 2021
Published: 7 December 2021

Publisher's Note: MDPI stays neutral with regard to jurisdictional claims in published maps and institutional affiliations.

Copyright: © 2021 by the authors. Licensee MDPI, Basel, Switzerland. This article is an open access article distributed under the terms and conditions of the Creative Commons Attribution (CC BY) license (https://creativecommons.org/licenses/by/4.0/).

Abstract: In the era of globalization and rapid economic growth, affecting most world economies, increased production and consumption are leading to higher levels of energy production and consumption. The growing demand for energy means that energy resources from conventional sources are not sufficient; moreover, its production generates high costs and contributes to the emission of greenhouse gases and waste. In view of the above, many countries have opted to implement an energy transformation. The energy transition allows the transition from an energy system based on conventional fuels to an energy system based mainly on renewable energy (RE) and low-emission sources. In the EU, the development of a "green economy" has become a strategic goal in the fight against climate change. The development of RE offers the possibility to improve the energy security of a given country and the entire EU. New, innovative technologies of RE also increase the attractiveness and competitiveness of the economies of the Member States. In line with the EU strategy, the activities carried out aim to achieve a situation in which, in 2050, the activities of economies will not endanger the environment. The main purpose of this article was the assessment of the RE sector in the Pomerania region in the context of energy transformation. To achieve this goal, PEST analysis regarding the functioning of the RE sector in the selected Polish region was used and the potential of the RE sector was determined using GIS tools on the basis of physical conditions. The article presents the research hypothesis that the RE sector within the Pomerania Voivodeship possesses appropriate energy potential, which will allow this Voivodeship to become an energy self-sufficient region based on the use of these energy sources (according to EU strategy). The implementation of the goal set in the article allowed for the verification of the research hypothesis, where the determined energy potential from the RE sector would cover the Voivodeship's needs due to the use of electricity and heat. The conducted research shows that the RE sector in these regions has high energy potential to meet the criteria outlined in EU legal documents and to implement them successfully within the intended period.

Keywords: renewable energy; energy transition; decarbonization strategy; energy potential; PEST analysis; Pomerania Voivodeship

1. Introduction

In the current era of globalization and economic growth, increased production and consumption leads to higher energy production levels. Energy usage plays a substantial role in globalization processes, thus leading to unsustainability through the use of natural resources on a large scale [1,2]. The continuous increase in energy demand means that the most frequently used energy sources are no longer sufficient. Energy from non-renewable sources has become more expensive, harmful for the environment, and thus less profitable [3,4]. This situation creates the perfect conditions for an intensive energy transformation in developing countries [5,6]. The shift to a "green economy" has become a global phenomenon and is based on the effective consumption of energy and the gradual replacement of fossil fuels with clean energy sources [7]. This process is understood as a transformation from the current energy system using fossil fuels to an energy system based mainly on low-emission and renewable sources. It includes the gradual replacement of exhaustible hydrocarbons and uranium fuel by renewable energy (RE) in almost all sectors of the economy (transport, industry, energy, heating, construction, agriculture, etc.).

For the EU countries, the transformation of the energy system has become an important strategic goal, the implementation of which will allow for increased energy security and the improved investment attractiveness of the economies of the Member States [1,2]. This transformation, however, is being processed through the institutional system first [8–11]. The future structure of electricity generation is determined in the context of trends in the EU climate policies [12]. Poland, which entered the EU in 2004, adapted and integrated its institutional model [13]. New member states such as Poland have made a commitment to reduce the emission of greenhouse gases by ratifying the Kyoto Protocol and by participating in the climate policy of the EU [14]. The EU is fighting climate change through very ambitious policies. Currently, the EU is on track to reduce gas emissions by at least 55% by 2030 ("2030 Climate Target Plan"). By 2050, Europe aims to become the first climate-neutral continent ("European Climate Law"). These are the short-term goals, and the long-term aim within Europe is to become a climate-neutral economy with net-zero greenhouse gas emissions. Among many key targets for 2030, the most important are as follows: at least 40% cuts in greenhouse gas emissions (from 1990 levels), at least 32% share for RE, and at least 32.5% improvement in energy efficiency. The 2030 package includes a higher renewables target and new rules to support the expansion of renewables. The EU increased its greenhouse gases emissions reduction target for 2030 from 40% to 55% and the RE target from 32% to 40%. This means that the EU will need 451 GW of wind power capacity by 2030, an increase from the current value of 180 GW. The long-term strategy is in line with the EU's commitment to global climate action under the European Green Deal and Paris Agreement (a global framework to avoid dangerous climate change by limiting global warming). Both of these acts seek to address climate change and environmental degradation. From a long-term perspective, the EU is to be transformed into a modern, resource-efficient, and competitive economy, ensuring no net emissions of greenhouse gases by 2050, economic growth decoupled from resource use, and no person and no place left behind.

European integration processes, amplified by overall globalization in the background, have led to improved socio-economic and institutional situations in many states [15–21]. With time, the improved condition of the economies allowed the introduction of new, innovative RE projects [22–26]. Besides RE projects, since the early 2000s, many private investors and businesses have become the additional "engine" supporting the rapid growth of this market in Eastern Europe [26,27]. EU programs are usually organized on either commercial or hybrid models, subsidized by the government or EU (with national or regional range). Thanks to such programs, and their national and regional institutions, new consumption attitudes in the field of RE have been popularized, and many new jobs have been created [28–30]. All these actions favor change in the structure of the labor market in Poland and in the EU in general—we can observe increased employment mobility [31–38]. The RE labor sector in Poland is one in which employment has been increasing for the last two decades [39–41]. It is expected that, during the COVID-19 pandemic, the political and economic environment may not support the short-term expansion of the RE sector, but, from a long-term perspective, positive trends are forecasted [42,43]. Moreover, further growth in total energy consumption is predicted, representing a huge challenge for the RE market as well [44–46].

It should also be emphasized that the new EU climate policies and the rapid development of innovative RE technologies are implemented at the level of both countries and regions [47–49]. The RE market transformation should be based on the diversification of energy sources—they are not evenly geographically accessible [50]. Taking into account all of the aforementioned facts, we conclude that it is worth analyzing the regional potential first (at the Voivodeship level) and then evaluating individual sources of RE in light of EU requirements.

If we consider the energy sector in Poland, this country is only at the beginning of the energy transformation—most of the region's energy is still obtained from hard coal and lignite. At the end of September 2021, the installed capacity in all generation sources in Poland amounted to 53,274.6 MW. In conventional power plants, the country has 36,585.1 MW, and in RES installations, 15,276 MW. Only in September, 15,216.4 GWh of energy was produced in Poland, of which 12,775.4 GWh was generated by the conventional segment of the energy sector, and RES installations represented 2385.4 GWh [51].

The Pomerania Voivodeship is located is located in the northern part of the country and borders the Baltic Sea. It includes the following geographical units: Pobrzeże Południowobałtyckie, Lakeland Pomorskie, and Żuławy. Its total area is 18,310 km^2. The population of the Voivodeship in 2019 amounted to 2,343,928 people and shows an upward trend. The population density was 128 people/km^2. Most of the population lives in urbanized areas (63.5%) [52].

For these reasons, the aim of this study was to examine the current state, potential, and development prospects of the RE sector in the Pomerania (Pomorskie in Polish) Voivodeship in the context of energy transformation. To achieve this goal, PEST analysis regarding the development of RE in the Pomerania Voivodeship was used and the potential of the RE sector was determined using GIS tools on the basis of physical conditions. The article presents the research hypothesis that the Pomerania Voivodeship has an appropriate level of RE potential, which will allow this Voivodeship to become an energy self-sufficient region in line with the EU's short- and long-term strategy based on the use of RE. The implementation of the goal set in the article allowed for the verification of the research hypothesis regarding whether the potential amount of RE electricity would cover the Voivodeship's needs. Moreover, the results obtained confirm that the requirements of both the 2030 and 2050 EU acts are expected to be satisfied in terms of RE and the RE market's transformation.

2. Assessment of Development Potential of RE in Pomerania Voivodeship—PEST Analysis

For the assessment of the potential of the Pomeranian Voivodeship in light of the energy transformation and development of RE, we use PEST analysis. PEST analysis belongs to the group of methods used to study the macro-environment.

The main elements of the macro-environment are (Figure 1) [53]:

- political: including political stability, RE policy, legislative environment;
- economic: including economic situation in the world and in the Pomerania Voivodeship, the labor market, interest rates;
- social: including demography, knowledge about RE, structure of availability of personnel and human resources;
- technological: including the innovativeness of the RE sector, transfer of technologies and techniques (Figure 1).

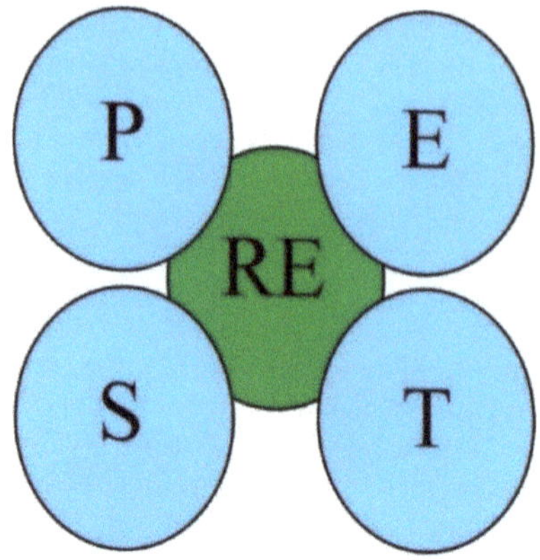

Figure 1. Utilized PEST analysis factors for RE macro-environment: P—political, E—economic, S—social, T—technological (author's own work).

In the presented work, it was considered advisable to subject the diagnosed factors to scoring, depending on the extent to which they support the RE sector. In the utilized scale ranging from 1 to 5, depending on the degree of their support for the development of renewable energy, individual points were assigned as follows:

- 1—very unconducive factor;
- 2—unconducive factor;
- 3—factor without influence;
- 4—contributing factor;
- 5—a very conducive factor.

At the same time, a formula for averaging the assessments of selected factors was adopted, assuming that this impact on its development means:

- below 2.00 points—the macro-environment is very hostile;
- 2.00–2.99 points—hostile macroenvironment;
- 3.00–3.49—neutral macroenvironment;
- 3.50–4.49—friendly macro-environment;
- 4.50–5.00—very friendly macroenvironment [53].

This PEST analysis was performed in line with the available literature as well as the current political, economic, social, and technological situation in the Pomerania region and Poland. The weights were counted from 1.00 to 5.00. To determine the weights, a team of experts was appointed, consisting of individuals from the Nicolaus Copernicus University, Gdańsk University of Technology, and the University of East Finland. In particular, the authors performed the PEST analysis using the weighting performed by a panel of experts in the fields of Economics, Chemistry, Biology, and Geography.

2.1. PEST Analysis—Political Environment

Table 1 contains the political determinants of the RE in the Pomerania region. The political factors of renewable energy sources in the Pomeranian Voivodeship generally result from the system and the political system in which it operates.

Table 1. The political environment of RE sources in the Pomerania Voivodeship (author's own work).

	Factor	Strengthening the RE Sector
1.	Political system and its stability	3.50
2.	Public administration system and its efficiency	2.00
3.	RE policy at the Voivodeship/state level	3.00
4.	Programs supporting the development of RE	4.50
5.	Documents on the use of the environment and GHG emissions	4.00
6.	A strong conventional energy lobby	1.00
7.	Membership in the EU (Poland's obligations to achieve a certain level of RE share)	5.00
	Overall assessment	3.29

The constitution law [54] states that Poland is a parliamentary republic and realizes the principles of independence and national sovereignty of a democratic state ruled by law. The current political system in the Republic of Poland can be considered relatively stable. The development of the research and science sphere, including RE, is positively influenced by a fairly friendly administrative order.

The administration in Poland is an integral part of the country, a vehicle of remembrance of the state's tradition and endowed with the privilege of legislative initiative. Moreover, the preamble to the Constitution of the Republic of Poland [52] defines a very important goal related to the need to improve the functioning of the public sector in the Republic of Poland. Despite many efforts, the public administration does not function efficiently in Poland—it is not conducive to the development of RE sources, which is often emphasized by the owners of RE installations.

As indicated by respondents in surveys [53–55], the policy in the field of RE at the state/Voivodeship level is not very conducive to the development of RE. The "Distance Act" [56] stopped the development of aeroenergy (wind energy) in Poland. The RE Act [57] is frequently amended, which discourages potential investors. For every 1 kWh of energy introduced to the grid, the prosumer can receive only 0.8 kWh of energy at any time.

The Polish energy policy, implemented until 2040, represents an opportunity for the development of RE in the Pomerania Voivodeship [58]. Poland's energy policy until 2040 places a strong emphasis on three main pillars. The first is a fair energy transition, the second is the construction of a parallel, emission-free energy system, and the third is good air quality.

When deciding to purchase/build an RE installation, it is worth using a subsidy/loan/financial relief. First, it is necessary to check whether there are funds for RE in a given town, commune, poviat, or Voivodeship and whether there are funds for investment in the national program or the EU program. For example, the "My Electricity" program allows one to obtain funding for the building of small photovoltaic installations (between 2 kW and 10 kW) in the form of a subsidy for up to 50% of the costs, but this is no more than PLN 5000 (in 2021, PLN 3000) for one project. Figure 2 shows the results of the "My Electricity 2" program [59]. The green color indicates the amount of co-financing granted to PV installation projects, while the black color indicates the installation power in a given Voivodeship (kW).

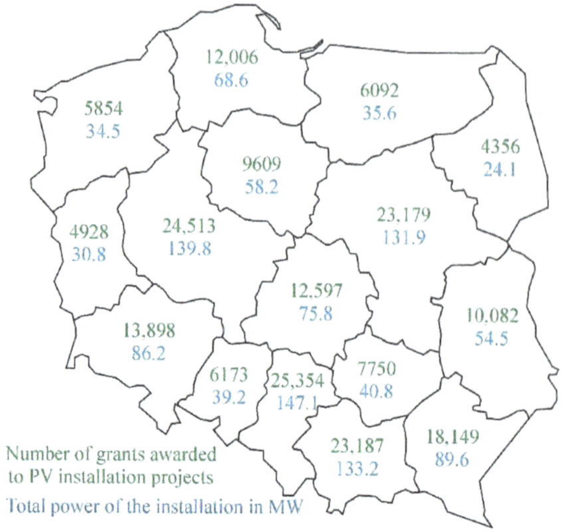

Figure 2. Results of the "My Electricity 2" program (own study from [59]).

In Poland, the Environmental Protection Law [60] describes legal regulations related to the use of the environment, including CHG emissions. According to the act, the principles of sustainable development constitute the basis for the preparation and updating of the national spatial development concept, Pomerania Voivodeship development strategies, and Voivodeship spatial development plans. The development of RE in the Pomerania Voivodeship/Poland is hindered by the strong lobby for conventional energy, mainly the coal lobby. Associated with mines, coal-fired power plants, and trade unions, they openly oppose the development of RE in Poland [55].

In order to ensure institutional "maturity" in Poland, this country is a member of the EU and other international organizations related to the law. Globalization processes positively influence the development of RE. This allows for open functioning in the European Research Area as part of the conducted research. This also allows for joint research projects and the free movement of employees between universities [55].

2.2. *PEST Analysis—Economic Environment*

Table 2 shows the economic factors of RE in the Pomerania region. In the case of EU Member States, Poland is characterized by economic growth. This was brought to a halt due to the COVID-19 pandemic. Figure 3 shows the dynamics of the volume and added value of Poland in the years 2016–2020 [61]. The world economy has practically recovered from the crisis that started in 2008 and is now struggling with the economic problems resulting from the pandemic. The prices of energy carriers remain at an average level; in October 2021, the price of crude oil was around USD 83–84/barrel [62], the price of hard coal was USD 233/Mg [63], and the price of natural gas was USD 5.30/million bt [64]. These prices are higher than 2 years ago [55], which means that energy from RE sources is becoming increasingly competitive in relation to energy from fossil fuels.

Table 2. The economic factors of RE in the Pomerania Voivodeship (author's own work).

	Factor	Strengthening the RE Sector
1.	Socio-economic development and economic growth	3.50
2.	The economic situation in the world, world fuel and energy prices, energy resources in the world	3.50
3.	Labor market (including in the RE sector), unemployment level	3.50
4.	Own funds for financing RE installations	2.00
5.	availability of loans and credit, interest rates,	4.00
6.	Economic support (auctions and green certificates)	3.00
7.	Investor activity, entrepreneurship, and innovation	4.00
8.	EU membership—globalization	5.00
	Overall assessment	3.56

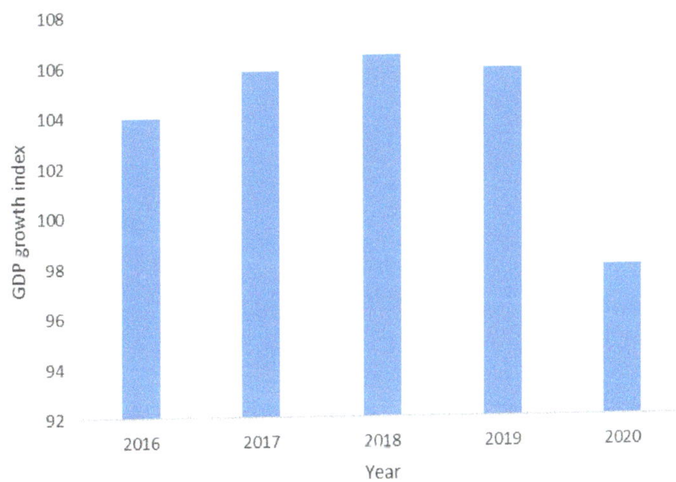

Figure 3. GDP growth index in the years 2016–2020 in Poland (own study from [59]).

An important element of the macro-environment of RE in the Pomeranian Voivodeship is the situation of the labor market: employment structure, unemployment rate, and population migrations. Official data [65] indicate that the unemployment rate in Poland at the end of August 2021 was 6.4% (5.6% in the Pomerania Voivodeship). Moreover, rural areas with relatively high unemployment offer excellent conditions for the development of renewable energy sources. A 20% rate of unemployment, which was measured in 2004/2005, and accession to the EU resulted in intensified emigration processes in Poland and in the Pomeranian Voivodeship. Currently, over 2 million Poles live abroad. Inflation in Poland was declining until the outbreak of the COVID-19 pandemic. Inflation has increased since then. The prices of consumer goods and services in August 2021, compared to the same month last year, were increased by 5.5%. Compared to the previous month, the prices of goods and services increased by 0.5% [66].

It is worth emphasizing that employment in the RE sector is growing year by year, which is important for the Polish economy. Today, almost 90,000 people are employed in the photovoltaics sector alone [67]. It is estimated that 150,000 people in Poland are employed within the RE sector. Regarding RE installations, although they are becoming cheaper every year, their price is still quite "prohibitive", especially for individual investors. Very few take into account that the operating costs of a heat pump are much lower than those of a coal-fired furnace (and the environmental and climate costs are even higher). Investors in

RE in Poland and in the Pomerania Voivodeship may apply for co-financing/loans from both EU funds and national funds [55].

The system of green certificates has been in use since 1 October 2005, based on the amended Energy Law [68]. The rapid development of the technology of biomass co-firing with coal led to an oversupply of green certificates on the market and a reduction in their prices from around PLN 300/MWh in 2012 to around PLN 25/MWh in mid-2017 and PLN 213/MWh in October 2021. Such low certificate prices have a negative impact on the financial liquidity of many RE installations in Poland. Designing and introducing the auction support system was a revolutionary change in the area of supporting electricity generation from RE sources. This system ensures the competitiveness of entities participating in the auctions, which translates into the most favorable electricity prices and, consequently, has an impact on the costs of electricity offered to end users. The auction is won by the entrepreneurs who offer energy at the lowest selling price [69].

In Poland, interest rates are currently at a record low, but they have grown in the past. The Lombard rate is 1.00%, the reference rate is 0.50%, the rediscount rate is 0.51%, and the deposit rate is 0.00% [70]. Polish RE investors are characterized by high innovation, entrepreneurship, and activity. They very often contribute their own ideas to improve the "conventional" technology, thanks to which the process of, e.g., methane fermentation has been improved. Increasingly, investors are developing two or more types of RE, e.g., biogas plants and PV panels. The accession to the EU and the opening of the borders allowed Poland to make an economic "leap". Since then, many trade agreements have been concluded, and there has been free movement of technologies and services. For several years, many Polish research centers have been conducting joint research projects with other research centers in the EU. The latest achievements of EU technology are more often implemented in Poland.

2.3. *PEST Analysis—Social Environment*

Table 3 presents the social environment of RE in the Pomerania Voivodeship in the context of its development potential.

Table 3. Social environment of RE in the Pomerania Voivodeship (author's own work).

	Factor	Strengthening the RE Sector
1.	Demographic situation	3.00
2.	Level of education	3.50
3.	Knowledge about RE	3.00
4.	Social acceptance of RE	4.00
5.	Social acceptance for building a green economy	3.00
6.	Human resources and employee availability	3.00
7.	The impact of RE development on jobs	4.00
8.	EU membership	4.50
	Overall assessment	3.50

The population of Poland has been steadily declining for a decade. At the end of 2020, there were 38,265 million Poles, which is 273,000 fewer than in 2011 (it is worth mentioning that, for example, Gdynia has 244,000 inhabitants) [71]. The current data are even more concerning, as 75,869 people in Poland had died from COVID-19 by 11 October 2021 [72]. In the Pomerania Voivodeship, the population at the end of 2020 was 2347 thousand [71]. The Polish society is aging and the population is set to decline in the future—in 2020, the fertility rate was 1.38 [67]. It is assumed that if the fertility rate fluctuates in the range of 2.10–2.15, we can talk about the so-called simple replacement of generations. This means that, in the assumed situation, each woman of reproductive age gives birth to slightly more than two children on average. Such a state is desirable in order to be able to ensure an optimal size of the workforce in the future, and thus not to overburden the pension system.

The unfavorable demographic situation in the Pomerania Voivodeship and in Poland has a very significant impact on the activity of the regional labor market. The related problems include the limited mobility of employment and the reduced ability to flexibly respond to changes in the labor market. A large group of the unemployed are people with higher education. Local authorities should support entrepreneurship by creating favorable employment conditions in micro- and small enterprises, which, inter alia, produce energy from RE sources [55].

Graduates of secondary schools (general and vocational) occupy a dominant position in the structure of the population according to the level of education; currently, the percentage of such people is 29.0%. Since 1995, the share of people with higher education in the 24–64 age group has increased from 9.7% to 21.2% [67]. Internet access, RE promotion (seminars, conferences, shows), and Poles traveling abroad have contributed to a gradual increase in knowledge of and support for RE. Secondary school students indicated that they learn about RE from the Internet, television and newspapers, and, to a lesser extent, from the curriculum content at school. Most of them assess their knowledge of RE as not very high. It is worth noting that teachers judge their knowledge of RE as good or very good, while information on RE is obtained mainly from the Internet and television [55]. Polish society has agreed to build a low-carbon economy, provided that it does not significantly increase the prices of energy and consumer goods. The respondents to this survey almost unanimously believed that the further development of renewable energy sources would have a positive impact on reducing the emission of pollutants into the environment [55].

Until recently, there were no fields of study related to RE offered by universities. Around 40 Polish universities have introduced to their offer fields of study/subjects related to RE sources. For example, at the Gdansk University of Technology, one can study "Green technologies and monitoring" [55].

The rapidly developing RE sector is looking for new employees, most often in rural areas, i.e., areas with a high unemployment rate. The technology of agricultural biogas plants fits perfectly into this scheme, thanks to which unemployment in Northern Poland may drop. Membership in the EU ensures the free movement of new technologies, as well as scientists and employees, in the RE sector.

2.4. PEST Analysis—Technological Environment

In recent years, the RE sector has been developing rapidly in Poland and around the world. Moreover, numerous studies have been conducted to increase the durability and efficiency of RE installations. These studies, as well as the increasing demand, make it possible to reduce/maintain the prices of RE installations [55].

Table 4 presents the technological environment of RE in the Pomerania Voivodeship in the context of its development potential. It needs to be highlighted that Polish power blocks are obsolete. Due to their poor technical condition, shutdowns are anticipated in the future. On 1 January 2018, the "Adamów" power plant was shut down. The electricity transmission network is poorly developed. It should be emphasized also that the Pomerania Voivodeship has excellent conditions for the development of RE—so it is worth developing small, local RE installations that will transmit energy almost without grid losses [55].

In Poland, there is little cooperation between industry and science. This also applies to RE. This cooperation is necessary to develop modern and cheap RE technologies. An opportunity for the development of this sector is the creation of numerous cluster initiatives, based on the endogenous potential of a given region. The efficiency of RE installations is equal to and sometimes exceeds that of conventional energy installations. Hydroenergy power plants obtain energy with an efficiency of up to 90%, while agricultural biogas plants, cogenerating electricity and heat, achieve efficiency of up to 85% [55].

Table 4. The technological environment of RE in the Pomerania Voivodeship (author's own work).

	Factor	Strengthening the RE Sector
1.	Innovativeness of the economy and the RE	4.50
2.	Energy infrastructure	2.50
3.	The power grid in the Voivodeship	2.00
4.	Economy–science cooperation	2.00
5.	RE efficiency	4.00
6.	RE potential in the Voivodeship	4.50
7.	EU membership	4.50
	Overall assessment	3.43

The RE potential calculated in Section 3 in the Pomerania Voivodeship is high. The amount of electricity would cover the Voivodeship's electricity needs, and 53% of its heat needs. Membership in the EU is an opportunity for the further economic development of Poland, including the RE sector. Cooperation and knowledge exchange take place within the European Research Area. The information society is developing.

3. Evaluation of Renewable Energy in the Pomerania Voivodeship—Physical Aspects

3.1. Use of Biomass in the Pomerania Voivodeship

Two biomass power plants with a total capacity of 2350 MW are operating in the Pomerania Voivodeship [73]. The area of energy plantations in the Voivodeship currently amounts to nearly 1.7 thousand ha [74]. Four species of plants are cultivated: three woody species—willow, poplar, and birch—and one species of grass from the genus Miscanthus. The crop structure is dominated by fast-growing poplar, amounting to 68%, with willow constituting 20% and Miscanthus constituting 12%, while birch is grown in negligible amounts. There are 40 wood biomass boiler houses with a total capacity of nearly 105 MW in the Pomerania Voivodeship. These are mainly wood-fired boiler houses; there are a few with pellets, and the boiler house at IP Kwidzyn operates with waste from the pulp and paper industry [75]. The number of straw-fired boiler houses in the Pomerania Voivodeship ranges from 25 to 30. Previously, boilers based on various Danish technological solutions were used; after being depleted, they were then replaced with more effective boilers fired with wood chips and pellets [75].

In the CHP plant in Lębork, waste generated in sawmills or furniture factories is the substrate used for the production of heat and electricity. The choice of biomass is not accidental. In the vicinity of Lębork, there are many farms (which can supply plant waste or switch to high-energy crops) and wood processing plants (which also produce wood waste). The plant has three chippers for biomass homogenization (Figure 4) [76]. Thanks to these activities, the old heating plant has reduced the combustion of coal, which had a negative impact on the environment and the health of the inhabitants.

According to the data of the Agricultural Market Agency, in the Pomerania Voivodeship, in Malbork, there is a plant producing biodiesel in the amount of 159 million dm^3, and in Goszyn, there is a plant producing bioethanol in the amount of 32 million dm^3. In turn, Gdańsk produces:

- liquid biocarbon in the amount of 286 million dm^3;
- biopropane in the amount of 32 million dm^3;
- biohydrogen in the amount of 1 120 million dm^3 [77].

Figure 4. Chipper for biomass homogenization in Lębork (photo: B. Igliński).

There are 22 biogas plants in operation in the Voivodeship, with a total capacity of 21,843 MW (data from the Energy Regulatory Office [78]; 20 biogas plants according to the Pomerania Spatial Planning Office [75]). According to the register of agricultural biogas producers, 10 agricultural biogas plants operate in the Pomerania Voivodeship (Table 5).

Table 5. Characteristics of agricultural biogas plants in the Pomerania Voivodeship (author's own study from: [75,78]).

City	Substrates	Efficiency (mln m^3 Biogas/Year)	Power (MW)		Annual Production (GWh/Year)	
			Electricity	Heat	Electricity	Heat
Koczała	Slurry, maize silage, straw	9.200	2.126	2.206	18.000	19.500
Pawłówko	Slurry, maize slaughterhouse waste, glycerin	3.810	0.946	0.420	3.000	3.900
Płaszczyca	Slurry, maize silage, waste plant mass, waste from processing plant products	2.900	0.625	0.600	5.300	5.900
Uniechówek	Slurry, maize silage	4.500	1.063	1.081	8.380	8.696
Darżyno	Potato waste, maize silage, miscellaneous waste, slurry	10.000	2.400	2.400	18.500	12.000
Lębork	Potato peelings, sewage sludge	3.500	1.200	1.251	9.328	no data
Tragamin	Corn silage, grass silage, manure, haylage, bio-waste	3.500	0.800	5.540	4.860	no data
Jaromierz	Agricultural substrates, potato pulp, bio-waste	5.000	0.999	8.360	8.670	no data
Płaszczyna	Distillation broth, corn silage, grain chaff	11.000	2.000	no data	no data	no data
Niedamowo	Pig slurry, vegetable and food waste	1.222	0.330 (possible extension)	0.365	no data	no data

3.2. Biomass Potential in the Pomerania Voivodeship

3.2.1. Amount of Energy from Wood Harvested from Forests and Wood Industry

The technical potential of waste solid biomass and biogas from wastes in the Pomerania Voivodeship was calculated. The main source of energy is waste. In order to calculate the available energy, the following assumptions were made:

- 15% of wood harvested in the forest is "waste wood" [55];
- the calorific value of biomass from forests is 7 GJ/m^3 (in the Voivodeship, pine is the dominant species) [79];
- the efficiency of obtaining energy is 80% (electricity from solid biomass will be obtained with an efficiency of 30%, and heat with an efficiency of 50%).

Formula (1) shows the annual amount of energy that can be obtained from wood harvested directly, i.e., logging in forests:

$$E_{WW} = 0.8 \cdot 0.15 \cdot I_W \cdot W_W, \quad (1)$$

where: E_{WW}—energy from wood generated in the process of logging (PJ/year), I_W—amount of wood harvested annually in forests (3.23 million m^3/year), W_W—calorific value of biomass (7 GJ/m^3).

The amount of energy that can be produced from waste wood from forests in the Pomerania Voivodeship is 2.71 PJ/year (283 GWh/year of electricity and 1.69 PJ/year of heat). In order to estimate the annual energy possible from waste wood from the wood industry, the same assumptions were made, but it was assumed that industrial and post-use waste wood constitutes 25% of that harvested [53].

Formula (2) shows the amount of energy that can be obtained from wood in the woodworking industry:

$$E_{wi} = 0.8 \cdot 0.25 \cdot I_W \cdot W_W, \quad (2)$$

where: E_{wi}—energy from wood generated in the wood industry (PJ/year), I_W—amount of wood harvested annually in forests (3.75 million m^3/year), W_W—calorific value of biomass (7 GJ/m^3).

The amount of energy that can be produced from waste wood from the wood industry in the Pomerania Voivodeship is 4.52 PJ/year (472 GWh/year of electricity and 2.92 PJ/year of heat).

A total of 40% of biomass from forests can be allocated to energy purposes (formula (3)):

$$E_W = 0.8 \cdot 0.4 \cdot I_W \cdot W_W, \quad (3)$$

where: E_W—energy from wood generated during logging and in the wood industry (PJ/year), I_W—amount of biomass harvested annually in the Voivodeship's forests (3.75 million m^3/year), W_W—calorific value of biomass (7 GJ/m^3).

The obtained amount of energy that can be produced every year from waste wood in the Pomerania Voivodeship is 7.23 PJ/year (755 GWh/year of electricity and 4.51 PJ/year of heat).

3.2.2. The Amount of Energy from Waste Biomass from Orchards

Orchard wood waste comes from both maintenance (e.g., cutting branches) and cleaning. The amount of wood produced annually during maintenance works varies depending on the age and species of trees—on average, it amounts to 7 Mg/(ha·year) [80]. As a result of the grubbing up of orchards, it is possible to obtain approx. 60 Mg/ha in the case of modern, low-growing plantations (15 years) and 80 Mg/ha of biomass in the case of older plantations (30 years) [80]; annually, this amounts to approximately 3.5 Mg/(ha·year). In order to calculate the available energy, the following assumptions were made:

- 30% of wood harvested in orchards can be used for energy purposes [55];
- grubbing up produces 3.5 Mg/(ha·year) of wood [80];

- 7 Mg/(ha·year) of wood is produced as a result of maintenance work [80];
- the calorific value of wood of fruit trees is, on average, 11.5 GJ/Mg [55,80];
- the efficiency of obtaining energy is 80% (electricity from solid biomass will be obtained with an efficiency of 30%, and heat with an efficiency of 50%).

Formula (4) shows the amount of energy that can be obtained from waste orchard wood:

$$E_o = 0.8 \cdot 0.3 \cdot (K_g + K_c) \cdot P_o \cdot W_o, \quad (4)$$

where: E_o—energy from orchard wood waste (PJ/year), K_g—amount of wood harvested annually as a result of grubbing up one hectare of orchard (3.5 Mg/(ha·year)), K_c—the amount of wood harvested annually as a result of cultivation works per hectare of the orchard (7 Mg/(ha·year), P_o—orchard area (2.0 thousand ha), W_o—calorific value of biomass from orchards (11.5 GJ/Mg).

The obtained amount of energy that can be produced every year from waste wood from orchards in the Pomerania Voivodeship is 0.058 PJ/year (6 GWh/year of electricity and 0.04 PJ/year of heat).

3.2.3. The Amount of Energy from Surplus Straw

In recent years, the agriculture of the Voivodeship has been dominated by the cultivation of cereals: rye, wheat, barley, oats, triticale, and cereal mixtures. According to the data of the Central Statistical Office, the annual weight of cereal grain harvested in the Pomerania Voivodeship is 874 thousand Mg/year of wheat, 181 thousand Mg/year of rye, 126 thousand Mg/year of barley, 66 thousand Mg/year of oats, and 222 thousand Mg/year of triticale [81,82].

In order to calculate the available energy, the following assumptions were made:

- the grain/straw ratio is, respectively, 0.8 for wheat, 1.4 for rye, 0.9 for barley, 1.05 for oats, and 0.95 for triticale [4];
- 30% of produced straw is surplus that can be used for energy purposes [55];
- the calorific value of straw (with a moisture content of approximately 20%) is, on average, 15 GJ/Mg [83,84];
- the efficiency of obtaining energy is 80% (electricity from solid biomass will be obtained with an efficiency of 30%, and heat with an efficiency of 50%).

Formula (5) shows the amount of energy that can be obtained from straw:

$$E_s = 0.8 \cdot 0.3 \cdot (Z_w/w_w + Z_r/w_r + Z_b/w_b + Z_o/w_o + Z_t/w_t) \cdot W_s, \quad (5)$$

where: E_s—energy from straw (PJ/year), W_s—the calorific value of straw (15 GJ/Mg) [4], Z_w, Z_r, Z_b, Z_o, Z_t—annual crops of cereal grains (wheat, rye, barley, oats, triticale) (million Mg/year), w_w, w_r, w_b, w_o, w_t—grain/straw ratio, respectively, 0.8, 1.4, 0.9, 1.05, 0.95 [53].

The obtained amount of energy that can be produced every year from surplus straw in the Pomerania Voivodeship is 5.97 PJ/year (622 GWh/year of electricity and 3.73 PJ/year of heat).

3.2.4. The Amount of Energy from Hay from Unused Pastures and Meadows

The Pomerania Voivodeship has a large area of pastures and meadows; according to the data of the Central Statistical Office, this amounts to 22 thousand ha and 106 thousand ha, respectively [83]. Due to the reduction in the number of livestock (mainly cattle), as well as the change in the animal nutrition system, most of the meadows and pastures are not used. In order to calculate the available energy, the following assumptions were made:

- 15% of pastures and meadows can be used to grow hay for energy purposes [55];
- annual weight of biomass harvested per hectare of pastures is 3.6 Mg/(ha·year), and 4.9 Mg/(ha·year) of meadows [80];
- hay's calorific value is 14 GJ/Mg [85];

- the efficiency of obtaining energy is 80% (electricity from solid biomass will be obtained with an efficiency of 30%, and heat with an efficiency of 50%).

Formula (6) shows the amount of energy that can be obtained from hay from unused pastures and meadows (PJ/year):

$$E_h = 0.8 \cdot 0.15 \cdot (z_p \cdot P_p + z_m \cdot P_m) \cdot W_h, \quad (6)$$

where: E_h—energy from hay from pastures and meadows (PJ/year), z_p, z_m—weight of hay harvested annually per hectare from pastures and meadows (3.6 Mg/(ha·year) and 4.9 Mg/(ha·year), respectively), P_m, P_p—area of pastures and meadows (million ha), W_h—hay's calorific value (14 GJ/Mg).

The obtained amount of energy that can be produced every year from hay from unused meadows and pastures in the Pomerania Voivodeship is 1.0 PJ/year (104 GWh/year of electricity and 0.63 PJ/year of heat).

3.2.5. The Amount of Energy from the Salix Viminalis Cultivated on Wasteland, Fallow Land, and 5% of the Cultivated Area

The Pomerania Voivodeship has a large area of wasteland (post-industrial and degraded areas) and fallow areas; according to the data of the Central Statistical Office, this amounts to 41.2 thousand ha and 12.6 thousand ha, respectively [83]. This land can be used for the production of energy crops: grasses (e.g., *Miscantus*), cereals (straw and bioethanol), trees (e.g., *Populus* L.), or shrubs (e.g., *Salix viminalis*). In this study, it was assumed that this would be *Salix viminalis*; it is a native species [86]. *Salix viminalis* tolerates the agroclimatic conditions practically all over Poland very well [82]. It was additionally assumed that 5% of the agricultural land area, i.e., 34.9 thousand ha, would be used for the production of bioenergy.

In order to calculate the available energy, the following assumptions were made:

- 20% of wasteland, 50% of fallow land, and 5% of agricultural land can be used for growing basket willow for energy purposes [55];
- the annual yield of *Salix viminalis* per hectare is 8 Mg/(ha·year), both on fallow land and wasteland, and on arable land, it is 16 Mg/(ha·year) [80];
- the calorific value of *Salix viminalis* is 19 GJ/Mg [87];
- the efficiency of obtaining energy is 80% (electricity from solid biomass will be obtained with an efficiency of 30%, and heat with an efficiency of 50%).

Formula (7) shows the amount of energy that can be obtained from the *Salix viminalis* grown on unused fallow land and wasteland:

$$E_b = 0.8 \cdot [Q_f \cdot (0.2 \cdot P_w + 0.5 \cdot P_f) + Q_a \cdot 0.05 \cdot P_a] \cdot W_w, \quad (7)$$

where: E_b—energy from basket willow grown on wasteland, fallow land, and agricultural land (TJ/year), Q_f—annual yield of basket willow per hectare on fallow and barren fields (8 Mg/(ha·year)), Q_a—annual yield of *Salix viminalis* per hectare on arable land (16 Mg/(ha·year)), P_f, P_w, P_a—area of wasteland, fallow land, and arable land (thous. ha), W_w—calorific value of biomass of *Salix viminalis* (19 GJ/Mg).

The obtained amount of energy that can be produced every year from basket willow grown on fallow land, wasteland, and agricultural land in the Pomerania Voivodeship is 10.3 PJ (1073 GWh/year of electricity and 6.44 PJ/year of heat).

3.2.6. Biogas Production from Waste

The Amount of Energy from Biogas from Animal Slurry and Bird Manure

It was assumed that biogas in the Pomerania Voivodeship would be produced from liquid manure and bird droppings, as well as from municipal waste and sewage sludge. According to the data of the Central Statistical Office, the total number of cattle, pigs, and poultry was 219,000, 772,000, and 6,499,000, respectively [83]. In order to calculate the available energy, the following assumptions were made:

- coefficients for converting animals into "large livestock units" (LLU) (500 kg) are 0.8 for cattle, 0.2 for pigs, and 0.004 for poultry [55];
- the average weight of animal slurry or bird droppings produced by an LLU is 44.9 kg/day for cattle, 43.5 kg/day for pigs, and 26.8 kg/day for poultry;
- biogas yield from slurry is 0.050 m^3/kg, from pig slurry is 0.055 m^3/kg, and from bird manure is 0.140 m^3/kg [88];
- biogas from animal slurry or bird droppings contains 60% of methane, with a calorific value of 35.73 MJ/m^3 [88];
- the technical potential of utilization biogas is 20% of the theoretical potential;
- the efficiency of obtaining energy is 80% (electricity from biogas will be obtained with an efficiency of 35%, and heat with an efficiency of 45%).

Formula (8) shows the annual amount of energy that can be obtained from biogas obtained from animal slurry or bird manure:

$$E_{bg} = 0.8 \cdot 0.2 \cdot 0.6 \cdot (0.8 \cdot N_c \cdot I_c \cdot U_c + 0.2 \cdot N_p \cdot I_p \cdot U_p + 0.004 \cdot N_{po} \cdot I_{po} \cdot U_{po}) \cdot W_m, \quad (8)$$

where: E_{bg}—energy from biogas obtained from animal slurry or bird manure (TJ/year), N_c, N_p, N_{po}—number of cattle, pigs, poultry (million heads), I_c, I_p, I_{po}—annual weight of animal slurry or bird droppings from a large unit count of cattle (16.4 Mg/year), pigs (15.9 Mg/year), poultry (9.8 Mg/year), U_c, U_p, U_{po}—biogas yield from cattle slurry (50 m^3/Mg), from pig slurry (55 m^3/Mg), from bird manure (140 m^3/Mg) [4], W_m—methane's calorific value (35.73 MJ/m^3).

The obtained amount of energy that can be produced every year from the utilization of biogas from animal slurry or bird manure in the Pomerania Voivodeship is 1.08 PJ (112 GWh/year of electricity and 0.68 PJ/year of heat).

The Amount of Energy from "Landfill" Biogas

The Pomerania region generated an annual mass of municipal waste amounting to 869 thousand Mg/year (data of the Central Statistical Office, Households and Public Utility Facilities [84]). Unfortunately, only approximately 20% of the theoretical potential is represented by biogas from municipal waste.

The potential energy calculation was based on the following assumptions, assuming that the biogas yield from municipal waste is 100 m^3/Mg [55]:

- biogas from municipal waste contains 55% methane, with a calorific value of 35.73 MJ/m^3 [88];
- the technical potential of the utilization of biogas accounts for 20% of the theoretical potential;
- the efficiency of obtaining energy is 80% (electricity from biogas will be obtained with an efficiency of 35%, and heat with an efficiency of 45%).

Formula (9) shows the amount of energy that can be obtained from biogas obtained from the biodegradable fraction of municipal waste:

$$E_{bo} = 0.8 \cdot 0.2 \cdot 0.55 \cdot N_b \cdot U_b \cdot W_m, \quad (9)$$

where: E_{bo}—energy from biogas from the biodegradable fraction of municipal waste (TJ/year), N_b—mass of biodegradable fraction of municipal waste (million Mg/year), U_b—biogas yield from the biodegradable fraction of municipal waste (100 m^3/Mg) [53], W_m—methane calorific value (35.73 MJ/m^3).

The amount of energy that can be obtained annually from biogas from the biodegradable fraction of municipal waste in the Pomerania Voivodeship is 0.273 PJ/year (28 GWh/year of electricity and 0.154 PJ/year of heat).

The Amount of Energy from Biogas in Wastewater Treatment Plants

In the Pomerania Voivodeship, 44.8 million m^3 of municipal wastewater is treated annually [89]. It was assumed that 50% of the sewage flowing into the treatment plant would obtain sludge (constituting 1% of the sewage) and that 1 m^3 of sludge can be used to obtain 15 m^3 of biogas [88].

In order to calculate the available energy, the following assumptions were made:

- 50% of municipal wastewater will be used to obtain biogas [55];
- the volume of sewage sludge is 1% of the incoming municipal sewage [88];
- biogas yield from sewage sludge is 15 m^3/m^3 = 15 [88];
- biogas from sewage sludge contains 60% methane, with a calorific value of 35.73 MJ/m^3 [90];
- the efficiency of obtaining energy is 80% (electricity from biogas will be obtained with an efficiency of 35%, and heat with an efficiency of 45%).

Formula (10) shows the amount of energy that can be obtained from biogas obtained from sewage sludge:

$$E_{bs} = 0.8 \cdot 0.5 \cdot 0.01 \cdot 0.6 \cdot V_{bs} \cdot U_{bs} \cdot W_m, \quad (10)$$

where: E_{bs}—energy from utilization biogas from sewage sludge (TJ/year), V_{bs}—annual volume of municipal wastewater flowing into the treatment plant (million m^3/year), U_{bs}—biogas yield from sewage sludge (15 m^3/m^3) [88], W_m—methane calorific value (35.73 MJ/m^3).

In total, 3.4 million m^3 of biogas can be obtained in the Pomerania Voivodeship, i.e., approximately 0.13 PJ/year of energy (13 GWh/year of electricity and 0.07 PJ/year of heat).

3.3. Wind Energy in the Pomerania Voivodeship

As shown in the wind speed map at the height of 140 m (Figure 5), it can be stated that the Pomerania Voivodeship provides very favorable conditions for wind energy generation. The Voivodeship is especially well-suited for the development of wind energy—not only on land but also offshore. Offshore wind energy may become a flywheel for companies from the region, including shipyards, which already today produce components for the offshore industry [91].

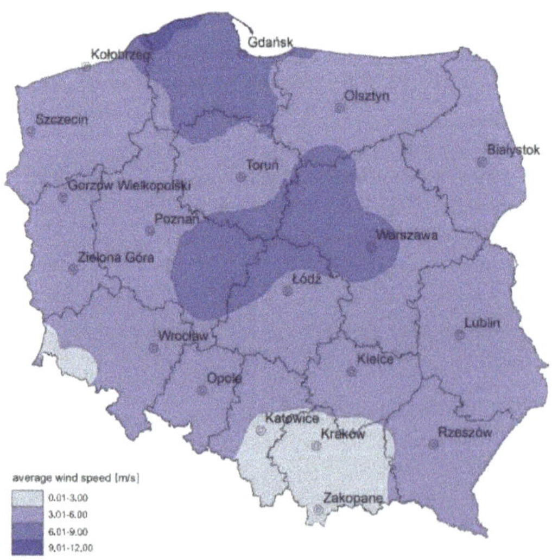

Figure 5. Wind speed analysis. Height of 140 m (own study from [92]).

The total capacity of wind farms in the Pomerania Voivodeship is 786 MW. In the draft program for the development of offshore wind energy and maritime industry, taking into account the available area of the Polish exclusive economic zone (2000 km^2 by 2030), wind conditions, productivity, and installed power density (6 MW/km^2), the theoretical potential was estimated at the level of 12 GW, with a generation potential of approximately 48–56 TWh. On the other hand, the technical potential of offshore energy was estimated to reach 7.4 GW by 2030 (Figure 6) [93].

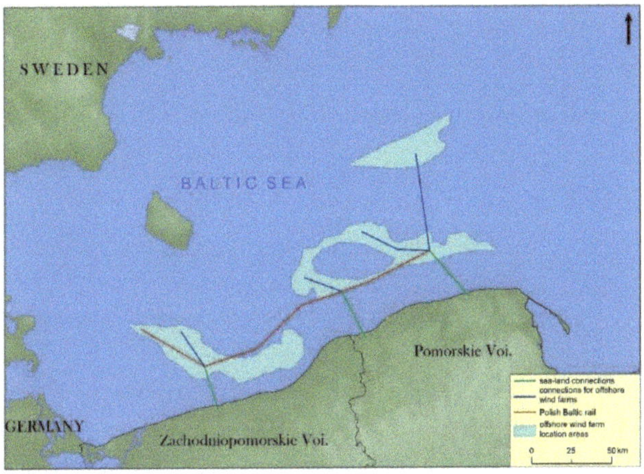

Figure 6. Potential locations of offshore wind farms in Poland (own study from [93]).

The Polska Grupa Energetyczna (Polish Energy Group) and Danish Ørsted have signed an investment agreement aimed at the development, construction, and operation of two offshore aeroenergy projects in the Baltic Sea, with a total capacity of approximately 2.5 GW. These are the Baltica-3 Wind Power Plant, with a capacity of over 1 GW, and the Baltica-2 Wind Power Plant, with a capacity of approximately 1.5 GW. Baltica-2 and Baltica-3 are eligible for participation in 2021 in the first phase of the operation of the offshore wind support scheme in Poland for wind farms with a total capacity of 5.9 GW. This system is a result of the new act to promote electricity generation in offshore wind farms, which was announced on 3 February 2021 in the Journal of Laws [94].

The technical potential of wind energy in the Pomerania Voivodeship was calculated. For this purpose, it was assumed that 140 m turbines will be erected, which means that the height of a turbine with a blade will be 215 m (h). The "Distance Act" [56] requires that the distance from residential buildings, areas of natural protection, and promotional forest complexes should be 10 h, i.e., 2150 m. The calculations performed indicate that the total area of the available areas is 60 km^2, i.e., only 0.3% of the area of the Pomerania Voivodeship (Figure 7). The amount of energy generated from aeroenergy in the Pomerania Voivodeship is 5.68 PJ, or 1.56 TWh. It is possible to make these assumptions early on, and this is the result of the "Distance Act" [56]. Its relaxation llows this system to produce many times more energy.

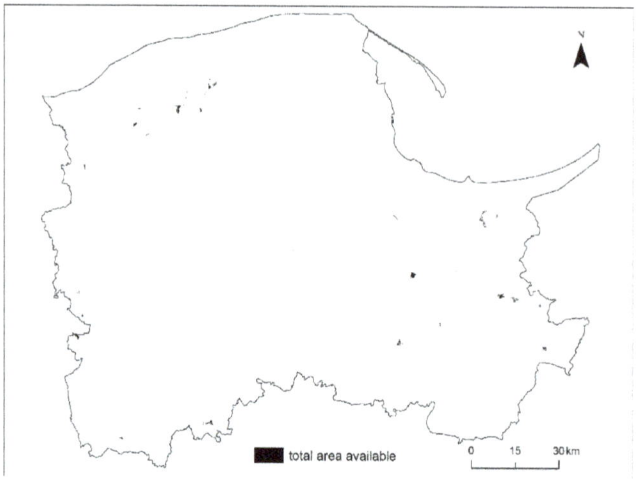

Figure 7. Area available for the development of aeroenergy in the Pomerania Voivodeship (own study).

3.4. The Potential of Hydropower in the Pomerania Voivodeship

The Pomerania region lies on the outskirts of two rivers—the Odra and the Vistula. As part of the Vistula River, which covers 86% of the province, it is possible to distinguish two hydrographic systems—Kashubian and Delta. According to Energy Regulatory Office data [78], there are 108 hydropower plants in the Pomerania Voivodeship, with a capacity of 34 MW. Considering that g = 9.81 m/s^2 and ρ = 1000 kg/m^3 and then = 0.85, we obtain:

$$P = 9.81 \cdot \theta \cdot H \cdot \eta \quad (11)$$

where:
—the capacity of a water turbine, or the volume of a stream of water flowing through the turbine within 1 s (m^3/s);
H—head (m);
—the efficiency of the water turbine, the gear, and the generator (Korolewski and Ligocki 2004).

Assuming efficiency at the level of 85%, we obtain the following model:

$$P = 8.34 \cdot H \cdot \theta \quad (12)$$

Assuming that the electronics will be able to operate for 6000 h/year at full power, i.e., 21,600,000 s/year, the annual amount of electricity from a given hydroelectric power plant E_{el}, for the output year (after conversion Wh to MWh), can be calculated as follows [4]:

$$E_k = 21.6 \cdot P \quad (13)$$

In the Pomerania Voivodeship, there are a number of facilities that can be used to produce electricity. These provide equally inexhaustible levels of water, serving in the past to fulfil energy targets, as well as for the identification of objects present in water during land reclamation. Based on the Equation (13) and data of the National Water Management Board on the levels of leveling and flooding (clearer, overflows) of water on the general furnaces, the theoretical power of electric and electric water, which can be calculated, is calculated.

In the Pomerania Voivodeship, the number of existing dams is:

- 154 dams on which a small hydropower plant (SHP) with a capacity of less than 5 kW can be placed;

- 95 dams on which a 5–10 kW SHP can be placed;
- 88 dams on which a 10–20 kW SHP can be placed;
- 67 dams on which a 20–50 kW SHP can be placed;
- 35 dams on which a 50–100 kW SHP can be placed;
- 48 dams on which one can place an SHP with a capacity of 100–500 kW;
- 4 dams on which an SHP with a capacity of over 500 kW can be placed (Figure 8).

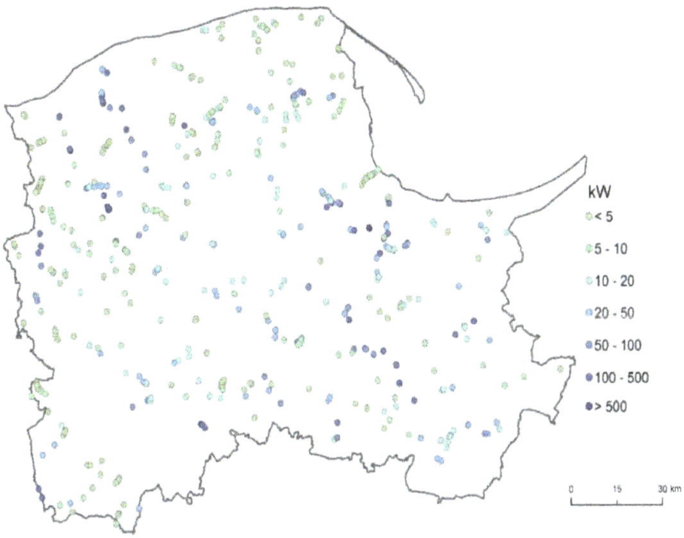

Figure 8. The location of the dams and the power of small hydroenergy plants that can be built in the Pomerania Voivodeship (own study from [95]).

The total theoretical electric power of all dams is 40.1 MW. This potential can, of course, be increased by building new water dams.

3.5. *Solar Energy in the Pomerania Voivodeship*

3.5.1. PV Potential on Roofs

The Pomerania Voivodeship has average/good conditions for the development of solar energy. At the end of 2020, there were already 89 photovoltaic (PV) farms in the region, with a total capacity of 59.2 MW [78]. There are also thousands of prosumer installations and thousands of solar collectors.

It was assumed that 10% of roofs in the Pomerania Voivodeship will be covered by PV. The value of electricity that can be obtained from the PV panels can be obtained from the following formula [55]:

$$E_r = 0.1 \cdot 0.15 \cdot P_d \cdot U \cdot N, \tag{14}$$

where: E_r—electricity that can be obtained annually from PV on roofs (J/year), P_d—roof area (62.3 km^2) (m^2), U—insolation (h/year), converted to (s/year), N—insolation (W/m2).

The received annual amount of energy is 0.231 TWh (833 TJ).

3.5.2. PV in Wasteland

The Pomerania Voivodeship has a large area of wasteland, with a total area of 41.2 thousand ha [80]. It was assumed that 20% of the wasteland in the Voivodeship will be subjected to reclamation in the "energy direction", where PV farms will be erected. The calculations took into account the insolation and solar irradiance of individual areas of Poland (data from the Institute of Meteorology and Water Management), and then the amount of electricity that can be obtained from PV panels with an efficiency of 15% was calculated, using the following formula [55]:

$$E_w = 0.20 \cdot 0.15 \cdot Pn \cdot I \cdot S, \quad (15)$$

where: E_w—electricity that can be obtained annually from PV in wasteland (J/year), P_w—wasteland area (m^2), I—insolation (h/year), converted to (s/year), N—solar irradiance (W/m^2).

The amount of electricity generated from PV will be 1.14 TWh (4120 TJ).

3.5.3. PV on Roads

The total length of public roads with hard surfaces in the Pomerania Voivodeship is 14,547 km [96]. It has been assumed that it is technically possible to place 10-m-wide photovoltaic panels on 10% of the road length. We calculated how much electricity can be obtained with PV panels with an efficiency of 15%, using the following formula:

$$E_r = 0.10 \cdot 0.15 \cdot D_r \cdot U \cdot N, \quad (16)$$

where: E_r—electricity that can be obtained annually from PV (J/year), D_a—length of roads with hard surface (m), U—insolation (h/year), converted to (s/year), N—insolation (W/m^2).

The calculated amount of obtainable electricity is 8112 TJ, i.e., 2254 TWh. The electricity produced can be used to power traffic lights, roadside bars, restaurants, etc. The estimated total amount of energy that can be obtained from PV in the Pomerania Voivodeship is 3625 TWh. Of course, this potential can be increased by making greater use of, for example, roofs or closed landfills.

3.6. Geothermal and Heat Pumps in the Pomerania Voivodeship

Figure 9 shows geological wells with a depth of at least 500 m in the Pomerania Voivodeship. Some of them could be used in the future to build heating plants and/or geothermal thermal springs [97]. In general, the Pomerania Voivodeship does not have sufficient conditions for the development of "deep" geothermal energy. However, heat pumps can be used on a large scale in this region.

Figure 9. Geological drilling with a depth of at least 500 m in the Pomerania Voivodeship (own study from [97]).

It was assumed that heat pumps will be installed in schools, kindergartens, accommodation, and cultural facilities. It was assumed that every fifth facility with a building area of 1500 m^2 will be heated on average, and the heat pump's heating power will be 70 W/m^2. The power of the heat pump P is the product of the area and the unit power of 70 W/m^2 [55]. Assuming an operating time of 2000 h a year and a heat pump's coefficient of performance (COP) = 3.8, the annual amount of heat produced can be calculated as follows:

$$E = P \cdot COP \cdot t, \tag{17}$$

where: E—heat production (MJ/year), P—power of heat pumps (MW), COP—coefficient of performance, t—annual operation time of the heat pump.

The calculated amount of energy for heat pumps in schools in the Voivodeship is 1708 PJ.

In 2020, there were 1555 tourist accommodation establishments in the Pomerania Voivodeship (as a result of the COVID-19 pandemic, this figure has decreased by 103 less compared to 2019), including hotels, motels, and guest houses [92]. It was assumed that 20% of such facilities will have a heat pump, the average area of a single facility is 1000 m^2, and the heat pump's heating power is also 70 W/m^2 [53]. Assuming the operation of heat pumps for 2000 h a year, on the basis of formula (16), it was calculated that the annual heat production in the Voivodeship will be 0.614 PJ.

In 2019, a total of 330 cultural facilities operated in the voivodship [98]. It was assumed that 20% of such facilities will have a heat pump, the average area of a single facility is 500 m^2, and the heat pump's heating power is also 70 W/m^2 [53]. Assuming the operation of heat pumps for 1500 h a year, based on formula 17 it was calculated that the annual heat production in cultural facilities would amount to 0.052 PJ.

4. Discussion

The political environment of RE science and research can be classified as relatively neutral (3.29 points) in the Pomerania Voivodeship. The main opportunity for the development of RE is represented, above all, by Poland's participation in the structures of the EU and active participation in the European Research Area. The main threats to the sphere of science and research in RE include extensive and complicated laws and low effectiveness in the practical implementation of the assumptions of the innovation policy.

The economic environment can be considered relatively conducive to the development of RE (3.56 points). One opportunity for development is, first of all, membership in the EU, which allows for financing research and investments in RE. Additionally, another opportunity is the relative stability of the economic situation in the Pomerania Voivodeship, as well as the innovativeness and activity of investors. Risks include high investment costs and fluctuating green certificate prices.

The social factor can be classified as positive to the RE sphere (3.50 points). The main opportunity is, above all, membership in the EU and the use of integration processes to educate society. Transnational partnership and investments in human capital are being developed. The main threats include the maintenance of the low reputation of science in society, the aging of society, and the demographic decline.

The technological factor can be considered neutral (3.43 points). The main opportunity for development is primarily the high efficiency of RE installations, as well as the transfer of technologies and techniques. The greatest threats are the poor condition of the power grid in Poland, little cooperation between science and the economy, and too few domestic producers of RE installations.

In summary, the PEST analysis shows that there are opportunities for the further development of RE in the Pomerania Voivodeship. Friendly legal regulations, greater financial subsidies, public education, and the development of domestic producers of RE installations may contribute to a significant reduction in GHG emissions and the greater energy independence of Poland.

The idea of the balanced development of the Pomerania Voivodeship is based on the rational production of energy from natural resources and renewable sources. Equilibrium (eco-development) has made it possible to meet the economic, social, and environmental needs of the next generation without compromising on the ability to meet the needs of future generations. The Pomerania region already ranks third in Poland in terms of the generation of electricity from renewable sources. In quantified terms, the use of RE in the Pomerania Voivodeship appears satisfactory, although the production of "green energy" still shows room for development [99]. Bioenergy, PV, and aeroenergy are characterized by the greatest potential regarding RE in the Pomerania Voivodeship. The potential of wind energy is significant in the case of the "Distance Act" agreement. The construction and operation of offshore wind farms will be a great opportunity for the economic development of the Pomerania Voivodeship. Then, the Voivodeship will become the main exporter of energy in Poland.

The Pomerania Voivodeship can become energy-independent in the future. The potential of RE electricity is greater than needed, and it is necessary to increase the amount of biomass for heating needs, and to install more solar collectors and heat pumps, as well improving the energy efficiency. RE may become an important factor for the economic development of the Pomerania Voivodeship. Today, there is a great demand in Poland for heat pumps and solar panels. Their production and assembly should be carried out by Polish companies. The development of RE in the Voivodeship is also an opportunity for cheaper electricity and heat. Each new RE installation/power plant brings huge financial benefits for the Voivodeship. Various types of taxes are transferred to the Voivodeship's coffers. A new RE plant means new jobs, resulting in a decline in unemployment in the voivodship.

Table 6 shows the amount of energy that can be obtained from RE in the Pomerania region. The energy mix should consist of bioenergy, aeroenergy, and PV. It should be emphasized that the energy mix will allow the energy self-sufficiency of the Pomerania region. When investing in RE, one should also consider strategies to increase energy efficiency.

Table 6. Annual potential electricity and heat from renewable sources in the Pomerania Voivodeship (author's own study).

RE Sector	Electricity (TWh/Year)	Heat (PJ/Year)
Bioenergy	3.03	16.17
Aeroenergy	1.56	
Hydroenergy	0.15	
PV	3.63	
Heat pumps		2.37
Summary	8.37	18.56

In the Pomerania region, in 2018, 35 PJ of heat and 7.5 TWh of electricity were consumed. According to the authors' calculations, the obtainable amount of energy from RE would be 8.37 TWh, in which the total amount of heat 18.56 PJ (Table 6). The calculated amount of electricity would cover the entirely of the Voivodeship's electricity needs, and 53% of its heat requirement [100]. Heat consumption should be reduced through thermal modernization measures. After thermal modernization, a building can consume up to 60–70% less energy than before modernization. Thus, passive, zero-energy, or even plus-energy buildings are becoming increasingly popular.

5. Summary

The focus of this article concerned the problem of the energy transformation of the EU countries, which is understood as the transition to an economy that produces and consumes largely energy from renewable sources. It was emphasized that the strategic goal of the EU's energy policy is the transition to a "green economy". According to the EU strategy, the activities carried out by 2050 are expected to lead to the improvement of the energy sector towards neutrality to the climate and the environment. Transforming the economies of the EU Member States into environmentally neutral economies requires huge effort in changing and adapting many governmental/institutional policies, business strategies, and consumer behavior. It is a great challenge for the EU to set up coordinated strategies at national and local levels to reduce energy consumption and waste to zero by 2050 and become the first climate-neutral continent. Even the short-term goals (2030) proposed in the EU strategy must be based on coordination within all four aspects: institutional, economic, environmental, and physical aspects. We performed our study at the regional level in terms of RE and energy transformation in several steps.

The PEST analysis to study the macro-environment of the Pomerania Voivodeship was performed first. The authors found that the policy in the field of RE at the Voivodeship level is not very conducive to the renewable energy sector. The RE Act [55] is quite often amended and this discourages potential investors; however, the Polish energy policy until 2040 offers an opportunity for the development of RE in the Pomerania Voivodeship. Despite the unfavorable institutional and demographics in the Pomerania Voivodeship, we evaluated the economic environment of RE in the Pomerania Voivodeship in the context of its development potential quite positively (3.56 out of 5). The PEST analysis showed that RE sector has a chance for further development in the Pomerania Voivodeship.

Next, the authors evaluated the physical potential of RE in the region. On the basis of the conducted analyses, it was found that the Pomerania Voivodeship offers very favorable conditions for wind energy and solar energy but does not have these conditions for "deep" geothermal energy (however, heat pumps can be used on a large scale here). We can conclude that the Pomerania Voivodeship displays all the necessary physical conditions for the production of energy from renewable sources.

Our goal was to examine the situation of the RE sector in the Pomerania region in the context of energy transformation. The Pomerania Voivodeship has an appropriate level of RE potential and may become an energy self-sufficient region. The estimated amount of RE electricity may also cover the Voivodeship's needs. The implementation of the goal allowed for the verification of this hypothesis, which confirms that all the EU's Member States' and regional governments' efforts are reasonable and the SDGs are achievable.

In summary, we can state that the calculated amount of electricity would cover fully the Voivodeship's electricity needs, and 53% of its heat needs. The vision in both the short- (2030) and long-term (2050) acts is to reach the level of 100% RE-based heating and cooling in Europe, which is achievable. The authors provide evidence at the level of one region (Pomeranian Voivodeship) that both EU strategies may be realistic and achievable under certain conditions. Thanks to the results obtained in one region, we know that there is significant potential to increase the RE share for the whole EU's heating and cooling sector and, at the same time, a higher refurbishment rate is also possible. Efficient and technologically smarter buildings should be the basis of Europe's decarbonization, but more effort, however, should be devoted towards renovating the current building stock. Reaching climate targets without decarbonizing them seems impossible. We conclude that, under certain conditions, and only with the EU, country-, and region-level institutional support, it will be possible to decarbonize EU building stock by 2050.

The obtained results indicate the necessity to extend the research of the energy sector in Poland to the entire territory of the country; then, SWOT analysis, PEST analysis, and analysis of the RE potential in terms of electricity and heat production can be performed. The conclusions drawn from the research would allow us to answer the questions of whether Poland is ready for the energy transformation based on the RE sources, which barriers should be overcome, what to pay attention to, and the opportunities and threats related to this process. The limitations of the research carried out in the article should also be emphasized. They are related to the differentiation of each region in Europe in terms of the analyzed conditions. The PEST analysis carried out and the GIS methods used constitute a universal tool for assessing the potential of the energy sector in the given region. However, it should be remembered that the analysis carried out must be based on detailed knowledge of the above-mentioned aspects of the region, cooperation with an expert panel, and the authors' extensive scientific experience in the energy sector. It should also be emphasized that it is difficult to obtain data at such a low level of aggregation as the Voivodeship.

Author Contributions: Conceptualization, B.I., K.F. and M.B.P.; methodology, B.I., K.F. and M.B.P.; software, M.S. and M.B.P.; validation, B.I., K.F., M.B.P., U.K., M.S., A.Z. and K.N.; formal analysis, B.I., M.S. and M.B.P.; investigation, B.I., K.F., M.B.P., U.K., M.S., A.Z. and K.N.; resources, B.I., K.F., M.B.P., U.K., M.S., A.Z. and K.N.; data curation, M.B.P. and B.I.; writing—original draft preparation B.I., K.F. and M.B.P.; writing—review and editing, B.I., K.F., M.B.P., U.K., M.S., A.Z. and K.N.; visualization, B.I., K.F., M.B.P. and M.S.; supervision, B.I. and M.B.P.; project administration, B.I. and M.B.P. All authors have read and agreed to the published version of the manuscript.

Funding: This research received no external funding.

Institutional Review Board Statement: Not applicable.

Informed Consent Statement: Not applicable.

Data Availability Statement: Data available on request due to restrictions. The data presented in this study are available on request from the corresponding author.

Conflicts of Interest: The authors declare no conflict of interest.

References

1. Overland, I. Energy: The missing link in globalization. *Energ. Res. Soc. Sci.* **2016**, *14*, 122–130. [CrossRef]
2. Rees, W.E. Globalization, trade and migration: Undermining sustainability. *Ecol. Econ.* **2006**, *59*, 220–225. [CrossRef]
3. Młynarski, T. Unia Europejska w procesie transformacji energetycznej. *Krak. Studia Międzynarodowe* **2019**, *1*, 31–44.
4. Igliński, B.; Skrzatek, M.; Iwański, P.; Krukowski, K. *Energia Odnawialna w Województwie Warmińsko-Mazurskim*; Nicolaus Copernicus University: Toruń, Poland, 2020.
5. Lin, M.-X.; Liou, H.M.; Chou, K.T. National energy transition framework toward SDG7 with legal reforms and policy bundles: The case of Taiwan and its comparison with Japan. *Energies* **2020**, *13*, 1387. [CrossRef]
6. Pietrzak, M.B.; Igliński, B.; Kujawski, W.; Iwański, P. Energy transition in Poland—assessment of the renewable energy sector. *Energies* **2021**, *14*, 2046. [CrossRef]
7. Chovancová, J.; Tej, J. Decoupling economic growth from greenhouse gas emissions: The case of the energy sector in V4 countries. *Equilib. Q. J. Econ. Econ. Policy* **2020**, *15*, 235–251. [CrossRef]
8. Gajdos, A.; Arendt, L.; Balcerzak, A.P.; Pietrzak, M.B. Future trends of labour market polarisation in Poland. The Perspective of 2025. *Trans. Bus. Economy* **2020**, *19*, 114–135.
9. Bhattacharya, M.; Churchill, S.A.; Paramati, S.R. The dynamic impact of renewable energy and institutions on economic output and CO_2 emissions across regions. *Renew. Energ.* **2017**, *111*, 157–167. [CrossRef]
10. Saidi, H.; El Montasser, G.; Ajmi, A.N. The role of institutions in the renewable energy-Ggrowth Nnexus in the MENA region: A panel cointegration approach. *Environ. Model Assess* **2020**, *25*, 259–276. [CrossRef]
11. Schoenefeld, J.J.; Knodt, M. Softening the surface but hardening the core? Governing renewable energy in the EU. *West Europ. Pol.* **2021**, *44*, 49–71. [CrossRef]
12. Gawlik, L. The Polish power industry in energy transformation process. *Miner. Econ.* **2018**, *31*, 229–237. [CrossRef]
13. Zielenkiewicz, M. Institutional environment in the context of development of sustainable society in the European Union countries. Equilibrium. *Quart. J. Econ. Econ. Pol.* **2014**, *9*, 21–37.
14. Grosse, T.G. Low carbon economy policy in Poland: An example of the impact of europeanisation. *Equilibrium. Q. J. Econ. Econ. Policy* **2011**, *6*, 9–39. [CrossRef]
15. Cheba, K.; Szopik-Depczyńska, K. Multidimensional comparative analysis of the competitive capacity of the European Union countries and geographical regions. *Oecon. Copernic.* **2017**, *8*, 487–504. [CrossRef]
16. Szopik-Depczyńska, K.; Kędzierska-Szczepaniak, A.; Szczepaniak, K.; Cheba, K.; Gajda, W.; Ioppolo, G. Innovation in sustainable development: An investigation of the EU context using 2030 agenda indicators. *Land Use Pol.* **2018**, *79*, 251–262. [CrossRef]
17. Szopik-Depczyńska, K.; Cheba, K.; Bąk, I.; Stajniak, M.; Simboli, A.; Ioppolo, G. The study of relationship in a hierarchical structure of EU sustainable development indicators. *Ecol. Indic.* **2018**, *90*, 120–131. [CrossRef]
18. Kijek, T.; Matras-Bolibok, A. The relationship between TFP and innovation performance: Evidence from EU regions. *Equilibrium. Q. J. Econ. Econ. Policy* **2019**, *14*, 695–709. [CrossRef]
19. Kijek, A.; Matras-Bolibok, A. Technological convergence across European regions. *Equilibrium. Q. J. Econ. Econ. Policy* **2020**, *15*, 295–313. [CrossRef]
20. Ginevičius, R. Multi-criteria assessment of socioeconomic systems' conditions based on hierarchically structured indicator systems. *Econom. Soc.* **2020**, *13*, 256–266. [CrossRef] [PubMed]

21. Androniceanu, A.M.; Kinnunen, J.; Georgescu, I.; Androniceanu, A. A multidimensional approach to competitiveness, innovation and well-being in the EU using canonical correlation analysis. *J. Compet.* **2020**, *12*, 5–21. [CrossRef]
22. Pao, H.-T.; Fu, H.C. Renewable energy, non-renewable energy and economic growth in Brazil. *Renew. Sustain. Energ. Rev.* **2013**, *25*, 381–392. [CrossRef]
23. Gross, R.; Leach, M.; Bauen, A. Progress in renewable energy. *Environ. Int.* **2003**, *29*, 105–122. [CrossRef]
24. Marinaș, M.-C.; Dinu, M.; Socol, A.-G.; Socol, C. Renewable energy consumption and economic growth. Causality relationship in Central and Eastern European countries. *PLoS ONE* **2018**, *13*, e0202951. [CrossRef]
25. Belke, A.; Dreger, C.; Dobnik, F. Energy consumption and economic growth-new insights into the cointegration relationship. *Energ. Econom.* **2010**, *33*, 782–789. [CrossRef]
26. Zinecker, M.; Skalická, M.; Balcerzak, A.P.; Pietrzak, M.B. Business angels in the Czech Republic: Characteristics and a classification with policy implications. *Econ. Res.-Kenosha Istraživanja* **2021**, 1–26. [CrossRef]
27. Zinecker, M.; Skalická, M.; Balcerzak, A.P.; Pietrzak, M.B. Identifying the impact of external environment on business angel activity. *Econ. Res.-Ekonomska Istraživanja* **2021**, 1–23. [CrossRef]
28. Caruso, G.; Colantonio, E.; Gattone, S.A. Relationships between renewable energy consumption, social factors, and health: A panel vector auto regression analysis of a cluster of 12 EU countries. *Sustainability* **2020**, *12*, 2915. [CrossRef]
29. Marra, A.; Colantonio, E. The path to renewable energy consumption in the European Union through drivers and barriers: A panel vector autoregressive approach. *Socio-Econ. Plan. Sci.* **2021**, *76*, 100958. [CrossRef]
30. Piekut, M. The consumption of renewable energy sources (RES) by the European Union households between 2004 and 2019. *Energies* **2021**, *14*, 5560. [CrossRef]
31. Zinecker, M.; Doubravský, K.; Balcerzak, A.P.; Pietrzak, M.B.; Dohnal, M. The Covid-19 disease and policy response to mitigate the economic impact in the EU: An exploratory study based on qualitative trend analysis. *Technol. Econ. Dev. Econ.* **2021**, *27*, 742–762. [CrossRef]
32. Rollnik-Sadowska, E.; Dąbrowska, E. Cluster analysis of effectiveness of labour market policy in the European Union. *Oeconomia Copernic.* **2018**, *9*, 143–158. [CrossRef]
33. Chocholatá, M.; Furková, A. The analysis of employment rates in the context of spatial connectivity of the EU regions. *Equilibrium. Q. J. Econ. Econ. Policy* **2018**, *13*, 181–213. [CrossRef]
34. Lewandowska-Gwarda, K. Female unemployment and its determinants in Poland in 2016 from the spatial perspective. *Oeconomia Copernicana* **2018**, *9*, 183–204. [CrossRef]
35. Fragkos, P.; Paroussos, L. Employment creation in EU related to renewables expansion. *Appl. Energy* **2018**, *230*, 935–945. [CrossRef]
36. Duscha, V.; Fougeyrollas, A.; Nathani, C.; Pfaff, M.; Ragwitz, M.; Resch, G.; Schade, W.; Breitschopf, B.; Walz, R. Renewable energy deployment in Europe up to 2030 and the aim of a triple dividend. *Energy Policy* **2016**, *95*, 314–323. [CrossRef]
37. Comodi, G.; Cioccolanti, L.; Mahkamov, K.; Penlington, R.; Lapuerta, M.; Hernandez, J.J.; Lora, E.E.S.; Venturini, O.; Cobas, V.R.M.; Palacio, J.C.E.; et al. Analysis of labour market needs for engineers with enhanced knowledge in renewable energy in some European and Latin-American Countries. *Energy Procedia* **2019**, *158*, 1135–1140. [CrossRef]
38. Pietrzak, M.B.; Balcerzak, A.P.; Gajdos, A.; Arendt, Ł. Entrepreneurial environment at regional level: The case of Polish path towards sustainable socio-economic development. *Entrep. Sustain. Issues* **2017**, *5*, 190–203. [CrossRef]
39. Markandya, A.; Arto, I.; Eguino, M.G.; Román, M.V. Towards a green energy economy? Tracking the employment effects of low-carbon technologies in the European Union. *Appl. Energy* **2016**, *179*, 1342–1350. [CrossRef]
40. World Bank Group. *Poland Energy Transition: The Path to Sustainability in the Electricity and Heating Sector*; World Bank: Washington, DC, USA, 2018.
41. Meluzín, T.; Balcerzak, A.P.; Pietrzak, M.B.; Zinecker, M.; Doubravský, K. The impact of rumours related to political and macroeconomic uncertainty on IPO success. Evidence from a qualitative model. *Transform. Bus. Econ.* **2018**, *2017*, 148–169.
42. Mamat, R.; Sani, M.S.; Khoerunnisa, F.; Kadarohman, A. Target and demand for renewable energy across 10 ASEAN countries by 2040. *Electr. J.* **2019**, *32*, 106670. [CrossRef]
43. Gielen, D.; Boshell, F.; Saygin, D.; Bazilian, M.D.; Wagner, N.; Gorini, R. The role of renewable energy in the global energy transformation. *Energy Strategy Rev.* **2019**, *24*, 38–50. [CrossRef]
44. Ma, J.; Oppong, A.; Acheampong, K.N.; Abruquah, L.A. Forecasting renewable energy consumption under zero assumptions. *Sustainability* **2018**, *10*, 576. [CrossRef]
45. Farrell, M. EU policy towards other regions: Policy learning in the external promotion of regional integration. *J. Eur. Public Policy* **2009**, *16*, 1165–1184. [CrossRef]
46. Scorza, F.; Santopietro, L. A systemic perspective for the sustainable energy and climate action plan (SECAP). *Eur. Plan. Stud.* **2021**. [CrossRef]
47. Musiał, W.; Zioło, M.; Luty, L.; Musiał, K. Energy policy of European Union member states in the context of renewable energy sources development. *Energies* **2021**, *14*, 2864. [CrossRef]
48. Aslani, A.; Naaranoja, M.; Helo, P.; Antila, E.; Hiltunen, E. Energy diversification in Finland: Achievements and potential of renewable energy development. *Int. J. Sustain. Energy* **2013**, *32*, 504–514. [CrossRef]
49. Seriño, M.N.V. Diversification of nonhydro renewable energy sources in developing countries. *Energy Ecol. Environ.* **2018**, *3*, 317–329. [CrossRef]

50. Xianguo, L. Diversification and localization of energy systems for sustainable development and energy security. *Energy Policy* **2005**, *33*, 2237–2243. [CrossRef]
51. Energy Market Agency. Available online: www.are.waw.pl (accessed on 10 November 2021).
52. Pomerania Voivodeship Data. Available online: https://gdansk.stat.gov.pl/dane-o-wojewodztwo/wojewodztwo-1057/warunki-naturalne-i-ochrona-srodowiska (accessed on 10 November 2021).
53. Igliński, B.; Iglińska, A.; Cichosz, M.; Kujawski, W.; Buczkowski, R. Renewable energy production in the Łódzkie Voivodeship. The PEST analysis of the RES in the voivodeship and in Poland. *Renew. Sustain. Energy Rev.* **2016**, *58*, 737–750. [CrossRef]
54. Polish Constitution. NS. Of Laws of 1997, No. 78, item 483. Available online: http://isap.sejm.gov.pl/isap.nsf/download.xsp/WDU19970780483/U/D19970483Lj.pdf (accessed on 15 October 2021).
55. Igliński, B. *Badanie Sektora Energii Odnawialnej w Polsce: Potencjał Techniczny, Badania Ankietowe, Analiza SWOT, Analiza PEST*; Nicolaus Copernicus University in Torun: Toruń, Poland, 2019.
56. Distance Act. NS. Of Laws of 2016, item 961. Available online: https://isap.sejm.gov.pl/isap.nsf/download.xsp/WDU20160000961/U/D20160961Lj.pdf (accessed on 16 October 2021).
57. RE Act. NS. Of Laws of 2015, item 478. Available online: https://isap.sejm.gov.pl/isap.nsf/download.xsp/WDU20150000478/U/D20150478Lj.pdf (accessed on 16 October 2021).
58. Polish Energy Strategy. Available online: www.gov.pl/web/klimat/polityka-energetyczna-polski (accessed on 29 September 2021).
59. My Electricity Program. Available online: https://mojprad.gov.pl (accessed on 24 September 2021).
60. Environmental Law. NS. Of Laws 2001 No. 62, item 627. Available online: https://isap.sejm.gov.pl/isap.nsf/download.xsp/WDU20010620627/U/D20010627Lj.pdf (accessed on 26 September 2021).
61. Central Statistical Office. Quarterly Accounts of Gross Domestic Product in 2016–2020; Warszawa, Poland. 2021. Available online: https://stat.gov.pl/obszary-tematyczne/rachunki-narodowe/kwartalne-rachunki-narodowe (accessed on 10 October 2021).
62. Raw Materials Quotations, Crude Oil. Available online: www.bankier.pl/inwestowanie/profile/quote.html?symbol=ROPA (accessed on 12 October 2021).
63. Coal Price. Available online: www.wnp.pl/gornictwo/notowania/ceny_wegla (accessed on 12 October 2021).
64. Natural Gas Price. Available online: www.bankier.pl/inwestowanie/profile/quote.html?symbol=GAZ-ZIEMNY (accessed on 12 October 2021).
65. Central Statistical Office. Registered Unemployment in the First Quarter of 2021; Warszawa, Poland. 2021. Available online: https://stat.gov.pl/obszary-tematyczne/rynek-pracy/bezrobocie-rejestrowane (accessed on 13 October 2021).
66. Central Statistical Office. Price Indices of Consumer Goods and Services in April 2021, Warszawa, Poland. 2021. Available online: https://stat.gov.pl/obszary-tematyczne/ceny-handel/wskazniki-cen/wskazniki-cen-towarow-i-uslug-konsumpcyjnych-w-kwietniu-2021-roku,2,114.html (accessed on 14 October 2021).
67. Institute for Renewable Energy. Photovoltaic Market in Poland, Warszawa, Poland. 2021. Available online: https://ieo.pl/pl/aktualnosci/1538-rynek-fotowoltaiki-w-polsce-2021 (accessed on 14 October 2021).
68. The Energy Law Act of April 10, 1997, Journal of Laws 1997, No. 2017, Item 220. Available online: https://isap.sejm.gov.pl/isap.nsf/download.xsp/WDU19970540348/U/D19970348Lj.pdf (accessed on 22 September 2021).
69. Auction System. Available online: www.ure.gov.pl/pl/urzad/informacje-ogolne/aktualnosci/8739,System-aukcyjny-dla-odnawialnych-zrodel-energii-ma-5-lat.html (accessed on 22 September 2021).
70. National Bank of Poland. Basic Interest Rates. Available online: www.nbp.pl/home.aspx?f=/dzienne/stopy.htm (accessed on 22 September 2021).
71. Central Statistical Office. Demographic Yearbook; Warszawa, Poland. 2020. Available online: https://stat.gov.pl/obszary-tematyczne/roczniki-statystyczne/roczniki-statystyczne/rocznik-demograficzny-2021,3,15.html (accessed on 11 October 2021).
72. Coronavirus Information. Available online: https://koronawirus.abczdrowie.pl (accessed on 12 October 2021).
73. Igliński, B.; Cichosz, M.; Skrzatek, M.; Buczkowski, R. Potencjał techniczny odpadowej biomasy stałej na cele energetyczne w Polsce. *Inżynieria Ochr. Sr.* **2019**, *2*, 109–118.
74. Available online: https://podr.pl (accessed on 11 October 2021).
75. Pomorskie Biuro Planowania Regionalnego. *Założenia Przestrzenne Rozwoju Odnawialnych Źródeł Energii w Województwie Pomorskim*; Pomeranian Regional Planning Office: Słupsk, Poland, 2015.
76. Available online: www.mpec.lebork.pl/pl (accessed on 14 September 2021).
77. Agricultural Market Agency. Register of Biofuel and Biocomponent Producers. Available online: www.kowr.gov.pl/uploads/pliki/DI/Biokomponenty_i_biopaliwa/W/Rejestr_wytworcow_2021.03.26.pdf (accessed on 9 September 2021).
78. The Energy Regulatory Office. Renewable Energy Sources. Available online: www.ure.gov.pl/pl/sekcja/456,Odnawialne-Zrodla-Energii.html (accessed on 8 September 2021).
79. State Forest Holding. State Forests, Report on the State of Forests in Poland; Warszawa, Poland. 2020. Available online: https://www.lasy.gov.pl/pl/informacje/publikacje/informacje-statystyczne-i-raporty/raport-o-stanie-lasow/raport-o-lasach-2020.pdf/view (accessed on 14 September 2021).
80. Jasiulewicz, M. *Potencjał Biomasy w Polsce*; Wydawnictwo Politechniki Koszalińskiej: Koszalin, Poland, 2010.

81. Central Statistical Office. Statistical Yearbook of Voivodeships; Warszawa, Poland. 2020. Available online: https://stat.gov.pl/obszary-tematyczne/roczniki-statystyczne/roczniki-statystyczne/rocznik-statystyczny-wojewodztw-2020,4,15.html (accessed on 10 September 2021).
82. Gradziuk, P. Ekonomiczne i ekologiczne aspekty wykorzystania słomy na cele energetyczne w lokalnych systemach grzewczych. *Acta Agrophysica* **2006**, *8*, 591–601.
83. Central Statistical Office. Statistical Yearbook of Agriculture 2020; Warszawa, Poland. 2021. Available online: https://stat.gov.pl/obszary-tematyczne/roczniki-statystyczne/roczniki-statystyczne/rocznik-statystyczny-rolnictwa-2020,6,14.html (accessed on 10 September 2021).
84. Owczuk, M.; Kołodziejczyk, K. Ocena możliwości wykorzystania słomy i wytłoków z lnicznika siewnego jako alternatywnego surowca energetycznego. *Chemik* **2011**, *6*, 537–542.
85. Terlikowski, J. Biomasa z trwałych użytków zielonych jako źródło energii odnawialnej. *Probl. Inżynierii Rol.* **2021**, *1*, 43–49.
86. Sobczyk, W. Plonowanie wierzby wiciowej-w świetle badań. *Polityka Energetyczna* **2007**, *10*, 547–555.
87. Krzyżaniak, M.; Szczukowski, S.; Tworkowski, J.; Stolarski, M.J. Właściwości energetyczne i skład chemiczny biomasy wierzby pozyskiwanej z systemu Eko-Salix. *Zesz. Probl. Postępów Nauk. Rol.* **2015**, *581*, 21–28.
88. Igliński, B.; Iglińska, A.; Kujawski, W.; Buczkowski, R.; Cichosz, M. Bioenergy in Poland. *Renew. Sustain. Energy Rev.* **2011**, *6*, 2999–3007. [CrossRef]
89. Central Statistical Office. Environmental Protection 2020; Warszawa, Poland. 2020. Available online: https://stat.gov.pl/obszary-tematyczne/srodowisko-energia/srodowisko/ochrona-srodowiska-2020,1,21.html (accessed on 11 September 2021).
90. Haynes, W.M. (Ed.) *CRC Handbook of Chemistry and Physics*, 97th ed.; CRC Press: Boca Raton, FL, USA, 2016.
91. Polish Wind Energy Association. Onshore Wind Energy in Poland TPA Poland; 2021. Available online: http://psew.pl/wp-content/uploads/2021/05/Raport_Ladowa-energetyka-wiatrowa-w-Polsce_2021-05-11.pdf (accessed on 12 September 2021).
92. Meteorological Materials Obtained from the Institute of Meteorology and Water Management, Warsaw (Excel Files).
93. Offshore in Poland. Available online: www.mir.gov.pl/Gospodarka_Morska/Pozwolenia/Documents/Mapa_potencjalnych_miejsc_farmy_wiatrowe.pdf (accessed on 2 October 2021).
94. Polish Wind Association Data. Available online: http://psew.pl/pge-i-orsted-podpisaly-umowe-o-utworzeniu-joint-venture-w-zakresie-budowy-farm-wiatrowych-na-baltyku (accessed on 3 October 2021).
95. The State Water Holding Polish Waters. Available online: www.kzgw.gov.pl (accessed on 4 October 2021).
96. Central Statistical Office. Transport. Maritime Economy; Szczecin, Poland; 2020. Available online: https://stat.gov.pl/obszary-tematyczne/transport-i-lacznosc/transport/gospodarka-morska-w-polsce-w-2020-roku,7,18.html (accessed on 4 October 2021).
97. Polish Geological Institute Data. Available online: www.pgi.gov.pl/dane-geologiczne/geologiczne-bazy-danych.html (accessed on 18 September 2021).
98. Central Statistical Office in Gdańsk. Tourism in the Pomerania Voivodeship in 2020; Gdańsk, Poland. 2021. Available online: https://gdansk.stat.gov.pl/opracowania-biezace/opracowania-sygnalne/sport-turystyka/turystyka-w-wojewodztwie-pomorskim-w-2020-r-,5,14.html (accessed on 5 October 2021).
99. Sejmik of the Pomerania Voivodeship. Pomerania Voivodeship Development Strategy 2030; Gdańsk, Poland. 2021. Available online: https://pomorskie.eu/wp-content/uploads/2021/04/Za%C5%82%C4%85cznik-do-uchwa%C5%82y_SWP_376_XXXI_21_SRWP2030_120421.pdf (accessed on 6 October 2021).
100. Central Statistical Office. Fuel and Energy Consumption in 2019, Warszawa, Poland. 2020. Available online: https://stat.gov.pl/obszary-tematyczne/srodowisko-energia/energia/zuzycie-paliw-i-nosnikow-energii-w-2019-roku,6,14.html (accessed on 6 October 2021).

energies

MDPI

Article

How to Meet the Green Deal Objectives—Is It Possible to Obtain 100% RES at the Regional Level in the EU?

Bartłomiej Igliński [1], Michał Bernard Pietrzak [2,*], Urszula Kiełkowska [1], Mateusz Skrzatek [1], Artur Gajdos [3], Anas Zyadin [4] and Karthikeyan Natarajan [4]

1 Faculty of Chemistry, Nicolaus Copernicus University in Toruń, Gagarina 11, 87-100 Toruń, Poland; iglinski@chem.umk.pl (B.I.); ulak@chem.umk.pl (U.K.); teusz@doktorant.umk.pl (M.S.)
2 Faculty of Management and Economics, Gdańsk University of Technology, Narutowicza 11/12, 80-233 Gdańsk, Poland
3 Faculty of Economics and Sociology, University of Lodz, POW 3/5, 90-255 Łódź, Poland; artur.gajdos@uni.lodz.pl
4 Faculty of Science and Forestry, School of Forest Sciences, University of Eastern Finland, P.O. Box 111, 801014 Joensuu, Finland; anas.zyadin@uef.fi (A.Z.); kartheikeyan@uef.fi (K.N.)
* Correspondence: michal.pietrzak@zie.pg.gda.pl

Abstract: The subject matter discussed in the article concerns the problem of the energy transformation of the European Union (EU) countries. In the case of the EU, the energy transformation has specific characteristics due to formal legal and institutional provisions. This means that the member states are obliged to implement the adopted Community Energy Strategy, which was defined under the European Green Deal. According to the EU policy, all member states are to have climate-neutral and zero-gas-emission economies by 2050. The energy transformation is to be largely based on the diversification of the energy sources used, with a dominance of renewable energy sources (RES). The article presents a research problem, where the question was asked whether achieving climate-neutral energy independence based solely on RES is possible at the regional level. It seems that the positive answer is an important argument in the discussion about the possibility of all member states achieving the goals set under the European Green Deal. Additionally, stating the possibility of energy independence from RES for a selected region is an important argument to promote just and bottom-up initiatives in order to carry out the energy transformation more effectively. The assessment of the energy potential from renewable sources at the regional level was carried out on the example of a selected NUTS 2 region, the Greater Poland Voivodeship in Poland. The main objective of the study is to analyze the possibility of obtaining independence from RES by the selected Greater Poland Voivodeship. The implementation of the objective consists of determining the energy potential from RES in the Voivodeship under study on the basis of the methods of a geographic information system (GIS). GIS methods were selected due to the fact that they allow for the spatial positioning of point, line, and surface structures in relation to the potential of RES, thus ensuring high accuracy of the obtained estimates. The analysis carried out in the study shows that the technical potential of RES in the Greater Poland Voivodeship is higher than the current electricity and heat usage. This means that by focusing solely on RES in the region, the Greater Poland Voivodeship can fully meet the energy demand thanks to its green resources. It should be emphasized that the Greater Poland Voivodeship is one of the coal-dependent regions in Poland that has already prepared a structured plan of just transformation. A locally and bottom-up prepared strategy assumes the conversion of the region from a "Coal Energy Region" into a "Green Energy Valley" in which economic development will be strictly connected to RES energy independence.

Keywords: renewable energy; sustainable development; Green Deal objectives; sustainable development goals; GIS method; European Union; Greater Poland Voivodeship

Citation: Igliński, B.; Pietrzak, M.B.; Kiełkowska, U.; Skrzatek, M.; Gajdos, A.; Zyadin, A.; Natarajan, K. How to Meet the Green Deal Objectives—Is It Possible to Obtain 100% RES at the Regional Level in the EU?. *Energies* **2022**, *15*, 2296. https://doi.org/10.3390/en15062296

Academic Editors: Adrian Ilinca, K.T. Chau, Abdelali El Aroudi and Alessandro Mauro

Received: 25 October 2021
Accepted: 16 March 2022
Published: 21 March 2022

Publisher's Note: MDPI stays neutral with regard to jurisdictional claims in published maps and institutional affiliations.

Copyright: © 2022 by the authors. Licensee MDPI, Basel, Switzerland. This article is an open access article distributed under the terms and conditions of the Creative Commons Attribution (CC BY) license (https://creativecommons.org/licenses/by/4.0/).

1. Introduction

For nearly thirty years, systematic growth of globalization and internationalization processes have been observed in the world [1–3]. These processes, through penetration into most markets, have resulted in intensive economic growth in most of the world's economies and a related increase in the level of human development [4–9]. It should be emphasized that along with the increase in ties between economies, there was an unprecedented increase in investment, innovation, production, and significant institutional changes, both at the national and international level [10–16]. Significant changes in production have resulted in an increase in energy demand in all world economies [17,18]. The existing resources of energy from non-renewable sources are slowly becoming insufficient, and their use is becoming more and more expensive and has an increasingly negative impact on the natural environment. All this, combined with the change in consumer patterns and changes in the labor markets [19–27], has influenced the energy transformation currently being carried out in most countries [28–32]. The subject of the article is part of the problem of the energy transformation of the EU countries. It should be emphasized that the energy transformation in the community was strengthened institutionally, where all countries are obliged to follow the directions of the community climate policy. The short-term and long-term measures taken under the European Green Deal are to lead to a situation where all member states are to have climate-neutral economies with zero gas emissions [33,34].

One of the leading tools for an effective energy transformation is to diversify the energy sources used, with RES being the dominant ones [35]. This means that for most countries, the energy transition in EU member states refers to the shift from one dominant energy to energy based on RES [35–38]. The article poses a research problem in the form of whether achieving climate-neutral energy independence under the European Green Deal is possible at the regional level. The answer to this question seems to be very important. A positive answer will mean that all member states will be able to achieve all the objectives of the European Green Deal with a properly applied economic development strategy. The analysis of energy independence at the regional level was carried out on the example of a selected NUTS 2 region, the Greater Poland Voivodeship in Poland. The analysis was carried out in two aspects. The first aspect concerns the analysis of the current energy situation in the voivodeship in the context of the bottom-up initiative of energy transformation [39,40]. The second aspect concerns the assessment of the energy potential from RES.

Poland is a country where the majority of electric power and heat is generated with hard coal and brown coal (lignite) combustion. The limited amount of natural resources, increasing prices of fossil fuels, and environmental pollution are the major reasons for the use of RES. RES cannot be exhausted, are locally available for each country and, thus, do not lead to any economic or military conflict [41].

Currently, the greatest amount of renewable energy is generated by the following voivodeships: Kuyavian–Pomeranian [42], West Pomeranian [43] Pomeranian [44], Greater Poland [45], and Łódzkie [46]. The total RES capacity in Poland is about 9 GW (Figure 1). The Greater Poland Voivodeship, located in central Poland, has great potential to develop all types of RES [47].

Poland is obliged by the EC to add RES to the energy mix. In 2020, in Poland, the coal-based share in the energy mix went below 70% for the first time. The EC's Renewable Energy Directives show the main goals and acts that emphasized the necessity to turn to a more sustainable and environmental economy. The directives have established the specific RES shares in the energy mix as targets. The EC's Renewable Energy Directive from 2009 predefined three goals to be achieved by 2020—reduce greenhouse gas emissions by at least 20%; (2) increase the share of renewable energy to at least 20% of consumption; (3) achieve energy savings of 20% or more. Additionally, all EU countries should achieve a 10% share of renewable energy in their transport sector [48]. This EC Renewable Energy Directive was revised in 2018, establishing the new 32% renewable energy (share in energy mix) target by 2030 [49]. The proposed EC Renewable Energy Directive from 2021 established a 40% renewable energy target by 2030. The "Energy Roadmap" adopted by the EC in 2011 presents a strategic plan leading to the transition to a competitive low-carbon economy by 2050, with the goal of reducing CO_2 emissions by 96–99% compared to 1990 [50].

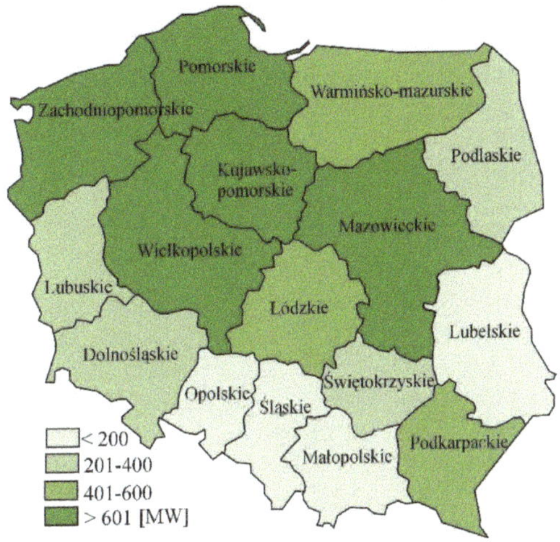

Figure 1. Capacity of RES in Poland (own study based on [47]).

The formal energy transformation in the EU was mainly a top-down initiative in the beginning. This means that the regions followed the formally established national and EC directions. The EC's "Strategy on Adaptation to Climate Change", introduced in 2013, includes the actions that must also be taken at the local, regional, and national levels in order to counteract the effects of climate change [51]. The most recent EU Strategy on Adaptation to Climate Change, established in 2021 empowered the local-level adaptation approach, focusing on individual citizens. The national response to the renewable energy directives were three documents established by the Polish government. Poland's Energy Policy until 2030, constituted in 2009, underlines the commitment to develop the use of RES [52]. Poland's National Energy and Climate Plan for years 2021–2030 [53], prepared as a fulfillment of the obligation set out in Regulation (EU) 2018/1999 of the European Parliament and of the Council of 11 December 2018 on the Governance of the Energy Union and Climate Action, established a 21–23% renewable energy target by 2030 (a 10-year plan). Prepared in 2021, Poland's Energy Policy until 2040 [54] predefines a 32% renewable energy target to be obtained by 2040, claiming that "The local level is the bedrock of adaptation".

In this paper, we underline the necessity of creating a well-organized and prepared bottom-up initiative of energy transformation, which starts within regions. Such an approach opens up more possibilities for energy transformation in the regions and automatically contributes to the Polish energy transformation. This does not mean that the RES regional policy is inconsistent with the Polish government policy and EU policy. It only underlines the necessity of performing RES transformation at the regional level more independently and equitably, taking into account the local context, such as the people's awareness of climate change, the fairness of the decision-making process, and trust in the local government authorities, which includes the overall evaluation of costs, risks, and benefits of RES in the region. This approach should be integrated with the national and EU RES objectives and primarily enhance the quality of life in the region.

Attention should be paid to the role of business angels and sustainable start-ups, which can be translated into the effective implementation of the next stages of the energy transition and development of the RES sector [55–57]. Obtaining financing for successful sustainable start-ups, which have established their market position, is also possible in the capital market, where an important issue is the choice of an appropriate time for the company to enter the capital market [58–61].

At the local level, energy justice plays a key role in shaping the local community's approach to energy transformation. Energy justice is defined in Poland as providing new development opportunities to regions and communities most affected by the negative effects of the transformation in connection to the low-carbon energy transition, providing new jobs and building new industries that participate in transforming the energy sector. The low-emission energy transformation envisaged in Poland's Energy Policy until 2040 plans and initiates modernization changes for the entire economy, guaranteeing energy security and ensuring a fair distribution of costs and protection of the most vulnerable social groups [62]. According to this document, the energy transformation will be

- Fair;
- participatory, local, bottom-up;
- focused on modernization and innovation;
- stimulating economic development, efficiency, and competitiveness.

According to Carley and Konisky [39], "the comprehensive energy justice framework can be said to include energy availability and access, affordability, due process, accountability and transparency, and both inter- and intra-generational equity". Such an approach points out that justice energy transformation comprises the fair distribution of benefits and burdens in the community, fair and transparent energy decision-making processes and procedures, understanding the historical context and related inequalities (and acts to reconcile them), and restorative justice treated as a tool to correct injustices [39]. The importance of consequentiality and trust in institutions on willingness-to-pay estimates towards the expansion of renewable energy is specifically important at the local level and has been analyzed and underlined in Germany by Oehlman and Meyerhoff [63]. Interpersonal and social trust in authorities has been proven to be related to community opinion on renewable energy in the United Kingdom. Energy justice and energy security are the base of an efficient and fair energy transformation in the region, as they shape the local community environment and opinions on renewable energies. This means that the community should have confirmation that the regional energy security and equity will be maintained. This includes maintaining a decent lifestyle and trust in the regional authorities and their decision-making processes.

A study performed by Rogers et. al. [64] examined the rural community perceptions of local renewable energy. The results stated that in the United Kingdom, "the community-based renewable energy projects, with high levels of public participation, are more likely to be accepted by the public than top-down development of large-scale schemes and may bring additional benefits such as increased engagement with sustainable energy issues."

The studies performed by Zoellner et al. [65], Upham, and Shackley [66] demonstrated the roles of justice and trust in local authorities in the shaping of the public acceptance of renewable energy developments in Germany and the United Kingdom (accordingly).

The EC European Green Deal Plan includes the European Green Deal Investment Plan (EGDIP), which specifies the Just Transition Mechanism. The mechanism comprises the tools ensuring a fair and just transition to a green economy on the regional level and supports citizens of the regions (that are most impacted by the transition). The Greater Poland Voivodeship is one of six coal regions in Poland, and the first one that has an elaborated primary version of the Territorial Just Transition Plan with a specifically defined concept of the Just Transformation of the Eastern Greater Poland Voivodeship. This concept reinforced a strong bottom-up content-related emphasis in preparation for the energy transformation plan in the region, taking into account the local social and economic environment. The transition plan and the concept of just transformation have been prepared and will be followed with the collaboration with and the mutual support of public and private entities, the science sector, and the local community. Both documents were considered by public consultation. The plan and concept for the Greater Poland Voivodeship is to maintain the energy characteristic of the region, become the national leader of the green transformation in Poland, and achieve climate neutrality as early as 2040. What is more, the Greater Poland Voivodeship aims to create a well-known brand for the region, called the "Greater Poland Voivodeship Energy Valley" (based on RES) in the document. To do so, the region has to be RES independent.

Fuel cells are devices that convert the chemical energy of fuel and oxidants into electricity. All types of fuel cells, unlike traditional methods, generate electricity without burning fuel and oxidants. This allows for avoiding the emissions of harmful compounds, including nitrogen oxides, sulfur oxides, hydrocarbons (which cause the formation of holes in the ozone), and carbon oxides. In modern fuel cells, the most frequently used fuel is hydrogen (H_2), while the oxidant is oxygen (O_2), supplied to the device in its pure form or together with atmospheric air. However, this does not mean that no other fuels are used in fuel cells. Currently, intensive research is underway on cells powered directly by methanol CH_3OH and carbon (in various forms). The fuel may also be CH_4 methane, HCOOH formic acid, N_2H_4 hydrazine, and NH_3 ammonia. Hydrogen is the fuel of choice for most cells due to its high reactivity in the presence of suitable catalysts, the possibility of producing it from hydrocarbons, and its high energy density when stored in a liquid form under high pressure at a low temperature. Unfortunately, although hydrogen is one of the most popular elements on Earth, it is mainly found in chemicals, primarily water. Hydrogen can be obtained from water by electrolysis, but unfortunately, a significant amount of energy must be invested in the process. Therefore, other sources of hydrogen have been sought and tested. In addition to obtaining hydrogen, an additional problem is its storage. The storage and transport of hydrogen require prior compression to a certain pressure or reduction to a liquid form [67–69].

The principle of the operation of fuel cells is the same as in galvanic cells, that is, batteries. However, unlike galvanic cells, fuel and oxidants are supplied externally in fuel cells. Thanks to this, they do not have such a limited time of use as traditional batteries. Fuel cells consist of two electrodes—an anode and cathode. The electrodes are separated by an electrolyte in liquid or solid form. Hydrogen (pure or as a component of air) is fed continuously to the anode, while oxygen (pure or as a component of air) is fed, also continuously, to the cathode. The ions must flow freely between the electrodes. In fuel cells, electrons reach the cathode, bypassing the electrolyte, through an external electrical circuit, making the cell a source of electromotive force. The said electrolyte is responsible for the transport of ions between the electrodes. For this reason, it must be a good ion conductor and at the same time, the weakest electron conductor. If the electrolyte did not meet even one of the requirements, the entire cell would not be able to function properly. The chemical reaction that takes place in the cell consists of breaking down the hydrogen into a proton and an electron at the anode, and then joining the reactants at the cathode. Electrochemical processes are accompanied by the flow of an electron from the anode to the cathode, bypassing the impermeable membrane. The electrochemical reaction of hydrogen and oxygen produces electricity, water, and heat [67–69].

In the future, fuel cells will be an integral part of the hydrogen fuel industry. Fuel cells are capable of supplying enough energy to meet the global energy needs. These technologies are very efficient and safe for the environment [67–69].

The main aim of the study was to analyze the possibilities of obtaining the RES independence by the Greater Poland Voivodeship. Achieving the goal consisted of determining the potential of energy for the selected NUTS 2 region, which was determined with the assumption of using only renewable sources. The claim of RES independence is an important factor to continue the process of introducing just and bottom-up coal region energy transformation opportunities.

In order to calculate the potential of renewable energy in the Greater Poland Voivodeship, the authors used GIS methods. Geographic information systems enable the spatial positioning of point, linear, and surface structures in relation to the potential of RES; these could respectively be geological boreholes, watercourses, or in the case of the surface, the total area of land excluded from a wind turbine location. As a result of including these structures into the coordinate system, it is possible to know the conditions of their positioning as well as of the surrounding area, such as the natural environment or technical infrastructure. In addition, the database and calculated parameters for each structure enable the measurement of the RES potential for the analyzed area.

The authors wanted to show that the selected region and, consequently, the entire country, in which energy is produced mainly from coal, has enough RES potential to cover 100% of its energy needs.

The authors used the geographical information system (GIS) method to calculate the technical potential of RES in the Greater Poland Voivodeship. For this purpose, meteorological data on wind speeds in the region were obtained. These data were extrapolated for a height of 100 m (this is the most common height of turbines in the voivodeship). We were the first to compare the technical potential of wind energy for the "10H" and "5H" distances (which will probably be introduced in Poland).

In the case of solar energy, both solar conditions and the availability of roofs were taken into account (roof installations do not take up space).

In the case of biomass, it was assumed that only waste biomass would be used for energy purposes. It was assumed that water power in the Voivodeship should develop in terms of the already existing dams, such as locks or weirs.

Geothermal energy in the Voivodeship should be based on the already existing boreholes (a map has been drawn up), primarily for heat production.

The conducted research is undoubtedly innovative in this sense. Implementation of the goal will allow for the assessment of the possibility of achieving the goals of the European Green Deal at the regional level, which can undoubtedly be translated into the possibility of achieving these goals both at the national level and at the EU level. The energy mix of renewable energy should be in the Greater Poland Voivodeship as well as in all of Poland; only in this way can 100% RES be achieved with the sustainable development of the economy.

2. Methodology—GIS in RES Research

GIS is increasingly used in RES sector research, including determining the potential of renewable energy. It allows you to calculate the potential of RES in a given region or country with great accuracy. Many papers have been published about using GIS for RES research over the last few years. The combination of a GIS and tools or multicriteria decision-making (MCDM) methods were used by Sánchez-Lozano et al. for performing an evaluation of the optimal placement of photovoltaic solar power plants in the area of Cartagena (Region of Murcia), Spain. An excellent analysis tool that allows for the creation of an extensive cartographic and alphanumeric database that will later be used by multi-criteria methodologies to solve problems simply and promote the use of multiple criteria is generated by the combination GIS-MCDM.

A suitable site selection for solar farms using GIS in the Karapinar region, Turkey was determined by Uyan [70]. The research showed that 15.38% (928.18 km^2) of the study area had low suitability, 14.38% (867.83 km^2) was moderately suitable, 15.98% (964.39 km^2) as suitable, and 13.92% (840.07 km^2) as the most suitable for solar farms.

Jahangiri et al. [71] showed that the eastern, central, and southwestern parts of Iran, the south of Oman, almost all parts of Iraq and Yemen, some northern and eastern parts of Egypt, the south of Jordan and Israel, as well as a small region in the southeast of Turkey are particularly ideal for setting up solar–wind power stations.

As far as Poland is concerned, there have not been many scientific studies so far in which the RES potential has been calculated using the GIS method. Sliz-Szkliniarz and Vogt [72] calculated the wind energy and biogas potential in the Kuyavian–Pomeranian Voivodeship. The application of a GIS-based approach showed that the Kuyavian–Pomeranian Voivodeship could be used for wind energy production to a great extent because the major technical potential remains untapped. By excluding the infrastructural and ecological-related barriers, the area of almost 7500 km^2 remains suitable for wind collection. The potential for biogas production could meet the demand of 442 GWh of heat, 368 GWh of electricity, or 98 Mm^3 of methane, based on the assumed biogas feedstock mix [73].

Rozakis et al. [74] determined the straw potential of Poland using the GIS method. In the article, the actual production of straw was modeled on a scale of local districts as well as the needs of its local use and the possibility of the redistribution of excessive quantities to regions with a deficit of straw based on statistical data from the Polish Central Statistical Office. The results showed that the straw surplus should be used in the energy sector along with its geographical distribution. The detailed results at the municipal level indicated an excess capacity for biomass co-firing by the plant and areas to be fulfilled by additional biomass sources, such as biomass from energy plantations and forests.

The current 100% RES solution in Portugal is in favor of wind and hydro energy [67]. Wind power should be introduced by the use of large reversible or pumped hydropower plants and could be achieved by installing bigger wind turbines and storage systems. It has become possible to combine these storage systems with a transport system. Hydrogen and batteries could become a storage solution for large future systems once the technology progresses [75].

In the paper by [76], a 100% RES system for Macedonia in the year 2050 was designed. The results of the analyses showed that a 100% renewable energy system in Macedonia is possible. However, to achieve this goal, a large share of biomass, solar power, and wind power, as well as different storage technologies are needed. The analysis showed that at the moment, half of the renewable energy system seems much more likely than a 100% renewable energy system. With the additional energy efficiency steps, which will lead to a decrease in consumption, and with the installation of new generation capacities, the achievement of this goal is possible.

For this article, the technical potential of wind energy, solar energy, biomass, hydropower, and geothermal energy in the Greater Poland Voivodeship was calculated. The authors wanted to know if the RES technical potential, counted with the use of the GIS method, will allow for 100% coverage of the energy needs of the Greater Poland Voivodship. More attention was devoted to wind energy, because the distance law, which inhibited the development of wind energy in Poland, was in force. In analyzing the aforementioned articles, one can draw the conclusion that in almost every country, there are good conditions for the development of one, and most often several, renewable energy sectors. In addition, in Poland, many regions can cover their energy needs only from RES.

3. The Description of the Greater Poland Voivodeship

The Greater Poland Voivodeship is located in the central-western part of Poland, with Poznań being its capital city (Figure 2). The surface area of the voivodeship region is 29,826 km^2 (second largest in the country after the Mazowieckie Voivodeship), which is equal to 9.53% of Poland's area. The number of inhabitants of the Greater Poland Voivodeship is 3.47 million people. In terms of land management, the area of the Greater Poland Voivodeship is dominated by arable lands, which make up 65.4% of the area, whilst forests make up 25.7% [45].

The climate of the Greater Poland Voivodeship belongs to the temperate climate zone, where oceanic and continental influences converge. Despite a highly developed river network and an abundance of lakes, the voivodeship has limited water resources. This results from a low amount of atmospheric precipitation (the annual precipitation is from 480 mm to 600 mm), as well as insufficient investments into anthropogenic retention. The hydrological situation is worsened by open-cast brown coal mining. Water shortages in the Greater Poland Voivodeship are estimated at about 350 million m^3 [45].

The dominant types of soil in the Greater Poland Voivodeship include podzolic and rusty soils, which make up 60% of the area, as well as lessive and brown soils at 20%. The remaining types are mainly wetland soil (pseudogley soils, gley-podzolic soils, half-bog peat soils, alluvial soils). The Greater Poland Voivodeship farmers boast the highest agricultural output, both market and global, in the country. The region comes second in terms of global and market plant production and first in terms of global and market animal production. The agriculture of the Greater Poland Voivodeship takes a dominant position when it comes to slaughter animal production, accounting for 22% of the whole country, of which pork production reaches nearly 26% whilst that of beef is 18.5%. The region also produces the highest amount of grain and sugar beet as well as a significant amount of rapeseed. In addition, the area covered by ground vegetable crops is higher than the domestic average (Figure 2) [45].

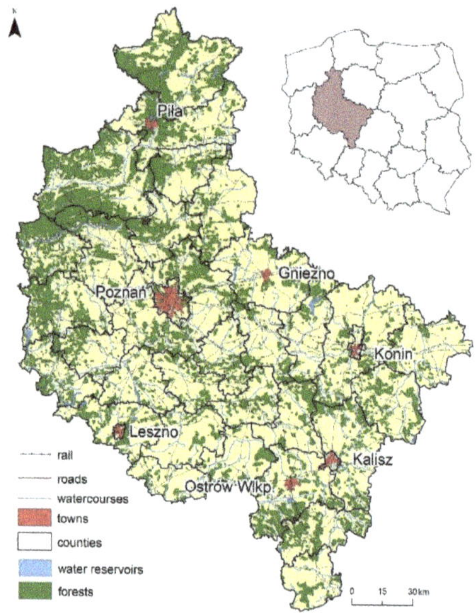

Figure 2. The physico-geographical map of the Greater Poland Voivodeship—watercourses and reservoirs (own study).

4. Wind Energy Potential in the Greater Poland Voivodeship

In order to calculate the wind speed in the Greater Poland Voivodeship, the so-called inverse distance interpolation of data provided by the Institute of Meteorology and Water Management in Warsaw was carried out using the mean monthly wind speed and the mean monthly wind directions (8 directions) from years 1990–2014. The obtained mean wind speed values at the basic height, that is, 10 m (v_p), were used to calculate mean wind speed values at the height of a rotor, that is, 100 m (v_1), according to the following formula:

$$v_1 = v_p(h/h_o)^k \tag{1}$$

where v_1 is the mean wind speed at a height of 100 m (m/s), v_p is the mean wind speed at a height of 10 m (m/s), h is the height of the rotor (in this case, 100 m), h_o is the basic height (in this case, 10 m), k is the exponent, k = 0.14–0.30 [53], and was assumed to be 0.22.

Substituting

$$v_1 = v_p(100/10)^{0.22} = 1.66\ v_p \tag{2}$$

The obtained results are displayed graphically in Figure 3. When analyzing Figure 3, it can be concluded that the best conditions for locating wind turbines in the Greater Poland Voivodeship region are in the southeast whilst the worst are in the west.

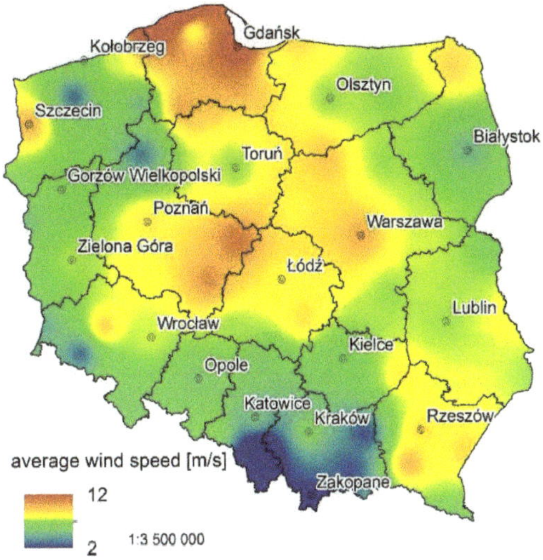

Figure 3. Average wind speed at a height of 100 m (own calculations).

According to the current regulations (the Wind Energy Investments Act, 2016), wind turbines in Poland must be placed away from other structures at a distance of 10 times (10*h*) their height (including the blades). Assuming the height is 140 m, the buffer will be 1400 m. Work is currently underway on the amendment of the RES Act; the "distance act" is to be amended, so a buffer of 700 m (5*h*) was also adopted in our calculations.

In the Greater Poland Voivodeship, the built-up area, along with the buffer zone of 700 m, reaches 16,655 km^2 (Figure 4) and 6609 km^2 for a buffer zone of 1400 m.

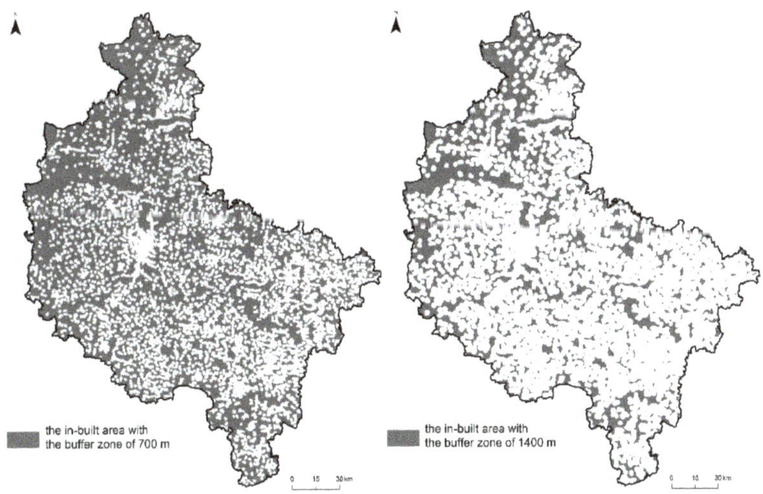

Figure 4. Area available for wind energy development—the built-up area with the buffer zones of 700 m and 1400 m in the Greater Poland Voivodeship (own calculations).

The area of the Greater Poland Voivodeship covered by flood plain areas is 6350 km^2 for a buffer zone of 700 m and 1260 km^2 for a buffer zone of 1400 m (Figure 5).

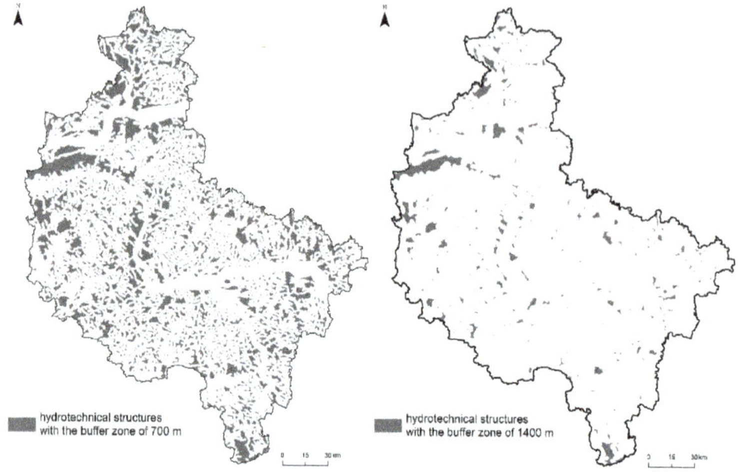

Figure 5. Area available for wind energy development—flood plains with the buffer zones of 700 m and 1400 m in the Greater Poland Voivodeship (own calculations).

The area of the Greater Poland Voivodeship covered by protected areas together with a buffer zone of 700 m is 12,910 km^2, and with a buffer zone of 1400 m, is 10,397 km^2 (Figure 6).

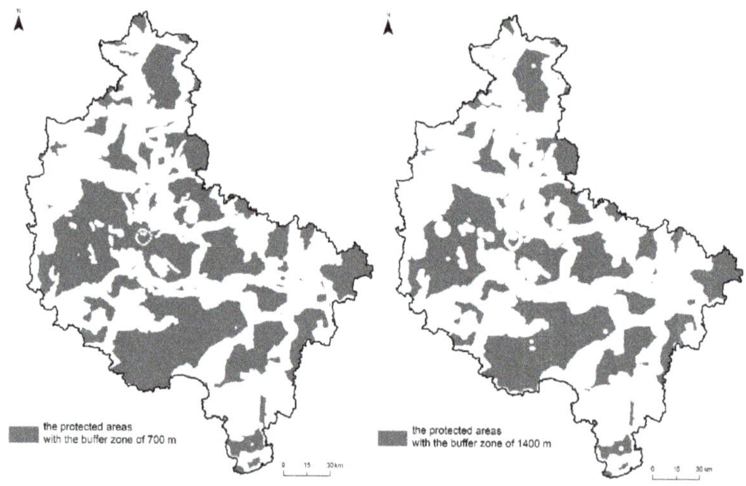

Figure 6. Area available for wind energy development—protected areas with the buffer zones of 700 m and 1400 m in the Greater Poland Voivodeship (own calculations).

The area of the Greater Poland Voivodeship covered by forest with a buffer zone of 700 m is 12,012 km^2, and it is 6394 km^2 for a buffer zone of 1400 m (Figure 7).

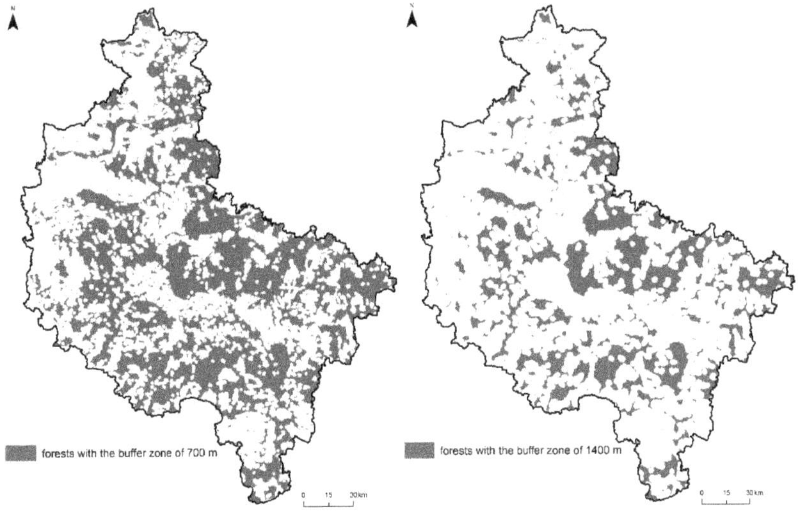

Figure 7. Area available for wind energy development—forest areas with the buffer zones of 700 m and 1400 m in the Greater Poland Voivodeship (own calculations).

In total, the available terrain for wind energy for a buffer zone of 700 m is 210 km^2 and for a buffer zone of 1400 m, only 2 km^2 (Figure 8).

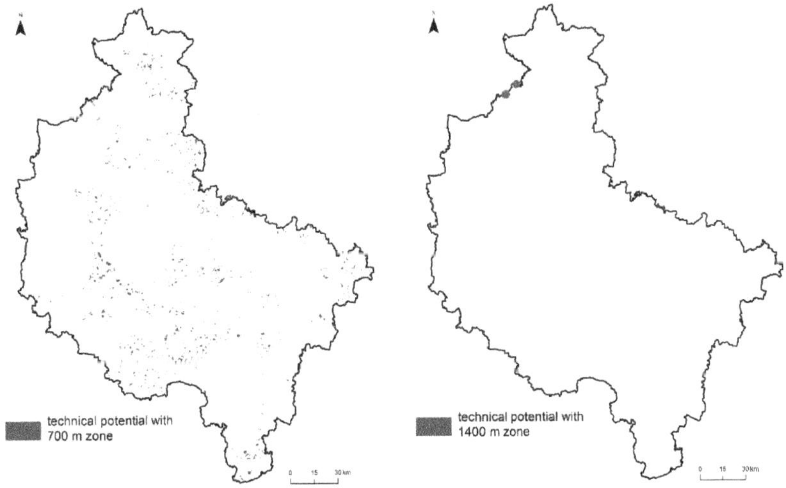

Figure 8. Area available for wind energy development with the buffer zone of 700 m and 1400 m in the Greater Poland Voivodeship (own calculations).

The area of the Greater Poland Voivodeship that could be technically designated for the construction of wind turbines is 210 km^2 (buffer zone of 700 m). It was assumed that the operating wind turbines will have a diameter r = 50 m. In order to ensure that turbines do not interfere with one another, it was assumed that the distance between the turbines should be 5r, whilst in the direction towards the wind, it should be 8r. It was assumed that the wind turbines will be positioned in a network of rectangles, and each of the rectangles will have the dimensions $5r \times 8r$. The area taken up by one wind turbine is 100,000 m^2 (0.1 km^2). Therefore, in an area of 377 km^2, it is technically possible to position 2100 wind turbines. This means that the technical potential of the wind power sector in the Greater

Poland Voivodeship is 4.2 GW. It was assumed that the amount of energy generated by one 2 MW wind turbine is 4.2 GWh. Thus, the total amount of electricity is 8,5 TWh, which is 65% of the electricity currently used in the Greater Poland Voivodeship [77].

5. Solar Energy Potential in the Greater Poland Voivodeship

The annual solar radiation energy per area unit on a horizontal plane in Poland reaches between 950 and 1250 kWh/m^2, whilst in the Greater Poland Voivodeship, it reaches about 1050–1150 kWh/m^2 (Figure 9). About 80% of this value is generated during the six months of the spring and summer seasons (April–September) [78,79]. It needs to be mentioned that solar operation in the summer is prolonged up to 16 h per 24-h period, whilst in winter, it shortens down to 8 h per 24-h period [79–81].

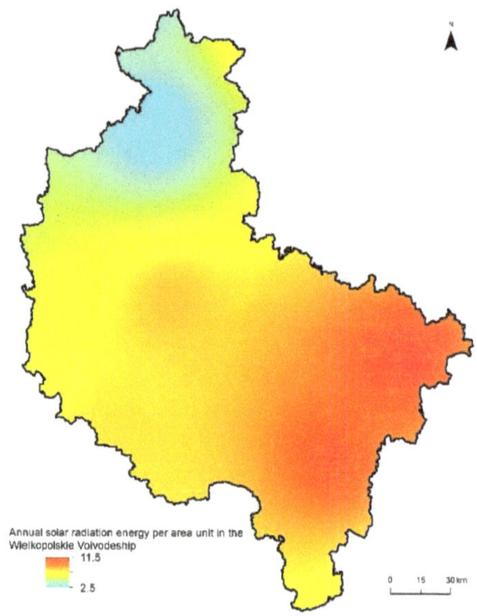

Figure 9. Annual solar radiation energy per area unit in the Greater Poland Voivodeship (own calculations).

The annual values of insolation in the Greater Poland Voivodeship vary from 1250 h in years with the highest cloud cover to 2000 h in the sunniest years. The long-term mean is 1600 h, which is similar to the long-term mean of insolation for all of Poland [77]. The analysis of variations in annual insolation carried out for Poznań over a 25-year period, 1990–2014, indicates that the insolation value has been regularly increasing by 11 h/year, on average.

Photovoltaic panels can be installed in many places where they take up a relatively small space or where typical economic activity cannot be undertaken (e.g., around airports) [77]. In the Greater Poland Voivodeship, such places could be the roofs of buildings, closed waste site grounds, and post-industrial wasteland. In our investigation, we assumed that 5% of the roofs of buildings could be designated for photovoltaic panels and 2% of the roofs could be used for solar collectors. The gross covered area in the Greater Poland Voivodeship is 1648 km^2 (Figure 10) [77].

Figure 10. The gross covered area in the Greater Poland Voivodeship (own calculations).

In addition, if we assume that the roof area of a typical building is approximately equal to the gross covered area—that is, according to the Polish Standard [77], the area of the vertical projection of the building in a completion state, determined by projecting onto the area surface all of its external walls (in reality, it could be slightly larger, as roofs are often sloping and not horizontal)—then the available areas for photovoltaic panels and collectors in the Greater Poland Voivodeship are 82.4 km^2 and 33.0 km^2, respectively. Based on the data [75], it can be assumed that from 1 m^2, at least 120 kWh of electric power could be generated per year; thus, the total amount of electric power would be 9.84 TWh. The amount of heat obtained from 1 m^2 of a collector was assumed to be at the level of 650 kWh annually, which, combined with the whole area of the Greater Poland Voivodeship, amounts to 21.45 PJ.

6. The Biomass Energy Potential in the Greater Poland Voivodeship

6.1. The Technical Potential of Waste Wood

The Greater Poland Voivodeship is known as "the breadbasket of Poland". This Voivodeship region is where a significant part of food production takes place. Agricultural production generates considerable amounts of waste, which, together with forest waste, could be used to generate energy [82]. Forestation in the Greater Poland Voivodeship is 25.7% and lower than the average for Poland, which is 29.4% [73]. If we assume that 15% of wood obtained directly from forests could be used for energy purposes (parts of bark, slash, and more coarse waste wood generated during logging), then 450,000 m^3 of waste wood per year could be obtained in the Greater Poland Voivodeship.

Significant amounts of waste [83] and used timber would be at least the same as the amount of waste wood obtained from forests, that is, 450,000 m^3 (technical potential). In total, the amount of waste forest biomass for energy purposes can be estimated to be 900,000 m^3 per year. If we assume that 1 m^3 of wood weighs 600 kg, then the weight of waste wood would be 540,000 tons per year.

The area covered by orchards in the Greater Poland Voivodeship is 16,400 ha [82]. The timber from orchards is obtained from both felling and maintenance work. As a result of the felling of orchards, it is technically possible to obtain about 80 tons of biomass per 1 ha in the case of older plantations (about 30 years old) and about 60 tons/ha in the case of modern dwarf plantations (about 15 years old). The amount of biomass created during maintenance work varies depending on the variety and age of the trees, from 4 to 10 tons/ha [64]. Assuming that due to the felling of orchards, 3.5 tons of biomass can be obtained per 1 ha per year, and that as a result of maintenance work, 7 tons of biomass is created per 1 ha per year, then it is possible to obtain 125,000 tons of waste biomass from orchards in the Greater Poland Voivodeship. In total, the waste wood in the Greater Poland Voivodeship amounts to 600,000 tons per year (technical potential).

6.2. The Technical Potential of Straw and Hay from Unused Meadows, Pastures, and Energy Crops

In 2020, the Greater Poland Voivodeship produced about 770,000 tons of wheat straw, 310,000 tons of rye straw, 180,000 tons of barley straw, 330,000 tons of oats straw, and 50,000 tons of triticale straw [77]. The total amount of straw is 1,640,000 tons. Assuming that the technical potential of straw is 30%, then 492,000 tons of straw could be used for energy purposes. In the Greater Poland Voivodeship, there are 234,500 ha of meadows and 17,400 ha of pastures. The amount of hay harvested from meadows amounts to 4.9 tons per 1 ha/year, whilst from pastures, this amount is 3.6 tons per 1 ha/year [83]. Assuming that the technical potential of hay from meadows and pastures that could be used is 30%, then the amount of hay that could be used for energy purposes is 367,000 tons per year.

In the Greater Poland Voivodeship, there are 33,000 ha of fallow land and 36,000 ha of idle land [83]. These areas could be cultivated or reclaimed using energy crops. Our plant of choice is the common osier, which has been described in previous papers. This plant is used for both energy production and land reclamation. It could be assumed that it is viable to cultivate 50% of fallows and 25% of idle land; this means that there is an available area of 25,500 ha. The agriculture in the Greater Poland Voivodeship is at the highest level compared to other parts of the country. This is why, despite fallows and idle land being designated for cultivation, it can be assumed that the average biomass crop would be 20 tons of dry mass per 1 ha per year. This means that there would be a crop of about 510,000 tons of biomass per year.

6.3. The Technical Potential of Agricultural Biogas

The technical potential of agricultural biogas in the Greater Poland Voivodeship was calculated, taking into consideration the conversion factor of livestock heads into livestock units (*LSU*; 500 kg) [84]. For cattle, the conversion rate is 0.8, for pigs—0.2, and for poultry—0.004. The mean amount of slurry per 1 *LSU* is 44.9 kg for cattle, 43.5 kg for pigs, and 26.8 kg for poultry. The number of heads was taken from the data of the Central Statistical Office [77]. The construction of biogas plants using slurry and/or poultry manure is technically and economically viable on farms with a livestock number of at least 100 heads of cattle, 500 heads of pigs, and 5000 heads of poultry. Thus, the technical potential of agricultural biogas from animal droppings in the Greater Poland Voivodeship should be estimated at 25% of the theoretical potential [83]. It was assumed that the production of biogas from 1 ton of cattle slurry was 50 m^3, from pig slurry—55 m^3, and from poultry manure—140 m^3. The amount of biogas that could be obtained in the Voivodeship is 40 million m^3.

In the Greater Poland Voivodeship, maize is grown for consumption purposes and as farm livestock fodder. After the maize knobs have been harvested for consumption purposes, what is left in the field are stalks and leaves, which could be used to produce biogas. It was assumed that it is technically possible to obtain biogas from 30% of sown plants, whilst straw constitutes 62% of the dry mass of the whole plant. In 2015, the area where maize was cultivated for its grain in the Greater Poland Voivodeship reached 3,288,000 ha [83]. Assuming the grain crop yielded 9 tons, whilst 1 ton of biomass can produce 90 m^3 of biogas, then the volume of biogas would be 1.3 billion m^3. Due to the agricultural character of the Greater Poland Voivodeship, the opportunities to build agricultural biogas plants in the Greater Poland Voivodeship are great [85,86].

6.4. The Technical Potential of Biogas from Municipal Waste and Sewage Treatment Plants

The amount of municipal waste generated in households and public-use buildings in the Greater Poland Voivodeship reaches 1.1 million tons, of which more than a half consists of biodegradable waste [87]. The technical potential of biogas from municipal waste can be estimated at the level of 40% of the theoretical potential. Assuming that 90 m^3 of biogas can be produced from 1 ton of waste, then it is possible to obtain nearly 20 million m^3 of biogas from municipal waste per year in the Greater Poland Voivodeship.

In the Greater Poland Voivodeship, 21.4 million m^3 of municipal sewage is treated [87]. Assuming that from 50% of effluents coming to the plant sludge can be obtained (sludge amounts to 1% of effluents) and that 1 m^3 of sludge produces 15 m^3 of biogas, then 1.6 million m^3 of biogas could be generated in the Voivodeship.

6.5. The Total Technical Potential of Bioenergy in the Voivodeship

The Greater Poland Voivodeship has a very high potential to produce biomass for energy purposes. Table 1 shows the amounts of electric power and heat (in cogeneration) that could be produced from solid biomass. In total, it comes to about 25.6 PJ, which includes 2837 GWh of electric power and 12.8 PJ of heat (Table 1). The development of biomass-based energy generation would contribute a few hundred new work places in agriculture, the transport industry, companies dealing with biomass processing, such as briquette and pellet production, and eventually, in new boiler plants and heat and power stations.

Table 1. Electricity and heat production from solid biomass in the Greater Poland Voivodeship (own calculations).

	Biomass (Thousand tons)	Calorific Value (MJ/kg)	Amount of Energy (80% Efficiency) (PJ)	Amount of Electric Power (30% Efficiency) (GWh)	Amount of Heat (50% Efficiency) (PJ)
Wooden waste	540	14	6.0	667	3.0
Waste from orchards	125	16	1.6	178	0.8
Straw	492	15	5.9	656	3.0
Hay	367	15	4.5	496	2.2
Energy crops	510	18.5	7.6	840	3.8
Total			**25.6**	**2837**	**12.8**

Table 2 demonstrates the amounts of electric power and heat (in cogeneration) that could be obtained from biogas in the Greater Poland Voivodeship. In total, it reaches about 23.8 PJ, which includes 3.1 GWh of electric power and 12.6 TJ of heat (Table 2). In the Greater Poland Voivodeship, about 6 TWh of electric power could be obtained from solid waste biomass and biogas, which is nearly equivalent to 50% of the currently used energy [71]. On the other hand, the amount of heat that could be generated is 25.4 PJ, which amounts to 88% of the currently used heat.

Table 2. Electric power and heat production from biogas in the Greater Poland Voivodeship (own calculations).

Source of Biogas	Biogas (Million m^3)	CH_4 Content (%)	CH_4 Volume (Million m^3)	Amount of Energy (85% Efficiency) (TJ)	Amount of Electric Power (40% Efficiency) (GWh)	Amount of Heat (45% Efficiency) (TJ)
Agriculture (manure, maize straw)	1.3	60	780	23.7	3.1	12.5
Municipal waste	10	50	5.0	152	20	80
Sewage sludge	1.6	55	0.9	27	4	14
Total			**786**	**23.8**	**3.1**	**12.6**

7. The Water Energy Potential in the Greater Poland Voivodeship

Until recently, mills and small hydropower plants could be frequently sighted in the Greater Poland Voivodeship. Apart from energy generation, they had been responsible for water retention for a few hundred years [88]. The drainage system had been closely linked, in a unique symbiosis one could say, to barrages [89]. Unfortunately, after World War II, most of the small hydropower stations fell into decline. In recent years, however, there has been a growing interest among investors in the hydropower sector in the Greater Poland Voivodeship [89–92].

Calculations of the technical potential of hydropower in the Greater Poland Voivodship with the use of already existing damming facilities were carried out. In river water power plants, electricity is obtained from kinetic energy, and especially from potential energy. Assuming the water density ρ = 1000 kg/m^3 and an efficiency of 85%, a formula [92,93] can be obtained, as follows:

$$P = 8.34\Theta \cdot H \tag{3}$$

where P is the hydroelectric power (MW), Θ is the flow (m^3/s), and H is the spades (height).

Assuming that the full power plant will operate 6000 h a year, the amount of energy from the E_k hydroelectric power plant will be as follows:

$$E_k = 21.6 \cdot P \tag{4}$$

The data Θ and H were obtained from the National Water Management Board [66].

Figure 11 was created using the data provided by the Geodesic and Cartographic Documentation Centre [94]. Within the area of the Voivodeship, there are 1229 hydrotechnical structures—10 dams, 11 sluice locks, and 1208 weirs.

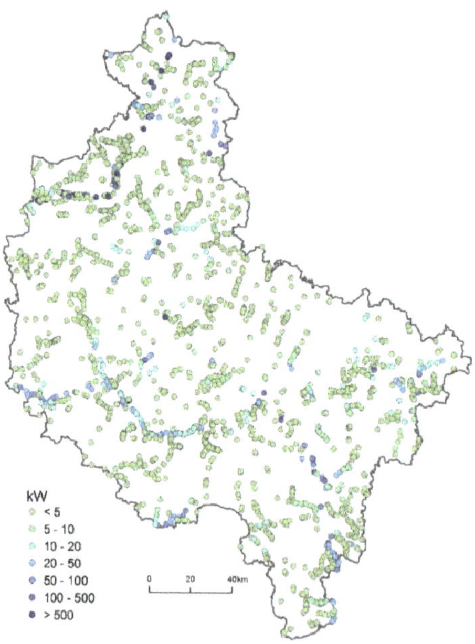

Figure 11. Hydrotechnical structures in the Greater Poland Voivodeship (own calculations).

It can be assumed that at least half of all the hydrotechnical structures in the Greater Poland Voivodeship could be used to generate electric power. If the same supposition as above was followed, then the potential would be a total of 123 MW, whilst the power generation would reach 615 GWh. Clean water is becoming a rarity in the Greater Poland Voivodeship. Thus, more care has been given to its purification in recent years; some of the sewage treatment plants could be used to generate electricity, the way it is already happening in Minsk (Belarus) [92].

In the Greater Poland Voivodeship, there are 425 sewage treatment plants [87]. As already mentioned, some of them could be used to generate electric power—there must be an adequate difference between the water levels and the amount of treated water. It can be assumed that from a technical standpoint, hydropower plants need to be erected at the sewage treatment plants with a flow of at least 1 million m^3 per year; 32 facilities meet this criterion (Figure 12). As an estimate, each of these could produce about 400–700 MWh of electric power. In total, this would provide about 8–12 GWh. It may not be a huge amount, but the main purpose of treating sewage is met anyway and electricity generation would just be a by-product.

Figure 12. Location of sewage treatment plants with a minimum capacity of 1 million m^3 per year (own calculations).

8. The Geothermal Energy Potential in the Greater Poland Voivodeship

In the Greater Poland Voivodeship, the density of the terrestrial heat flow varies from 70 to 100 mW/m^2 (Figure 13) [95,96]. We know this is a result of a high number of exploratory bore holes, which were created after World War II, mainly during exploration for oil and natural gas resources. Using already existing bore holes results in significantly lower investment costs in the geothermal energy sector [95,96].

The data from the Central Geological Database [97] were used to create Figure 14, which shows bore holes of a minimum depth of 500 m in the Greater Poland Voivodeship.

The Greater Poland Voivodeship has good conditions not only for heat generation but also for electricity production [98]. In our study, it was assumed that binary geothermal heat and power plants could be implemented; their parameters are collated in Table 3. The construction of four binary heat and power plants will facilitate the production of 65.65 TJ of heat and 8.55 GWh of electric power per year. Medical and tourist centers are to be built next to the geothermal plants. For example, on the Island of Pociejewo in Konin, the geothermal water reaches a temperature of 97 °C and mineralization of 70 g/dm^3 at a depth of 2600 m. According to the initial physico-chemical investigation, this is a highly mineralized sodium chloride type of water, which contains large amounts of chloride, sodium, magnesium, and calcium ions as well as many microelements. It meets all the parameters of medicinal water. The architectural development plan of the island in Konin contains the geothermal plant Pociejewo, "The Ecological Town Salon", including a complex of 15, mainly indoor, swimming pools, 50 health and wellness salons, an exclusive hotel, and an indoor sports hall [99].

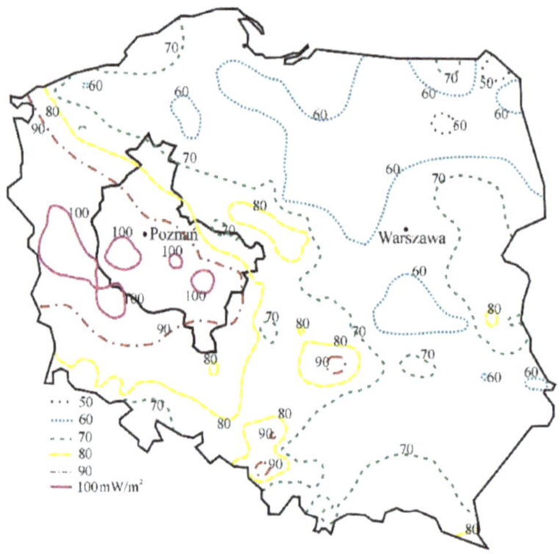

Figure 13. The map of terrestrial heat flow density for the Greater Poland Voivodeship in comparison to the whole of Poland (own study from [95,96]).

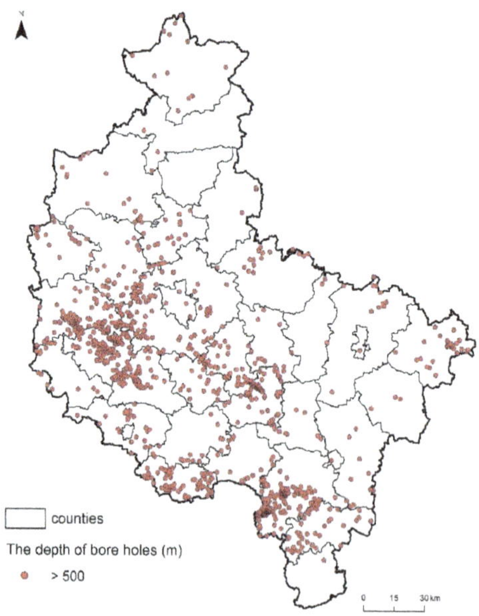

Figure 14. The bore holes of a minimum depth of 500 m in the Greater Poland Voivodeship (own study from [97]).

Table 3. Parameters of binary geothermal heat and power plants in the Greater Poland Voivodeship (net power value, taking into account own energy usage and loss) [98].

Locality	Depth of the Production Bore Hole (m)	Water Temperature at the Outlet (°C)	Total Thermal Power (MW)	Net Amount of Generated Heat (TJ/Year)	Total Electric Power (MW)	Net Amount of Generated Electric Power (GWh/Year)
Koło	4000	118	28.3	64.7	1.05	3.16
Turek	3550	98	2.45	0.25	0.45	1.70
Konin	2897	98	3.23	0.35	0.43	1.15
Ślesin	3358	98	0.30	0.35	0.45	2.54
Total	-	-	-	**65.65**	-	**8.55**

9. Total Technical Potential of the Renewable Energy in the Greater Poland Voivodeship

The implementation of the goal set in the article allowed for the designation of the technical potential of the renewable energy sector in the Greater Poland Voivodeship using GIS methods. The obtained results indicate that the amount of available electricity and heat is higher than the energy demand of the Voivodeship (Table 4). As for electric power, it is possible to generate 23.34 TWh, which is 1.8 times more than the demand. As for heat, it is about 47 PJ, which is 1.6 times more than the demand. A surplus of energy could be sent to other voivodeship regions. In addition, our accurate calculations show that it is possible to produce 100% RES in other regions in Poland. This means that the Greater Poland Voivodeship may be energy independent in the future, where both electricity and heat would come entirely from renewable energy sources.

Table 4. Technical potential of RES in the Greater Poland Voivodeship (own calculations).

RES	Amount of Electric Power (TWh)	Amount of Heat (PJ)
Wind Energy	8.50	
Solar Energy	9.84	21.45
Bioenergy	5.99	25.40
Hydropower	0.63	
Geothermal Energy	0.01	0.01
Total	**23.34**	**46.86**
Demand in the Voivodeship	**12.48**	**27.89**
Oversupply of energy (ratio)	**1.87**	**1.68**

Source: own calculations.

This is an important argument in the further process of introducing just and bottom-up initiatives that would enable the region's coal-based energy transformation. Additionally, the development of RES contributes to increasing the energy security of the society and economy, which are powered by a network of smaller and safer power plants, and leads to stronger local communities.

10. Discussion and Summary

The national coal regions have constituted the foundation of economic growth and prosperity at the local and national levels in the EU for decades. The XIXth and XXth centuries show a huge relation between coal mining in six coal regions in Poland (including the eastern part of the Greater Poland Voivodeship), the Ruhr region of Germany, Western Macedonia in Greece, and the southwest region of Bulgaria, and the economic and social development. A great value in performing energy transformation is establishing one region as an example for others. It is said that an example of just energy transformation is the Ruhr region of Germany. The transformation started there earlier and is much more ahead of the curve. The industrial sector of this German region has been restructured and is open for high-tech, tourism, and cultural services. The successful departure from coal mining has its foundation in focusing on the local strategy and authorities. The recommendations and transferable solutions reaccessioned on the Ruhr region transformation cover (1) the use of the bottom-up approach, (2) focusing on regional disparities, (2) establishing local value chains connected to the reindustrialization, and (3) empowering supra-regional competitiveness.

A transition to RES that is in line with the energy strategy of the EU is inevitable. As it has been mentioned, the comprehensive energy justice framework includes many issues, the first point of which is energy availability and access. RES may create a base of energy independence in coal regions, enabling full green energy availability and access for the local community. The analysis carried out in the study showed that by focusing only on RES resources in the region, the Greater Poland Voivodeship may fully meet the demand for energy with its green resources. The obtained results allow for concluding that in the area of the selected region (Greater Poland Voivodeship), it is possible to achieve climate-neutral energy independence under the European Green Deal. This means that despite the diversity of the energy potential of NUTS 2 regions in the EU, the energy transformation towards RES is justified at the regional level. This, in turn, is a good sign that all member states of the community will be able to achieve the objectives of the European Green Deal.

What is even more noticeable and economically important, the Greater Poland Voivodeship may remain energy competitive with other regions, as it may produce more energy than is needed. This means that this region may maintain its energy characteristic and economic power. The Greater Poland Voivodeship has high RES potential, and introducing a fair and just bottom-up strategy may convert the coal energy region to a green energy valley and further obtain the national and EU objectives. As the RES potential is far higher than the needs of the region, open-minded reindustrialization may bring large benefits. RES availability and energy transformation may lead to a diversified production centralization in the region, which, in conjunction with the west–east transition route, may even become a geopolitical hub, empowering the region and even national competitiveness and growth.

Moving towards RES is an enormous challenge, especially for local communities in coal regions. Local and family traditions, individual life choices, and a sense of security are based on coal mining in many situations. Juliette de Grandpré, Senior Policy Advisor, WWF Germany, said that "We need to make sure Europe's transition towards a net-zero emissions economy does not happen at the expense of these regions" [100]. Above and beyond the RES benefits, the Ruhr region is still dealing with a high unemployment rate, double that in other German regions. This is what the Greater Poland Voivodeship wants to avoid. To do so, the local community has to be aware of the region's RES potential, the large possibilities generated by RES, and be sure that the region maintains energy security and that it can be competitive and economically strong. The key roles played here are the local authorities, their financial and educative support, decision-making transparency, trust, equity, and of course, understanding and empathy. This may create a "green" and open-minded community that is able to meet challenges.

In this study, we showed that the Greater Poland Voivodeship has opportunities and resources to become an energy-independent region and become a Green Energy Valley in the near future. This energy independence is planned to be obtained under a fair, just, bottom-up, locally and community-oriented energy transformation. Performing such a region transformation is a milestone in achieving energy neutrality in Poland. There is only one way to move toward RES, and local support is indispensable. The possibility of the Greater Poland Voivodeship to achieve RES independence is an unmediated fact, but the details of how to, in practice, effectively support the local community while maintaining regional differences, still remain under discussion.

Author Contributions: Conceptualization, B.I. and M.B.P.; methodology, B.I. and M.B.P.; software, M.S. and M.B.P.; validation, B.I., M.B.P., U.K., M.S., A.G., A.Z. and K.N.; formal analysis, B.I., M.S. and M.B.P.; investigation, B.I., M.B.P., U.K., M.S., A.G., A.Z. and K.N.; resources, B.I., M.B.P., U.K., M.S., A.Z. and K.N.; data curation, M.B.P. and B.I.; writing—original draft preparation B.I. and M.B.P.; writing—review and editing, B.I., M.B.P., U.K., M.S., A.G., A.Z. and K.N.; visualization, B.I., M.B.P. and M.S.; supervision, B.I. and M.B.P.; project administration B.I. and M.B.P. All authors have read and agreed to the published version of the manuscript.

Funding: This research received no external funding.

Data Availability Statement: Data available on request due to restrictions. The data presented in this study are available on request from the corresponding author.

Conflicts of Interest: The authors declare no conflict of interest.

References

1. Hussain, H.I.; Szczepańska-Woszczyna, K.; Kamarudin, F.; Anwar, N.A.M.; Saudi, M.H.M. Unboxing the black box on the dimensions of social globalisation and the efficiency of microfinance institutions in Asia. *Oeconomia Copernic.* **2021**, *12*, 557–592. [CrossRef]
2. Rees, W.E. Globalization, trade and migration: Undermining sustainability. *Ecol. Econom.* **2006**, *2*, 220–225. [CrossRef]
3. Bartożewski, K.; Trąpczyński, P. Is internationalisation beneficial for novice internationalisers? The performance effects of firm-specific advantages, internationalisation degree and firm size revisited. *Oeconomia Copernic.* **2021**, *12*, 53–75. [CrossRef]
4. Szopik-Depczyńska, K.; Cheba, K.; Bąk, I.; Stajniak, M.; Simboli, A.; Ioppolo, G. The study of relationship in a hierarchical structure of EU sustainable development indicators. *Ecol. Indic.* **2018**, *90*, 120–131. [CrossRef]
5. Skare, M.; Porada-Rochoń, M. Financial and economic development link in transitional economies: A spectral Granger causality analysis 1991–2017. *Oeconomia Copernic.* **2019**, *1*, 7–35. [CrossRef]
6. Sánchez-López, C.; Aceytuno, M.T.; De Paz-Bañez, M.A. Inequality and globalisation: Analysis of European countries. *Econ. Sociol.* **2019**, *4*, 84–100. [CrossRef] [PubMed]
7. Simionescu, M.; Lazányi, K.; Sopková, G.; Dobeš, K.; Balcerzak, A.P. Determinants of economic growth in V4 countries and Romania. *J. Comp.* **2017**, *1*, 103–113. [CrossRef]
8. Zinecker, M.; Doubravský, K.; Balcerzak, A.P.; Pietrzak, M.B.; Dohnal, M. The COVID-19 disease and policy response to mitigate the economic impact in the EU: An exploratory study based on qualitative trend analysis. *Technol. Econ. Dev. Econ.* **2021**, *27*, 742–762. [CrossRef]
9. Wosiek, M. Rural-urban divide in human capital in Poland after 1988. *Oeconomia Copernic.* **2020**, *11*, 183–201. [CrossRef]
10. Szopik-Depczyńska, K.; Kędzierska-Szczepaniak, A.; Szczepaniak, K.; Cheba, K.; Gajda, W.; Ioppolo, G. Innovation in sustainable development: An investigation of the EU context using 2030 agenda indicators. *Land Use Policy* **2018**, *79*, 251–262. [CrossRef]
11. Roszko-Wójtowicz, E.; Grzelak, M.M. Macroeconomic stability and the level of competitiveness in EU member states: A comparative dynamic approach. *Oeconomia Copernic.* **2020**, *11*, 657–688. [CrossRef]
12. Prokop, V.; Kotkova Striteska, M.; Stejskal, J. Fostering Czech firms? innovation performance through efficient cooperation. *Oeconomia Copernic.* **2021**, *12*, 671–700. [CrossRef]
13. Markauskas, M.; Baliute, A. Technological progress spillover effect in Lithuanian manufacturing industry. *Equilib. Q. J. Econ. Econ. Policy* **2021**, *16*, 783–806. [CrossRef]
14. Nowak, P. Cooperation of enterprises in innovative activities on the example of Polish regions. *Equilib. Q. J. Econ. Econ. Policy* **2021**, *16*, 839–857. [CrossRef]
15. Balcerzak, A.P. Quality of institutions in the European Union countries. Application of TOPSIS based on entropy measure for objective weighting. *Acta Polytech. Hung.* **2020**, *1*, 101–122. [CrossRef]
16. Zielenkiewicz, M. Institutional environment in the context of development of sustainable society in the European Union countries. Equilibrium. *Quart. J. Econ. Econ. Policy* **2014**, *9*, 21–37. [CrossRef]
17. Belke, A.; Dreger, C.; Dobnik, F. Energy consumption and economic growth–new insights into the cointegration relationship. *Energy Econom.* **2010**, *33*, 782–789. [CrossRef]
18. Piekut, M. The consumption of renewable energy sources (RES) by the European Union households between 2004 and 2019. *Energies* **2021**, *14*, 5560. [CrossRef]
19. Jankiewicz, M.; Pietrzak, M.B. Assessment of trends in the share of expenditure on services and food in the Visegrad group member states. *Int. J. Bus. Soc.* **2020**, *2*, 977–996. [CrossRef]
20. Zadykowicz, A.; Chmielewski, K.J.; Siemieniako, D. Proactive customer orientation and joint learning capabilities in collaborative machine to machine innovation technology development: The case study of automotive equipment manufacturer. *Oeconomia Copernic.* **2020**, *11*, 531–547. [CrossRef]
21. Wosiek, M. Unemployment and new firm formation: Evidence from Polish industries at the regional level. *Equilib. Q. J. Econ. Econ. Policy* **2021**, *16*, 765–782. [CrossRef]
22. Fragkos, P.; Paroussos, L. Employment creation in EU related to renewables expansion. *Appl. Energy* **2018**, *230*, 935–945. [CrossRef]

23. Gajdos, A.; Arendt, L.; Balcerzak, A.P.; Pietrzak, M.B. Future trends of labour market polarisation in Poland–The perspective of 2025. *Transform. Bus. Econ.* **2020**, *3*, 114–135.
24. Dmytrów, K.; Bieszk-Stolorz, B. Comparison of changes in the labour markets of post-communist countries with other EU member states. *Equilib. Q. J. Econ. Econ. Policy* **2021**, *16*, 741–764. [CrossRef]
25. Pietrzak, M.B.; Balcerzak, A.P.; Gajdos, A.; Arendt, Ł. Entrepreneurial environment at regional level: The case of Polish path towards sustainable socio-economic development. *Entrep. Sustain. Issues* **2017**, *5*, 190–203. [CrossRef]
26. Svabova, L.; Tesarova, E.N.; Durica, M.; Strakova, L. Evaluation of the impacts of the COVID-19 pandemic on the development of the unemployment rate in Slovakia: Counterfactual before-after comparison. *Equilib. Q. J. Econ. Econ. Policy* **2021**, *16*, 261–284. [CrossRef]
27. Markandya, A.; Arto, I.; Eguino, M.G.; Román, M.V. Towards a green energy economy? Tracking the employment effects of low-carbon technologies in the European Union. *Appl. Energy* **2016**, *179*, 1342–1350. [CrossRef]
28. Pao, H.-T.; Fu, H.-C. Renewable energy, non-renewable energy and economic growth in Brazil. *Renew. Sustain. Energy Rev.* **2013**, *25*, 381–392. [CrossRef]
29. Pietrzak, M.B.; Igliński, B.; Kujawski, W.; Iwański, P. Energy transition in Poland–assessment of the renewable energy sector. *Energies* **2021**, *14*, 2046. [CrossRef]
30. Chovancová, J.; Tej, J. Decoupling economic growth from greenhouse gas emissions: The case of the energy sector in V4 countries. *Equilib. Q. J. Econom. Econom. Policy.* **2020**, *15*, 235–251. [CrossRef]
31. Lin, M.-X.; Liou, H.M.; Chou, K.T. National energy transition framework toward SDG7 with legal reforms and policy bundles: The case of Taiwan and its comparison with Japan. *Energies* **2020**, *13*, 1387. [CrossRef]
32. Grosse, T.G. Low carbon economy policy in Poland: An example of the impact of Europeanisation. *Q. J. Econom. Econom. Policy.* **2011**, *1*, 9–39. [CrossRef]
33. 2019 Europe Sustainable Development Report. In *Sustainable Development Solutions Network and Institute for European Environmental Policy*; Institute for European Environmental Policy: Brussels, Belgium, 2019.
34. Musiał, W.; Zioło, M.; Luty, L.; Musiał, K. Energy policy of European Union member states in the context of renewable energy sources development. *Energies* **2021**, *14*, 2864. [CrossRef]
35. Igliński, B.; Skrzatek, M.; Kujawski, W.; Cichosz, M.; Buczkowski, R. SWOT analysis of renewable Energy sector in Mazowieckie Voivodeship: Current progress, prospects and policy implications. *Env. Dev. Sustain.* **2021**, *24*, 77–111. [CrossRef]
36. Overland, I. Energy: The missing link in globalization. *Energy Res. Soc. Sci.* **2016**, *14*, 122–130. [CrossRef]
37. Kochanek, E. Evaluation of energy transition scenarios in Poland. *Energies* **2021**, *14*, 6058. [CrossRef]
38. Kałążny, A.; Morawski, W. Taxation of assets used to generate energy–in the context of the transformation of the Polish Energy sector from coal energy to low-emission Energy. *Energies* **2021**, *14*, 4587. [CrossRef]
39. Carley, S.; Konisky, D.M. The justice and equity implications of the clean energy transition. *Nat. Energy* **2020**, *5*, 569–577. [CrossRef]
40. David, M.; Schönborn, S. Bottom-up energy transition narratives: Linking the global with the local? A comparison of three German renewable Co-Ops. *Sustainability* **2018**, *10*, 924. [CrossRef]
41. Budzianowski, W.M.; Gomes, J.F.P. Perspectives for low-carbon production until 2030: Lessons learned from the comparison of local context in Poland and Portugal. *Energy Sources Part B Econ. Plan. Policy* **2016**, *6*, 534–541. [CrossRef]
42. Igliński, B.; Kujawski, W.; Buczkowski, R.; Cichosz, M. Renewable energy in the Kujawsko-Pomorskie Voivodeship (Poland). *Renew. Sustain. Energ. Rev.* **2010**, *4*, 1336–1341. [CrossRef]
43. Igliński, B.; Buczkowski, R.; Cichosz, M.; Piechota, G.; Kujawski, W.; Plaskacz, M. Renewable energy production in the Zachodniopomorskie Voivodeship (Poland). *Renew. Sustain. Energy Rev.* **2013**, *27*, 768–777. [CrossRef]
44. Igliński, B.; Piechota, G.; Iglińska, A.; Cichosz, M.; Buczkowski, R. The study on the SWOT analysis of renewable energy sector on the example of the Pomorskie Voivodeship (Poland). *Clean Technol. Environ. Policy* **2015**, *1*, 45–61. [CrossRef]
45. Igliński, B.; Buczkowski, R.; Iglińska, A.; Cichosz, M.; Plaskacz-Dziuba, M. SWOT analysis of the renewable energy sector in Poland: Case study of Wielkopolskie region. *J. Power Technol.* **2015**, *2*, 143–157.
46. Igliński, B.; Iglińska, A.; Cichosz, M.; Kujawski, W.; Buczkowski, R. Renewable energy production in the Łódzkie Voivodeship: The PEST analysis of the RES in the Voivodeship and in Poland. *Renew. Sustain. Energy Rev.* **2016**, *58*, 737–750. [CrossRef]
47. Biuletyn Informacji Publicznej Urzędu Regulacji Energetyki. Available online: www.ure.gov.pl (accessed on 25 September 2021).
48. European Union. Directive 2009/28/EC of the European Parliament and of the Council of 23 April 2009 on the promotion of the use of energy from renewable sources and amending and subsequently repealing Directives 2001/77/EC and 2003/30/EC. *Off. J. Eur. Union* **2009**, *5*, 2009.
49. Directive (EU) 2018/2001 of the European Parliament and of the Council of 11 December 2018 on the Promotion of the Use of Energy from Renewable Sources PE/48/2018/REV/1. Available online: https://eur-lex.europa.eu/legal-content/EN/TXT/?uri=uriserv:OJ.L_.2018.328.01.0082.01.ENG (accessed on 30 September 2021).
50. Energy Roadmap 2050. In *European Commission Staff Working Paper*; European Commission: Brussels, Belgium, 2011; Part 2/2; p. 36.
51. Caravella, S.; Costantini, V.; Crespi, F. Mission-oriented policies and technological sovereignty: The case of climate mitigation technologies. *Energies* **2021**, *14*, 6854. [CrossRef]
52. Poland's Energy Policy until 2030. Available online: www.cire.pl/artykuly/materialy-problemowe/45029-polityka-energetyczna-polski-do-2030-roku (accessed on 28 September 2021).

53. Poland's National Energy and Climate Plan for Years 2021–2030. Available online: https://www.gov.pl/web/klimat/national-energy-and-climate-plan-for-the-years-2021-2030 (accessed on 29 September 2021).
54. Poland's Energy Policy until 2040. Available online: www.gov.pl/web/klimat/polityka-energetyczna-polski (accessed on 29 October 2021).
55. Zinecker, M.; Skalická, M.; Balcerzak, A.P.; Pietrzak, M.B. Business angels in the Czech Republic: Characteristics and a classification with policy implications. *Econ. Res.-Kenosha Istraživanja* **2021**, 1–26. [CrossRef]
56. Gorączkowska, J. Enterprise innovation in technology incubators and university business incubators in the context of Polish industry. *Oeconomia Copernic.* **2020**, *11*, 799–817. [CrossRef]
57. Zinecker, M.; Skalická, M.; Balcerzak, A.P.; Pietrzak, M.B. Identifying the impact of external environment on business angel activity. *Econ. Res.-Ekonomska Istraživanja* **2021**, 1–23. [CrossRef]
58. Meluzín, T.; Zinecker, M.; Balcerzak, A.P.; Pietrzak, M.B.; Doubravský, K. Institutional Settings and their Impact on the IPO Activity: An Exploratory Study Based on Qualitative Modelling. *Acta Polytech. Hung.* **2021**, *18*, 215–235. [CrossRef]
59. Meluzín, T.; Balcerzak, A.P.; Pietrzak, M.B.; Zinecker, M.; Doubravský, K. The impact of rumours related to political and macroeconomic uncertainty on IPO success: Evidence from a qualitative model. *Transform. Bus. Econ.* **2018**, *2017*, 148–169.
60. Meluzín, T.; Zinecker, M.; Balcerzak, A.P.; Doubravský, K.; Pietrzak, M.B.; Dohnal, M. The timing of initial public offerings: Non-numerical model based on qualitative trends. *J. Bus. Econ. Manag.* **2018**, *19*, 63–79. [CrossRef]
61. Meluzín, T.; Zinecker, M.; Balcerzak, A.P.; Pietrzak, M.B. Why do companies stay private? Determinants for IPO candidates to consider in Poland and the Czech Republic. *East. Eur. Econ.* **2018**, *56*, 471–503. [CrossRef]
62. Oehlmann, M.; Meyerhoff, J. Stated preferences towards renewable energy alternatives in Germany–do the consequentiality of the survey and trust in institutions matter? *J. Environ. Econ. Policy* **2017**, *1*, 1–16. [CrossRef]
63. Rogers, J.; Simmons, E.; Convery, I.; Weatherall, A. Public perceptions of community-based renewable energy projects. *Energy Policy* **2008**, *36*, 4217–4226. [CrossRef]
64. Zoellner, J.; Schweizer-Ries, P.; Wemheuer, C. Public acceptance of renewable energies: Results from case studies in Germany. *Energy Policy* **2008**, *36*, 4136–4141. [CrossRef]
65. Upham, P.; Shackley, S. The case of a proposed 21.5MWe biomass gasifier in Winkleigh, Devon: Implications for governance of renewable energy planning. *Energy Policy* **2006**, *34*, 2161–2172. [CrossRef]
66. Sánchez-Lozano, J.M.; Teruel-Solano, J.; Soto-Elvira, P.L.; García-Cascales, M.S. Geographical Information Systems (GIS) and Multi-Criteria Decision Making (MCDM) methods for the evaluation of solar farms locations: Case study in south-eastern Spain. *Appl. Energy* **2013**, *24*, 544–556. [CrossRef]
67. Xu, Q.; Guo, Z.; Xia, L.; He, Q.; Li, Z.; Bello, I.T.; Zheng, K.; Ni, M. A comprehensive review of solid oxide fuel cells operating on various promising alternative fuels. *Energy Conv. Manag.* **2022**, *253*, 115174. [CrossRef]
68. Huang, Y.; Xiao, X.; Kang, H.; Lv, J.; Zeng, R.; Shen, J. Thermal management of polymer electrolyte membrane fuel cells: A critical review of heat transfer mechanisms, cooling approaches, and advanced cooling analysis. *Energy Conv. Manag.* **2022**, *254*, 115221. [CrossRef]
69. Ghanem, R.S.; Nousch, L.; Richter, M. Modelling of a grid-independent set-up of a PV/SOFC micro-CHP system combined with seasonal energy storage for residential applications. *Energy* **2022**, *15*, 1388. [CrossRef]
70. Uyan, M. GIS based solar farms site selection using analytic hierarchy process (AHP) in Karapinar region, Konya/Turkey. *Renew. Sustain. Energy Rev.* **2013**, *28*, 11–17. [CrossRef]
71. Jahangiri, M.; Ghaderi, R.; Haghani, A.; Nematollahi, O. Finding the best location for establishment of solar-wind power stations in Middle-East using GIS: A review. *Renew. Sustain. Energy Rev.* **2016**, *66*, 38–52. [CrossRef]
72. Sliz-Szkliniarz, B.; Vogt, J. GIS-based approach for evaluation of wind energy potential: A case study for the Kujawsko-Pomorskie Voivodeship. *Renew. Sustain. Energy Rev.* **2011**, *15*, 1696–1707. [CrossRef]
73. Sliz-Szkliniarz, B.; Vogt, J. A GIS-based approach for evaluating the potential of biogas production from livestock manure and crops at a regional scale: A case study for Kujawsko-Pomorskie Voivodeship. *Renew. Sustain. Energy Rev.* **2012**, *16*, 752–763. [CrossRef]
74. Rozakis, S.; Kremmydas, D.; Pudełko, R.; Borzecka, M. Straw potential for energy purposes in Poland and optimal allocation to major co-firing power plants. *Biomass Bioenergy* **2013**, *58*, 275–285. [CrossRef]
75. Ozawa, A.; Kudoh, Y. Assessing uncertainties of life-cycle CO_2 emissions using hydrogen energy for power generation. *Energies* **2021**, *14*, 6943. [CrossRef]
76. Ćosić, B.; Krajačić, G.; Duić, N. A 100% renewable energy system in the year 2050: The case of Macedonia. *Energy* **2012**, *48*, 80–87. [CrossRef]
77. Główny Urząd Statystyczny. *Zużycie Paliw i Nośników Energii w 2019 Roku*; GUS: Warsaw, Poland, 2019.
78. Igliński, B.; Cichosz, M.; Kujawski, M.; Plaskacz-Dziuba, M.; Buczkowski, R. Helioeneregy in Poland–Current state, surveys and prospects. *Renew. Sustain. Energy Rev.* **2016**, *58*, 862–870. [CrossRef]
79. Rataj, M.; Berniak-Woźny, J.; Plebańska, M. Poland as the leader in terms of photovoltaic market growth dynamics–behind the scenes. *Energies* **2021**, *14*, 6987. [CrossRef]
80. Gnatowska, R.; Moryń-Lucharczyk, E. The place of photovoltaics in Poland's energy mix. *Energies* **2021**, *14*, 1471. [CrossRef]
81. Igliński, B.; Iglińska, A.; Kujawski, W.; Buczkowski, R.; Cichosz, M. Bioenergy in Poland. *Renew. Sustain. Energy Rev.* **2011**, *6*, 2999–3007. [CrossRef]

82. Główny Urząd Statystyczny. *Rocznik Statystyczny Leśnictwa 2020*; GUS: Warsaw, Poland, 2020.
83. Główny Urząd Statystyczny. *Rocznik Statystyczny Rolnictwa 2020*; GUS: Warsaw, Poland, 2020.
84. Jasiulewicz, M. *Potencjał Biomasy w Polsce*; University of Technology in Koszalin Publishing: Koszalin, Poland, 2010.
85. Igliński, B.; Buczkowski, R.; Iglińska, A.; Cichosz, M.; Piechota, G.; Kujawski, W. Agricultural biogas plants in Poland: Investment process, economical and environmental aspects, biogas potential. *Renew. Sustain. Energy Rev.* **2012**, *7*, 4890–4900. [CrossRef]
86. Piechota, G.; Igliński, B. Biomethane in Poland–current status, potential, perspective and development. *Energies* **2021**, *14*, 1517. [CrossRef]
87. Główny Urząd Statystyczny. *Ochrona Środowiska 2020*; GUS: Warsaw, Poland, 2020.
88. Kasiulis, E.; Punys, P.; Kvaraciejus, A.; Dumbrauskas, A.; Jurevičius, L. Small hydropower in the Baltic States–current status and potential for future development. *Energies* **2020**, *13*, 6731. [CrossRef]
89. Kishore, T.S.; Patro, E.R.; Harish, V.S.K.V.; Haghighi, A.T. A comprehensive study on the recent progress and trends in development of small hydropower projects. *Energies* **2021**, *14*, 2882. [CrossRef]
90. Manzano-Agugliaro, F.; Taher, M.; Zapata-Sierra, A.; Juaidi, A.; Montoya, F.G. An overview of research and energy evolution for small hydropower in Europe. *Energ Convers. Manag.* **2017**, *75*, 476–489. [CrossRef]
91. Restor Hydro. Available online: www.restor-hydro.eu/en/tools/mills-map (accessed on 21 November 2020).
92. Igliński, B. Hydro energy in Poland: The history, current state, potential, SWOT analysis, environmental aspects. *Intern. J. Energy Water Resour.* **2019**, *3*, 61–72. [CrossRef]
93. Date obtained from National Water Management Board. 2020.
94. Data obtained from Geodesic and Cartographic Documentation Centre. 2020.
95. Szewczyk, J.; Gientka, D. Terrestrial heat flow density in Poland–A new approach. *Geol. Q.* **2009**, *1*, 125–140.
96. Igliński, B.; Buczkowski, R.; Kujawski, W.; Cichosz, M.; Piechota, G. Geoenergy in Poland. *Renew. Sustain. Energy Rev.* **2012**, *5*, 2545–2557. [CrossRef]
97. Data obtained from Central Geological Database. 2020.
98. Bujakowski, W.; Tomaszewska, B. *Atlas Wykorzystania Wód Termalnych do Skojarzonej Produkcji Energii Elektrycznej i Cieplnej przy Zastosowaniu Układów Binarnych w Polsce*; Jak: Kraków, Poland, 2014.
99. Gospodarka Konin. Wyspa Pociejewo. Available online: http://gospodarka.konin.pl/wyspa-Pociejewo.htmlx (accessed on 1 October 2021).
100. WWF. EU Coal Regions Share Just Transition Needs Despite Their Differences. Available online: https://wwf.panda.org/wwf_news/?359877/EU-coal-regions-share-just-transition-needs-despite-their-differences (accessed on 1 October 2021).

energies

MDPI

Article

A Simulation Model of Power Demand Management by Manufacturing Enterprises under the Conditions of Energy Sector Transformation

Justyna Smagowicz [1], Cezary Szwed [1,*], Dawid Dąbal [2] and Pavel Scholz [3]

1 Faculty of Management, Warsaw University of Technology, 02-524 Warsaw, Poland; justyna.smagowicz@pw.edu.pl
2 InterMarium Ltd., 31-548 Kraków, Poland; dawid.dabal@flexsim.pl
3 Faculty of Mechanical Engineering, Czech Technical University in Prague, 166 07 Prague, Czech Republic; pavel.scholz@cvut.cz
* Correspondence: cezary.szwed@pw.edu.pl; Tel.: +48-22-234-84-32

Abstract: This paper addresses electricity consumption management in manufacturing enterprises. The research aims to provide manufacturing enterprises with an effective tool to control electricity costs. Recently, some factors have been observed to affect the rapid changes in the operating conditions of enterprises. These include the transformation of the power sector toward renewable energy, the disruption of supply chains resulting from a coronavirus pandemic, political crises, and process automation. A method for the analysis and management of electricity consumption in enterprises based on simulation modeling is proposed. The simulation model contains predefined objects representing physical system elements and the data processing algorithm. The production order execution time, energy consumption, employee overtime, and machine load are included in the model. The results show that it is possible to determine the level of power available for the process completion and its influence on the production volume and realization time. In the studied case, when the available power was reduced by half, there was an increase in order execution time of nearly 25 percent and an increase in energy consumption of nearly 15 percent. The method can be used in the operational activities of enterprises as well as extended to different types of production processes.

Keywords: electricity consumption; simulation modeling; production management; manufacturing company

Citation: Smagowicz, J.; Szwed, C.; Dąbal, D.; Scholz, P. A Simulation Model of Power Demand Management by Manufacturing Enterprises under the Conditions of Energy Sector Transformation. *Energies* **2022**, *15*, 3013. https://doi.org/10.3390/en15093013

Academic Editors: Bartlomiej Iglinski and Michał Bernard Pietrzak

Received: 25 March 2022
Accepted: 18 April 2022
Published: 20 April 2022

Publisher's Note: MDPI stays neutral with regard to jurisdictional claims in published maps and institutional affiliations.

Copyright: © 2022 by the authors. Licensee MDPI, Basel, Switzerland. This article is an open access article distributed under the terms and conditions of the Creative Commons Attribution (CC BY) license (https://creativecommons.org/licenses/by/4.0/).

1. Introduction

1.1. General Factors Influencing Energy Price Increases

The primary resource necessary for the operation of practically any economy is electricity. Electricity is supplied from generation units via transmission and distribution grids to the customers. The generating sources are primarily large system power plants. Renewable sources have still not become dominant, despite significant political pressure in recent decades and massive subsidies to support their installation and operation.

The significant increase in the number of small distributed sources has also not changed the structure of the power systems. The amount of renewable energy supplied to consumers has increased, but so have the problems of grid operators responsible for the security of supply and balancing the power system resources. So far, the main visible effect of the transformation of the electricity sector has been an increase in electricity prices for consumers. The electricity costs are distributed differently among various customer groups depending on political and regulatory decisions in individual countries.

In the past few years, additional factors driving up the operating costs of manufacturing companies have been the coronavirus pandemic and local and global political crises. The coronavirus pandemic led to the disruption of supply chains for materials, components, and products. In the operation of manufacturing companies, this meant that there was an additional factor causing significant changes in electricity consumption. When production stops, electricity consumption drops temporarily. The contracted energy is not collected. When production is resumed, consumption rises significantly to make up for the earlier failure to produce. Politic crises can lead to a sharp increase in the price of primary fuels for electricity production. In the long run, they can also lead to changes in the energy policy of individual countries (i.e., the replacement of one primary fuel by another, usually more expensive one). This may mean an increase in the cost of electricity supply, which applies to both municipal consumers and manufacturing companies.

The automation and robotization of production processes is another factor leading to an increase in the energy costs of manufacturing companies. For several decades, along with the development and growth in the efficiency of processes, their energy intensity has also increased. With the noticeable trend of rising energy prices, the problem of reducing energy consumption has become significant for most companies. The undertaken actions aim to reduce the operating costs of enterprises, or at least stop their growth. However, it turns out that only a part of the electricity supplied to the enterprise is used directly for production processes. As in the case of thermal power plants, a significant part of this energy is wasted on thermal losses [1]. Companies are trying to reduce energy losses, and these actions should be aimed in different directions. On one hand, to reduce thermal losses of energy, and on the other hand, to lead to the best possible use of energy that goes into the production process.

1.2. Determinants of Energy Consumption Management in Enterprises

With the development of automation and robotics, electricity is becoming a key resource for the execution of the production process, and its lack has already led and will lead in the future to the interruption/suspension of ongoing processes. Energy as a resource is used in the entire production area, starting from ensuring the operation of machinery, through running transport and storage systems, to ensuring proper working environmental conditions. This includes artificial lighting, maintaining a constant temperature, ventilation, etc. [2].

There are three basic levels of research identified in the literature, depending on the details of the analysis. The first level consists of a single workstation, usually a machine. The second is a production cell, a controlled cluster of interconnected machines. The third is the production system, a cluster of many production cells and areas that allow for handling activities within the entire enterprise. For example, it may include material and product storage areas and social areas. At each of these levels, a specific set of resources must be identified and the extent of electricity consumption must be determined [3,4].

For selected resources, consumption over time can be constant or variable. This factor is an important criterion when deciding on the running of technological operations as well as their financial implications for the company. The possibility of variable power consumption is more flexible for the entire process and allows for the operations to be adjusted to the company's capabilities. Constant power consumption in time usually means that power is drawn when the resource is working. This situation, however, requires a decision as to when the resource should be switched on, whether it is possible to ensure its continuous operation, and what this means for the company.

Another important aspect is the use of different technologies in the implementation of production processes that have different characteristics. For example, this may be related to the necessity of performing certain activities in a certain sequence or procedure. It can also be related to the maintenance of certain parameters of machine operation (temperature, humidity, pollution level, noise level). It turns out that even in the case of variable power consumption by a given resource, the company may not have such high flexibility. Due to the nature of the operation/technology used, the resource will be forced to maintain its operating parameters, consuming an adequate amount of power, incommensurately greater than the operation performed. For example, ensuring a high temperature of the furnace or dryer during operation usually means a high power consumption throughout the entire period of operation of the machine, not only during the processing of the material/product, but also while waiting for the delivery of material to operate.

1.3. Objectives and the Paper Content

The main subject of this research was the impact of the occurrence of power limitation on the production system. The production system is a deliberately designed and organized system, within which the available resources are used to produce specific products that meet the consumer needs [5]. This system is based on the execution of the production process, using four key resources. The resources include material resources, human resources, information resources, and capital resources. In the last few years, more and more attention has been paid to including the energy component in the definition of the production system. This component is one of the inputs to the manufacturing process. Electricity makes it possible to perform technological operations on machines as well as to ensure the functioning of the entire production environment.

It is likely that soon, both the increase in electricity prices and the reduction in the availability of electricity may constitute one of the main challenges for manufacturing companies. In this context, this paper aims to propose a method that would enable manufacturing companies to analyze electricity consumption in rapidly changing production conditions. In particular, the method should answer the question of how the constrained availability of power or large differences in electricity prices between different times can be compensated by changing the production plan. A change in the production plan is understood as an extension of the execution time for a given production batch, moving the execution to another time, or determining which part of the production batch will not be completed on the previously planned date. This type of approach has not been considered in the literature thus far. The main conclusion of this paper is that with the help of available simulation modeling programs, it is possible to analyze electricity consumption in production processes and draw practical benefits for the company. An additional conclusion is the possibility of developing and applying the proposed method to various production processes, which requires further research.

2. Literature Review

The literature review is divided into several key topics concerning the addressed problem. These topics include power sector and market characteristics, energy management and energy consumption in manufacturing companies, energy management system, and simulation modeling in energy management. The literature studied on each topic can be seen in Table 1, and only the most relevant literature is cited in the text.

Table 1. Literature review structure and resources.

Literature Review Topic		Literature
Power sector and market characteristics	Poland	[6–8]
	Czech Republic	[9–15]
	Both countries	[9,16,17]
Energy management in manufacturing companies	General perspective	[6,18–27]
	Systematic approach	[25–33]
	Optimization areas	[1,21,24,26,27,31,34–40]
Energy consumption in manufacturing companies	Consumption levels	[3,4,41,42]
	Consumption analysis	[1,41–49]
Energy management system	Structure	[50–58]
	Specialized software	[50,58–65]

Table 1. *Cont.*

Literature Review Topic		Literature
Simulation modeling in energy management	Areas/sectors of use	[26,66–75]
	Level of machine	[66,76]
	Level of plant	[66,69–75,77,78]
	Level of plant with supporting technology	[26,66,68,79]
	Integration of parameters	[26,66–75,77,78,80]
	Analysis, optimization, strategy validation, and other benefits	[2,26,66–75,77,78,81,82]

2.1. Power Sector and Market Characteristics

The transformation of the power sector in some countries is mainly concerned with increasing the share of renewable energy in total energy consumption, which is mostly taking place in European Union countries. This transformation of the sector is driven by EU policy objectives and affects both the power sector and the functioning of the energy market. The directions and pace of changes vary between the EU countries. The biggest challenges are faced by countries where the share of energy produced from coal is the highest. Such countries include Poland and the Czech Republic, which have different amounts of energy production from coal and declare different energy policies.

Poland declares its support for mining and the use of coal, especially because changing the fuel mix would require huge expenditure. The reason would not only be the need for new technologies, but also the necessity to cope with the general transformation of the economy and its social effects. At the same time, energy security must be ensured. The entire transformation should therefore be based primarily on the modernization of existing coal-fired power plants (where cost-effective), resulting in greater efficiency and cost savings. Moreover, the use of renewable energy sources is expected to increase, although this type of energy is not significantly used in Poland [6]. Achieving the coal reduction target will be a challenge, though studies undertaken suggest that it should be achievable [7]. However, Poland's final energy mix will depend heavily on the impact of various environmental policy factors on the energy industry [8].

The Czech Republic is currently an exporter of electricity and is reported to be the seventh-largest exporter of electricity in the last decade. However, unlike Poland, the energy policy update enforces a clearer shift away from coal, also taking into account the potential social and economic impacts [9]. In terms of electricity production, an important role in the replacement of coal will be played, especially by the greater use of RES and nuclear energy. In the field of heating, this will be the use of natural gas [10,11].

However, the plans to move away from coal are probably too ambitious. Replacing coal with nuclear power has become increasingly problematic in recent years. Decisions on the construction of new units are being delayed or cancelled, and the global construction of nuclear power plants is running behind schedule. The commissioning of some new units is planned just before the official end of life of the current units [12].

There is also a relatively high level of opposition to the construction of reservoirs with hydroelectric power plants (primarily for water supply) and wind power plants. Similarly, the landscape conditions are not entirely ideal for these sources to be used to a greater extent in the Czech Republic, especially for hydropower plants [13]. The greater development of RES is undoubtedly hindered by the reduction in state aid [14] and contradictory public opinion regarding these resources [15].

The whole energy situation in both countries is extremely problematic and there is potential talk of instability of the electricity grid in the whole region [9]. In principle, it is possible to expect a scenario in which the slow reduction in coal use leads to an increase in energy prices (due to the purchase of emission allowances, the maintenance of a strategic reserve, or the import of coal) or a scenario in which there is a shortage of resources for electricity and heat production. The transformation of energy production must therefore go hand in hand with the transformation of the electricity market. Given that there is a certain similarity between the Czech and Polish energy markets [16], it can be assumed that the development and transformation in both markets could follow a similar pattern.

The key is the creation of a dual-energy market, where, in addition to buying and selling the physical electricity consumed, the readiness to supply energy is also purchased to ensure energy security. This is because it is necessary to hold a strategic reserve in conventional sources in the case of outages or the instability of electricity supply from RES (problems in the case of no wind or cloudy skies). However, this solution does not correspond to the idea of green electricity, is economically expensive, and is also not potentially feasible where there are not enough energy sources available (e.g., because they have been decommissioned due to high emissions). In this context, the responsibility for the stability of supply is beginning to shift much more from the manufacturers and distributors to the customers themselves. An important role should be played here by the DSR (demand-side response) service to reduce energy consumption on the customer's side, without which ambitious energy goals will not be achieved [17].

2.2. Energy Management in Manufacturing Companies

Within DSR, unlike other tools, the customer takes care of power management, as it is necessary to know the current consumption and the possibilities of reducing it for quality DSR operation (see Energy management below). It will no longer only be necessary to adjust production according to the current consumption, but thanks to the financial reward for "non-consumption" (or, conversely, "over-consumption"), it will be possible to influence the behavior of the consumers themselves. Thus, demand will be managed using a pricing policy, and, paradoxically, there may be situations where consumers consume more, but at a lower cost [18]. Moreover, as it turns out, DSR and management through a clear pricing policy alone can have a much greater impact on RES uptake than other factors such as the customers' environmental sentiment [19]. This fact, on the other hand, is not so surprising, because in general, the reasons for increasing energy efficiency, which is closely linked to consumption, are mainly organizational (e.g., education, know-how), managerial (e.g., competence), or economic (e.g., cost reduction) [20]. One of the ways to deal with grid instability could also be so-called energy clusters or hubs, which would have more predictable energy consumption patterns and thus enable a more efficient use of RES [6,21].

However, ensuring energy-efficient production and implementing DSR in the manufacturing sector is significantly more complicated than, for example, in the residential and service sectors [21,22], given the difficulty of matching energy optimization with production requirements [23,24]. A large number of factors are involved in the whole process (see below). Energy management should help companies cope with these requirements [25–27]. Scientific articles have mentioned several important areas to focus on in this context.

First and foremost is the importance of setting up a framework—a systematic approach using ISO 50001 and ANSI MSE 2000 (United States) [25,28,29]. Or also other standards (e.g., EN 16231) [26,30,31]. Next is respecting all key steps of the energy management process. This includes, for example, the setting of KPIs, which is important, among other things, in terms of the actual optimization of energy consumption and the measurability of the effect [26,27,32]. Furthermore, the mapping and analysis of the whole system, focusing on energy management (not only electricity, but also gas, water, and raw materials), on load curves and the possibilities of influencing them to shift the load from the peaks to the valleys as well as on the characteristics of the production program (batch sizes, production processes). This step should be implemented from the whole production plant down to the machine components. Various tools such as energy audits, value stream mapping, Sankey diagrams, etc., can be used for this purpose [26,29,33].

Without process analysis or knowledge of KPIs, it is not possible to address energy optimization, which includes activities such as process modification, modeling, planning, scheduling, forecasting, etc., to ensure energy-efficient production [26,27]. Energy optimization focuses on several key areas such as [21]:

- Production technology and its actual selection in terms of the design and operating characteristics of machines and their components, tools, and other equipment including their maintenance [26,31,34].
- The material and design of the products. Furthermore, the manufacturing process itself, for example, in terms of pressing using sequential tooling on several presses [26,34].
- The organization, planning, and scheduling of production [26]. Important in this context is the focus on the selection of specific machines (e.g., regarding their energy efficiency, reliability, or utilization to limit machine downtime, regarding batch sizes, etc.) [35–37].
- Educating workers on machine operation and maintenance, operating modes, etc. [24,26,32].
- Operation of supporting technologies and equipment in terms of heating and cooling of buildings, water heating, etc.

Finally, it is of course necessary to simultaneously respect the energy (e.g., energy availability, regulations, prices), operational (e.g., current operation of some machines), safety, and other constraints (e.g., quality) [24,34]. In some cases, these can often be subtle limitations (e.g., switching off machines can compromise their thermal stability) [38].

As some of the factors may be variable [30] such as the electricity consumption of the technologies depending on the type of product or the electricity available from the grid, the optimization needs to focus on different scenarios taking these factors into account [38].

Moreover, it should be noted that only part of the electricity supplied to the enterprise is used for the realization of production processes [1]. Part of the energy is transferred to powering other devices (not directly related to technological operations) or the realization of auxiliary processes. According to the concept of lean manufacturing, these activities are called non-value-added but necessary operations. Their execution supports the realization of the basic process (i.e., it is necessary/required), but they do not translate directly into manufacturing a product for the customer. Especially in the engineering sector, the state of "processing" reaches a high percent share in energy consumption, however, depending on the technologies used in the industry, this share may not be significant/dominant [39,40].

2.3. Energy Consumption in Manufacturing Companies

The study of energy consumption in manufacturing companies focuses on identifying and analyzing the elements that contribute to power consumption. Each element is characterized by a different energy consumption profile. In a manufacturing company, the number of elements to be surveyed is significantly high, therefore, depending on the purpose of the survey, it is necessary to divide the company into smaller elements, for which the survey is simpler.

It is possible to study individual elements, with their impact on the examined area, and then add/sum up other elements, thus extending the scope of the research, the examined energy consumption. In the literature, these ranges are called levels of energy testing. The most common is the division into three to six levels. Research in this range has been conducted, for example, by [3,4,41].

Based on these, four levels of energy consumption testing have been identified:

- Machine level—A single machine or tool is tested, which is used to perform production operations.
- Multi-machine level—Logical organization of devices in the system in the form of a production cell (production line, socket); the devices perform the assigned operations in series or parallel.
- Factory level—a separate system of interconnected devices is examined.

- Multi-factory level—differentiated manufacturing companies that are in a relationship with each other due to the joint performance of activities, generating synergy effects are subject to examination.

Determining the energy consumption levels is easiest to study for the lowest level (machine level), and at subsequent levels, add energy consumption profiles for additional resources included in the analysis. For example, examining energy consumption at the multi-machine level would consist of overlaying the energy consumption profiles of each machine and adding up the results for each moment/period of analysis [42–44]. Testing the energy consumption of a single machine is a complex task [45–47] and requires the definition of a consumption profile for each machine. The consumption profile is closely linked to the various states that the machine can be in (process operation state, standby state, etc.). In each of these states, the power consumption of the machine is different and the total energy consumption depends on the duration of each state. Research on the identification of possible states to be reached by machines has been the subject of many publications, for example, [41,45,48], especially in the engineering industry. Based on the research presented, nine states were identified: Power off, Shut down, Warm-up, Power on, Start-up, Stand by, Production, Maintenance, Failure. The publications by [1,48,49] identified the status/characteristics of the values taken in each state by defining them as constant or stochastic (or variable). This aspect is important for the detailed analysis of power consumption at a given time and the study of parameters, conditioning the power consumption.

2.4. Energy Management Systems

Another important issue in the context of this research is the systems supporting the management of energy consumption by enterprises. An energy management system is an interconnected set of elements, devices, and tools (hardware and software) used for monitoring, predicting, controlling, and optimizing energy consumption. Such systems have so far been mainly used by large enterprises, but with the development of modern technologies, digital twins, and the concept of Industry 4.0, they will become and are becoming much more accessible to smaller companies [50–52].

The whole system consists of several layers, the number of which varies according to the detail of the view of the issue. The first layer consists of the end devices (machines, equipment, but also energy sources), where the energy consumption data are collected [53,54]. The second layer is the communication and integration layer and is used to send data from devices in the first layer and to transmit commands from higher layers in the direction of the end devices [53,54]. The third layer consists of a server with a database and possibly other tools (machine learning, prediction engine, etc.) or a cloud environment [51,53,55]. The fourth layer is then functional (information and control), and consists of either specialized energy management software (Siemens Simatic, Wattics, EnergyCAP, ProntoForms) or other less sophisticated tools for prediction and data visualization such as Python and Microsoft Power BI [51,53,55]. Within the previous two layers, they can be integrated with other enterprise systems designed for production management such as the ERP, MES, APS, or simulation tools [55,56]. The fifth layer is the control logic itself in the form of objectives, strategies, and rules to ensure energy-efficient operation, respectively, the optimization of energy consumption, respecting production requirements and various other constraints [56–58].

Depending on the specific solution, specialized energy management software can have different functions [59–61]. Key functions include [62–65].

- Visualization and analysis of historical data related to KPIs, different types of energy data, but also weather and environmental conditions, production, seasonal effects, costs, and energy prices.
- Monitoring of current energy consumption and other energy data including various automatic alerts.
- Prediction and forecasting of energy demand or energy prices.

- Regular reporting focusing on KPIs and other data.
- Control of energy consumption, production, and possibly energy storage.
- Integration with production planning and scheduling tools to optimize consumption.

However, to effectively manage energy consumption, respectively, ensure energy-efficient operation of production using energy management software, it is necessary to understand the optimal rules and strategies for the management itself, taking into account different conditions (available energy, energy prices, etc.). Despite the relatively broad functionality, energy management systems do not allow for the testing of various scenarios of future events related to energy limitations and the selection of the solution best suited to the needs of enterprises. Simulation modeling offers such possibilities.

2.5. Simulation Modeling in Energy Management

Simulation modeling is a suitable tool to design and verify various rules and strategies for power and energy management, to find the optimal production setup in terms of energy consumption, and also for the analysis of energy consumption itself. Its use is quite widespread:

- Especially in the automotive industry [26,66,67];
- In the field of battery production (especially for automobiles) [68,69];
- In energy-intensive industries such as plastics and metal forming, the paper industry, or the foundry industry [66,70,71]; and
- There are also studies from the mechanical engineering industry in general [72–75].

In these sectors, simulation models are created for both systems consisting of a single machine or a multistage machine [66,76] and for entire production plants [69,77,78], sometimes including, for example, the operation of buildings or other supporting technologies [26,66,68,79].

The main motive for using simulation modeling is almost always to analyze energy consumption and reduce costs or energy intensity to resource depletion or global warming. To reduce consumption, the following approaches can be identified primarily, namely, optimization of machine control functions, optimization of parameters in production, and the design of energy-efficient production at the beginning of its design. However, individual approaches and models are quite specific and focused on addressing a particular problem and need [26,70,77]. In this context, their wider use is thus rather limited and a comprehensive view is usually missing—the greater integration and interaction of various factors and parameters such as production equipment operation, process parameters, building operation, price tariffs, etc. For example, if parameters such as operation and management of individual machines, production batches, efficiency, and number of operating machines, failure rates, building operation, etc. are integrated into the model in great detail, parameters such as price tariffs or constraints on the energy supply side are no longer integrated into the model [68,69,78]. On the other hand, if parameters such as price tariffs or supply-side constraints are already integrated, other parameters are not integrated to a sufficient extent or in sufficient detail [71,72,75]. Another situation is that models only integrate certain types of energy such as electricity, for example, [69,72,73].

One reason why other different factors and parameters are not integrated into the models may be that simulation tools still provide their users with limited options in some respects, despite some progress in recent years [80]. However, the key point remains that simulation modeling allows companies to analyze and optimize consumption [26,69,70] as well as validate different energy optimization strategies [2,81]. Moreover, without having, for example, APS software, which they often cannot even buy for financial reasons, it is also important that, unlike APS software, simulation modeling allows for the inclusion of uncertainty in the model (e.g., in the form of machine failure rates or production quality) [68,70,77]. Another benefit is that simulation modeling enables "what if" analyses [78,82].

3. Materials and Methods

A comprehensive study of the energy consumption of a production system was based on the identification of all elements of the system that consume electricity. The exclusion of any element will result in an incomplete view of the analysis and subsequently in inappropriate decisions being made by the enterprise. The literature review shows that depending on the purpose of the study and the analysis capabilities, the approaches will vary. In this paper, a multi-stage research method is presented to show how to study energy consumption in an enterprise using simulation modeling. The research method is shown in Figure 1. The subsequent stages of the method are described below.

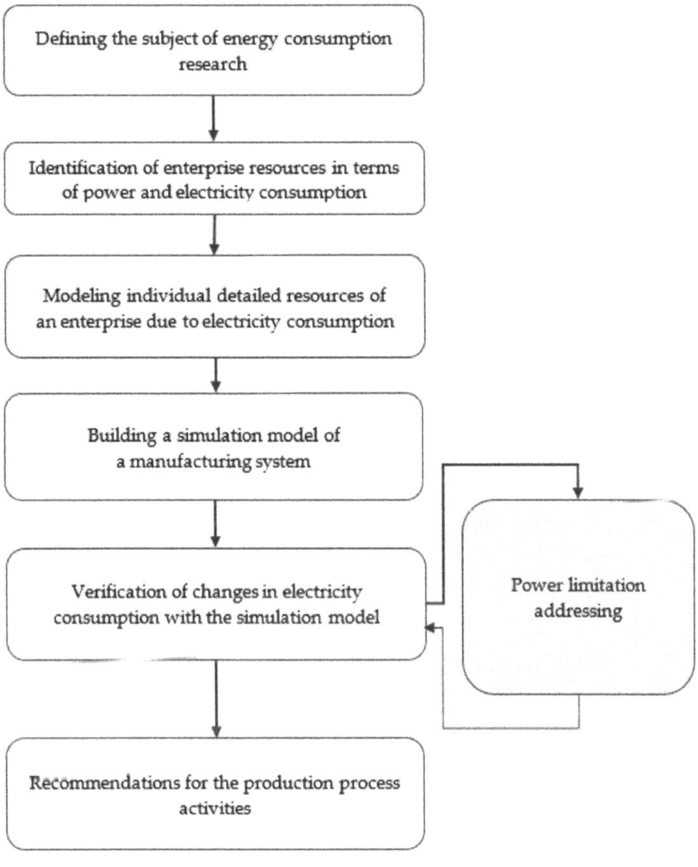

Figure 1. Research method.

Stage 1. Defining the subject of energy consumption research. The purpose of the analysis to be performed will influence the scope of the work undertaken. From the literature survey conducted, there are several approaches to studying energy consumption, related to either a tight or wide scope of the study. In the tight scope, the analysis deals with energy consumption on a single machine, where the study will focus on that single facility and the instrumentation/equipment used on that machine. In the case when the analysis concerns the entire production cell (production line, production slot), tests will be conducted for each machine (including equipment) and the transport operations related to the flow of material between objects (machines, buffers between stations, place of collection of materials/parts or storage of finished products). Transport operations include belt feeders, forklifts, etc.

A comprehensive approach assumes the identification of the impact of the environment in which the studied production system operates. Therefore, the analysis of the entire production system should not only take into account the individual energy consumption of individual production cells (located in a given area), the service system (transport, picking materials, parts, semi-finished or finished goods), but also the environment of the given production system (lighting, air conditioning, social facilities, computers for processing production orders). In selected studies, further levels of analysis appeared concerning energy consumption within interconnected production systems or the entire supply chain. Due to the research topic undertaken in this paper, the authors decided to include the first three levels indicated in the research method.

The specific scope of the research carried out influences the subsequent activities performed within the method—limiting or extending the scope of these activities.

Stage 2. Identification of enterprise resources in terms of power and electricity consumption. The subject specified in stage 1 sets the level at which the electricity consumption analysis will be performed. Based on this, it is clear how detailed the enterprise resources should be analyzed. The classification of enterprise resources includes physical resources, information resources, human resources, and financial resources. In the framework of each group, it is necessary to identify all objects used in the implementation of the production process that are associated with energy consumption. The largest number of objects can be found in the material resources, where the following are distinguished: machines, tools, possible means of transport, or equipment of the entire production system. In other resources, the range of identified objects will be tighter and will be more limited to the level of analysis performed (machine level, production cell level, or production system level).

As part of information resources, the knowledge that employees have should be identified. Knowledge itself is not associated with energy consumption, but the way it is collected, processed, and the process of exchanging knowledge among employees (for example, through computers, tablets, and cell phones) is.

Within human resources, employees who perform tasks in the production area under study should be identified. Similar to information resources, workers affect energy consumption by performing activities related to/involving machine operation, connecting/disconnecting tooling, or operating logistics equipment. Therefore, the identified activities must be classified appropriately to the levels of analysis performed. For example, devices used to collect and process data on the process flow or readers recording working time should be analyzed only at the level of examining a production cell or the whole production system.

The final resource is financial resources. Similar to the previously described information and human resources, the nature of financial resources does not involve/directly cause energy consumption. However, the activities related to recording or managing these resources using electronic equipment (computers, tablets) are associated with energy consumption. The identification of each resource at the listed levels of energy use analysis is shown in Table 2.

Table 2. Identification of resources corresponding to energy consumption levels.

	Machine Level	Production Cell Level	Production System Level
Physical resources	Machine (including equipment)	Machine (including equipment) Buffers Material Handling System	Machine (including equipment) Buffers Material handling system Equipment of the entire production system Social facilities

Table 2. *Cont.*

	Machine Level	Production Cell Level	Production System Level
Information resources	Electronic equipment (computers, tablets)	Electronic equipment (computers, tablets)	Electronic equipment (computers, tablets)
Human resources		Electronic equipment (computers, tablets)	Electronic equipment (computers, tablets) Employee time recording software
Financial resources			Electronic equipment (computers, tablets)

Stage 3. Modeling individual detailed resources of an enterprise due to electricity consumption. It is necessary to identify the power and energy consumption profile of the enterprise, having determined the enterprise resources to be studied. For this purpose, for each identified resource, the parameters characterizing the consumption profile are determined. The range of parameters studied is complex and should take into account the standard nature of the work of a given resource. Additionally, it should include changes of this nature depending on the changing factors of the surrounding environment. Therefore, these parameters can be divided into internal factors (related to the operation of the machine—the operating states of the analyzed object) and external factors. If there are several resources with similar parameters working at the same time, the resources can be aggregated and studied as a whole. A summary of the key factors is presented in Table 3.

Table 3. Parameters characterizing the power and energy consumption profile.

	List of Factors	Impact
Internal factors	Power Off	The machine does not consume power or consumes very little power.
	Warm-Up/Fast-Warm Up	The machine consumes additional power to set the operating parameters at the right level. In the case of the "Fast Warm-Up" option, the power consumption is even higher.
	Idle	The machine consumes a nominal power assigned to the standby mode in readiness for the next technological operation (maintaining the set machine parameters).
	Processing	The machine consumes a nominal power in the course of a technological operation.
	Stand by	The machine consumes a nominal power assigned to the machine's ready state to enter "Idle" mode.
	Failure	The machine consumes reduced power if the disturbance prevents or restricts operation. The machine may consume an increased power where the disturbance degrades the performance of the machine.
	Maintenance	The machine consumes varying power depending on the maintenance and repair work performed.

Table 3. *Cont.*

	List of Factors	Impact
External factors	Time of the day	In the production area, there is an increased energy consumption in the afternoon and at night due to the need to switch on artificial lights.
	Season	In the production area, there is increased energy consumption in winter due to the need for heating. In summer, on the other hand, increased energy consumption can result from the need to switch on air conditioning.
	Thermal losses of associated machines	In a production area, the increased/reduced energy consumption must be considered to the influence of other machines in the area. For example, a machine that generates a lot of heat during operation may require additional cooling equipment. This would be particularly advisable in summer when the temperature in the production hall can also be high. Similarly in the winter period (lower temperatures), the cooling equipment may be used less often due to the lower temperature on the production floor.

Stage 4. Building a simulation model of a manufacturing system. Based on the specified level of analysis of energy consumption (machine, production cell, or production system), the objects that should be placed in the simulation model should be identified. The list of objects should be as presented in Table 1. For each level, the resources enabling the realization of the production process have been defined, divided into the material, personnel, financial, and information resources. Within the framework of the simulation model being built, it is possible to represent material and personnel resources using objects occurring in the selected computer modeling program. In the next step, it is necessary to identify the connections occurring between these objects. For each identified object, it is necessary to determine the set of parameters, characterizing the data on the course of the production process (duration of a technological operation, setup time, the size of the production batch) and data on energy consumption (normative for different states of the machine). The simulation model should be validated to confirm the correctness of the representation of the tested real system.

The developed simulation model makes it possible to carry out several analyses of a given production area without the need to interfere in the real process. Particularly in the case of the analysis of power consumption by the equipment, carrying out tests on the real system would be difficult due to the necessity of stopping the production of objects in progress, the possibility of failure to the objects consuming energy from the system, or the entire power system in the company. An additional advantage of using computer simulation is the possibility of conducting tests in a short time considering many different variants of solutions through one built model—changing only the values of selected variables. The more data are entered into the model, the more analyses (with a wider scope) will be possible to conduct.

Stage 5. Verification of changes in electricity consumption with the simulation model. Power limitation addressing. In a general case, the subject of research can be the analysis of the impact of the power limitation (by a set value) on the analyzed production system. The power limitation may be of soft or hard character (constraint). The soft constraint may result from high electricity prices in a certain part of the planning period (e.g., during selected hours of the work shift). Exceeding this constraint is possible and affects the high cost of implementing the production process. Maintaining this constraint makes it necessary to shift production to other hours (e.g., night hours), when the energy price is acceptable. The hard case of the capacity constraint may be caused by the limitation of supply by the power system. This constraint cannot be exceeded. Any non-executed production must be postponed to another period. In any case, at the start of the operation, the production system has a certain level of available capacity. Then, during the execution of the process, the available power is reduced in the specified time. It is necessary to identify objects on which it is possible to change the state of power consumption. This should be conducted in the period with the introduced constraint, according to Table 2.

The reduction in power consumption can be realized in many ways, depending on the possibilities and needs of the implemented production process. A general (optimal or sub-optimal) procedure for power reduction should be the subject of further research. In the general case, a mathematical model of the problem and an optimization algorithm for its solution should be developed. In the practical case, a heuristic approach can be used. In such an approach, different action scenarios can be proposed and investigated. The scenarios can be based on the sequential start-up of operations, keeping selected machines in operation, keeping a bottleneck in operation, and in worst cases, stopping production. The criterion for each scenario should be to not exceed the limit of available power.

Stage 6. Recommendations for the production process activities. Based on the results obtained from the experiments, it is necessary to determine the impact of different strategies for dealing with the occurrence of power limitations. The simulation carried out allows one to determine the change in the studied parameters in terms of measurable and non-measurable quantities. For each investigated state, it is necessary to determine quantities such as the production schedule, the graph of power consumption, the duration of the production process, the value of energy consumption, and the load of machines involved in the implemented production process. The data obtained from the simulation model are only the proposals of different strategies of action by the enterprise and should be evaluated by the enterprise because of the possibility of implementing each strategy in the production process and the effects it will have on the enterprise. The results of the evaluation should be recommendations for further actions in the production process.

The indicated characteristics such as process duration and energy consumption are measurable parameters that allow for a simple comparative analysis of each strategy. At the same time, they may require a more in-depth analysis of a given parameter. For example, the increase in the duration of the production process causes both delays in delivering the product to the customer and affects the increased wear and tear of machinery/equipment or the need to pay compensation to employees. Repeated allocation of overtime to employees will have an impact on lowering the motivation of employees to work, the overtiredness of employees, thus consequently lowering the work efficiency or increasing the number of mistakes made by employees.

The increased exploitation of machinery leads to faster deterioration of its components, which means the implementation of more frequent repairs and maintenance and the need to stop production in a given production area. It is also possible for more frequent failures in the system to occur, which can be particularly severe when they occur at the moment of realization of the production process just after lifting the limitation of power consumption—at the moment of increased realization of the operation. These characteristics are difficult to introduce to the simulation model, focusing on the analysis of the selected production area. These will be taken into account as additional quantities analyzed in the given treatment strategies. The analyzed parameters include the amount of employee overtime, energy consumption by other resources necessary for the implementation of the production process (lighting of workspaces, social rooms, corridors, ventilation/space heating) as well as increased operation of machinery/equipment. In summary, it is necessary to use practical data from the enterprise to determine recommendations. In terms of production management, this is primarily an assessment of whether it is possible to implement the production process in a manner deviating from the established ones. This applies to changing the order of some technological operations, the implementation of only selected technological operations, a different order of starting and ending operations, the possibility of extending the operation of machinery, and the availability of an adequate number of workers, etc. The above information should be collected, analyzed, and evaluated by an authorized production manager. Alternatively, it can be collected in the form of a procedure to be followed in the production process in the case of a limitation of available capacity. The procedure will have a specific form for each company. As far as enterprise control is concerned, the choice of the recommended strategy should first be assessed based on the simulation results by comparing the values of the determined process parameters. On this basis, information on the cost of implementing the selected strategy and the possible lost benefits in the case of its abandonment can be determined. As a result, a decision should be made on whether the selected strategy will be implemented.

4. Results

The method characterized in the previous section was used to conduct computational experiments. From the wide spectrum of possible energy consumption management problems to be studied, the case of hard constraint power limiting was selected for the study. The analysis involved the level of the production cell. The purpose of the research was to verify the completeness and consistency of the proposed method. The possibility of building and including real production conditions in the simulation model in terms of power and electricity consumption was investigated. Based on the simulation experiments, different scenarios for responding to the situation of available power limitation were defined and investigated.

4.1. *Defining the Subject of Energy Consumption Research*

The subject under study was the production cell of a selected enterprise in the mechanical industry. The production line produces two types of products (Product 1, Product 2), which are elements of the electric motor of production machines. The production cell consists of four machines and two belt feeders. Each production workstation is equipped with a buffer where materials, parts, and semi-finished products used for the production process are stored. An operator is assigned to each workstation, who operates the machine (loads/unloads semi-finished products) and controls the machine parameters during the process.

4.2. *Identification of Enterprise Resources in Terms of Power and Electricity Consumption*

In the next step, all resources of the production line that consume power during the production process were identified (Table 4). The power consumption for buffers and belt feeders was constant at any time during the process, while for machines, it changed depending on the machine state.

Table 4. Resources of the production line.

Resource Type	Individual Resource	Type of Power Consumption	Permitted Operating States
Machine	Machine 1 Machine 2 Machine 3 Machine 4	Variable	Power Off Warm Up Processing Idle
Buffer	Buffer 1 (for M1) Buffer 2 (for M2) Buffer 3 (for M3) Buffer 4 (for M4)	Constant (0)	Power Off
Conveyor	Belt feeder 1 Belt feeder 2	Constant	Power Off Processing

4.3. Modeling Individual Detailed Resources of an Enterprise Due to Electricity Consumption

In the next step, a simulation model of the investigated production process was developed. The simulation model consisted of resources whose power consumption during the process realization was variable. For the remaining resources, power consumption was constant in each of the assumed states. In the "Power Off" state, the power consumption was 0 kW, while in the "Processing" state, it was at the set level for the operation of the resource. For example, in the case of buffers in the process under study, the power consumption was at the level of 0 kW, because they are a place for depositing products. They do not require the use of special equipment in the form of cooling/heating/keeping the movement of products, which would consume energy. For each resource, there is a profile of power consumption, which in the case of machinery is more complicated. This required introducing rules of power consumption in each of the states of the machine as well as defining the possibility of switching between the states. Characteristics of the production process in the form of unit power consumption by the machine as well as the duration of individual operations (for two manufactured products) are presented in Table 5.

Table 5. Characteristics of the implemented production process.

Factor	Possible States	M1	M2	M3	M4
Power consumption [kW]	Power Off (P1)	0	0	0	0
	Power Off (P2)	0	0	0	0
	Warm Up (P1)	1	2	1	2
	Warm Up (P2)	2	1	2	1
	Processing (P1)	6	9	10	12
	Processing (P2)	6	9	10	12
	Idle (P1)	3	4	4	4
	Idle (P2)	3	4	4	4
Time of operations [sec.]	Warm Up (P1)	60	60	60	60
	Warm Up (P2)	30	30	30	30
	Processing (P1)	28	47	55	28
	Processing (P2)	25	45	63	24

The duration of operations was given for two states of the machine (Warm-Up, Processing) because the duration of the other two states (Power Off, Idle) was calculated as part of the procedure for selecting operations on machines implemented in the simulation model. The "Power Off" state occurs when the work in the examined production cell is switched off—outside the set calendar of working hours from 6:00 a.m. to 2:00 p.m. The "Idle" state occurs on machines when the machine is waiting for a technological operation to be performed. The waiting time varies for different machines and also depends on the adopted strategy for the implementation of the production schedule.

4.4. Building a Simulation Model of a Manufacturing System

In a simulation model, the course of the investigated manufacturing process was mapped, taking into account the resources whose power consumption varies in time. To build the simulation model, data on the duration of individual operations and the volume of energy consumption for each state on specific machines were used (Table 5). Simulations were performed for a specific set of production orders scheduled during one working shift of 8 h. The simulation model developed in FlexSim 20.1.3 software is shown in Figure 2.

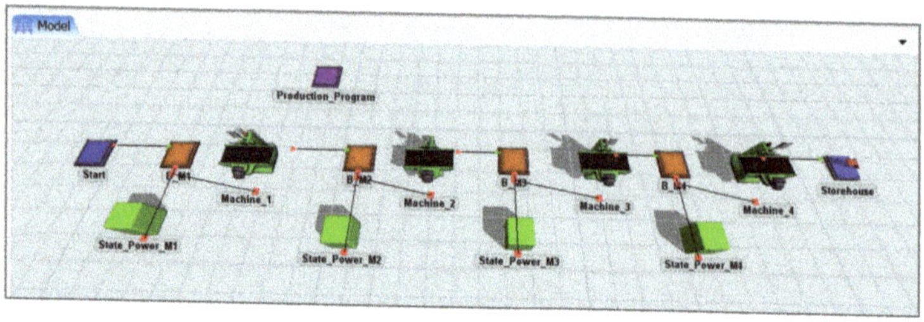

Figure 2. Simulation model of the production process.

Within the framework of the model, the logic of assigning tasks to be performed in the production process was developed. This logic is particularly important when there is a limitation of power consumption for a given production cell. The logic developed in the *ProcessFlow* tool of FlexSim program is presented in Figure 3.

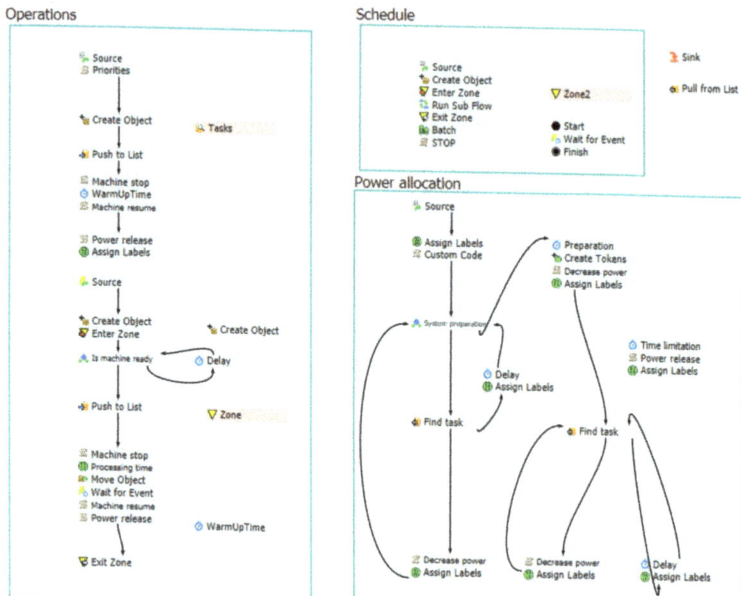

Figure 3. Data processing logic in the simulation model.

The data processing logic of the model was built on three elements. The first is the production schedule; the second is a flow of operations when an order is started on a machine; and the third is an assignment of power to a specific machine. At each iteration of the process, the state of the used power is examined, and then the power that can be distributed in the production cell to the remaining machines is determined (based on the difference between the available power and the power already used in the system). The power must be distributed to those machines that have made a request, and the decision to allocate power to each machine is made based on the established priorities in the production flow (order of request, order of flow in the process, etc.).

4.5. Verification of Changes in Electricity Consumption with the Simulation Model

The experiments analyzed the possible strategies for the company to deal with power curtailment in the assumed time interval. The analysis was performed at two levels. In the first, different variants of reducing peak power by 5, 10, 15, or 20 kW were analyzed.

In the second, different handling scenarios and their impact on the implementation of the production process were examined. Within each experiment, the following factors were investigated:

- Process schedule;
- Machine load schedule;
- Diagram of power consumption for the production cell;
- Process execution time; and
- Level of electricity consumption within the execution of the assumed production program.

The first step of the analysis was to simulate the production process for the base state, without the occurrence of the constraint. The obtained results make it possible to identify the points where, in case of the occurrence of a power constraint, it will be necessary to make changes in the production schedule. The power consumption graph with the production schedule is shown in Figure 4.

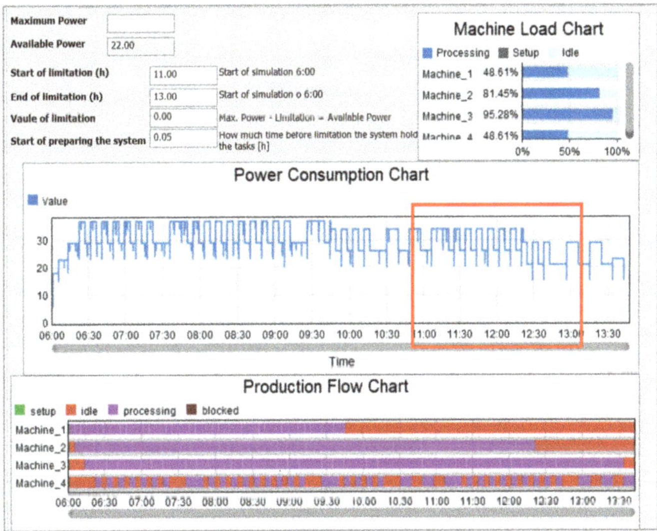

Figure 4. Power consumption and production schedule for the base state.

In the base state, the total process execution time was 462.6 min (i.e., it was executed during one work shift). One worker was assigned to each machine, therefore, their availability was 480 min, taking into account a 20-min work break. The work break was assigned to each station at a different time of the production schedule, therefore, it did not affect the working restrictions of the machines. The productivity of the machines varied from 48.61% (M1, M4) to 95.28% (M3). At the same time, the presented machine load chart showed that machine M3 was the bottleneck of the process, according to the concept of theory of constraints. The production schedule showed that on three workstations (M1, M2, M3), the course of production in the absence of restrictions on power consumption took place without restrictions/interruptions. The only breaks/restrictions occurred at station M4, where they were caused by waiting for the machine to run off/flow of semi-finished products from machine M3. While waiting, the machine operates in "Idle" mode. The factors of the production process for the base state are shown in Table 6.

If a constraint occurs from 11:00 a.m. to 1:00 p.m., part of the production must be reduced due to exceeding the available power consumption volume. The company can solve this problem according to the four proposed action scenarios:

1. Execution of operations in a sequential manner, starting from the last link of the production process and switching on the remaining links;

2. Realization of production with the assumption of maintaining work for selected machines (M2, M4) because of the applied production technology;
3. Subordination of the process under the maximum use of the bottleneck following the concept of theory of constraints; and
4. Disabling the execution of the production process for the period of introduction of the constraint or setting all machines in the work state "Idle".

Table 6. Production process factors for the base state.

Factor	Base State
Process execution time [min.]	462.6
Total electricity consumption [kWh]	231.96
Employees working time [min.]	480
Maximum machine load [%]	M3 95.28
Minimum machine load [%]	M1/M4 48.61

4.5.1. Scenario 1

When the power limitation occur, operations on machine M1 are completed. On the other machines, operations are started or the machines are set to the "Idle" state. Although the job allocation priority was set to the last operation on machine M4, it was not set to run continuously. Operations must be run sequentially. This means that products from machine M2 are passed to machine M3, and then machine M4. This allows for subsequent production orders to be executed. It is important to note that the start of production of a given production batch must be completed and only then can the machine operate in the "Idle" state. Figure 5 shows the power consumption, depending on the production schedule, for four variants of the size of power limitations. The red box indicates the times of power limitation. There was a clear decrease in the frequency and diversity of the performed technological operations.

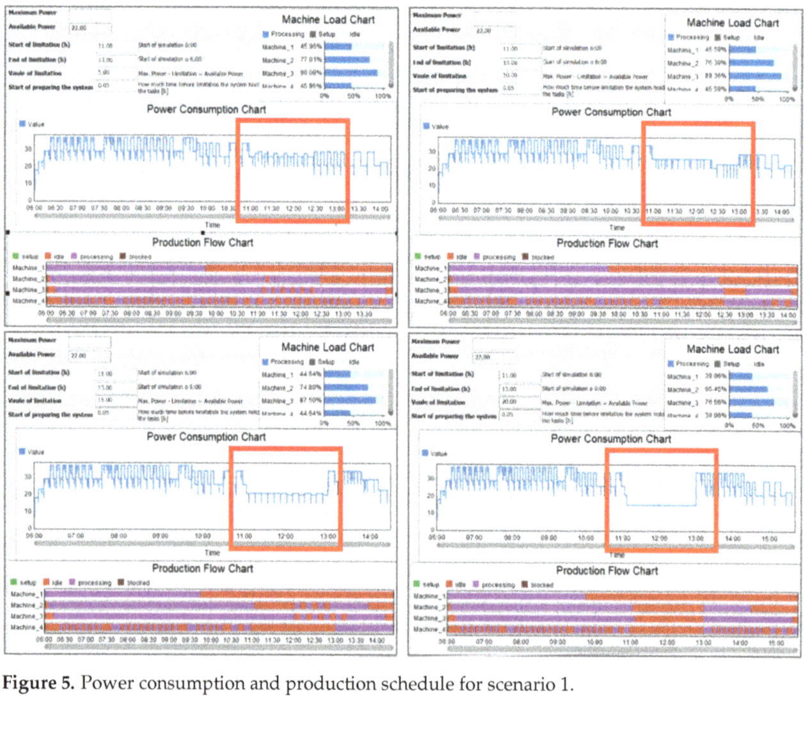

Figure 5. Power consumption and production schedule for scenario 1.

Table 7 shows the factors of the production process for different variants of power limitation.

Table 7. Production process factors for scenario 1.

Factor	Variant 1	Variant 2	Variant 3	Variant 4
Process execution time [min.]	489.6	493.2	504	576
Total electricity consumption [kWh]	238.5	239.63	242.25	260.25
Overtime for employees, per workstation [min.]	9.6	13.2	24	96
Maximum machine load [%]	M3 90.09	M3 89.36	M3 87.50	M3 76.56
Minimum machine load [%]	M1/M4 45.96	M1/M4 45.59	M1/M4 44.64	M1/M4 39.06

Increasing the amount of power limitation (decreasing the available power) increased the production process time by 27 min (with the 5 kW available power reduction) to 113.4 min (with the 20 kW reduction). Increasing the process realization increased the electricity consumption in the production cell due to the increased time the machines were in the "Idle" state. The power consumption increased from 6.54 (variant 1) to 28.29 (variant 4) with respect to the base state (Table 6). As the reduction in power consumption increased, the productivity of the machines used in the process decreased. It is because of the lower possible value of power consumption that the selected machines cannot run but have to be put into the "Idle" state. The maximum utilization of machine M3 decreased from a value of 90.09% (with the 5 kW available power reduction) to a value of 76.56% (with the 20 kW available power reduction).

4.5.2. Scenario 2

When the power limitation occurs, operations on machine M1 are completed. On the remaining machines, individual operations are started or the machines are set to the "Idle" state. In the production cell, two machines (M2 and M4) should constantly be in the "Processing" state (i.e., the machine setting parameters should enable the start or execution of the production process at any time). Due to the characteristics of the process, meeting this condition for machine M4 was not possible. However, for machine M2, it was possible in two variants, a reduction in power consumption by 5 or 10 kWh, while in other cases, it was again impossible. Table 8 presents the factors of the production process for different power limitations.

Table 8. Production process factors for scenario 2.

Factor	Variant 1	Variant 2	Variant 3	Variant 4
Process execution time [min.]	492	493.2	541.2	576
Total electricity consumption [kWh]	239.21	239.63	251.55	260.25
Overtime for employees, per workstation [min.]	12	13.2	61.2	96
Maximum machine load [%]	M3—89.66	M3—89.36	M3—87.50	M3—76.56
Minimum machine load [%]	M1/M4 45.75	M1/M4 45.59	M1/M4 44.64	M1/M4 39.06

Increasing the amount of power limitations influenced the increase in the time execution of the production process by 29.4 min (with the 5 kW available power reduction) up to 113.4 min (with the 20 kW available power reduction). Increased process realization was associated with increased electricity consumption in the production cell—from 7.26 (variant 1) to 28.29 (variant 4) with respect to the base state (Table 6). Again, with the increase in the limitation of power consumption, the efficiency of individual machines used in the realization of the production process decreased. The maximum utilization of machine M3 decreased from the value of 89.66% (with the 5 kW available power reduction) to 76.56% (with the 5 kW available power reduction).

4.5.3. Scenario 3

When a limitation occurs, operations on machine M1 are completed. On the remaining machines, individual operations are started or the machines are set to "Idle". In this scenario, the production was subject to maximum utilization of the bottleneck of the process (i.e., machine M3). Machine M3 should therefore be run continuously to ensure that its productivity is at the highest level. Note that the power allocation priors for a given machine should also be set on the machines upstream of the bottleneck. This means that the bottleneck is continuously provided with semi-finished products for production. Continuity of operation of machine M3 was maintained in two variants with a power reduction of 5 or 10 kW. In the other variants, there was an interruption to the production operations. With the 20 kW available power reduction, all production operations on machines were stopped and only the readiness of machines to work was maintained. Figure 6 presents power consumption, depending on the realized production schedule, for four variants of the size of the power consumption limitation. The red box indicates the times of power limitation. There was a clear decrease in the frequency and diversity of technological operations.

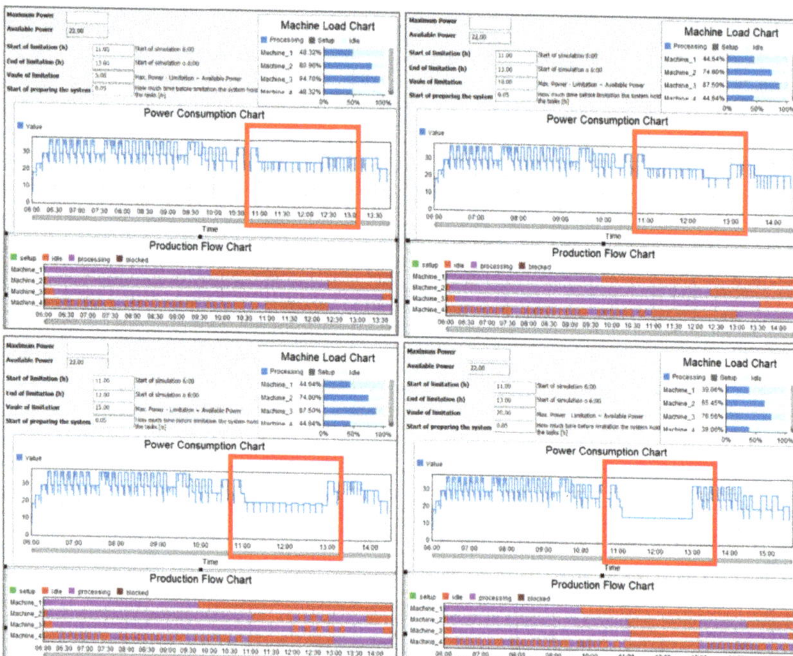

Figure 6. Power consumption and production schedule for scenario 3.

Table 9 shows the factors of the production process for different power limitations.

Table 9. Production process factors for scenario 3.

Factor	Variant 1	Variant 2	Variant 3	Variant 4
Process execution time [min.]	465.6	504	504	576
Total electricity consumption [kWh]	232.67	242.25	242.25	260.25
Overtime for employees, per workstation [min.]	0	24	24	96
Maximum machine load [%]	M3—94.70	M3—87.50	M3—87.50	M3—76.56
Minimum machine load [%]	M1/M4 48.32	M1/M4 44.54	M1/M4 44.64	M1/M4 39.06

Increasing the amount of the power constraint increased the process execution time by 3 min (with the 5 kW available power reduction) to 113.4 min (with the 20 kW available power reduction). Of all the scenarios presented, this one had the shortest process execution extension. It also resulted in the lowest increase in electricity consumption in the production cell, which ranged from 0.71 kWh (variant 1) to 28.29 kWh (variant 4) with respect to the base state (Table 6). With the increase in the reduction in power consumption, the efficiency of individual machines decreased. However, the efficiency values for the first few variants were significantly higher than in the previous scenarios. The maximum utilization of machine M3 decreased from the value of 94.70% (with the 5 kW available power reduction) to the value of 76.56% (with the 20 kW available power reduction).

4.5.4. Scenario 4

When the constraint occurs, operations on machine M1 have been completed and the machine is in the "Idle" state. However, in the considered scenario, the power constraint was so high (with the 30 kW available power reduction) that no machine could complete the production process. Therefore, all machines entered the "Idle" state and waited for the release of the power limitation to continue the technological operations. Table 10 shows the production process factors associated with the power limitation.

Table 10. Production process factors for scenario 4.

Factor	Maximum Limitation Variant
Process execution time [min.]	576
Total electricity consumption [kWh]	260.25
Overtime for employees, per workstation [min.]	96
Maximum machine load [%]	M3—76.56
Minimum machine load [%]	M1/M4 39.06

The results obtained in the scenario of waiting for the release of the power limitation to continue the production process are also presented in each previous scenario when the power curtailment was 20 kW. Likewise in this scenario, there was a significant increase in the duration of the production process of 113.4 min, which was simultaneously associated with an increased electricity consumption of 260.25 kWh. The production rates in the form of efficiency of the machines used in the process were also at a low level and ranged from 39.06% (for machines M1 and M4) to 76.56% (for machine M3).

4.6. Recommendations for the Production Process Activities

The simulations examined four key event scenarios possible for the enterprise. The results were analyzed and provided to the enterprise with recommendations. The best results in terms of reducing production order, lead time extension, and energy consumption were obtained in scenario 3. This solution subordinates power allocation to resources, prioritizing the bottleneck of the process, identified from the point of view of the production organization. The company's management has to make a decision taking into account both criteria from the production organization (shorter order processing time, less load on machines and employees, no overtime) and energy consumption analysis (reduction of energy consumption). These factors have a direct impact on reducing the costs of business. The paper did not take into account any classified information (e.g., legal provisions) that may affect the decisions made by the management, hence the interpretation of the results obtained from the simulation led to the establishment of recommendations for the enterprise.

At the same time, as part of the model building and validation process, the introduction of appropriate values of process parameters was checked and verified, corresponding to the given assumptions, requirements, and limitations of the tested production process.

5. Discussion

The discussion of the results is conducted in two areas: the first area was the analysis of the quantitative results obtained from the simulation experiments; and the second area was the evaluation of the proposed method in the context of applications for manufacturing companies and applications for conducting further research of the problem.

In the first area, a simulation model of the problem was developed in FlexSim software. All practically important elements of the problem concerning the management of power and electricity consumption in the investigated production process were represented in the model. Research scenarios were prepared, adequate to the needs of solving the problem. Simulation experiments were carried out according to the assumptions. Each experiment provided the expected quantitative results. Analysis of the results showed that the best results in terms of managing power consumption when a constraint occurred were obtained in scenario 3. In this scenario, production was subordinated to the bottleneck of the process. By increasing the throughput of the process and continuity of material flow (continuity of work) at the bottleneck, the lowest electricity consumption in the whole system was obtained (Table 9).

When limiting the available power by 20 kW, a stoppage of the production process was obtained in every scenario studied. All machines were running in the "Idle" state (i.e., waiting for the possibility of carrying out operations). The question to be analyzed was whether it is worthwhile keeping the machines in the "Idle" state. It is also possible to switch machines to the "Power Off" state. Then, when the restriction is removed, all machines must be restarted. The answer to this question depends on whether the electricity consumption in the "Warm Up" state covers the electricity consumption in the "Idle" state. In addition, it has to be assessed whether stopping and starting the machines increases their running costs and makes them wear out faster. In addition to electricity consumption, the need for increased overtime costs for employees must also be considered. These issues should be considered together when deciding how to respond to a reduction in an available capacity for a production process.

In the second area of discussion, it can be concluded that the proposed method was successful. The required data were collected, models of power consumption by the enterprise resources were prepared, experiments were carried out, and practical conclusions were formulated. However, it should be noted that for the analyzed production process, the authors of the paper managed to adequately prepare the research environment.

From a practical point of view, it is the ability to prepare the research environment within the enterprise that determines the ability to manage power and electricity consumption. For the company, this means that good internal preparation is required beforehand. Above all, the machinery and equipment must be metered. For decision-making, the measurements must be available virtually online. For possible long-term analyses, the measurements should be stored in databases, which requires that the company has qualified personnel. In addition, the company must have computer programs for simulations. Such programs require the purchase of licenses and the training of personnel to build models and conduct simulation experiments, which are costly and time-consuming activities. Other elements, indirectly related to the analyzed problem, are the type of manufactured products, the industry in which the company operates, and the technologies used. Not in every case is it possible to take sufficiently flexible actions to be able to react to problems related to the limitation of available power or the high prices of electricity.

On a general level, to mitigate these challenges, manufacturing companies can take the following actions:

- Assess how energy price increases affect the company's cost increases;
- Assess how energy supply interruptions, if any, affect production losses;
- Estimate investment expenditures for the purchase of energy consumption metering and software and hardware; and
- In the case of a decision to implement the solution, acquire appropriate specialists or services.

Implementation of the above activities requires special care and the assessment of risks associated with their operational execution. In practice, the management of power and energy consumption in most manufacturing enterprises, especially those for which energy costs are not significant, requires measures of a very basic nature. Thus far, the perception of energy as a public good has not been conducive to both the building managers' awareness of possible problems and to prepare enterprises in the technical layer. There is also a lack of work in the literature to help identify and better understand the reasons for this situation. The assessment of the convergence of the obtained results with the knowledge in enterprises on power and electricity consumption management requires further research.

In the research area, the formulated problem seems to be extremely interesting. The proposed method should be developed quantitatively and qualitatively. Simulation experiments conducted quantitatively require the construction of formal mathematic models of individual production resources. Additionally, optimization models should be formulated, taking into account the various decision-making criteria, not only in the field of power and electricity costs. The investigated problem has a multi-criteria character. Such models and possibly dedicated algorithms should help decision-makers in enterprises to make well-reasoned business decisions.

The advantage of the applied method is that it supports the decision-making process in manufacturing companies. Currently, in many of them, the decision-making process is based on simple quantitative data, intuition, and the experience of production managers as they do not have efficient and effective supporting tools at their disposal.

The method allows one to verify the possibility of carrying out the adopted production schedule in specific conditions of the power possessed in a given production cell, which also enables determining the adverse effects of reducing power consumption in a given time. Simulation modeling enables quick verification of different event scenarios in the scope of the occurring power reduction. It also provides a basis on which to assess the consequences of using different solutions in response to the limitation. Carrying out the simulation takes a short time, and after its execution, a detailed production schedule is obtained. Other data are also available concerning the extension of the order execution time and the amount of machine load. Depending on the model built and the adopted characteristics/parameters of the production process, these data can be varied and dedicated to a specific company. This information is taken into account in the built mathematical model, which shows the relationships between variables. Then, the described relations are translated into a simulation model.

It can be concluded that the main advantage of the proposed solution is the possibility of the experimental verification of different energy consumption management strategies. Complex systems do not have such functionality. Simulation modeling software is intuitive and easy to install, does not require large hardware resources, and can be used in both large and small enterprises. Simulation models are scalable and their functions can be defined by users. A certain limitation in their use is their difficult integration with other enterprise systems (no API).

In qualitative terms, it is important to note that the problem under study is related to a much broader research area that concerns the use of modern information technologies including the field of communication IoT systems and data analysis AI algorithms. Manufacturing companies strive to build digital twins of the production systems they own. Regardless of the choice of a given software or tools, the application of the presented method will have a versatile character.

The changes should be considered in the context of the functioning of power markets, which primarily concerns a new definition of products and services in these markets and the necessity of the instantaneous valuation of these products.

6. Conclusions

Simulation modeling can be used to effectively study power and electricity consumption in production processes. Such studies provide potential cost reductions in manufacturing companies. Due to the flexibility of available software, the proposed approach can be applied to different processes.

This paper draws attention to the broad context of the research to be conducted. Apart from the obvious problem of rising electricity prices, one has to take into account possible future limitations in power availability. The very issue of differentiating between power and electricity can be a problem for production companies. Considering electricity as a commodity and not as a public good would help to better balance the power system resources, but this would require changes necessary in the rules of the energy markets.

Finally, it should be noted that the problem under consideration may not seem very urgent at present. According to the authors, this is a mistaken impression. The current pandemic crisis, political crises that are likely to recur, and the drive to implement Industry 4.0 solutions may very quickly verify such a view. The time to discuss and prepare solutions for these challenges is now.

Author Contributions: Conceptualization, C.S. and J.S.; Methodology, J.S., P.S. and C.S.; Software, J.S. and D.D.; Validation, J.S.; Formal analysis, J.S. and C.S.; Investigation, J.S.; Resources, J.S.; Data curation, J.S. and D.D.; Writing—original draft preparation, J.S., P.S. and C.S.; Writing—review and editing, J.S. and C.S.; Visualization, D.D.; Supervision, C.S.; Project administration, C.S. All authors have read and agreed to the published version of the manuscript.

Funding: This research was funded by the Polish National Agency for Academic Exchange, under grant no. PPI/APM/2018/1/00047, entitled "Industry 4.0 in Production and Aeronautical Engineering" (International Academic Partnerships Program). The APC was funded by the Polish National Agency for Academic Exchange.

Institutional Review Board Statement: Not applicable.

Informed Consent Statement: Not applicable.

Data Availability Statement: Data are contained within the paper.

Acknowledgments: Technical support: InterMarium Ltd., Kraków, Poland.

Conflicts of Interest: The authors declare no conflict of interest.

References

1. Gutowski, T.; Dahmus, J.; Thiriez, A. Electrical energy requirements for manufacturing processes. In Proceedings of the 13th CIRP International Conference on Life Cycle Engineering, Leuven, Belgium, 31 May–2 June 2006; pp. 1–5.
2. Delgado-Gomes, V.; Oliveira-Lima, J.A.; Martins, J.F. Energy consumption awareness in manufacturing and production system. *Int. J. Comput. Integr. Manuf.* **2017**, *30*, 84–95. [CrossRef]
3. Duflou, J.R.; Sutherland, J.W.; Dornfeld, D.; Herrmann, C.; Jeswiet, J.; Kara, S.; Hauschild, M.; Kellens, K. Towards energy and resource efficient manufacturing: A processes and systems approach. *CIRP Ann.* **2012**, *61*, 587–609. [CrossRef]
4. Apostolos, F.; Alexios, P.; Georgios, P.; Panagiotis, S.; George, C. Energy Efficiency of Manufacturing Processes: A Critical Review. *Procedia Cirp* **2013**, *7*, 628–633. [CrossRef]
5. Durlik, I. *Organizacja i Zarządzanie Produkcją*; PWE: Warsaw, Poland, 2004; p. 31.
6. Gawlik, L. The Polish Power Industry in Energy Transformation Process. *Miner. Econ.* **2018**, *31*, 229–237. [CrossRef]
7. Li, Y.; Zhang, H.; Kang, Y. Will Poland Fulfill Its Coal Commitment by 2030 An Answer Based on a Novel Time Series Prediction Method. *Energy Rep.* **2020**, *6*, 1760–1767. [CrossRef]
8. Kaszyński, P.; Kamiński, J. Coal Demand and Environmental Regulations: A Case Study of the Polish Power Sector. *Energies* **2020**, *13*, 1521. [CrossRef]
9. Dvořák, J.; Wittlingerová, Z.; Vochozka, M.; Stehel, V.; Maroušková, A. Updated Energy Policy of the Czech Republic May Result in Instability of the Electricity Grid in Central Europe. *Clean Technol. Environ. Policy* **2018**, *20*, 41–52. [CrossRef]
10. Sivek, M.; Jirásek, J.; Kavina, P.; Vojnarová, M.; Kurková, T.; Bašová, A. Divorce after Hundreds of Years of Marriage: Prospects for Coal Mining in the Czech Republic with Regard to the European Union. *Energy Policy* **2020**, *142*, 111524. [CrossRef]
11. Burchart-Korol, D.; Pustejovska, P.; Blaut, A.; Jursova, S.; Korol, J. Comparative Life Cycle Assessment of Current and Future Electricity Generation Systems in the Czech Republic and Poland. *Int. J. Life Cycle Assess.* **2018**, *23*, 2165–2177. [CrossRef]
12. Zahájeno Detailní Bezpečnostní Posouzení Uchazečů o Dodávku Nového Jaderného Bloku v Dukovanech | MPO. Available online: https://www.mpo.cz/cz/rozcestnik/pro-media/tiskove-zpravy/zahajeno-detailni-bezpecnostni-posouzeni-uchazecu-o-dodavku-noveho-jaderneho-bloku-v-dukovanech---260344/ (accessed on 26 October 2021).
13. Macháč, J.; Zaňková, L. Renewables—To Build or Not? Czech Approach to Impact Assessment of Renewable Energy Sources with an Emphasis on Municipality Perspective. *Land* **2020**, *9*, 497. [CrossRef]

14. Gürtler, K.; Postpischil, R.; Quitzow, R. The Dismantling of Renewable Energy Policies: The Cases of Spain and the Czech Republic. *Energy Policy* **2019**, *133*, 110881. [CrossRef]
15. Čábelková, I.; Strielkowski, W.; Firsova, I.; Korovushkina, M. Public Acceptance of Renewable Energy Sources: A Case Study from the Czech Republic. *Energies* **2020**, *13*, 1742. [CrossRef]
16. Bluszcz, A.; Manowska, A. The Use of Hierarchical Agglomeration Methods in Assessing the Polish Energy Market. *Energies* **2021**, *14*, 3958. [CrossRef]
17. Tucki, K.; Orynycz, O.; Wasiak, A.; Świć, A.; Dybaś, W. Capacity Market Implementation in Poland: Analysis of a Survey on Consequences for the Electricity Market and for Energy Management. *Energies* **2019**, *12*, 839. [CrossRef]
18. Zator, S.; Skomudek, W. Impact of DSM on Energy Management in a Single-Family House with a Heat Pump and Photovoltaic Installation. *Energies* **2020**, *13*, 5476. [CrossRef]
19. Dudek, D.; Lipowski, M.; Bondos, I. Changing Energy Supplier on the Market with a Strong Position of Incumbent Suppliers—Polish Example. *Energies* **2021**, *14*, 3933. [CrossRef]
20. Solnørdal, M.T.; Foss, L. Closing the Energy Efficiency Gap—A Systematic Review of Empirical Articles on Drivers to Energy Efficiency in Manufacturing Firms. *Energies* **2018**, *11*, 518. [CrossRef]
21. Mohammadi, M.; Noorollahi, Y.; Mohammadi-ivatloo, B.; Hosseinzadeh, M.; Yousefi, H.; Khorasani, S.T. Optimal Management of Energy Hubs and Smart Energy Hubs—A Review. *Renew. Sustain. Energy Rev.* **2018**, *89*, 33–50. [CrossRef]
22. Pei, W.; Ma, X.; Deng, W.; Chen, X.; Sun, H.; Li, D. Industrial Multi-Energy and Production Management Scheme in Cyber-Physical Environments: A Case Study in a Battery Manufacturing Plant. *IET Cyber-Phys. Syst. Theory Appl.* **2019**, *4*, 13–21. [CrossRef]
23. Harjunkoski, I.; Maravelias, C.T.; Bongers, P.; Castro, P.M.; Engell, S.; Grossmann, I.E.; Hooker, J.; Méndez, C.; Sand, G.; Wassick, J. Scope for Industrial Applications of Production Scheduling Models and Solution Methods. *Comput. Chem. Eng.* **2014**, *62*, 161–193. [CrossRef]
24. Ramin, D.; Spinelli, S.; Brusaferri, A. Demand-Side Management via Optimal Production Scheduling in Power-Intensive Industries: The Case of Metal Casting Process. *Appl. Energy* **2018**, *225*, 622–636. [CrossRef]
25. Merkert, L.; Harjunkoski, I.; Isaksson, A.; Säynevirta, S.; Saarela, A.; Sand, G. Scheduling and Energy—Industrial Challenges and Opportunities. *Comput. Chem. Eng.* **2015**, *72*, 183–198. [CrossRef]
26. Hacksteiner, M.; Fuchs, G.; Bleicher, F. Strategic Energy Management in Mechanical Series Production: An Industrial Use-Case. *Procedia Manuf.* **2019**, *33*, 59–66. [CrossRef]
27. Cannata, A.; Taisch, M. Introducing Energy Performances in Production Management: Towards Energy Efficient Manufacturing. In *Advances in Production Management Systems. New Challenges, New Approaches. APMS 2009. IFIP Advances in Information and Communication Technology*; Vallespir, B., Alix, T., Eds.; Springer: Berlin/Heidelberg, Germany, 2010; Volume 338, pp. 168–175. Available online: https://hal.inria.fr/hal-01055856/document (accessed on 15 September 2021)ISBN 978 3 642 16357-9.
28. Javied, T.; Deutsch, M.; Franke, J. A Model for Integrating Energy Management in Lean Production. *Procedia CIRP* **2019**, *84*, 357–361. [CrossRef]
29. Hassan Khattak, S.; Oates, M.; Greenough, R. Towards Improved Energy and Resource Management in Manufacturing. *Energies* **2018**, *11*, 1006. [CrossRef]
30. Castrillón-Mendoza, R.; Rey-Hernández, J.M.; Rey-Martínez, F.J. Industrial Decarbonization by a New Energy-Baseline Methodology. Case Study. *Sustainability* **2020**, *12*, 1960. [CrossRef]
31. Gordić, D.; Babić, M.; Jovičić, N.; Šušteršič, V.; Končalović, D.; Jelić, D. Development of Energy Management System—Case Study of Serbian Car Manufacturer. *Energy Convers. Manag.* **2010**, *51*, 2783–2790. [CrossRef]
32. Thollander, P.; Johansson, M. Energy Management in Industry—Success Factors and Way Forward. Available online: https://www.diva-portal.org/smash/get/diva2:1151593/FULLTEXT02.pdf (accessed on 27 October 2021).
33. Saleem, M.U.; Usman, M.R.; Shakir, M. Design, Implementation, and Deployment of an IoT Based Smart Energy Management System. *IEEE Access* **2021**, *9*, 59649–59664. [CrossRef]
34. Gao, M.; He, K.; Li, L.; Wang, Q.; Liu, C. A Review on Energy Consumption, Energy Efficiency and Energy Saving of Metal Forming Processes from Different Hierarchies. *Processes* **2019**, *7*, 357. [CrossRef]
35. Xenos, D.P.; Mohd Noor, I.; Matloubi, M.; Cicciotti, M.; Haugen, T.; Thornhill, N.F. Demand-Side Management and Optimal Operation of Industrial Electricity Consumers: An Example of an Energy-Intensive Chemical Plant. *Appl. Energy* **2016**, *182*, 418–433. [CrossRef]
36. Wang, Y.; Li, L. Time-of-Use Based Electricity Demand Response for Sustainable Manufacturing Systems. *Energy* **2013**, *63*, 233–244. [CrossRef]
37. Halmschlager, V.; Hofmann, R. Assessing the Potential of Combined Production and Energy Management in Industrial Energy Hubs—Analysis of a Chipboard Production Plant. *Energy* **2021**, *226*, 120415. [CrossRef]
38. Jin, L.; Zhang, C.; Fei, X. Realizing Energy Savings in Integrated Process Planning and Scheduling. *Processes* **2019**, *7*, 120. [CrossRef]
39. Kordonowy, D.N. A Power Assessment of Machining Tools, Massachusetts Institute of Technology. Bachelor's Thesis, Department of Mechanical, Cambridge, MA, USA, 2001.
40. Apostolos, F.; Panagiotis, S.; Konstantinos, S.; George, C. Energy Efficiency Assessment of Laser Drilling Process. *Phys. Procedia* **2012**, *39*, 776–783. [CrossRef]

41. Herrmann, C.; Thiede, S. Process chain simulation to foster energy efficiency in manufacturing. *CIRP J. Manuf. Sci. Technol.* **2009**, *1*, 221–229. [CrossRef]
42. Rahimifard, S.; Seow, Y.; Childs, T. Minimising Embodied Product Energy to support energy efficient manufacturing. *CIRP Ann.* **2010**, *59*, 25–28. [CrossRef]
43. Heilala, J.; Vatanen, S.; Tonteri, H.; Montonen, J.; Lind, S.; Johansson, B.; Stahre, J. Simulationbased sustainable manufacturing system design. In Proceedings of the 40th Winter Simulation Conference, Miami, FL, USA, 7–10 December 2008; pp. 1922–1930.
44. Johansson, B.J.; Skoogh, A.; Mani, M.; Leong, S.K. Discrete event simulation as requirements specification for sustainable manufacturing systems design. In Proceedings of the PerMIS'09 Conference, Gaithersburg, MD, USA, 21 September–23 October 2009.
45. Dietmair, A.; Verl, A. A generic energy consumption model for decision making and energy efficiency optimization in manufacturing. *Int. J. Sustain. Eng.* **2009**, *2*, 123–133. [CrossRef]
46. Mouzon, G.; Yildirim, M.B.; Twomey, J. Operational methods for minimization of energy consumption of manufacturing equipment. *Int. J. Prod. Res.* **2007**, *45*, 4247–4271. [CrossRef]
47. Mori, M.; Fujishima, M.; Inamasu, Y.; Oda, Y. A study on energy efficiency improvement for machine tools. *CIRP Ann.* **2011**, *60*, 145–148. [CrossRef]
48. Weinert, N.; Chiotellis, S.; Seliger, G. Methodology for planning and operating energyefficient production systems. *CIRP Ann.* **2011**, *60*, 41–44. [CrossRef]
49. Wang, J. A Multi-Granularity Model for Energy Consumption Smulation and Control of Discrete Manufacturing System. 2013. Available online: https://link.springer.com/chapter/10.1007/978-3-642-38391-5_112 (accessed on 14 September 2021).
50. Shrouf, F.; Ordieres, J.; Miragliotta, G. Smart Factories in Industry 4.0: A Review of the Concept and of Energy Management Approached in Production Based on the Internet of Things Paradigm. In Proceedings of the 2014 IEEE International Conference on Industrial Engineering and Engineering Management, Selangor, Malaysia, 9–12 December 2014; pp. 697–701. Available online: https://ieeexplore.ieee.org/stamp/stamp.jsp?tp=&arnumber=7058728 (accessed on 16 September 2021).
51. Steindl, G.; Stagl, M.; Kasper, L.; Kastner, W.; Hofmann, R. Generic Digital Twin Architecture for Industrial Energy Systems. *Appl. Sci.* **2020**, *10*, 8903. [CrossRef]
52. Nouiri, M.; Trentesaux, D.; Bekrar, A. Towards Energy Efficient Scheduling of Manufacturing Systems through Collaboration between Cyber Physical Production and Energy Systems. *Energies* **2019**, *12*, 4448. [CrossRef]
53. Santos, D.; Ferreira, J.C. IoT Power Monitoring System for Smart Environments. *Sustainability* **2019**, *11*, 5355. [CrossRef]
54. Saad, A.; Faddel, S.; Mohammed, O. IoT-Based Digital Twin for Energy Cyber-Physical Systems: Design and Implementation. *Energies* **2020**, *13*, 4762. [CrossRef]
55. Martirano, L.; Borghi, L.; Bua, F.; Cristaldi, L.; Grigis, G.; Lavecchia, C.; Liziero, M.; Mongioví, L.; Nastri, E.; Tironi, E. Energy Management Information Systems for Energy Efficiency. In Proceedings of the 2018 IEEE International Conference on Environment and Electrical Engineering and 2018 IEEE Industrial and Commercial Power Systems Europe (EEEIC/I CPS Europe), Palermo, Italy, 12–15 June 2018; pp. 1–7. Available online: https://ieeexplore.ieee.org/stamp/stamp.jsp?tp=&arnumber=8493712&tag=1 (accessed on 16 September 2021).
56. Shrouf, F.; Miragliotta, G. Energy Management Based on Internet of Things: Practices and Framework for Adoption in Production Management. *J. Clean. Prod.* **2015**, *100*, 235–246. [CrossRef]
57. Sucic, B.; Al-Mansour, F.; Pusnik, M.; Vuk, T. Context Sensitive Production Planning and Energy Management Approach in Energy Intensive Industries. *Energy* **2016**, *108*, 63–73. [CrossRef]
58. Bougain, S.; Gerhard, D.; Nigischer, C.; Uĝurlu, S. Towards Energy Management in Production Planning Software Based on Energy Consumption as a Planning Resource. *Procedia CIRP* **2015**, *26*, 139–144. [CrossRef]
59. Swords, B.; Coyle, E.; Norton, B. An Enterprise Energy-Information System. *Appl. Energy* **2008**, *85*, 61–69. [CrossRef]
60. Mařík, K.; Schindler, Z.; Stluka, P. Decision Support Tools for Advanced Energy Management. *Energy* **2008**, *33*, 858–873. [CrossRef]
61. Lee, D.; Cheng, C.-C. Energy Savings by Energy Management Systems: A Review. *Renew. Sustain. Energy Rev.* **2016**, *56*, 760–777. [CrossRef]
62. Industrial Energy Management | Rockwell Automation Finland. Available online: https://www.rockwellautomation.com/en-fi/capabilities/industrial-energy-management.html (accessed on 27 October 2021).
63. ABB Energy Management Software Solution for Industrial Plants. Available online: https://new.abb.com/cpm/energy-manager (accessed on 27 October 2021).
64. Energy Management. Available online: https://new.siemens.com/global/en/products/automation/industry-software/automation-software/energymanagement.html (accessed on 27 October 2021).
65. Energy Management & Monitoring System Software | Smart Meter Monitoring Platform. Available online: https://www.wattics.com/ (accessed on 27 October 2021).
66. Roemer, A.C.; Strassburger, S. A Review of Literature on Simulation-Based Optimization of the Energy Efficiency in Production. In Proceedings of the 2016 Winter Simulation Conference (WSC), Washington, DC, USA, 11–14 December 2016; pp. 1416–1427. Available online: http://simulation.su/uploads/files/default/2016-roemer-strassburger.pdf (accessed on 17 September 2021).
67. Thiede, S.; Kurle, D.; Herrmann, C. The Water–Energy Nexus in Manufacturing Systems: Framework and Systematic Improvement Approach. *CIRP Ann.* **2017**, *66*, 49–52. [CrossRef]
68. Weeber, M.; Wanner, J.; Schlegel, P.; Birke, K.P.; Sauer, A. Methodology for the Simulation Based Energy Efficiency Assessment of Battery Cell Manufacturing Systems. *Procedia Manuf.* **2020**, *43*, 32–39. [CrossRef]

69. Silva, G.V.; Thomitzek, M.; Abraham, T.; Herrmann, C.; Braunschweig, T. Simulation-Based Assessment of Energy Demand and Costs Associated with Production Scrap in the Battery Production. Available online: http://www.asim-fachtagung-spl.de/asim2021/papers/Proof_159.pdf (accessed on 14 September 2021).
70. Sobottka, T.; Kamhuber, F.; Heinzl, B. Simulation-Based Multi-Criteria Optimization of Parallel Heat Treatment Furnaces at a Casting Manufacturer. *J. Manuf. Mater. Processing* **2020**, *4*, 94. [CrossRef]
71. Dunkelberg, H.; Sondermann, M.; Meschede, H.; Hesselbach, J. Assessment of Flexibilisation Potential by Changing Energy Sources Using Monte Carlo Simulation. *Energies* **2019**, *12*, 711. [CrossRef]
72. Kim, S.; Meng, C.; Son, Y.-J. Simulation-Based Machine Shop Operations Scheduling System for Energy Cost Reduction. *Simul. Model. Pract. Theory* **2017**, *77*, 68–83. [CrossRef]
73. Meißner, M.; Myrzik, J.; Wiederkehr, P. Representation of Energy Efficiency Interdependencies of Manufacturing Processes on the Shop Floor Level. *Procedia CIRP* **2020**, *88*, 252–257. [CrossRef]
74. Delbrügger, T.; Meißner, M.; Wirtz, A.; Biermann, D.; Myrzik, J.; Rossmann, J.; Wiederkehr, P. Multi-Level Simulation Concept for Multidisciplinary Analysis and Optimization of Production Systems. *Int. J. Adv. Manuf. Technol.* **2019**, *103*, 3993–4012. [CrossRef]
75. Renna, P. Peak Electricity Demand Control of Manufacturing Systems by Gale-Shapley Algorithm with Discussion on Open Innovation Engineering. *J. Open Innov. Technol. Mark. Complex.* **2020**, *6*, 29. [CrossRef]
76. Scharmer, V.M.; Bröskamp, S.; Schulz, J.; Zaeh, M.F. Demand Planning Strategies for the Control of Energy Flexible Components of Machine Tools. *Procedia Manuf.* **2020**, *43*, 360–367. [CrossRef]
77. Abele, M.; Unterberger, E.; Friedl, T.; Carda, S.; Roth, S.; Hohmann, A.; Reinhart, G. Simulation-Based Evaluation of an Energy Oriented Production Planning System. *Procedia CIRP* **2020**, *88*, 246–251. [CrossRef]
78. Keshari, A.; Sonsale, A.N.; Sharma, B.K.; Pohekar, S.D. Discrete Event Simulation Approach for Energy Efficient Resource Management in Paper & Pulp Industry. *Procedia CIRP* **2018**, *78*, 2–7. [CrossRef]
79. Rodrigues, G.S.; Espíndola Ferreira, J.C.; Rocha, C.R. A Novel Method for Analysis and Optimization of Electric Energy Consumption in Manufacturing Processes. *Procedia Manuf.* **2018**, *17*, 1073–1081. [CrossRef]
80. Garwood, T.L.; Hughes, B.R.; Oates, M.R.; O'Connor, D.; Hughes, R. A Review of Energy Simulation Tools for the Manufacturing Sector. *Renew. Sustain. Energy Rev.* **2018**, *81*, 895–911. [CrossRef]
81. Mawson, V.J.; Hughes, B.R. The Development of Modelling Tools to Improve Energy Efficiency in Manufacturing Processes and Systems. *J. Manuf. Syst.* **2019**, *51*, 95–105. [CrossRef]
82. Roth, S.; Thimmel, M.; Fischer, J.; Schöpf, M.; Unterberger, E.; Braunreuther, S.; Buhl, H.U.; Reinhart, G. Simulation-Based Analysis of Energy Flexible Factories in a Regional Energy Supply System. *Procedia Manuf.* **2019**, *33*, 75–82. [CrossRef]

energies

MDPI

Article

Analyzing Wind Energy Potential Using Efficient Global Optimization: A Case Study for the City Gdańsk in Poland

Olgun Aydin [1,*], Bartłomiej Igliński [2], Krzysztof Krukowski [3] and Marek Siemiński [3]

1 Faculty of Management and Economics, Gdańsk University of Technology, Traugutta 79, 80-233 Gdańsk, Poland
2 Faculty of Chemistry, Nicolaus Copernicus University in Toruń, Gagarina 7, 87-100 Toruń, Poland; iglinski@chem.umk.pl
3 Faculty of Economic Sciences, University of Warmia and Mazury in Olsztyn, Oczapowskiego 2, 10-719 Olsztyn, Poland; kkruk@uwm.edu.pl (K.K.); markos@uwm.edu.pl (M.S.)
* Correspondence: olgun.aydin@pg.edu.pl

Abstract: Wind energy (WE), which is one of the renewable energy (RE) sources for generating electricity, has been making a significant contribution to obtaining clean and green energy in recent years. Fitting an appropriate statistical distribution to the wind speed (WS) data is crucial in analyzing and estimating WE potential. Once the best suitable statistical distribution for WS data is determined, WE potential and potential yield could be estimated with high accuracy. The main objective of this paper is to propose a novel approach for calculating wind energy potential. For this purpose, the Efficient Global Optimization (EGO) technique was proposed for fitting a statistical distribution to WS data and the performance of the technique was compared with genetic algorithm (GA), simulated annealing (SA), and differential evolution (DE). Performance metrics showed that EGO is providing better estimations compared with GA, SA, and DE. Based on Weibull parameters obtained by using EGO, potential WE and potential annual revenue were estimated for Gdańsk, which is the capital of Pomerania Voivodeship in Poland, in the case of having city-type wind turbines in the city center. Estimations for Gdańsk showed that city-type wind turbines might be helpful for producing electricity from WE in the city without being limited by constraints such as having a long distance between wind turbines and buildings. If such wind turbines were erected on the roofs of residential buildings, malls, or office buildings, there is a possibility that part of the electric energy needed for such buildings could be generated using WE. However, this topic should be further investigated from technical and financial perspectives.

Keywords: renewable energy; energy transition; wind energy; energy prices; efficient global optimization (EGO); Weibull distribution

Citation: Aydin, O.; Igliński, B.; Krukowski, K.; Siemiński, M. Analyzing Wind Energy Potential Using Efficient Global Optimization: A Case Study for the City Gdańsk in Poland. *Energies* **2022**, *15*, 3159. https://doi.org/10.3390/en15093159

Academic Editors: Ricardo J. Bessa and Davide Astolfi

Received: 16 February 2022
Accepted: 21 April 2022
Published: 26 April 2022

Publisher's Note: MDPI stays neutral with regard to jurisdictional claims in published maps and institutional affiliations.

Copyright: © 2022 by the authors. Licensee MDPI, Basel, Switzerland. This article is an open access article distributed under the terms and conditions of the Creative Commons Attribution (CC BY) license (https://creativecommons.org/licenses/by/4.0/).

1. Introduction

The subject of the article is the issue of WE production, which is an important source of RE in the world. Undoubtedly, the topic raised is relevant to the energy transition currently taking place in most of the world's economies [1–3]. The main objective of the energy transition is to move from the dominant role of fossil/nuclear fuels to the consideration of diverse RE sources, where WE, if the physical conditions are right, can represent a significant share of the RE portfolio [4–6]. Due to the systematic growth of global energy demand, the rational use of energy from renewable sources is one of the most important elements of sustainable development, bringing tangible effects for humanity as well as energy and ecology [7–10].

It should be emphasized that the production of RE and the implementation of energy transition has been possible for two decades at least. It turns out that it was only the occurrence of appropriate institutional, economic, and social changes in economies that allowed the real energy transformation of entire economies to begin, including state structures, local governments, businesses, and citizens [11–16]. Undoubtedly, most of the significant changes are due to the globalization processes that have been developing systematically for nearly 30 years. These processes have resulted in a significant increase in the interdependence between all markets [17–20]. This has contributed to a significant increase in the socio-economic level of countries [21–24]. Global economies are experiencing strong growth, which is being attributed to the increase in investment and the level of innovation [25–30]. In addition, we should notice the increase in the level of wealth of society, the emergence of new patterns of consumption, and changes in the labour market [31–39]. All this has contributed to the fact that both the production of RE from the national level to the level of the individual consumer and other activities related to the energy transition have found ground for implementation. In the case of the European Union (EU) member states, including Poland, the processes of energy transformation are intimately linked to the achievement of the Sustainable Development Goals [40–43]. It should be noted that many efforts also indicate that the effective implementation of energy transition processes requires bottom-up involvement and consideration of energy justice [44].

Significant development of RE in Poland took place at the beginning of the second decade of the 21st century when the process of implementing the third energy package and the implementation of the ambitious goals of the EU's climate policy started to be in charge under the commitments' so-called "20-20-20 in 2020", i.e., increasing energy efficiency by controlling energy consumption more, increasing the share of RE, and reducing carbon dioxide emissions [44]. RE sources are undoubtedly seen by European decision makers as a solution to reduce emissions [4]. RE sources are an alternative to primary and non-renewable hydrocarbon fuels. Although RE is characterized by the cyclical replenishment of resources in natural processes, the level of consumption of this type of energy as a primary energy source is still low [5,6]. As shown in Table 1, the capacity of renewable sources has been increasing since 2010 and showed a significant increase in 2020. However, the capacity is not at the desired level to be able to use RE as a primary energy source.

Table 1. Installed capacity of RE sources [45].

Type of RE Installation	Installed Capacity [MW]		
	2010	2015	2020
Biogas	82.884	212.497	255.699
Biomass	356.190	1122.670	1512.885
Photovoltaics	0.033	71.031	887.434
Wind Energy	1180.044	4582.036	6347.111
Hydropower	937.044	981.799	976.047
Total	2556.423	6970.033	9979.176

Total global wind capacity is currently up to 743 GW, helping the world avoid over 1.1 billion tons of CO_2 per year equivalent to the annual CO_2 emissions in South America. However, WE sources with the capacity of 180 GW should be activated each year in the world to avoid the worst effects of climate change. This means that industry and policymakers must act quickly to accelerate the switch from traditional energy sources to RE sources [46,47].

In spite of the termination of China's feed-in tariffs (FiT) and the planned phase-out of the United States' full-rate Production Tax Credit (PTC), the world's two largest economies increased their combined market share by 15 percent to 76 percent [46].

A record for onshore installations was also achieved in the Asia Pacific, North America, and Latin America regions in the calendar year 2020. According to the International Energy Agency, in these three regions a total of 74 GW of new onshore wind power was installed. This represents a 76 percent increase in capacity over the previous year. There was just a 0.6 percent year-over-year (YoY) rise in new onshore wind installations in Europe last year, which was due to the slow recovery of onshore wind installations in Germany. There were

8.2 GW onshore installations in Africa and the Middle East last year, which is almost the same as in 2019 [46].

The main objective of this paper is to propose an approach to the problem of WE generation that will ensure the highest efficiency of the energy generation process and the economic viability of this process. In order to achieve the objective, the authors developed the following added values: an overview in terms of development of WE in Poland and Pomerania Voivodeship, which is the area in Poland with significant potential when it comes to WE, was carried out; a novel approach was proposed for fitting a statistical distribution to WS data for estimating WE potential in a more efficient way; the discussion was conducted about the potential benefits of having a city-type wind turbine in the city.

The city-type wind turbine offers a possibility to install the turbines on the top of buildings in the city and, thanks to this, residents in each building could generate part of the electricity that they need. This could potentially make the transition to WE from traditional energy sources faster, as installing this type of turbine would not require big spaces, long-lasting investment planning, or any other limitations. For these purposes, within the scope of the study, WS data in an hourly format for almost the last seven years for Gdańsk (Poland) were obtained from a third-party provider. The two-parameter Weibull distribution (TPWD) was then fitted to WS with the help of Maximum Likelihood Estimation (MLE). Efficient Global Optimization (EGO) was used on top of MLE to find optimum parameters of TPWD. Moreover, performance of EGO was compared with performance of GA, SA, and DE, which are the algorithms that have been used by researchers in the literature to fit statistical distribution to WS. To compare the performances, root mean squared error (RMSE) and coefficient of determination (R^2) were used. Parameters were obtained for each month and annual data by using each technique. Comparisons were provided. By using parameters of TPWD for annual data, potential WE was estimated for Gdańsk, which is the capital of Pomerania Voivodeship in Poland, for the case of having city-type wind turbine in the city center of Gdańsk.

The construction of the paper is as follows: Section 2 focuses on a brief history of WE in general, the development of WE in Poland, and WE potential in Pomerania Voivodeship. Section 3 describes the methodology used in the study. Section 4 covers details about dataset used in the study. In Section 5, results are discussed, while Section 6 concludes the study and provides information about potential further research studies.

2. Development of the WE Market in Poland

2.1. WE: A Brief History

WE has been used by humans for a really long time, alongside sunlight, e.g., to dry agricultural crops. It is also worth remembering that important geographical discoveries were possible thanks to WE that "powered" sailing ships [47,48].

In early 2000 BCE, Egyptians used WE to propel their boats. The Code of Hammurabi (circa 1750 BCE) shows that WE was also used in Persia. In India, in the fourth century BCE, the first windmill was used for pumping water and already in the second century BCE in China windmills were used to irrigate farmland. At the beginning of our era, the first windmills were constructed in the Mediterranean countries [48].

The first European windmills appeared in England in the 9th century, in France in the 11th century, and in the 13th century they became popular in all Western Europe. The oldest image of a windmill in Europe is on the first page of an English manuscript from 1270. Originally, the windmill was a wooden "booth" that was rotated around a centrally located pole to set the wings to the wind. The revolution in the construction of windmills was made by the Dutch, who in 1390 introduced four-wing structures. The "Dutch" type windmills gained popularity in Europe in the 17th century [48].

The industry became more interested in wind power plants in the early 1980s. As an initiative of Danish power companies, a turbine with a capacity of 660 kW was developed. The following years were marked by the resolution of many technical problems related to the generator's construction, mechanical strength, and the selection of appropriate materials for the towers and rotor blades. In the last 20 years, a real "boom" in aero energy in the world has been happening [48].

The first Polish wind turbines were erected in the 1930s in Podkarpackie, a region in the south-east of Poland. Before the outbreak of World War II, 504 wind turbines were in operation in Poland. The first Polish wind potential map was published in 1958 in the book by Rynkowski entitled *Small Wind Farms*. The first wind turbine in Poland based on the new technology was erected in 1991 in Żarnowiec, a village in the north of Poland, as a replacement for the existing hydroelectric power plant. The first Polish wind farm (6 × 800 MW) was built in Barzowice in Pomerania Voivodeship in 2001 [48].

2.2. WE in Poland and Pomerania Voivodeship

Until 2016, WE was developing well in Poland (Table 2). As a result of the entry that went into force in 2016 regarding the act on investments in wind farms, there was a stagnation on new WE projects. The barrier is the inability to meet the requirement of a minimum distance of 10 × H (H = total height of the wind turbine with the blade in full elevation) from the buildings [49,50].

Table 2. Dynamics of the WE market in Poland [51].

Year	Installed Capacity of Onshore Wind Installations [GW]
2013	3.39
2014	3.83
2015	4.58
2016	5.81
2017	5.85
2018	5.86
2019	5.92
2020	6.35
2021	≈6.80

The progressive inclusion of the most advanced projects in 2018–2020 has resulted in an increase in new onshore wind farm capacity seen in late 2020 and 2021. As a result, the installed capacity potential increased to approximately 6.80 GW [52] and in the next two or three years it is planned to exceed 10 GW [52]. The government's announcements of distance regulation are likewise positive, with the expectation of another investment "boom" of 3–4 GW by 2025 [53].

The strategic objective is to maximize the potential of Polish onshore wind energy. By 2030–2035, the Polish Wind Energy Association (PWEA) [48] anticipates that Poland will be able to generate 22–24 GW of energy from wind [53]. Clean electrical energy derived from the most sustainable RE sources is important to maintain the Polish economy's international competitiveness. Every single additional gigawatt to wind farm capacity results in significant cost savings. It has a direct effect on the wholesale price of electricity, which has decreased by an average of more than PLN 20/MWh on the wholesale market since 2007. Poland's energy system appears to be defying global trends. Fossil fuels—hard coal and lignite—continue to account for a share of domestic output; nonetheless, the share of RE continues to expand. In 2020, coal's proportion in the energy mix fell below 70% for the first time in history. In 2020, over 28 TWh of electricity was generated from RE sources, including nearly 16 TWh from WE. Poland's energy production is becoming increasingly uncompetitive as CO_2 emissions and domestic coal costs continue to grow. WE is the most advantageous alternative to fossil fuel based energy production [54].

2.3. WE in the Pomerania Voivodeship

As shown in Figure 1, at a height of 140 m the Pomerania Voivodeship has exceptionally excellent WE conditions. The Voivodeship is particularly well-suited to the growth of WE, both on land and at sea. Offshore WE might serve as a propeller for regional businesses, such as shipyards, which already supply components for the offshore industry [55].

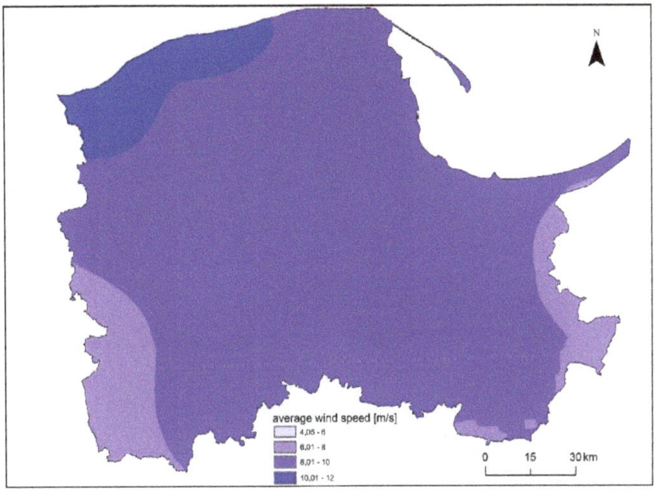

Figure 1. WS at a height of 140 m in Pomerania Voivodeship (created based on [55]).

WE capacity in Pomerania Voivodeship is 786 MW [56]. Taking into consideration the available area of the Polish exclusive economic zone (which is expected to grow to 2000 km^2 by 2030), wind conditions, productivity, and installed power density (6 MW/km^2), the theoretical potential is estimated to be 12 GW, with an approximate energy generation of 48–56 TWh, according to the draft program for the development of offshore wind energy and maritime industry.

The first wind farm in the Pomerania Voivodeship, with a capacity of 150 kW, was established in 1991 in Lisewo near Gniewino. In the same year, a 90 kW power plant was built in Swarzewo near Puck (currently closed). Since 2005, following Poland joining the EU when some legal barriers were removed, the number of investments in WE have started to increase [57].

The Energy Regulatory Office issued a license for the largest wind farm in Poland, which is located in the Pomerania and West-Pomerania voivodeships. The investment was carried out by the Potęgowo company belonging to the Israeli Mashav fund. Its power is 219 MW. The Potęgowo wind farm is located in the Słupsk and Sławno districts. It consists of 81 General Electric turbines with a capacity of 2.5 MW and 2.75 MW. For its construction, the investor received a loan from the European Bank in the amount of PLN 209 million. The total cost of the investment was PLN 1.25 billion. The farm also won an auction to supply electricity [51,58].

The Airport Wind Farm, which has a capacity of 90 MW, was officially inaugurated in the Voivodeship in 2015. PGE Energia Odnawialna S.A., a subsidiary of the PGE Capital Group, is the company that owns and operates the power plant. The Airport Wind Farm is the largest renewable energy investment made by the PGE Group since May 2014 [58].

The area available for the construction of wind farms in the Pomerania Voivodeship, including the buffer zone 2150 m from residential buildings, is 2716 km^2 (Figure 2).

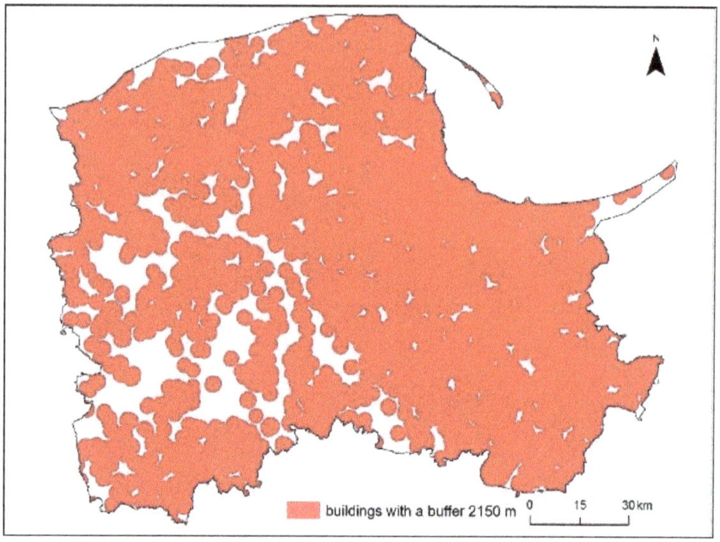

Figure 2. Buildings with a buffer of 2150 m (own elaboration).

The area available for the construction of wind farms in the Pomerania Voivodeship, including the buffer zone 2150 m from protected areas such as national parks, landscape parks, and nature reserves, is 2552 km^2 (Figure 3). The available area for wind farm construction in the Pomerania Voivodeship, including the buffer zone 200 m from forests (a condition regarding protecting bats), is 9568 km^2 (Figure 4).

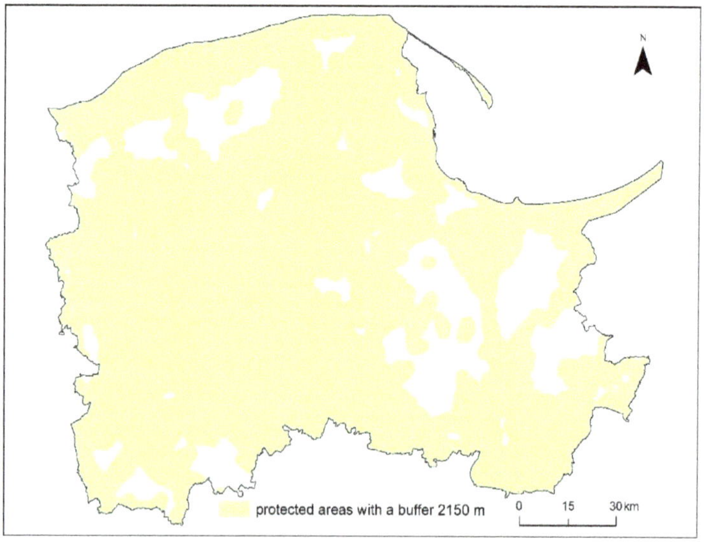

Figure 3. Protected areas with a buffer of 2150 m (own elaboration).

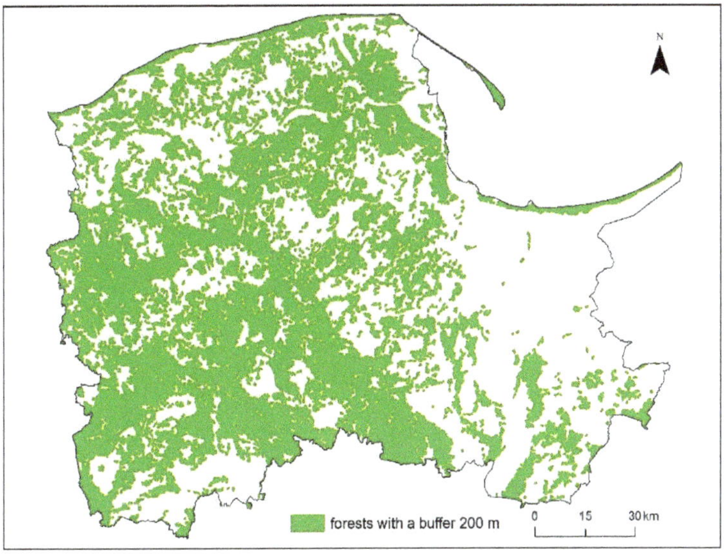

Figure 4. Forests with 200 m buffer (own elaboration).

The area available for the construction of wind farms in the Pomerania Voivodeship, including the hydrographic network and the 90 m buffer zone (propeller length 75 m increased by an additional 15 m) from the surface waters, is 3106 km^2 (Figure 5). The available area for the construction of wind farms in the Pomerania Voivodeship, including the infrastructure network and the 90 m buffer zone (propeller length 75 m increased by an additional 15 m), is 15,820 km^2 (Figure 6). Even after taking into consideration all the restricting factors, the accessible land area amounts to only 60 km^2 or less than 0.3 percent of the total land area of the Pomerania Voivodeship (Figure 7).

Figure 5. Water surfaces with 90 m buffer (own elaboration).

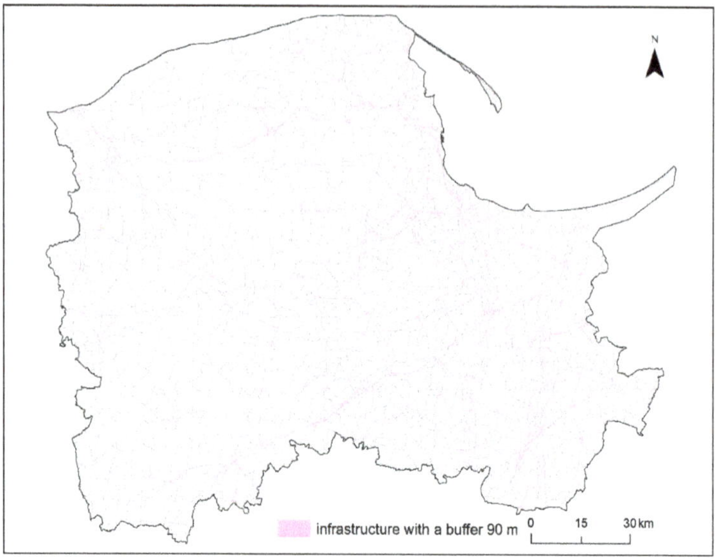

Figure 6. Infrastructure with a 90 m buffer (own elaboration).

Figure 7. Available area for constructing wind turbines in the Pomerania Voivodeship (own elaboration).

3. Methodology

In recent years, WE has made a substantial contribution to the production of clean and green energy. It is vital to be able to examine and estimate WE potential by fitting an appropriate statistical distribution to the WS data. Hourly WS data for Gdańsk over the last seven years were gathered from a third-party provider for this purpose as part of the study's scope. Then, using MLE, TPWD was fitted to the WS data. EGO, GA, SA, and DE were utilized to find the optimum TPWD parameters that maximize the likelihood function. Performance metrics were calculated to compare the performance of EGO with other methods. Following this, potential WE for Gdańsk, the capital of the Pomerania Voivodeship in Poland, was evaluated using TPWD parameters.

3.1. Parameter Estimation for Distribution of WS

Because of the intermittent nature of WS, it is necessary to understand and analyze the statistical properties of WS that have a substantial impact on WE and the design of power generators [56]. Probability distribution functions (PDF) are a way to describe how the random variables are likely to behave. The PDF can help to describe the change in WS over time. For the purpose of depicting WS patterns, several probability distributions such as Weibull, lognormal, gamma, Rayleigh, and mixed distributions are utilized, among others [52,59,60]. TPWD is widely used in the literature. The Weibull is flexible and is proven to fit WS data very well [52,61]. The parameters of the TPWD are shape and scale. An accurate assessment of the Weibull parameters is required to anticipate WE potential and understand WS characteristics. In order to determine the optimal parameters of the Weibull distribution (WD), researchers have developed a number of different ways over the years. The graphical method (GM), the moments method (MOM), the least-squares estimation (LSE), and (MLE) are the most frequently used methodologies [62,63].

Justus et al., proposed an approach [64] that employs mean and standard deviation of WS for estimating parameters of WS PDF. Stevens and Smulders used MLE to find parameters of WS PDF [63]. Jowder compared the empirical techniques to the graphical approaches and found that empirical techniques produce more accurate results [65]. For the parameter estimation, Akdag and Dinler proposed the power density factor and energy pattern factor [62]. The novel method was used for several locations in Turkey and the findings were compared with those produced using the GM and MLE methods. George compared five alternative approaches for calculating shape and scale parameters of the TPWD [66]. The maximum likelihood method outperformed among others. Chang examined six approaches for estimating the parameters of the WD: GM, MOM, empirical method (EM), MLE, modified MLE, and energy pattern factor/power density method (EPFPDM) [67].

Researchers also used the equivalent energy method to estimate parameters of the WD [68,69]. The performance of parameter estimation of the WD is also influenced by the sample size [70]. To predict Weibull parameters, probability-weighted moments based on the power density method (PWMBP) was used and PWMBP outperformed among other methods [71].

Aside from numerical approaches, a metaheuristic optimization algorithm can be used to estimate parameters. The parameters can be determined using various optimization algorithms. Chang used particle swarm optimization (PSO) to estimate parameters of the WD. PSO was used to estimate parameters using WS data collected from several climatic zones in Taiwan [68].

Wu et al. [72] proposed logistic distributions for assessing the WE potential in Inner Mongolia using maximum likelihood estimation. Using multi-objective moments, Usta et al., developed a novel approach for estimating the parameters of the WD [73]. Tosunoglu [74] focused on fitting several distributions to WS data for Turkey. MOM, MLE, and probability-weighted moments (PWMs) methods were applied. Chaurasiya et al. [75] applied nine numerical approaches for estimating the shape and scale parameters of the WD for calculating wind power in southern India. The results showed that shape and scale parameters have a significant impact on wind power calculations [76]. The least-squares method was applied to find the parameter of the WD [77,78]. For estimating the single and combined parameters of probability distributions, Alrashidi et al. [77] introduced a new metaheuristic optimization algorithm. Gungor et al. [79] explored the suitability of four different numerical approaches for estimating the WD parameters for WS data. Kumar et al. [80] concentrated on MLE using the differential evolution technique.

According to the reviewed literature, the TPWD is the most general distribution for representing WS distribution and assessing WE potential. To estimate the parameter, the researchers used a number of strategies to optimize the distribution's log-likelihood function. It is also noticed in the literature that researchers mostly use RMSE and R^2 for comparing performance of different optimization algorithms while estimating statistical distribution of WS. This study is primarily concerned with MLE and EGO.

3.2. Estimating Parameters of WD Using MLE

Modern estimation theory has application in a wide variety of fields, spanning from statistics to economics, engineering design, and many more. For a vast majority of applications, the estimation of an unknown parameter is required based on a collection of observations. Different parameter estimation methods can be found in the literature, the most common ones are GM, MLE, and MOM. Because of its theoretical capabilities, the MLE is often preferred over other methods.

The likelihood function is maximized by a set of parameters, which are MLE estimations. When fitting a distribution to the WS data, the TPWD is commonly used. The distribution function can be written as shown in Equation (11) [80].

$$f(x) = \left(\frac{k}{c}\right)\left(\frac{x}{c}\right)^{k-1} e^{-\left(\frac{x}{c}\right)^k},\ x \geq 0,\ c > 0,\ k > 0 \quad (1)$$

The WD likelihood function is as shown in Equation (1).

$$L = \prod_{i=!}^{N}\left(\frac{k}{c}\right)\left(\frac{x}{c}\right)^{k-1} e^{-\left(\frac{x}{c}\right)^k} \quad (2)$$

and its log-likelihood function will be:

$$\log(L) = Nlnk - Nclnc - \sum_{i=1}^{N}\left(\frac{x_i}{c}\right)^k + (k-1)\sum_{i=1}^{N} lnx_i \quad (3)$$

The EGO is used and compared with other techniques such as GA, SA, and DE for optimizing the log-likelihood function of the WD in this study. Detailed results are presented in the following section.

3.3. EGO

EGO is closely linked with kriging metamodeling. The EGO approach is focused on solving optimization problems in a low number of function evaluations and the approach offers clear stopping criteria based on expected improvement (*EI*). The EI function is produced based on the Kriging model. To get a new sampling point, the EI function is maximized. Then this new data point is added to the initial set. This process is repeated until the EI function value does not change significantly.

The Kriging model can be simply defined as shown in Equation (4), where $\mathbf{x}^{(i)} = \left(x_1^{(i)}, \ldots, x_k^{(n)}\right)$ and $\mathbf{y}^{(i)} = y\left(\mathbf{x}^{(i)}\right)$.

$$y\left(\mathbf{x}^{(i)}\right) = \mu + \epsilon\left(\mathbf{x}^{(i)}\right) \quad (4)$$

In this equation, μ is the mean of the stochastic process; $\epsilon\left(\mathbf{x}^{(i)}\right)$ is normally distributed independent error term with mean zero and variance σ^2. Correlation between $\epsilon\left(\mathbf{x}^{(i)}\right)$ and $\epsilon\left(\mathbf{x}^{(j)}\right)$ could be defined as shown in Equation (5) [81].

$$\text{Corr}[\epsilon(\mathbf{x}^{(i)}),\ \epsilon(\mathbf{x}^{(j)})] = \sum_{h=1}^{k}\theta_h\left|x_h^i - x_h^j\right|^{p_h},\ \theta_h \geq 0,\ p_h \in\ [1,2],\ i,j = (1,\ldots,n) \quad (5)$$

θ_h is importance measuring for the variable x_h and p_h is the smoothness parameter of the correlation function. μ and σ^2 are unknown. They can be estimated by using the parameters of the correlation function which are θ_h and p_h. For estimating the parameters, MLE is used. Likelihood function could be written as shown in Equation (6) [81]:

$$L = \frac{1}{(2\pi)^{\frac{n}{2}}(\sigma^2)^{\frac{n}{2}}|\mathbf{R}|^{\frac{1}{2}}} \exp\left[-\frac{(\mathbf{y}-1\mu)'\mathbf{R}^{-1}(\mathbf{y}-1\mu)}{2\sigma^2}\right] \tag{6}$$

where $\mathbf{y} = (y^{(i)}, \ldots, y^{(n)})$ is the n-vector for response values and 1 is a vector of ones. Since μ and σ^2 are unknown, estimations of μ and σ^2 could be calculated as shown in Equations (7) and (8).

$$\hat{\mu} = \frac{1'\mathbf{R}^{-1}\mathbf{y}}{1'\mathbf{R}^{-1}1} \tag{7}$$

$$\hat{\sigma}^2 = \frac{(\mathbf{y}-1\hat{\mu})'\mathbf{R}^{-1}(\mathbf{y}-1\hat{\mu})}{n} \tag{8}$$

By changing the Equations (7) and (8) with $\hat{\mu}$ and $\hat{\sigma}^2$ from the likelihood function, "concentrated likelihood function" is created. It depends only on θ_h and p_h. Denote that **r** gives the correlation between the error terms for **x***, which is not observed previously, and the error for **x**, which is observed previously. The correlation between those two could be written as shown in Equation (9).

$$\mathbf{r}(\mathbf{x}^*) \equiv \mathrm{Corr}[\epsilon(\mathbf{x}^*),\ \epsilon(\mathbf{x})]. \tag{9}$$

After having all the equations together, the Kriging model can be converted into the form shown in Equation (10).

$$\hat{y}(\mathbf{x}^*) = \hat{\mu} + \mathbf{r}'\mathbf{R}^{-1}(\mathbf{y}-1\hat{\mu}) \tag{10}$$

Following the process of creating the Kriging model, *EI* criteria is described as follows. Denote that the function $y = f(x)$, the improvement (I) over f_{min}, which is the minimum response value of $f(x)$. The improvement now can be defined as

$$I = \begin{cases} (f_{min} - y),\ y < f_{min} \\ 0,\ otherwise \end{cases}. \tag{11}$$

When y has normal distribution with $\hat{y}$ mean and s^2 variance, expected value of I can be calculated by following Equations (12) and (13).

$$E(I) = \int_{-\infty}^{f_{min}} (f_{min} - y)\phi(y)dy \tag{12}$$

Expected Improvement (*EI*) function can be shown as follows:

$$EI = (f_{min} - \hat{y})\Phi\left(\frac{f_{min} - \hat{y}}{s}\right) + s\phi\left(\frac{f_{min} - \hat{y}}{s}\right) \tag{13}$$

where Φ () is cumulative distribution function (CDF) and ϕ is PDF of a standard Normal distribution [81].

4. Data

The hourly WS dataset for Gdańsk (latitude: 54.352° N, longitude: 18.646° E, 10 m. height) over the last seven years between 1 January 2015 and 26 July 2021 was gathered from Open Weather Map [82]. Then, using MLE, the TPWD was fitted to the WS data for each month and the annual data. For the purpose of determining the optimal parameters of the WD that maximize the likelihood function, the SA, GA, DE, and EGO were applied and performance of the techniques was compared. Following the obtaining of the parameters, potential WE and wind power for Gdańsk, the capital of the Pomerania Voivodeship in Poland, were calculated.

The dataset contains 60,100 rows and 25 columns and it provides information about WS, minimum temperature, maximum temperature, pressure, wind angle, amount of rain, amount of snow, information about how the weather looks (rainy, snowy, etc.).

Table 3 shows summary statistics regarding monthly average of WS and minimum and maximum temperatures in Celciusº. As shown in the table, the warmest month for Gdańsk is August and the coldest month is January. Table 3 also presents the monthly average of WS (m/s) in Gdańsk. As shown in the table, the monthly average WS does not differ dramatically between months within a year. According to the table, it can be concluded that the months in which the average WS is higher than others are April, December, and May. The lowest WS average is observed in August.

Table 3. Average WS, minimum temperature, maximum temperature per month in Gdańsk between 1 January 2015 and 26 July 2021.

Month	Average WS (m/s)	Min. Temperature (Average—Celciusº)	Max. Temperature (Average—Celciusº)
January	2.72	−1.21	1.94
February	2.87	−0.58	2.77
March	3.15	2.33	5.39
April	3.51	6.15	9.61
May	3.17	10.66	14.39
June	3.02	15.65	19.24
July	3.12	16.76	19.96
August	2.38	17.17	20.85
September	2.68	13.29	16.52
October	2.77	8.29	11.25
November	2.83	4.27	6.86
December	3.19	1.86	4.48

Since obtaining the dataset from the third-party vendor was easy and quick and since the dataset contains hourly WS information for almost seven years, it was preferred to be used in the study. However, variables, including WS, in the dataset were collected by a single sensor located near the old town in Gdańsk. In conclusion to this, estimations for potential wind power were made only for this location. There was also no possibility to get WS information for different heights or for different parts of the city (e.g., parts of the city where long and tall buildings are located). For future studies, researchers plan to obtain datasets from different sources, such as local authorities or any other official sources, to be able to avoid the limitations mentioned above.

5. Results

As one of the goals of the study is fitting WS data to TPWD and estimating parameters of the distribution, an R package called "DiceOptim" was used for applying EGO, a "DEoptim" package was used for applying DE, a "GA" package was used for applying GA, and an "optimization" package was used for applying SA [83–87].

Table 4 shows the estimated value of shape (*k*) and scale (*c*) parameters of the TPWD using four different techniques. From the table, it can be concluded that there are no huge differences between the parameters estimated using the four different techniques. Figures 8 and 9 represent the histogram of observed wind speed and the estimated TPWD obtained using four techniques per each month of the year. Table 5 and Figures 10 and 11 show the performance of techniques based on two different metrics: RMSE and R^2. From Table 5 and Figures 10 and 11, it can be seen that EGO performs better than other techniques for estimating the parameters of TWPD; the larger the value of R^2 the better the performance of estimation as seen in Figure 10 that R^2 calculated for estimations.

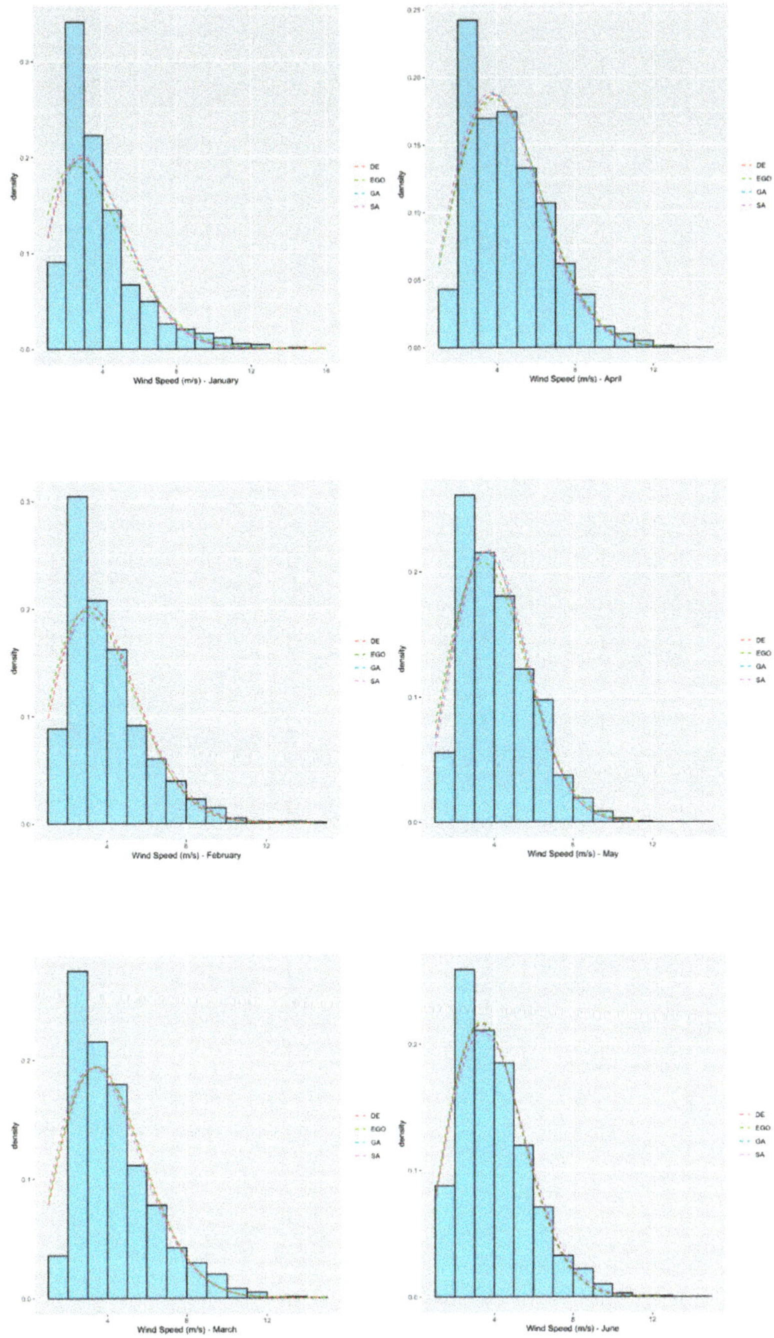

Figure 8. Monthly probability density function estimation and monthly histograms of observed WS data (from January to June) (own elaboration).

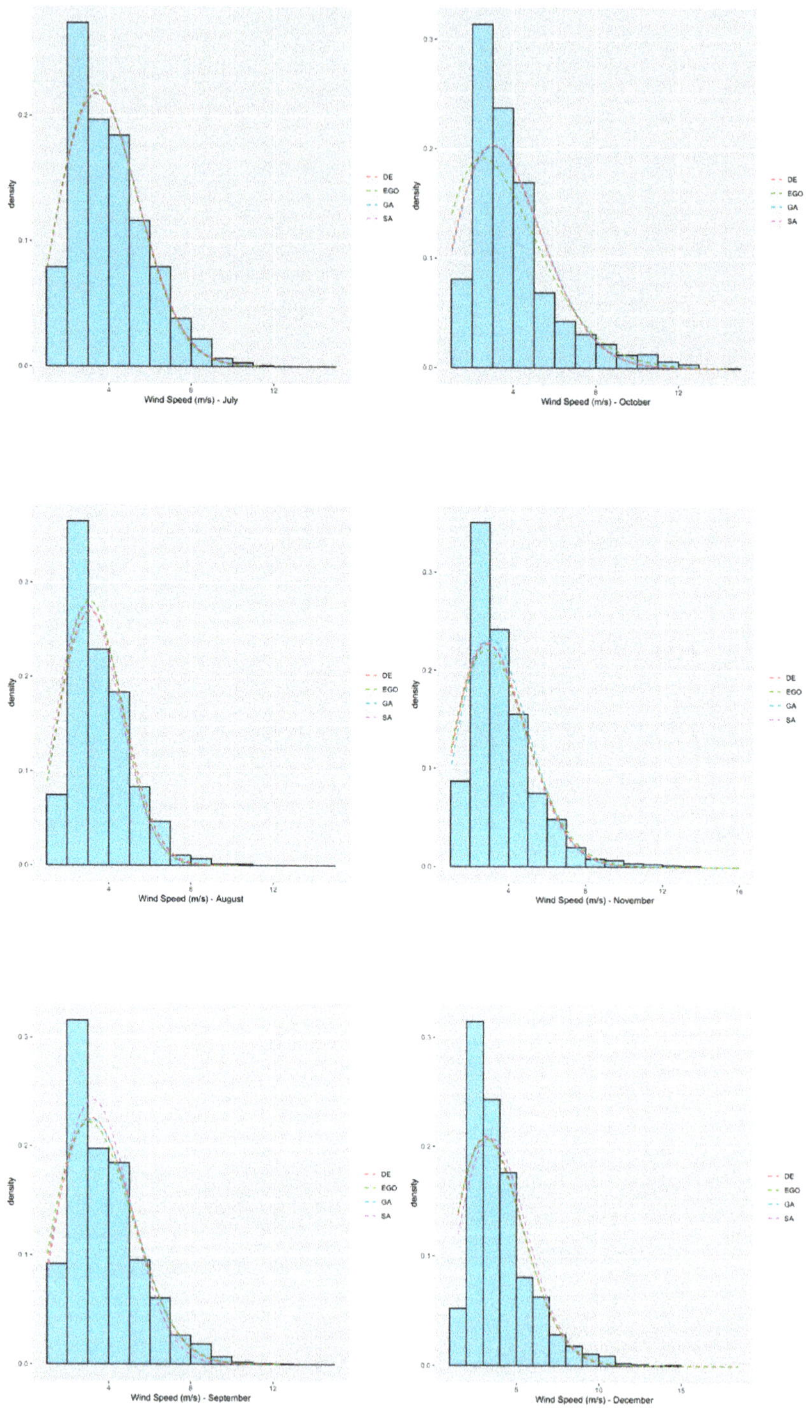

Figure 9. Monthly probability density function estimation and monthly histograms of observed WS data (from July to December) (own elaboration).

Table 4. Estimated TPWD parameters for monthly wind speed data.

Month	SA		GA		DE		EGO	
	k	*c*	*k*	*c*	*k*	*c*	*k*	*c*
January	1.93	4.15	1.91	4.23	1.93	4.21	1.72	4.11
February	2.00	4.42	2.06	4.32	2.06	4.33	1.96	4.28
March	2.06	4.51	2.15	4.64	2.15	4.64	2.12	4.60
April	2.20	4.86	2.26	4.98	2.24	4.98	2.22	5.02
May	2.44	4.56	2.38	4.50	2.36	4.48	2.23	4.46
June	2.25	4.44	2.27	4.35	2.27	4.34	2.26	4.31
July	2.31	4.35	2.30	4.35	2.30	4.35	2.32	4.31
August	2.48	3.64	2.53	3.78	2.53	3.78	2.59	3.70
September	2.41	4.04	2.24	4.13	2.26	4.13	2.17	4.09
October	2.02	4.25	2.00	4.25	2.00	4.22	1.72	4.11
November	2.09	3.86	2.21	3.96	2.11	3.91	2.04	3.94
December	2.24	4.45	2.13	4.28	2.13	4.28	2.09	4.27

Table 5. Performance comparison based on different metrics.

Month	SA		GA		DE		EGO	
	RMSE	R^2	RMSE	R^2	RMSE	R^2	RMSE	R^2
January	0.6512	0.9022	0.6496	0.9027	0.6616	0.8990	0.6136	0.9132
February	0.4967	0.9373	0.5130	0.9331	0.5007	0.9363	0.4650	0.9451
March	0.4762	0.9451	0.5178	0.9350	0.4835	0.9434	0.4758	0.9452
April	0.3828	0.9667	0.3830	0.9667	0.3856	0.9662	0.3802	0.9671
May	0.3760	0.9560	0.3732	0.9566	0.3658	0.9583	0.3665	0.9582
June	0.3632	0.9592	0.3605	0.9598	0.3358	0.9651	0.3363	0.9650
July	0.3713	0.9567	0.3672	0.9577	0.3703	0.9570	0.3599	0.9594
August	0.4569	0.8925	0.4526	0.8945	0.4487	0.8963	0.4539	0.8939
September	0.4913	0.9182	0.4471	0.9323	0.4664	0.9263	0.4234	0.9392
October	0.6639	0.8896	0.6802	0.8841	0.6649	0.8893	0.6229	0.9028
November	0.5996	0.8805	0.6262	0.8697	0.5745	0.8903	0.5636	0.8945
December	0.6441	0.8830	0.6111	0.8946	0.5934	0.9007	0.5602	0.9115

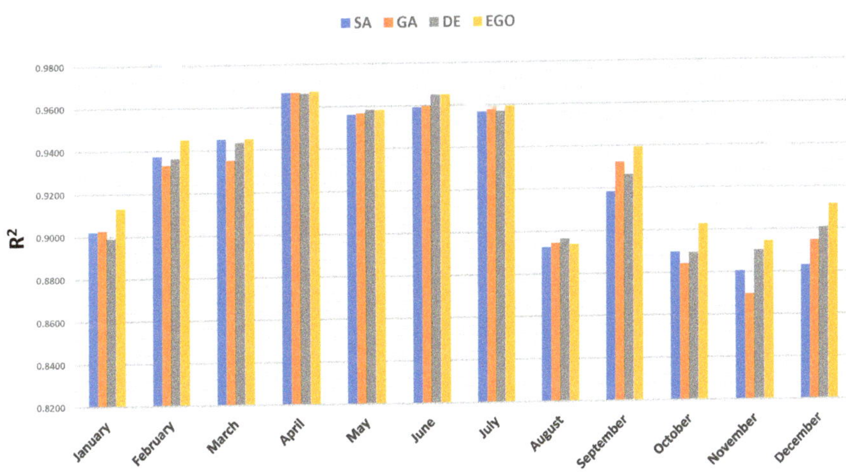

Figure 10. R-Square values for monthly parameter estimation per each technique (own elaboration).

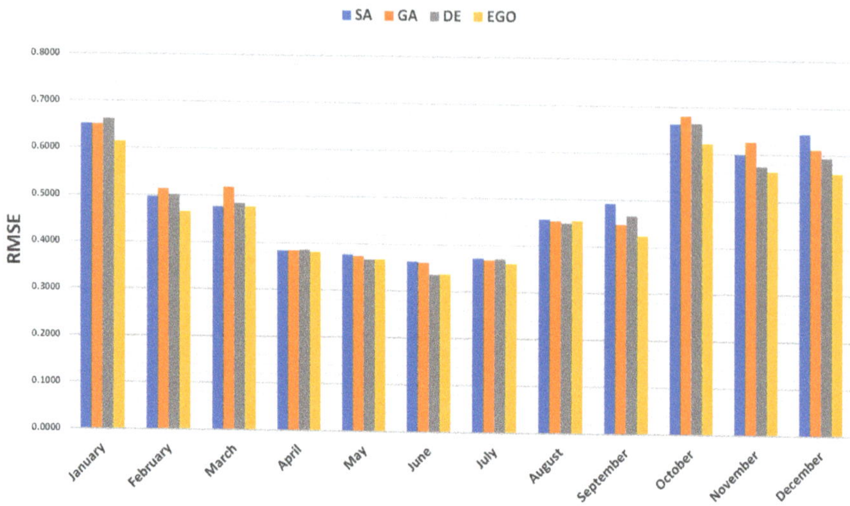

Figure 11. RMSE values for monthly parameter estimation per each technique (own elaboration).

RMSE is also one of the most common metrics to compare the techniques for distribution fitting; the smaller the value of RMSE the better the performance of estimation. Figure 11 shows that RMSE calculated for estimations based on EGO is lower than others for most of the months. Table 6 shows estimated value of shape (k) and scale (c) parameters of the TPWD for annual data using four different techniques. Figure 12 represents the histogram of observed wind speed and the estimated TPWD obtained using four techniques for annual data. From both Table 6 and Figure 12, it can be concluded that there are no huge differences between the parameters estimated using the four different techniques. Table 6 also shows the performance of techniques based on two different metrics: RMSE and R^2. From Table 6 and Figure 13, it can be seen that EGO provides the lowest RMSE and the highest R^2. In other words, EGO has the best performance among other techniques for estimating the parameters of TWPD for annual data.

In order to evaluate the wind energy potential, it is critical to estimate the TPWD parameters. EGO was utilized to estimate the parameters of TPWD in this study. EGO findings were compared with findings from the GA, SA, and DE algorithms. The EGO parameter estimation for TPWD yielded more precise outcomes. According to R^2 and RMSE, the EGO is superior to other algorithms.

Table 6. Performance comparison and parameter estimation for yearly data.

	Parameters		Metrics	
Technique	k	c	**RMSE**	R^2
SA	2.16	4.40	0.501242	0.9300
GA	2.14	4.33	0.486022	0.9342
DE	2.15	4.33	0.482129	0.9352
EGO	2.05	4.25	0.465032	0.9397

When the Weibull is chosen as PDF, the average wind power density per square meter is calculated as shown below [87,88]:

$$P_W = \frac{1}{2}\rho\bar{x}^3 \frac{\Gamma(1+\frac{3}{k})}{[\Gamma(1+\frac{1}{k})]^3} \quad (14)$$

Based on the estimated parameters of the WD, wind power density can be calculated by using the Equation (14), where $\overline{x}$ is the average wind speed, k is the shape parameter of the WD, and ρ is the standard air density, which is assumed to be equal to 1.225 kg/m^3 [87].

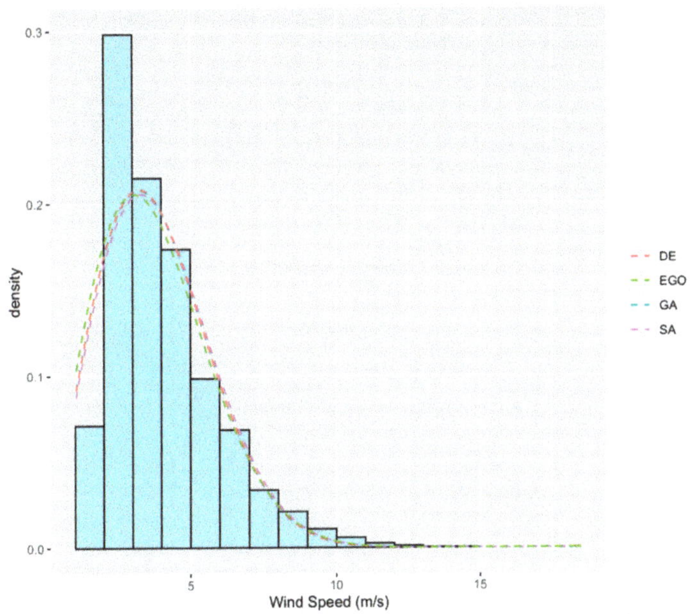

Figure 12. Probability density function estimation and histogram of observed WS for annual data (own elaboration).

k was estimated as equal to 2.05 and $\overline{x}$ is calculated as 3.81. Using Equation (14), wind power density can be calculated and it is equal to 62.89 W/m^2.

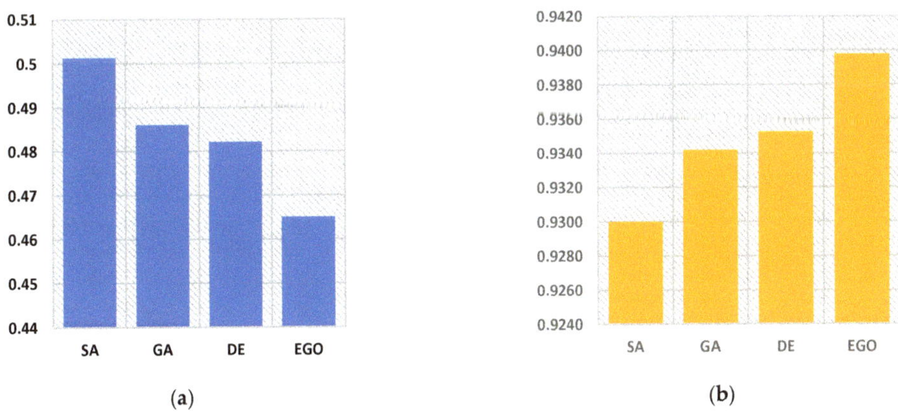

Figure 13. RMSE (**a**) and R^2 values (**b**) for parameter estimation using yearly data (own elaboration).

According to the *Small Wind Guidebook* provided by the U.S. Department of Energy, Office of Energy Efficiency and Renewable Energy, Skystream 3.7 is a type of wind turbine that can be used in urban areas [89]. The Skystream 3.7 is a wind turbine that turns wind into usable electricity for homes and small businesses. For households and smaller businesses, Skystream 3.7 may supply 40% or up to 90% of their total energy needs [88]. Taking this information into account, potential WE is calculated under the assumption of having Skystream 3.7 installed in Gdańsk city center.

WE is calculated as shown in Equation (15). To be able to calculate potential WE, information about swept area (SWA) and power coefficient (PC) of the wind tribune must be known [90].

$$P_E = P_W * PC * SA \quad (15)$$

SWA and maximum PC of Skystream 3.7 are 10.87 m^2 and 0.4, respectively [90–92]. P_E is calculated as 273.4457 kWh. Annually, P_E is calculated as 3281 kWh. According to Rachuneo.pl [93], energy prices per kWh in Poland range between 0.69 PLN and 0.78 PLN. By using this information, approximate annual revenue could be calculated between 2263 PLN and 2559 PLN.

6. Conclusions and Recommendations

For the purpose of analyzing and estimating WE potential, it is critical that WS data are fitted to a correct statistical distribution with high precision. Then, using the parameters of the statistical distribution, potential WE and wind power can be calculated. Within the scope of this study, wind power and WE potential were calculated for Gdańsk, the capital of Pomerania Voivodeship—one of the most important regions in Poland in terms of WE potential. Goals of the study are to propose a novel approach for estimating TPWD's parameters by using EGO and to shed a bit more light on the topic of potential benefits of having city-type wind turbines in a city. For these purposes, a dataset that contains hourly WS information for Gdańsk was used. In the following step, the TPWD was fitted to the monthly and annual WS data using MLE with EGO, SA, DE, and GA. Performance of the EGO was compared with other techniques using RMSE and R^2. Comparisons showed that EGO is providing more accurate estimations than other techniques. Using the parameters of the TPWD for annual data obtained by using EGO, potential WE and wind power for Gdańsk were calculated.

Based on the calculations, by having single Skystream 3.7 wind turbines in the city center of Gdańsk, 3281 kWh energy could be generated annually and this could bring revenue between 2263 PLN and 2559 PLN. These calculations revealed that city-type wind turbines might play an important role in generating electricity from WE. Erecting large wind turbines has limitations such as long distances between wind turbines and buildings according to the official regulations. If the city-type wind turbines were to be installed on the rooftops of residential buildings, shopping malls, or office buildings in Gdańsk city center, a portion of the electric energy needed by these buildings could be generated by using WE. However, payback periods and other potential limitations should be investigated.

The most important question is how to have widespread installation of smaller urban-type wind turbines. Two important directions should be emphasized here. The first direction is the development of entrepreneurship directed to the production of green energy. Undoubtedly, an important role is played here by business angels and the creation of sustainable startups. The creation of startups can most significantly translate into the widespread establishment of urban-type wind turbines by companies and consumer households [94–96]. Many startups should succeed in the market in this matter; then there is a need to transform the startup into a listed company in order to raise the necessary funds for development. In this case, it is important to decide the appropriate capital market and the timing of the market entry [97–100]. The second course of action is to focus on grassroot civic initiatives; adequately targeted activities at the local level can play a key role in the community's approach to wind energy and the widespread use of urban-type wind turbines for energy production [101–103].

For future studies, the authors would like to consider not only calculating potential wind energy for different parts of the city but also potential challenges when it comes to installing city-type wind turbines. In addition, they aim to expand the scope of the study by calculating wind energy potential for other cities in Pomerania Voivodeship.

Author Contributions: Conceptualization, O.A. and B.I.; methodology, O.A. and B.I.; software, O.A. and B.I.; validation, O.A. and B.I.; formal analysis, O.A. and B.I.; investigation, O.A., B.I., K.K. and M.S.; resources, O.A. and B.I.; data curation, O.A. and B.I.; writing—original draft preparation, O.A. and B.I.; writing—review and editing, O.A., B.I., K.K. and M.S.; visualization, O.A. and B.I.; supervision, B.I.; project administration, B.I. All authors have read and agreed to the published version of the manuscript.

Funding: This research received no external funding.

Institutional Review Board Statement: Not applicable.

Informed Consent Statement: Not applicable.

Data Availability Statement: Data available on request due to restrictions. The data presented in this study are available on request from the corresponding author.

Conflicts of Interest: The authors declare no conflict of interest.

References

1. Lin, M.-X.; Liou, H.M.; Chou, K.T. National Energy Transition Framework toward SDG7 with Legal Reforms and Policy Bundles: The Case of Taiwan and Its Comparison with Japan. *Energies* **2020**, *13*, 1387. [CrossRef]
2. Pietrzak, M.B.; Igliński, B.; Kujawski, W.; Iwański, P. Energy transition in Poland—Assessment of the renewable energy sector. *Energies* **2021**, *14*, 2046. [CrossRef]
3. Strunz, S. The German energy transition as a regime shift. *Ecol. Econ.* **2014**, *100*, 150–158. [CrossRef]
4. Bednarczyk, J.L.; Brzozowska-Rup, K.; Luściński, S. Determinants of the energy development based on renewable energy sources in Poland. *Energies* **2021**, *14*, 6762. [CrossRef]
5. Janowski, M. Generacja wiatrowa w KSE—diagnoza funkcjonowania w latach 2010–2018. *Elektroenerg. Współczesność I Rozw.* **2019**, *1*, 18–24.
6. Marks-Bielska, R.; Bielski, S.; Pik, K.; Kurowska, K. The importance of renewable energy sources in Poland's energy mix. *Energies* **2020**, *13*, 4624. [CrossRef]
7. Matuszewska-Janica, A.; Żebrowska-Suchodolska, D.; Ala-Karvia, U. Hozer-Koćmiel, M. Changes in electricity production from renewable energy sources in the European Union countries in 2005–2019. *Energies* **2021**, *14*, 6276. [CrossRef]
8. Chovancová, J.; Tej, J. Decoupling economic growth from greenhouse gas emissions: The case of the energy sector in V4 countries. *Equilib. Q. J. Econ. Econ. Policy* **2020**, *15*, 235–251. [CrossRef]
9. Grosse, T.G. Low Carbon Economy Policy in Poland: An Example of the Impact of Europeanisation. *Equilib. Quart. J. Econ. Econ. Policy* **2011**, *6*, 9–39. [CrossRef]
10. Piekut, M. The Consumption of Renewable Energy Sources (RES) by the European Union Households between 2004 and 2019. *Energies* **2021**, *14*, 5560. [CrossRef]
11. Henderson, J.; Anupama, S. The Energy Transition: Key Challenges for Incumbent and New Players in the Global Energy System, OIES Paper. Oxford Institute for Energy Studies. 2021. Available online: www.oxfordenergy.org/publications/the-energy-transition-key-challenges-for-incumbent-and-new-players-in-the-global-energy-system/ (accessed on 10 January 2022).
12. United Nations. Theme Report on Energy Transition. Toward the Achievement of SDG7 and Net-Zero Emission. 2021. Available online: https://www.un.org/sites/un2.un.org/files/2021-twg_2--062321.pdf (accessed on 10 January 2022).
13. Markandya, A.; Arto, I.; González-Eguino, M.; Román, M.V. Towards a green energy economy? Tracking the employment effects of low-carbon technologies in the European Union. *Appl. Energy* **2016**, *179*, 1342–1350. [CrossRef]
14. Gielen, D.; Boshell, F.; Saygin, D.; Bazilian, M.D.; Wagner, N.; Gorini, R. The role of renewable energy in the global energy transformation. *Energy Strat. Rev.* **2019**, *24*, 38–50. [CrossRef]
15. Ellabban, O.; Abu-Rub, H.; Blaabjerg, F. Renewable energy resources: Current status, future prospects and their enabling technology. *Renew. Sustain. Energy Rev.* **2014**, *39*, 748–764. [CrossRef]
16. Balcerzak, A.P. Quality of Institutions in the European Union countries. Application of TOPSIS Based on Entropy Measure for Objective Weighting. *Acta Polytech. Hung.* **2020**, *17*, 101–122. [CrossRef]
17. Overland, I. Energy: The missing link in globalization. *Energy Res. Soc. Sci.* **2016**, *14*, 122–130. [CrossRef]
18. Rees, W.E. Globalization, trade and migration: Undermining sustainability. *Ecol. Econ.* **2006**, *59*, 220–225. [CrossRef]
19. Dreher, A. Does globalization affect growth? Evidence from a new index of globalization. *Appl. Econ.* **2006**, *38*, 1091–1110. [CrossRef]
20. Sánchez-López, C.; Aceytuno, M.T.; De Paz-Bañez, M.A. Inequality and globalisation: Analysis of European countries. *Econ. Sociol.* **2019**, *4*, 84–100. [CrossRef]

21. Simionescu, M.; Lazányi, K.; Sopková, G.; Dobeš, K.; Balcerzak, A.P. University of Economics in Bratislava Determinants of Economic Growth in V4 Countries and Romania. *J. Compet.* **2017**, *9*, 103–116. [CrossRef]
22. Skare, M.; Porada-Rochoń, M. Financial and economic development link in transitional economies: A spectral Granger causality analysis 1991–2017. *Oecon. Copernic.* **2019**, *10*, 7–35. [CrossRef]
23. Ginevičius, R. Multi-criteria assessment of socioeconomic systems' conditions based on hierarchically structured indicator systems. *Econ. Sociol.* **2020**, *13*, 256–266. [CrossRef] [PubMed]
24. Zinecker, M.; Doubravský, K.; Balcerzak, A.P.; Pietrzak, M.B.; Dohnal, M. The COVID-19 disease and policy response to mitigate the economic impact in the EU: An exploratory study based on qualitative trend analysis. *Technol. Econ. Dev. Econ.* **2021**, *27*, 742–762. [CrossRef]
25. Roszko-Wójtowicz, E.; Grzelak, M.M. Macroeconomic stability and the level of competitiveness in EU member states: A comparative dynamic approach. *Oecon. Copernic.* **2020**, *11*, 657–688. [CrossRef]
26. Prokop, V.; Kotkova Striteska, M.; Stejskal, J. Fostering Czech firms? innovation performance through efficient cooperation. *Oecon. Copernic.* **2021**, *12*, 671–700. [CrossRef]
27. Markauskas, M.; Baliute, A. Technological progress spillover effect in Lithuanian manufacturing industry. *Equilib. Q. J. Econ. Econ. Policy* **2021**, *16*, 783–806. [CrossRef]
28. Androniceanu, A.-M.; Kinnunen, J.; Georgescu, I.; Androniceanu, A. A Multidimensional Approach to Competitiveness, Innovation and Well-Being in the EU Using Canonical Correlation Analysis. *J. Compet.* **2020**, *12*, 5–21. [CrossRef]
29. Cheba, K.; Szopik-Depczyńska, K. Multidimensional comparative analysis of the competitive capacity of the European Union countries and geographical regions. *Oecon. Copernic.* **2017**, *8*, 487–504. [CrossRef]
30. Nowak, P. Cooperation of enterprises in innovative activities on the example of Polish regions. *Equilib. Q. J. Econ. Econ. Policy* **2021**, *16*, 839–857. [CrossRef]
31. Jankiewicz, M.; Pietrzak, M.B. Assesment of Trends in the Share of Expenditure on Services and Food in the Visegrad Group Member States. *Int. J. Bus. Soc.* **2020**, *21*, 977–996. [CrossRef]
32. Piekut, M. Patterns of Energy Consumption in Polish One-Person Households. *Energies* **2020**, *13*, 5699. [CrossRef]
33. Zadykowicz, A.; Chmielewski, K.J.; Siemieniako, D. Proactive customer orientation and joint learning capabilities in collaborative machine to machine innovation technology development: The case study of automotive equipment manufacturer. *Oecon. Copernic.* **2020**, *11*, 531–547. [CrossRef]
34. Wosiek, M. Unemployment and new firm formation: Evidence from Polish industries at the regional level. *Equilib. Q. J. Econ. Econ. Policy* **2021**, *16*, 765–782. [CrossRef]
35. Fragkos, P.; Paroussos, L. Employment creation in EU related to renewables expansion. *Appl. Energy* **2018**, *230*, 935–945. [CrossRef]
36. Gajdos, A.; Arendt, L.; Balcerzak, A.P.; Pietrzak, M.B. Future trends of labour market polarisation in Poland—The perspective of 2025. *Transform. Bus. Econ.* **2020**, *3*, 114–135.
37. Dmytrów, K.; Bieszk-Stolorz, B. Comparison of changes in the labour markets of post-communist countries with other EU member states. *Equilib. Q. J. Econ. Econ. Policy* **2021**, *16*, 741–764. [CrossRef]
38. Pietrzak, M.B.; Balcerzak, A.P.; Gajdos, A.; Arendt, Ł. Entrepreneurial environment at regional level: The case of Polish path towards sustainable socio-economic development. *Entrep. Sustain. Issues* **2017**, *5*, 190–203. [CrossRef]
39. Svabova, L.; Tesarova, E.N.; Durica, M.; Strakova, L. Evaluation of the impacts of the COVID-19 pandemic on the development of the unemployment rate in Slovakia: Counterfactual before-after comparison. *Equilib. Q. J. Econ. Econ. Policy* **2021**, *16*, 261–284. [CrossRef]
40. European Commission. The European Green Deal. 2019. Available online: https://eur-lex.europa.eu/legal-content/EN/TXT/?uri=COM%3A2019%3A640%3AFIN (accessed on 10 January 2022).
41. Zielenkiewicz, M. Institutional Environment in the Context of Development of Sustainable Society in the European Union Countries. *Equilib. Quart. J. Econ. Econ. Policy* **2014**, *9*, 21–38. [CrossRef]
42. Szopik-Depczynska, K.; Cheba, K.; Bąk, I.; Stajniak, M.; Simboli, A.; Ioppolo, G. The study of relationship in a hierarchical structure of EU sustainable development indicators. *Ecol. Indic.* **2018**, *90*, 120–131. [CrossRef]
43. Szopik-Depczynska, K.; Kędzierska-Szczepaniak, A.; Szczepaniak, K.; Cheba, K.; Gajda, W.; Ioppolo, G. Innovation in sustainable development: An investigation of the EU context using 2030 agenda indicators. *Land Use Policy* **2018**, *79*, 251–262. [CrossRef]
44. Carley, S.; Konisky, D.M. The justice and equity implications of the clean energy transition. *Nat. Energy* **2020**, *5*, 569–577. [CrossRef]
45. Available online: www.ure.gov.pl/pl/oze/potencjal-krajowy-oze/5753,Moc-zainstalowana-MW.html (accessed on 15 September 2021).
46. REN21. *Renewables 2021: Global Status Report*; REN21: Paris, France, 2021.
47. Olabi, A.G.; Wilberforce, T.; Elsaid, K.; Salameh, T.; Sayed, E.T.; Husain, K.S.; Abdelkareem, M.A. Selection Guidelines for Wind Energy Technologies. *Energies* **2021**, *14*, 3244. [CrossRef]
48. Igliński, B.; Iglińska, A.; Koziński, G.; Skrzatek, M.; Buczkowski, R. Wind energy in Poland—History, current state, surveys, renewable energy sources Act, SWOT analysis. *Renew. Sustain. Energy Rev.* **2016**, *64*, 19–33. [CrossRef]
49. Ustawa z Dnia 20 Maja 2016 r. o Inwestycjach w Zakresie Elcktrowni Wiatrowych, Dz.U. poz. 961. 2016. Available online: https://isap.sejm.gov.pl/isap.nsf/DocDetails.xsp?id=WDU20160000961 (accessed on 25 October 2021).
50. Wyrobek, J.; Popławski, Ł.; Dzikuć, M. Analysis of financial problems of wind farms in Poland. *Energies* **2021**, *14*, 1239. [CrossRef]

51. Available online: www.ure.gov.pl/pl/sekcja/456,Odnawialne-Zrodla-Energii.html (accessed on 17 September 2021).
52. Carta, J.A.; Ramirez, P.; Velazquez, S. A review of wind speed probability distributions used in wind energy analysis: Case studies in the Canary Islands. *Renew. Sustain. Energy Rev.* **2009**, *13*, 933–955. [CrossRef]
53. Polish Wind Energy Association Official Web Page. Available online: http://psew.pl/en (accessed on 24 September 2021).
54. PSEW. Wkład Krajowych Dostawców w Rozwój Energetyki Wiatrowej na Lądzie i jej Wpływ na Polski Rynek Pracy do 2040 r. 2019. Available online: http://psew.pl/wp-content/uploads/2019/06/Analiza-WiseEuropa-Wp%C5%82yw-energetyki-wiatrowej-na-l%C4%85dzie-na-gospodark%C4%99-Polski.pdf (accessed on 10 October 2021).
55. Institute of Meteorology and Water Management: Warszawa, Poland, 2021; *unpublished work*.
56. Li, G.; Shi, J. Application of Bayesian model averaging in modeling long-term wind speed distributions. *Renew. Energy* **2010**, *35*, 1192–1202. [CrossRef]
57. Available online: https://zbkiks.ug.edu.pl/mk/kistowski_b_2_4.pdf (accessed on 11 October 2021).
58. Igliński, B.; Kiełkowska, U.; Pietrzak, M.B.; Skrzatek, M. *Energia Odnawialna w Województwie Pomorskim*; Nicolaus Copernicus University: Toruń, Poland, 2021.
59. Celik, A.N. A statistical analysis of wind power density based on the Weibull and Rayleigh models at the southern region of Turkey. *Renew. Energy* **2004**, *29*, 593–604. [CrossRef]
60. Elmahdy, E.E.; Aboutahoun, A.W. A new approach for parameter estimation of finite Weibull mixture distributions for reliability modeling. *Appl. Mathem. Model.* **2013**, *37*, 1800–1810. [CrossRef]
61. Mathew, S. *Wind Energy: Fundamentals, Resource Analysis and Economics*; Springer: Berlin/Heidelberg, Germany, 2006; Volume 1.
62. Akdag, S.A.; Dinler, A. A new method to estimate Weibull parameters for wind energy applications. *Energy Convers. Manag.* **2009**, *50*, 1761–1766. [CrossRef]
63. Stevens, M.J.M.; Smulders, P.T. The estimation of the parameters of the Weibull wind speed distribution for wind energy utilization purposes. *Wind Eng.* **1979**, *3*, 132–145.
64. Justus, C.; Hargraves, W.; Mikhail, A.; Graber, D. Methods for estimating wind speed frequency distributions. *J. Appl. Meteorol.* **1978**, *17*, 350–353. [CrossRef]
65. Jowder, F.A. Wind power analysis and site matching of wind turbine generators in Kingdom of Bahrain. *Appl. Energy* **2009**, *86*, 538–545. [CrossRef]
66. George, F. A comparison of shape and scale estimators of the two-parameter Weibull distribution. *J. Modern Appl. Stat. Methods* **2014**, *13*, 3. [CrossRef]
67. Chang, T.P. Performance comparison of six numerical methods in estimating Weibull parameters for wind energy application. *Appl. Energy* **2011**, *88*, 272–282. [CrossRef]
68. Chang, T.P. Wind energy assessment incorporating particle swarm optimization method. *Energy Convers. Manag.* **2011**, *52*, 1630–1637. [CrossRef]
69. Rocha, P.A.C.; De Sousa, R.C.; De Andrade, C.F.; da Silva, M.E.V. Comparison of seven numerical methods for determining Weibull parameters for wind energy generation in the northeast region of Brazil. *Appl. Energy* **2012**, *89*, 395–400. [CrossRef]
70. Arslan, T.; Bulut, Y.M.; Yavuz, A.A. Comparative study of numerical methods for determining Weibull parameters for wind energy potential. *Renew. Sustain. Energy Rev.* **2014**, *40*, 820–825. [CrossRef]
71. Usta, I. An innovative estimation method regarding Weibull parameters for wind energy applications. *Energy* **2016**, *106*, 301–314. [CrossRef]
72. Wu, J.; Wang, J.; Chi, D. Wind energy potential assessment for the site of Inner Mongolia in China. *Renew. Sustain. Energy Rev.* **2013**, *21*, 215–228. [CrossRef]
73. Usta, I.; Arik, I.; Yenilmez, I.; Kantar, Y.M. A new estimation approach based on moments for estimating Weibull parameters in wind power applications. *Energy Convers. Manag.* **2018**, *164*, 570–578. [CrossRef]
74. Tosunoglu, F. Accurate estimation of T year extreme wind speeds by considering different model selection criterions and different parameter estimation methods. *Energy* **2018**, *162*, 813–824. [CrossRef]
75. Chaurasiya, P.K.; Ahmed, S.; Warudkar, V. Study of different parameters estimation methods of Weibull distribution to determine wind power density using ground based Doppler SODAR instrument. *Alex. Eng. J.* **2018**, *57*, 2299–2311. [CrossRef]
76. Seo, S.; Oh, S.D.; Kwak, H.Y. Wind turbine power curve modeling using maximum likelihood estimation method. *Renew. Energy* **2019**, *136*, 1164–1169. [CrossRef]
77. Alrashidi, M.; Rahman, S.; Pipattanasomporn, M. Metaheuristic optimization algorithms to estimate statistical distribution parameters for characterizing wind speeds. *Renew. Energy* **2020**, *149*, 664–681. [CrossRef]
78. Guedes, K.S.; De Andrade, C.F.; Rocha, P.A.; Mangueira, R.S.; Moura, E.P. Performance analysis of metaheuristic optimization algorithms in estimating the parameters of several wind speed distributions. *Appl. Energy* **2020**, *268*, 114952. [CrossRef]
79. Gungor, A.; Gokcek, M.; Uçar, H.; Arabacı, E.; Akyüz, A. Analysis of wind energy potential and Weibull parameter estimation methods: A case study from Turkey. *Int. J. Environ. Sci. Technol.* **2020**, *17*, 1011–1020. [CrossRef]
80. Kumar, R.; Kumar, A. Application of differential evolution for wind speed distribution parameters estimation. *Wind Eng.* **2021**, *45*, 1544–1556. [CrossRef]
81. Aydin, L.; Aydin, O.; Artem, H.S.; Mert, A. Design of dimensionally stable composites using efficient global optimization method. *Proc. Inst. Mech. Eng. Part L J. Mater. Des. Appl.* **2019**, *233*, 156–168. [CrossRef]
82. Open Weather Map. Available online: https://openweathermap.org (accessed on 24 October 2021).

83. Roustant, O.; Ginsbourger, D.; Deville, Y. DiceKriging, DiceOptim: Two R Packages for the Analysis of Computer Experiments by Kriging-Based Metamodeling and Optimization. *J. Stat. Softw.* **2012**, *51*, 1–55. [CrossRef]
84. Scrucca, L. GA: A Package for Genetic Algorithms in R. Journal of Statistical Software. *Found. Open Access Stat.* **2013**, *53*, 1–37. [CrossRef]
85. Mullen, K.; Ardia, D.; Gil, D.L.; Windover, D.; Cline, J. DEoptim: An R package for global optimization by differential evolution. *J. Stat. Softw.* **2011**, *40*, 1–26. [CrossRef]
86. Husmann, K.; Lange, A.; Spiegel, E. The R Package Optimization: Flexible Global Optimization with Simulated-Annealing. 2017. Available online: https://cran.r-project.org/web/packages/optimization/vignettes/vignette_master.pdf (accessed on 12 March 2022).
87. Gökçek, M.; Bayülken, A.; Bekdemir, Ş. Investigation of wind characteristics and wind energy potential in Kirklareli, Turkey. *Renew. Energy* **2007**, *32*, 1739–1752. [CrossRef]
88. Saxena, B.K.; Rao, K.V.S. Estimation of wind power density at a wind farm site located in Western Rajasthan region of India. *Procedia Technol.* **2016**, *24*, 492–498. [CrossRef]
89. *WINDExchange: Small Wind Guidebook*; Wind Exchange, 18 November 2021. Available online: https://windexchange.energy.gov/small-wind-guidebook (accessed on 12 September 2021).
90. Royal Academy of Engineering. Wind Turbine Power Calculations. RWE Npower Renewables, Mechanical and Electrical Engineering Power Industry. 2010. Available online: https://www.raeng.org.uk/publications/other/23-wind-turbine (accessed on 10 November 2021).
91. Ray, A. *Performance Study of a Small Scale Wind Turbine*; Pennsylvania State University: State College, PA, USA, 2014.
92. Wind Power–Skystream 3.7 Specifications, Solardirect. Available online: https://www.solardirect.com/archives/wind-power/skystream/st-specifications.html (accessed on 15 January 2022).
93. Cena Prądu w Polsce w 2021 Roku. Ile Kosztuje 1 kWh Energii Elektrycznej? Available online: www.rachuneo.pl/cena-pradu#cena-pradu-pge (accessed on 18 November 2021).
94. Zinecker, M.; Skalická, M.; Balcerzak, A.P.; Pietrzak, M.B. Business angels in the Czech Republic: Characteristics and a classification with policy implications. *Econ. Res.-Kenosha Istraživanja* **2021**, *16*, 1–26. [CrossRef]
95. Gorączkowska, J. Enterprise innovation in technology incubators and university business incubators in the context of Polish industry. *Oecon. Copernic.* **2020**, *11*, 799–817. [CrossRef]
96. Zinecker, M.; Skalická, M.; Balcerzak, A.P.; Pietrzak, M.B. Identifying the impact of external environment on business angel activity. *Econ. Res.-Ekonomska Istraživanja* **2021**, *12*, 1–23. [CrossRef]
97. Meluzín, T.; Zinecker, M.; Balcerzak, A.P.; Pietrzak, M.B.; Doubravský, K. Institutional Settings and their Impact on the IPO Activity: An Exploratory Study Based on Qualitative Modelling. *Acta Polytech. Hung.* **2021**, *18*, 215–235. [CrossRef]
98. Meluzín, T.; Balcerzak, A.P.; Pietrzak, M.B.; Zinecker, M.; Doubravský, K. The impact of rumours related to political and macroeconomic uncertainty on IPO success: Evidence from a qualitative model. *Transform. Bus. Econ.* **2018**, *2017*, 148–169.
99. Meluzín, T.; Zinecker, M.; Balcerzak, A.P.; Doubravský, K.; Pietrzak, M.B.; Dohnal, M. The timing of initial public offerings: Non-numerical model based on qualitative trends. *J. Bus. Econ. Manag.* **2018**, *19*, 63–79. [CrossRef]
100. Meluzín, T.; Zinecker, M.; Balcerzak, A.P.; Pietrzak, M.B. Why do companies stay private? Determinants for IPO candidates to consider in Poland and the Czech Republic. *East. Eur. Econ.* **2018**, *56*, 471–503. [CrossRef]
101. Oehlmann, M.; Meyerhoff, J. Stated preferences towards renewable energy alternatives in Germany–do the consequentiality of the survey and trust in institutions matter? *J. Environ. Econ. Policy* **2016**, *1*, 1–16. [CrossRef]
102. Rogers, J.; Simmons, E.; Convery, I.; Weatherall, A. Public perceptions of community-based renewable energy projects. *Energy Policy* **2008**, *36*, 4217–4226. [CrossRef]
103. Zoellner, J.; Schweizer-Ries, P.; Wemheuer, C. Public acceptance of renewable energies: Results from case studies in Germany. *Energy Policy* **2008**, *36*, 4136–4141. [CrossRef]

MDPI
St. Alban-Anlage 66
4052 Basel
Switzerland
Tel. +41 61 683 77 34
Fax +41 61 302 89 18
www.mdpi.com

Energies Editorial Office
E-mail: energies@mdpi.com
www.mdpi.com/journal/energies

www.ingramcontent.com/pod-product-compliance
Lightning Source LLC
Chambersburg PA
CBHW061601220926
51850CB00005B/392